THE M&F SOLUTION

STUDENT RESOURCES

- Interactive eBook
- New Practice Quiz Generator
- Flashcards
- Games: Crossword Puzzles and Beat the Clock

- Videos
- Trackable Quizzes
- Trackable Interactives
- Review Cards

Students sign in at **www.cengagebrain.com**

INSTRUCTOR RESOURCES

- All Student Resources
- Engagement Tracker
- First Day of Class Instructions
- LMS Integration
- Instructor's Manual
- Test Bank
- PowerPoint® Slides
- Instructor Prep Cards

Instructors log in at **www.cengage.com/login**

Print

M&F3 delivers all the key terms and all the content for the **Marriage & Family** course through a visually engaging and easy-to-reference print experience.

CourseMate

CourseMate provides access to the full **M&F3** narrative, alongside a rich assortment of quizzing, flashcards, and interactive resources for convenient reading and studying.

© tele52/Shutterstock.com

CENGAGE
Learning®

M&F, Third Edition
David Knox

Vice President, General Manager, 4LTR Press
   and the Student Experience: Neil Marquardt

Product Director, 4LTR Press: Steven E. Joos

Product Manager: Steven E. Joos

Content/Media Developer: Sarah Dorger

Product Assistant: Mandira Jacob

Market Strategist: Elizabeth Rankin

Sr. Content Project Manager: Kim Kusnerak

Manufacturing Planner: Ron Montgomery

Production Service: MPS Limited

Sr. Art Director: Stacy Jenkins Shirley

Cover/Internal Designer: Ke Design,
   Mason, Ohio

Cover Image: ©Thinkstock, Stockbyte
   collection/George Doyle

   Monitor, back cover: ©A-R-T/Shutterstock

   Tablet, page i: ©tele52/Shutterstock

Intellectual Property

   Analyst: Deanna Ettinger

   Project Manager: Brittani Morgan

Vice President, General Manager, Social Science
   & Qualitative Business: Erin Joyner

Product Director: Marta Lee-Perriad

For product information and technology assistance, contact us at
Cengage Learning Customer & Sales Support, 1-800-354-9706

For permission to use material from this text or product,
submit all requests online at **www.cengage.com/permissions**
Further permissions questions can be emailed to
permissionrequest@cengage.com

Unless otherwise noted all items © Cengage Learning.

Library of Congress Control Number: 2014960339

ISBN: 978-1-305-40637-7

Cengage Learning
20 Channel Center Street
Boston, MA 02210
USA

Cengage Learning is a leading provider of customized learning solutions with office locations around the globe, including Singapore, the United Kingdom, Australia, Mexico, Brazil, and Japan. Locate your local office at:
www.cengage.com/global

Cengage Learning products are represented in Canada by Nelson Education, Ltd.

To learn more about Cengage Learning Solutions, visit **www.cengage.com**

Purchase any of our products at your local college store or at our preferred online store **www.cengagebrain.com**

Printed in the United States of America
Print Number: 01   Print Year: 2015

Brief Contents

© Phaendin/Shutterstock.com

Courtesy of Charlene Johnson

M&F

4LTR Press solutions are designed for today's learners through the continuous feedback of students like you. Tell us what you think about **M&F** and help us improve the learning experience for future students.

YOUR FEEDBACK MATTERS.

Contents

Rachel Calisto

3 Gender in Relationships 48

Courtesy of Brittany Bolen

4 Love and Relationship Development 68

Courtesy of Rachel Calisto

5 Communication and Technology in Relationships 96

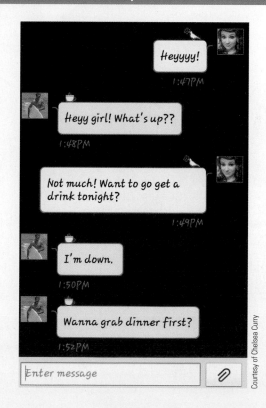

Courtesy of Chelsea Curry

6 Sexuality in Relationships 116

Hero Images/Getty Images

7 GLBTQ Relationships 134

Red Chopsticks/Getty Images

9 Money, Work, and Relationships 176

Rob Marmion/Shutterstock.com

10 Abuse in Relationships   190

Iofilolo/Getty Images

Courtesy of Chelsea Curry

12 Rearing Children 226

Courtesy of Brittany Bolen

13 Stress and Crisis in Relationships 246

Nicki Pardo/Getty Images

14 Divorce and Remarriage 266

Bacho/Shutterstock.com

15 The Later Years 292

Marriages and Families: An Introduction

"Enjoy the **little things** in life . . . For one day you'll look back and realize they **were the big things**."

—KURT VONNEGUT, WRITER

As the title of this chapter implies, there is no longer one definition or structure of "marriage" and "family" but various definitions and structures. No longer is marriage exclusively a heterosexual relationship but between persons of the same sex. And families are no longer two adults and children but single parent families headed by either a woman or man. In this chapter we embrace the diversity of marriages and families, identify how they are changing, suggest a choices framework (as well as other theoretical views) for marriage/family, and emphasize the need to be cautious about accepting the findings of research studies in marriage and family. We begin with the traditional conception of the term *marriage*.

1-1 Marriage

With all the talk of having no interest in getting married, enjoying singlehood, and pursuing one's education/career, "raising a family" remains one of the top values for undergraduates. In a nationwide study of 165,743 undergraduates in 234 colleges and universities, almost three-fourths (73%) identified raising a family as an essential

Rachel Calisto

objective (82% chose financial success as their top goal) (Eagan et al., 2013). In this chapter we review the definitions and types of marriage and family, various theoretical frameworks, and how researchers go about conducting M & F research so that we can be more informed about our own decisions.

Although young adults think of marriage as "love" and "commitment" (Muraco & Curran, 2012), the federal government regards **marriage** as a legal relationship that binds a couple together for the reproduction, physical care, and socialization of children. Each society works out its own details of what marriage is. In the United States, marriage is a legal contract between a heterosexual couple (although an increasing number of states are now recognize same-sex marriage) and the state in which they reside, that specifies the economic relationship between the couple (they become joint owners of their income and debt) and encourages sexual fidelity. The fine print of what marriage involves includes the following elements.

marriage a legal contract signed by a couple with the state in which they reside that regulates their economic and sexual relationship.

1-1a Elements of Marriage

Several elements comprise the meaning of marriage in the United States.

Legal Contract Marriage in our society is a legal contract into which two people of different sexes and legal age may enter when they are not already married to someone else. The age required to marry varies by state and is usually from 16 to 18 (most states set 17 or 18 as the requirement). In some states (e.g., Alabama) individuals can marry at age 14 with parental or judicial consent. In California, individuals can marry at any age with parental consent. The marriage license certifies that a legally empowered representative of the state perform the ceremony, often with two witnesses present. The marriage contract gives increased power to the state. Not only does the government dictate who may marry (e.g., persons of certain age, not currently married) but also the conditions of divorce (e.g., division of property, custody of children, and child support).

common-law marriage a marriage by mutual agreement between cohabitants without a marriage license or ceremony (recognized in some, but not all, states).

Under the laws of the state, the license means that spouses will jointly own all future property acquired and that each will share in the estate of the other. In most states, whatever the deceased spouse owns is legally transferred to the surviving spouse at the time of death. In the event of divorce and unless the couple has a prenuptial agreement, the property is usually divided equally regardless of the contribution of each partner. The license also implies the expectation of sexual fidelity in the marriage. Though less frequent because of no-fault divorce, infidelity is a legal ground for both divorce and alimony in some states.

The marriage license is also an economic authorization that entitles a spouse to receive payment from a health insurance company for medical bills if the partner is insured, to collect Social Security benefits at the death of one's spouse, and to inherit from the estate of the deceased. Spouses are also responsible for each other's debts.

Though the courts are reconsidering the definition of what constitutes a "family," the law is currently designed to protect spouses, not lovers or cohabitants. An exception is **common-law marriage**, in which a heterosexual couple cohabits and presents themselves as married; they will be regarded as legally married in those states that recognize such marriages. Common-law marriages exist in 14 states. Persons married by common law who move to a non-common-law state are recognized as being married in the state to which they move.

© iStock.com/Natalia Bratslavsky

TAKING CHANCES IN ROMANTIC RELATIONSHIPS

Marriage is about love and love is about making choices—some of them are risky such as moving in together after knowing each other for a short time, changing schools to be together, and forgoing condom usage thinking "this time won't end in a pregnancy." To assess the degree to which undergraduates take chances in their romantic relationships, 381 students completed a 64-item questionnaire posted on the Internet (Elliott et al., 2012). The majority of respondents were female (over 80%) and White (approximately 74%). Over half of the respondents (53%) described their relationship status as emotionally involved with one person, with 4% engaged or married. Of the various risk-taking behaviors identified on the questionnaire, eight were identified by 25% or more of the respondents as behaviors they had participated in. These eight are identified below.

Almost three-fourths (72%) of the sample self-identified as being a "person willing to take chances in my love relationship." However, only slightly over one-third of the respondents indicated that they considered themselves as risk takers in general. These percentages suggest that college students may be more likely to engage in risk-taking behavior in love relationships than in other areas of their lives. Both love and alcohol were identified as contexts for increasing one's vulnerability for taking chances in romantic relationships—60% and 66%, respectively. Both being in love and drinking alcohol (both love and alcohol may be viewed as drugs) gives one a sense of immunity from danger or allows one to deny danger.

Being both in love and drinking alcohol can increase one's vulnerability for taking chances in romantic relationships.

Most Frequent Risk-Taking Behaviors in a Romantic Relationship, N = 381

Risk-Taking Behavior	Percent
Unprotected sex	70
Being involved in a "friends with benefits" relationship	63
Broke up with a partner to explore alternatives	46
Had sex before feeling ready	41
Disconnected with friends because of partner	34
Maintained long-distance relationship (one year)	32
Cheated on partner	30
Lied to partner about being in love	28

Elliott et al., 2012.

Emotional Relationship Ninety-three percent of married adults in the United States point to love as their top reason for getting married (Cohn, 2013). Love is also an important reason for staying married. Forty-one percent of 4,730 undergraduates reported that they would divorce if they no longer loved their spouse (Hall & Knox, 2015).

American emphasis on love as a reason to marry is not shared throughout the world. Individuals in other cultures (e.g., India and Iran) do not require feelings of love to marry—love is expected to follow, not precede, marriage. In these countries, parental approval and similarity of religion, culture, education, and family background are considered more important criteria for marriage than love.

Sexual Monogamy Marital partners expect sexual fidelity. Almost two-thirds (66%) of 4,695 undergraduates agreed with the statement, "I would divorce a spouse who had an affair" (Hall & Knox, 2015).

Legal Responsibility for Children Although individuals marry for love and companionship, one of the most important reasons for the existence of marriage from the viewpoint of society is to legally bind a male and a female for the nurture and support of any children they may have. In our society, child rearing is the primary responsibility of the family, not the state.

> ## "If we don't **shape our kids**, they will be shaped by outside forces that don't care **what shape our kids are in.**"
>
> —LOUISE HART, PARENT EDUCATOR

Marriage is a relatively stable relationship that helps ensure that children will have adequate care and protection, will be socialized for productive roles in society, and will not become the burden of those who did not conceive them. Even at divorce, the legal obligation of the noncustodial parent to the child is maintained through child-support payments.

Announcement/Ceremony The legal binding of a couple is often preceded by an announcement in the local newspaper and then followed by a formal ceremony in a church or synagogue. The presence of parents, siblings, and friends at the wedding helps to verify the commitment of the partners to each other and helps marshal the social and economic support to launch the couple into married life. Most people in our society decide to marry.

When married people are compared with singles, the differences are strikingly in favor of the married (see Table 1.1 for the benefits of marriage and the liabilities of singlehood). The advantages of marriage over singlehood are true for first as well as subsequent marriages. However, just being married is not beneficial to all individuals. Being married is associated with obesity and spouses often do not sleep as well as singles since a spouse may snore or bed hog (Rauer, 2013).

1-1b Types of Marriage

Although we think of marriage in the United States as involving one man and one woman, other societies view marriage differently. **Polygamy** is a form of marriage involving more than two spouses. Polygamy occurs "throughout the world . . . and is found on all continents and among adherents of all world religions" (Zeitzen, 2008). There are three forms of polygamy: polygyny, polyandry, and pantagamy.

Polygyny **Polygyny** involves one husband and two or more wives and is practiced illegally in the United States by some religious fundamentalist groups. These groups are primarily in Arizona, New Mexico, and Utah (as well as Canada), and have splintered off from the Church of Jesus Christ of Latter-day Saints (commonly known as the Mormon Church). To be clear, the Mormon Church does not practice or condone polygyny (the church outlawed it in 1890). Those that split off from the Mormon Church represent only about 5% of Mormons in Utah. The largest offshoot is called the Fundamentalist Church of Jesus Christ of the Latter-day Saints (FLDS). Members of the group feel that the practice of polygyny is God's will. Although the practice is illegal, polygynous individuals are rarely prosecuted because a husband will have only one legal wife while the others will be married in a civil ceremony.

Polyandry The Buddhist Tibetans foster yet another brand of polygamy, referred to as **polyandry**, in which one wife has two or more (up to five) husbands. These husbands, who may be brothers, pool their resources to support one wife. Polyandry is a much less common form of polygamy than polygyny. The major reason for polyandry is economic. A family that cannot afford wives or marriages for each of its sons may find a wife for the eldest son only. Polyandry allows the younger brothers to also have sexual access to the one wife that the family is able to afford.

Polyamory **Polyamory** means multiple loves (poly = many; amorous = love) and is a lifestyle in which lovers embrace the idea of having multiple emotional and sexual partners. During the mid-1800s, the Oneida Community of Oneida, New York, embraced a form of

polygamy a generic term referring to a marriage involving more than two spouses.

polygyny a form of polygamy in which one husband has two or more wives.

polyandry a form of polygamy in which one wife has two or more husbands.

polyamory multiple loves (poly = many; amorous = love) and is a lifestyle in which lovers embrace the idea of having multiple emotional and sexual partners.

Table 1.1
Benefits of Marriage and the Liabilities of Singlehood

	Benefits of Marriage	Liabilities of Singlehood
Health	Spouses have fewer hospital admissions, see a physician more regularly, and are sick less often. They recover from illness/surgery more quickly.	Single people are hospitalized more often, have fewer medical checkups, and are sick more often.
Longevity	Spouses live longer than single people.	Single people die sooner than married people.
Happiness	Spouses report being happier than single people.	Single people report less happiness than married people.
Sexual satisfaction	Spouses report being more satisfied with their sex lives, both physically and emotionally.	Single people report being less satisfied with their sex lives, both physically and emotionally.
Money	Spouses have more economic resources than single people.	Single people have fewer economic resources than married people.
Lower expenses	Two can live more cheaply together than separately.	Cost is greater for two singles than one couple.
Drug use	Spouses have lower rates of drug use and abuse.	Single people have higher rates of drug use and abuse.
Connected	Spouses are connected to more individuals who provide a support system—partner, in-laws, etc.	Single people have fewer individuals upon whom they can rely for help.
Children	Rates of high school dropouts, teen pregnancies, and poverty are lower among children reared in two-parent homes.	Rates of high school dropouts, teen pregnancies, and poverty are higher among children reared by single parents.
History	Spouses develop a shared history across time with significant others.	Single people may lack continuity and commitment across time with significant others.
Crime	Spouses are less likely to be involved in crime.	Single people are more likely to be involved in crime.
Loneliness	Spouses are less likely to report loneliness.	Single people are more likely to report being lonely.

polyamory (complex marriage—every man was married to every woman). Today in Louisa, Virginia, half of the 100 members of Twin Oaks Intentional Community are polyamorous in that each partner may have several emotional or physical relationships with others at the same time. Although not legally married, these adults view themselves as emotionally bonded to each other and may even rear children together. Polyamory is not swinging, as polyamorous lovers are concerned about enduring, intimate relationships that include sex. A couple who has a polyamorous relationship often have an **open relationship**—a stable relationship in which the partners regard their own relationship as primary but agree that each may have emotional and physical relationships with others.

Pantagamy **Pantagamy** describes a group marriage in which each member of the group is "married" to the others. Also known as a *three-way marriage*, examples have existed in Brazil and the Netherlands whereby one male was "married" to two females. While these are not legal marriages, they reflect the diversity of lifestyle preferences and patterns. Theoretically, the arrangement could be of any sex, gender, and sexual orientation. The example in the Netherlands was of a heterosexual man "married" to two bisexual women.

The "one-size-fits-all" model of relationships and marriage is nonexistent. Individuals may be described as existing on a continuum from heterosexuality to homosexuality, from rural to urban dwellers, and from being single and living alone to being married and living in communes. Emotional relationships range from being close and loving to being distant and violent. Family diversity includes two parents (other or same-sex), single-parent families, blended families, families with adopted children, multigenerational families, extended families, and families representing different racial, religious, and ethnic backgrounds. *Diversity* is the term that accurately describes marriage and family relationships today.

open relationship a stable relationship in which the partners regard their own relationship as primary but agree that each may have emotional and physical relationships with others.

pantagamy a group marriage in which each member of the group is "married" to the others.

1-2 Family

Most people who marry choose to have children and become a family. However, the definition of what constitutes a family is sometimes unclear. This section examines how families are defined, their numerous types, and how marriages and families have changed in the past sixty years.

1-2a Definitions of Family

The U.S. Census Bureau defines **family** as a group of two or more people related by blood, marriage, or adoption. This definition has been challenged because it does not include foster families or long-term couples who live together. Marshall (2013) surveyed 105 faculty members from 19 Ph.D. marriage and family therapy programs and found no universal agreement on the definition of the family. Same-gender couples, children of same gender couples, and children with nonresidential parents were sometimes excluded from the definition of the family.

The answer to the question "Who is family?" is important because access to resources such as health care, Social Security, and retirement benefits is involved. Unless cohabitants are recognized by the state in which they reside as in a "domestic partnership," cohabitants are typically not viewed as "family" and are not accorded health benefits, Social Security, and retirement benefits of the partner. Indeed, the "live-in partner" may not be allowed to see the beloved in the hospital, which limits visitation to "family only."

The definition of who counts as family is being challenged. In some cases, families are being defined by function rather than by structure—what is the level of emotional and financial commitment and interdependence between the partners? How long have they lived together? Do the partners view themselves as a family?

Sociologically, a family is defined as a kinship system of all relatives living together or recognized as a social unit, including adopted people. The family is regarded as the basic social institution because of its important functions of procreation and socialization, and because it is found in some form in all societies.

Same-sex couples (e.g., Ellen DeGeneres and her partner) certainly define themselves as family. Increasingly, more states are recognizing marriages between same-sex individuals. Short of marriage, some states recognize committed gay relationships as **civil unions** (pair-bonded relationships given legal significance in terms of rights and privileges).

Although other states may not recognize same-sex marriages or civil unions (and thus people moving from these states to another state lose the privileges associated with marriage), over 24 cities and countries (including Canada) recognize some form of domestic partnership. **Domestic partnerships** are relationships in which cohabiting individuals are given some kind of official recognition by a city or corporation so as to receive partner benefits (e.g., health insurance). Disney recognizes domestic partnerships. Walmart offers benefits to same-sex partners. Domestic partnerships do not confer any federal recognition or benefits.

Some view their pets as part of their family. About 60% of Americans own a pet. In a Gallo Family Vineyard survey of 691 pet owners, 93% agreed that their pet was a part of the family (Payne & Bravo, 2013). Examples of treating pets like children include living only where there is a fenced-in backyard, feeding the pet a special diet, hanging a stocking and/or buying presents for the pet at Christmas, buying "clothes" for the pet, and leaving money in one's will for the care of the pet. Some pet owners buy accident insurance—Progressive© insurance covers pets. And pets are now the legal subject of divorce—the divorcing parties are granted custody and visitation rights to the animals of the couple (Gregory, 2010).

1-2b Types of Families

There are various types of families.

Family of Origin Also referred to as the **family of orientation**, this is the family into which you were born

family a group of two or more people related by blood, marriage, or adoption.

civil union a pair-bonded relationship given legal significance in terms of rights and privileges.

domestic partnership a relationship in which individuals who live together are emotionally and financially interdependent and are given some kind of official recognition by a city or corporation so as to receive partner benefits.

family of orientation the family of origin into which a person is born.

Courtesy of Caroline Schacht

or the family in which you were reared. It involves you, your parents, and your siblings. When you go to your parents' home for the holidays, you return to your **family of origin**. Your experiences in your family of origin have an impact on your own relationships. If you grew up in a loving intact family, you have a different set of expectations than if your parents were conflictual, divorced, and do not speak to each other today. Indeed, positive mother–father relationship quality is linked to children's outcomes. In a study of 773 parents, those reporting having stable and supportive relationships also reported fewer behavioral problems with their children who were ages 3 through 9. The researchers also found that marital relationship quality and children's behavioral problems were reciprocally related (Goldberg & Carlson, 2014).

Siblings in one's family of origin provide a profound influence on one another's behavior and emotional development and adjustment (McHale et al., 2012). Meinhold et al. (2006) noted that the relationship with one's siblings, particularly the sister–sister relationship, represents the most enduring relationship in a person's lifetime. Sisters who lived near one another and who did not have children reported the greatest amount of intimacy and contact.

Family of Procreation

The **family of procreation** represents the family that you will begin should you marry and have children. Of U.S. citizens living in the United States 65 years old and over, 96% have married and established their own family of procreation (*Statistical Abstract of the United States, 2012–2013*, Table 34). Across the life cycle, individuals move from the family of orientation to the family of procreation.

Nuclear Family

The **nuclear family** refers to either a family of origin or a family of procreation. In practice, this means that your nuclear family consists of you, your parents, and your siblings; or you, your spouse, and your children. Generally, one-parent households are not referred to as nuclear families. They are binuclear families if both parents are involved in the child's life or single-parent families if one parent is involved in the child's life and the other parent is totally out of the picture.

Is the Nuclear Family Universal?

Sociologist George Peter Murdock's classic study (1949) emphasized that the nuclear family is a "universal social grouping" found in all of the 250 societies he studied. The nuclear family channels sexual energy between two adult partners who reproduce and also cooperate in the care of offspring and their socialization to be productive members of society. "This universal social structure, produced through cultural evolution in every human society, as presumably the only feasible adjustment to a series of basic needs, forms a crucial part of the environment in which every individual grows to maturity" (p. 11).

The universality of the nuclear family has been questioned. In *Sex at Dawn*, Ryan and Jetha (2010) reviewed cross-cultural data and emphasized that the terms *marriage* and *family* do not have universal meanings. In some groups, adults have sexual relationships with various partners throughout their life and view themselves as mothers and fathers to all of the children in the community. Children in these villages view all adults as their mother and father.

Dr. Robert Bunger (2014) is a premier anthropologist. His reaction to the thesis of *Sex at Dawn* follows:

In my opinion the idea that everyone had sex with whomever and that all adults were parents of everyone's children is utter nonsense. Louis Henry Morgan in *Ancient Society* (published in the 19th century) suggested that early humans lived in a state of "primitive promiscuity" and the idea was taken up by Marx and Engels. I do not know of any society that actually lives that way. The Muria Ghond of India have a system whereby the young people, between puberty and marriage, live in a group marriage where everyone is allowed to have sex with everyone else of the other sex. At some point they drop out and settle into monogamous marriage. I do not think that there is any traditional society where group marriage for adults is the norm. I think that some communal movements like the Amana society tried group marriage but later gave it up.

Traditional, Modern, and Postmodern Family

Sociologists have identified three central concepts of the family. The **traditional family** is the two-parent nuclear family, with the husband as breadwinner and the wife as homemaker. The **modern family** is the dual-earner

family of origin the family into which an individual is born or reared, usually including a mother, father, and children.

family of procreation the family a person begins by getting married and having children.

nuclear family family consisting of an individual, his or her spouse, and his or her children, or of an individual and his or her parents and siblings.

traditional family the two-parent nuclear family with the husband as breadwinner and wife as homemaker.

modern family the dual-earner family, in which both spouses work outside the home.

Table 1.2

Differences between Marriage and the Family in the United States

Marriage	Family
Usually initiated by a formal ceremony.	Formal ceremony not essential.
Involves two people.	Usually involves more than two people.
Ages of the individuals tend to be similar.	Individuals represent more than one generation.
Individuals usually choose each other.	Members are born or adopted into the family.
Ends when spouse dies or is divorced.	Continues beyond the life of the individual.
Sex between spouses is expected and approved.	Sex between near kin is neither expected nor approved.
Requires a license.	No license needed to become a parent.
Procreation expected.	Consequence of procreation.
Spouses are focused on each other.	Focus changes with addition of children.
Spouses can voluntarily withdraw from marriage.	Parents cannot divorce themselves from obligations to children via divorce.
Money in unit is spent on the couple.	Money is used for the needs of children.
Recreation revolves around adults.	Recreation revolves around children.

Reprinted by permission of Dr. Lee Axelson.

postmodern family
nontraditional families emphasizing that a healthy family need not be heterosexual or have two parents.

binuclear family
family in which the members live in two households.

blended family a family created when two individuals marry and at least one of them brings a child or children from a previous relationship or marriage. Also referred to as a stepfamily.

extended family
the nuclear family or parts of it plus other relatives.

family, in which both spouses work outside the home. **Postmodern families** represent a departure from these models and include lesbian or gay couples and mothers who are single by choice (Silverstein & Auerbach, 2005).

Binuclear Family A **binuclear family** is a family in which the members live in two separate households. This family type is created when the parents of the children divorce and live separately, setting up two separate units, with the children remaining a part of each unit. Each of these units may also change again when the parents remarry and bring additional children into the respective units (**blended family**). Hence, the children may go from a nuclear family with both parents, to a binuclear unit with parents living in separate homes, to a

blended family when parents remarry and bring additional children into the respective units.

Extended Family The **extended family** includes not only the nuclear family (or parts of it) but other relatives as well. These relatives include grandparents, aunts, uncles, and cousins. An example of an extended family living together would be a husband and wife, their children, and the husband's parents (the children's grandparents). Asians are more likely than Anglo-Americans to live with their extended families. However, commitment to the elderly may be changing as a result of the westernization of Asian countries such as China, Japan, and Korea.

The terms *marriage* and *family* are often thought to be the same. Table 1.2 identifies the differences.

1-3 Changes in Marriage and Family

Whatever family we experience today was different previously and will change yet again. A look back at some changes in marriage and the family follow.

1-3a The Industrial Revolution and Family Change

The Industrial Revolution refers to the social and economic changes that occurred when machines and factories, rather than human labor, became the dominant mode for the production of goods. Industrialization occurred in the United States during the early- and mid-1800s and represents one of the most profound influences on the family.

Before industrialization, families functioned as an economic unit that produced goods and services for its own consumption. Parents and children worked together in or near the home to meet the survival needs of the family. As the United States became industrialized, more men and women left the home to sell their labor for wages. The family was no longer a self-sufficient unit that determined its work hours. Rather, employers determined where and when family members would work. Whereas children in preindustrialized America worked on farms and contributed to the economic survival of the family, children in industrialized America became economic liabilities rather than assets. Child labor laws and mandatory education removed children from the labor force and lengthened their dependence on parental support. Eventually, both parents had to work away from the home to support their children. The dual-income family had begun.

During the Industrial Revolution, urbanization occurred as cities were built around factories and families moved to the city to work in the factories. Living space in cities was crowded and expensive, which contributed to a decline in the birthrate and thus smaller families. The development of transportation systems during the Industrial Revolution made it possible for family members to travel to work sites away from the home and to move away from extended kin. With increased mobility, many extended families became separated into smaller nuclear family units consisting of parents and their children. As a result of parents leaving the home to earn wages and the absence of extended kin in or near the family household, children had less adult supervision and moral guidance. Unsupervised children roamed the streets, increasing the potential for crime and delinquency.

Industrialization also affected the role of the father in the family. Employment outside the home removed men from playing a primary role in child care and in other domestic activities. The contribution men made to the household became primarily economic.

Finally, the advent of industrialization, urbanization, and mobility is associated with the demise of

Families from familistic cultures such as China who immigrate to the United States soon discover that their norms, roles, and values are challenged.

familism (e.g., focus on what is important for the family) and the rise of **individualism** (focus on what it important for the individual). When family members functioned together as an economic unit, they were dependent on one another for survival and were concerned about what was good for the family. This familistic focus on the needs of the family has since shifted to a focus on self-fulfillment—individualism. Families from familistic cultures such as China who immigrate to the United States soon discover that their norms, roles, and values begin to alter in reference to the industrialized, urbanized, individualistic patterns and thinking. Individualism and the quest for personal fulfillment are thought to have contributed to high divorce rates, absent fathers, and parents spending less time with their children.

Hence, although the family is sometimes blamed for juvenile delinquency, violence, and divorce, it is more accurate to emphasize changing social norms and conditions of which the family is a part. When industrialization takes parents out of the home so that they can no longer be constant nurturers and supervisors, the likelihood of aberrant acts by their children/adolescents increases. One explanation for school violence is that absent, career-focused parents have failed to provide close supervision for their children.

1-3b Changes in the Last 65 Years

Enormous changes have occurred in marriage and the family since the 1950s. Table 1.3 reflects

familism philosophy in which decisions are made in reference to what is best for the family as a collective unit.

individualism philosophy in which decisions are made on the basis of what is best for the individual.

Table 1.3
Changes in Marriages and Families, 1950 and 2015

	1950	2015
Family relationship values	Strong values for marriage and the family. Individuals who wanted to remain single or child free are considered deviant, even pathological. Husband and wife should not be separated by jobs or careers.	Individuals who remain single or child free experience social understanding and sometimes encouragement. Single and childfree people are no longer considered deviant or pathological but are seen as self-actuating individuals with strong job or career commitments. Husbands and wives can be separated for reasons of job or career and live in a commuter marriage. Married women in large numbers have left the role of full-time mother and housewife to join the labor market.
Gender roles	Rigid gender roles, with men dominant and earning income while wives stay home, taking care of children.	Egalitarian gender roles with both spouses earning income. Greater involvement of men in fatherhood.
Sexual values	Marriage was regarded as the only appropriate context for intercourse in middle-class America. Living together is unacceptable, and a child born out of wedlock was stigmatized. Virginity is sometimes exchanged for marital commitment.	For many, concerns about safer sex have taken precedence over the marital context for sex. Virginity is rarely exchanged for anything. Living together is regarded as not only acceptable but sometimes preferable to marriage. For some, unmarried single parenthood is regarded as a lifestyle option. Hooking up is new courtship norm.
Homogamous mating	Strong social pressure exists to date and marry within one's own racial, ethnic, religious, and social class group. Emotional and legal attachments are heavily influenced by obligation to parents and kin.	Dating and mating have become more heterogamous, with more freedom to select a partner outside one's own racial, ethnic, religious, and social class group. Attachments are more often by choice.
Cultural silence on intimate relationships	Intimate relationships are not an appropriate subject for the media.	Individuals on talk shows, interviews, and magazine surveys are open about sexuality and relationships behind closed doors.
Divorce	Society strongly disapproves of divorce. Familistic values encouraged spouses to stay married for the children. Strong legal constraints keep couples together. Marriage is forever.	Divorce has replaced death as the endpoint of a majority of marriages. Less stigma is associated with divorce. Individualistic values lead spouses to seek personal happiness. No-fault divorce allows for easy severance. Marriage is tenuous. Increasing numbers of children are being reared in single-parent households apart from other relatives.
Familism versus individualism	Families are focused on the needs of children. Mothers stay home to ensure that the needs of their children are met. Adult concerns are less important.	Adult agenda of work and recreation has taken on increased importance, with less attention being given to children. Children are viewed as more sophisticated and capable of thinking as adults, which frees adults to pursue their own interests. Day care is used regularly.
Homosexuality	Same-sex emotional and sexual relationships are a culturally hidden phenomenon. Gay relationships are not socially recognized.	Supreme Court legalized same-sex marriage. Gay relationships are, increasingly, a culturally open phenomenon (e.g., television sitcoms, gay athletes).
Scientific scrutiny	Aside from Kinsey's, few studies are conducted on intimate relationships.	Acceptance of scientific study of marriage and intimate relationships.
Family housing	Husbands and wives live in same house.	Husbands and wives may "live apart together" (LAT), which means that, although they are emotionally and economically connected, they (by choice) maintain two households, houses, condos, or apartments.
Technology	Nonexistent except phone.	Use of iphones, texting, sexting, Facebook.

some of these changes. One of the most obvious changes is technology. Marriage relationships are initiated, developed, and maintained with cell phone technology. Individuals stay in contact with each other all day via text messaging. We will discuss the impact of technology on relationships in greater detail in Chapter 5 on communication and technology.

In spite of the persistent and dramatic changes in marriage and the family, marriage and the family continue to be resilient. Using this **marriage-resilience perspective**, changes in the institution of marriage are not viewed negatively nor are they indicative that marriage is in a state of decline. Indeed, these changes are thought to have "few negative consequences for adults, children, or the wider society" (Amato et al., 2007, p. 6).

Courtesy of E Fred Johnson, Jr.

A social exchange view of marital roles emphasizes that spouses negotiate the division of labor on the basis of exchange. For example, a man participates in child care in exchange for his wife earning an income, which relieves him of the total financial responsibility. Social exchange theorists also emphasize that power in relationships is the ability to influence, and avoid being influenced by, the partner.

Albert Einstein's second marriage to Elsa Einstein provides another example of exchange. "She was an efficient and lively woman, who was eager to serve and protect him . . . . He was pleased to be looked after . . . which allowed him to spend hours in a rather dreamy state, focusing more on the cosmos than on the world around him." (Isaacson, 2007, p. 247).

1-4 Theoretical Frameworks for Viewing Marriage and the Family

All **theoretical frameworks** are the same in that they provide a set of interrelated principles designed to explain a particular phenomenon and provide a point of view. In essence, theories are explanations. The more common frameworks follow.

1-4a Social Exchange Framework

The **social exchange framework** is one of the most commonly used theoretical perspectives in marriage and the family. The framework views interaction and choices in terms of cost and profit. It operates from a premise of **utilitarianism**—the theory that individuals rationally weigh the rewards and costs associated with behavioral choices. Vespa (2013) studied cohabitants age 50 and older and found that unhealthy but wealthy males were more likely to marry—they trade their wealth/agreement to marry for caregiving by a female who needs economic support/wants to be married.

1-4b Family Life Course Development Framework

The **family life course development** framework emphasizes the important role transitions of individuals that occur in different periods of life and in different social contexts. For example, young unmarried lovers may become cohabitants, then parents, grandparents, retirees, and widows. The family life cycle is a basic set of stages through which not all individuals pass (e.g., the childfree) and in which there is great diversity, particularly in regard to race and education (e.g., African Americans are less likely to marry; the highly educated are less likely to divorce) (Cherlin, 2010).

marriage-resilience perspective the view that changes in the institution of marriage are not indicative of a decline and do not have negative effects.

theoretical framework a set of interrelated principles designed to explain a particular phenomenon and to provide a point of view.

social exchange framework spouses exchange resources, and decisions are made on the basis of perceived profit and loss.

utilitarianism the doctrine holding that individuals rationally weigh the rewards and costs associated with behavioral choices.

family life course development the stages and process of how families change over time.

The family life course developmental framework has its basis in sociology (e.g., role transitions), whereas the **family life cycle** has its basis in psychology, which emphasizes the various developmental tasks family members face across time (e.g., marriage, childbearing, preschool, school-age children, teenagers). If developmental tasks at one stage are not accomplished, functioning in subsequent stages will be impaired. For example, one of the developmental tasks of early American marriage is to emotionally and financially separate from one's family of origin. If such separation from parents does not take place, independence as individuals and as a couple may be impaired.

1-4c Structure-Function Framework

The **structure-function framework** emphasizes how marriage and family contribute to society. Just as the human body is made up of different parts that work together for the good of the individual, society is made up of different institutions (e.g., family, religion, education, economics) that work together for the good of society. **Functionalists** (structure-function theorists) view the family as an institution with values, norms, and activities meant to provide stability for the larger society. Such stability depends on families performing various functions for society.

First, families serve to replenish society with socialized members. Because our society cannot continue to exist without new members, we must have some way of ensuring a continuing supply. However, just having new members is not enough. We need socialized members—those who can speak our language and know the norms and roles of our society. Girgis et al. (2011) emphasized that "societies rely on families to produce upright people who make for conscientious, law-abiding citizens lessening the demand for governmental policing and social services" (p. 245).

Disaster is the result for a child born into a family which does not function properly. Genie is a young girl who was discovered in the 1970s; she had been kept in isolation in one room in her California home for 12 years by her abusive father (James, 2008). She could barely walk and could not talk. Although provided intensive therapy at UCLA and the object of thousands of dollars of funded research, Genie progressed only slightly. Today, she is in her late 50s, institutionalized, and speechless. Her story illustrates the need for socialization; the legal bond of marriage and the obligation to nurture and socialize offspring help to ensure that this socialization will occur.

Second, marriage and the family promote the emotional stability of the respective spouses. Society cannot provide enough counselors to help us whenever we have emotional issues/problems. Marriage ideally provides in-residence counselors who are loving and caring partners with whom people share (and receive help for) their most difficult experiences.

Children also need people to love them and to give them a sense of belonging. This need can be fulfilled in a variety of family contexts (two-parent families, single-parent families, extended families). The affective function of the family is one of its major offerings. No other institution focuses so completely on meeting the emotional needs of its members as marriage and the family.

Third, families provide economic support for their members. Although modern families are no longer self-sufficient economic units, they provide food, shelter, and clothing for their members. One need only consider the homeless in our society to be reminded of this important function of the family.

In addition to the primary functions of replacement, emotional stability, and economic support, other functions of the family include the following:

- **Physical care.** Families provide the primary care for their infants, children, and aging parents. Other agencies (neonatal units, day care centers, assisted-living residences, shelters) may help, but the family remains the primary and recurring caretaker. Spouses also show concern about the physical health of each other by encouraging each other to take medications and to see a doctor.

- **Regulation of sexual behavior.** Spouses are expected to confine their sexual behavior to each other, which reduces the risk of having children who do not have socially and legally bonded parents, and of contracting or spreading sexually transmitted infections.

- **Status placement.** Being born into a family provides social placement of the individual in society. One's

family life cycle stages which identify the various challenges faced by members of a family across time.

structure-function framework emphasizes how marriage and family contribute to the larger society.

functionalists structural functionalist theorists who view the family as an institution with values, norms, and activities meant to provide stability for the larger society.

family of origin largely determines one's social class, religious affiliation, and future occupation. Baby Prince George Alexander Louis, son of Kate Middleton and Prince William of the royal family of Great Britain, was born into the upper class and is destined to be in politics by virtue of being born into a political family.

- **Social control.** Spouses in high-quality, durable marriages provide social control for each other that results in less criminal behavior. Parole boards often note that the best guarantee against recidivism is a spouse who expects the partner to get a job and avoid criminal behavior and who reinforces these behaviors.

Personal View: "I Was Stolen from My Family"

Rabbit Proof Fence is a 2002 film that tells the story of Aboriginal children who were taken by force from their parents by the Australian government between 1885 and 1969.

The government is an institution that has an enormous impact on family life. A dramatic example is the Australian government policy in regard to Aboriginal children. In Australia, between 1885 and 1969, between 50 and 100 thousand half caste (one White parent) Aboriginal children were taken by force from their parents by the Australian police. The rationale by the White society was that it wanted to convert these children to Christianity and to destroy their Aboriginal culture which was viewed as primitive and without value. The children were forced to walk or were taken by camel hundreds of miles away from their parents to church missions. (See the *Rabbit-Fence Proof* DVD.)

Bob Randall (2008) is one of the children who was taken by force from his parents at age 7, never to see them again. He was, literally, stolen from his family—physically taken from his mother and taken away, never to see her again. Of his experience, he wrote,

Instead of the wide open spaces of my desert home, we were housed in corrugated iron dormitories with rows and rows of bunk beds. After dinner we were bathed by the older women, put in clothing they called pajamas, and then tucked into one of the iron beds between the sheets. This

was a horrible experience for me. I couldn't stand the feel of the cloth touching my skin (p. 35).

The Australian government subsequently apologized for the laws and policies of successive parliaments and governments that inflicted profound grief, suffering, and loss on the Aborigines. He noted that the Aborigines continue to be marginalized and nothing has been done to compensate for the horror of taking children from their families.

Bob Randall visited the marriage and family class of the author, told of his being taken away from his mother, and sang "Brown Skin Baby (They Took Me Away)." A video of his singing and playing this song close to where he was taken away in Australia over 60 years ago has been posted to YouTube by the Global Oneness Project.

1-4d Conflict Framework

Conflict framework views individuals in relationships as competing for valuable resources (e.g., time, money, power). Conflict theorists recognize that family members have different goals and values that produce conflict. Adolescents want freedom (e.g., stay out all night with new love interest) while parents want their child to get a good night's sleep, not get pregnant, and stay on track in school.

Conflict theorists also view conflict not as good or bad but as a natural and normal part of relationships. They regard conflict as necessary for the change and growth of individuals, marriages, and families. Cohabitation relationships, marriages, and families all have the potential for conflict. Cohabitants are in conflict about commitment to marry, spouses are in conflict about the division of labor, and parents are in conflict with their children over rules such as curfew, chores, and homework. These three units may also be in conflict with other systems. For example, cohabitants are in conflict with the economic institution for health benefits for their partners. Similarly, employed parents are in conflict with their employers for flexible work hours, maternity or paternity benefits, and day care facilities.

Levent Konuk/Shutterstock.com

Conflict theory is also helpful in understanding choices in relationships with regard to mate selection and jealousy. Unmarried individuals in search of a partner are in competition with other unmarried individuals for the scarce resources of a desirable mate. Such conflict is particularly evident in the case of older women in competition for men. At age 85 and older, there are twice as many women (3.7 million) as there are men (1.8 million) (*Statistical Abstract of the United States, 2012–2013*, Table 7). Jealousy is also sometimes about scarce resources. People fear that their "one and only" will be stolen by someone else who has no partner. Thus wives are aware of how much time their husbands spend talking to the attractive newly divorced female at a social gathering.

conflict framework view that individuals in relationships compete for valuable resources.

symbolic interaction framework views marriage and families as symbolic worlds in which the various members give meaning to each other's behavior.

1-4e Symbolic Interaction Framework

The **symbolic interaction framework** views marriages and families as symbolic worlds in which the various members give meaning to one another's behavior. Human behavior can be understood only by the meaning attributed to behavior. Curran et al. (2010) assessed the meaning of marriage for 31 African Americans of different ages and found that the two most common meanings were commitment and love. Herbert Blumer (1969) used the term *symbolic interaction* to refer to the process of interpersonal interaction. Concepts inherent in this framework include the definition of the situation, the looking-glass self, and the self-fulfilling prophecy.

Definition of the Situation Two people who have just spotted each other at a party are constantly defining the situation and responding to those definitions. Is the glance from the other person (1) an invitation to approach, (2) an approach, or (3) a misinterpretation—was he or she looking at someone else? The definition used will affect subsequent interaction.

Getting married also has different definitions/meanings. For "marriage naturalists" it is an event that is a natural progression of a relationship (often begun in high school) and is expected of oneself, one's partner, and both of their families. Persons in rural areas more often have this view. In contrast, "marriage planners" are more metropolitan and view marriage as an event one "gets ready for" by completing one's college or graduate school education, establishing oneself in a job/career, and maturing emotionally and psychologically. These individuals may cohabit and have children before they decide to marry (Kefalas et al., 2011).

Looking-Glass Self The image people have of themselves is a reflection of what other people tell them about themselves (Cooley, 1964). People develop an idea of who they are by the way others act toward them. If no one looks at or speaks to them, they will begin to feel unsettled, according to Charles Cooley. Similarly, family members constantly hold up social mirrors for one another into which the respective members look for definitions of self.

G. H. Mead (1934), a classic symbolic interactionist, believed that people are not passive sponges but that they evaluate the perceived appraisals of others, accepting some opinions and not others. Although some parents teach their children that they are worthless, these children may reject the definition by believing in more positive social mirrors from friends, teachers, and lovers.

Self-Fulfilling Prophecy Once people define situations and the behaviors in which they are expected to engage, they are able to behave toward one another in predictable ways. Such predictability of behavior affects subsequent behavior. If you feel that your partner expects you to be faithful, your behavior is likely to conform to these expectations. The expectations thus create a self-fulfilling prophecy.

1-4f Family Systems Framework

The **family systems framework** views each member of the family as part of a system and the family as a unit that develops norms of interacting, which may be explicit (e.g., parents specify when their children must stop texting for the evening and complete homework) or implicit (e.g., spouses expect fidelity from each other). These rules serve various functions, such as the allocation of keeping the education of offspring on track and solidifying the emotional bond of the spouses.

Rules are most efficient if they are flexible (e.g., they should be adjusted over time in response to a child's growing competence). A rule about not leaving the yard when playing may be appropriate for a 4-year-old but inappropriate for a 16-year-old.

Family members also develop boundaries that define the individual and the group and separate one system or subsystem from another. A boundary may be physical, such as a closed bedroom door, or social, such as expectations that family problems will not be aired in public. Boundaries may also be emotional, such as communication, which maintains closeness or distance in a relationship. Some family systems are cold and abusive; others are warm and nurturing.

In addition to rules and boundaries, family systems have roles (leader, follower, scapegoat) for the respective family members. These roles may be shared by more than one person or may shift from person to person during an interaction or across time. In healthy families, individuals are allowed to alternate roles rather than being locked into one role. In problem families, one family member is often allocated the role of scapegoat, or the cause of all the family's problems (e.g., an alcoholic spouse).

Family systems may be open, in that they are receptive to information and interaction with the outside world, or closed, in that they feel threatened by such contact. The Amish have a closed family system and minimize contact with the outside world. Some communes also encourage minimal outside exposure. Twin Oaks Intentional Community of Louisa, Virginia, does not permit any of its almost 100 members to own a television or keep one in their room. Exposure to the negative drumbeat of the evening news is seen as harmful.

Holmes et al. (2013) used a family systems perspective to explain the transition of spouses and their marriage to parenthood. The researchers noted that it is the context that must be considered to understand changes. For example, having a daughter is associated with more conflict for fathers across time, and this impacts the interaction of the wife with her husband.

"Feminism is the radical notion that women are people."

—ANI DIFRANCO, AMERICAN SINGER

1-4g Feminist Framework

Although a **feminist framework** views marriage and family as contexts of inequality and oppression for women, there are 11 feminist perspectives, including lesbian feminism (emphasizing oppressive heterosexuality), psychoanalytic feminism (focusing on cultural domination of men's phallic-oriented ideas and repressed emotions), and standpoint feminism (stressing the neglect of women's perspective and experiences in the production of knowledge) (Lorber, 1998). Regardless of which feminist framework is being discussed, all feminist frameworks have the themes of inequality and oppression. Feminists seek equality in their relationships with their partners.

family systems framework views each member of the family as part of a system and the family as a unit that develops norms of interaction.

feminist framework views marriage and the family as contexts for inequality and oppression.

1-5 Choices in Relationships: View of the Text

While the previous theoretical frameworks are useful in understanding marriage and the family, in this text we encourage a proactive approach of taking charge of your life and making wise relationship choices. Making the right choices in your relationships, including marriage and family, is critical to your health, happiness, and sense of well-being. Your times of greatest elation and sadness will be in reference to your love relationships.

Although we have many choices to make in our society, among the most important are whether to marry, whom to marry, when to marry, whether to have children, whether to remain emotionally and sexually faithful to one's partner, and whether to protect oneself from sexually transmitted infections and unwanted pregnancy. Though structural and cultural influences are operative, a choices framework emphasizes that individuals have some control over their relationship destiny by making deliberate choices to initiate, nurture, or terminate intimate relationships.

"Things **do not** happen. Things are **made to** happen."

—JOHN F. KENNEDY, 35TH PRESIDENT

1-5a Facts about Choices in Relationships

The facts to keep in mind when making relationship choices include the following.

Not to Decide Is to Decide Not making a decision is a decision by default. If you are sexually active and decide not to use a condom, you have made a decision to increase your risk for contracting a sexually transmissible infection, including HIV. If you do not make a deliberate choice to end a relationship that is unfulfilling, abusive, or going nowhere, you have made a choice to continue in that relationship and have little chance of getting into a more positive and satisfying relationship. If you do not make a decision to be faithful to your partner, you have made a decision to be vulnerable to cheating.

Action Must Follow a Choice While the private life of Woody Allen has been the subject of public dismay (e.g., he married his long-time partner's adopted daughter), he is one of the few Hollywood directors who is given complete control over all aspects of his films. His success began with a decision to become a stand-up comedian and make a name for himself, then use this influence to launch his film career. His biographer writes, "Woody is nothing if he is not deliberate. Decisions may take a long while to be made, but once his mind is made up to do something, he devotes all his effort to it" (Lax 1991, p. 156). While in his twenties, Allen performed two to three shows a night to small, 50-person audiences six nights a week ($75 to $100 a week) for two years. He was beset with the fear of "going live," exhausted at the grueling schedule, and doubted the future (should he quit?). He persevered, pushed through another six months, got his break, and moved forward.

"In any moment of **decision**, the best thing you can do is the **right thing**, the **next best** thing is the **wrong thing**, and the **worst thing** you can do is **nothing**."

—THEODORE ROOSEVELT, 26TH PRESIDENT

Some Choices Require Correction Some of our choices, although appearing correct at the time that we make them, turn out to be disasters. Once we realize that a choice is having consistently negative consequences, we need to stop defending the old choice, reverse the position, make new choices, and move forward. Otherwise, one remains consistently locked into continued negative outcomes of "bad" choices. For example, choosing a partner who was loving and kind but who turns out to be abusive and dangerous requires correcting that choice. To stay in the abusive relationship will have predictable disastrous consequences. To make the decision to disengage and to move on opens the opportunity for a loving relationship with another partner. In the meantime, living alone may be a better alternative than living in a relationship in which you are abused and may end up dead. Other examples of making corrections involve ending dead or loveless relationships (perhaps after investing time and effort to improve the relationship or love feelings), changing jobs or career, and changing friends.

Choices Involve Trade-Offs By making one choice, you relinquish others. Every relationship choice you make will have a downside and an upside. If you decide to stay in a relationship that becomes a long-distance relationship, you are continuing involvement in a relationship that is obviously important to you. However, you may spend a lot of time alone and wonder if you made the right decision to continue the relationship. If you decide to marry, you will give up your freedom to pursue other emotional and/or sexual relationships, and you will also give up some of your control over how you spend your money—but you may also get a wonderful companion with whom to share life. Any partner that you select will also have characteristics that must be viewed as a trade-off. One woman noted of her man, "he doesn't do text messaging or email . . . he doesn't even know how to turn on a computer. But he knows how to build a house, plant a garden, and fix a car . . . trade-offs I'm willing to make."

Choices Include Selecting a Positive or Negative View As Thomas Edison progressed toward inventing the light bulb, he said, "I have not failed. I have found ten thousand ways that won't work." Ron Wayne, negotiating with Steve Jobs, was offered a 10% share in the computer giant Apple when it started up, but he was ambivalent since he would have to invest money that he feared he would lose and thus he did not invest. In early 2011, his 10% stake would have been worth approximately $2.6 billion. Later in life he said that he was not bitter and had made the best decision at the time (Isaacson, 2011, p. 65).

"**Nothing** is either **good or bad** but thinking makes it so."

—WILLIAM SHAKESPEARE, HAMLET

In spite of an unfortunate event in your life, you can choose to see the bright side. Regardless of your circumstances, you can opt for viewing a situation in positive terms. A breakup with a partner you have loved can be viewed as the end of your happiness or an opportunity to become involved in a new, more fulfilling relationship. The discovery of your partner cheating on you can be viewed as the end of the relationship or as an opportunity to examine your situation, to open up communication channels with your partner, and to develop a stronger connection. Discovering that you are infertile can be viewed as a catastrophe or as a challenge to face adversity with your partner. It is not the event but your view of it that determines its effect on you.

Most Choices Are Revocable; Some Are Not Most choices can be changed. For example, a person who has chosen to be sexually active with multiple partners can decide to be monogamous or to abstain from sexual relations in new relationships. People who have been unfaithful in the past can elect to be emotionally and sexually committed to a new partner.

Other choices are less revocable. For example, backing out of the role of parent is very difficult. Social pressure keeps most parents engaged, but the law (e.g., forced child support) is the backup legal incentive. Hence, the decision to have a child is usually irrevocable. Choosing to have unprotected sex may also result in a lifetime of coping with sexually transmitted infections.

> "Do you think we enjoy **hearing about** your brand-new **million-dollar home** when we **can barely afford** to eat Kraft Dinner sandwiches in our own **grimy little shoe boxes** and we're pushing thirty?"
>
> —DOUGLAS COUPLAND, FROM *GENERATION X: TALES FOR AN ACCELERATED CULTURE*

Choices of Generation Y Generations vary (see Table 1.4). Those in **Generation Y** (typically born between 1979 and 1984) are the children of the baby boomers. Numbering about 80 million, these Generation Yers (also known as the Millennial or Internet Generation) have been the focus of their parents' attention. They have been nurtured, coddled, and scheduled into day care to help them get ahead. The result is a generation of high self-esteem, self-absorbed individuals who believe they are "the best." Unlike their parents, who believe in paying one's dues, getting credentials, and sacrificing through hard work to achieve economic stability, Generation Yers focus on fun, enjoyment, and flexibility. They might choose a summer job at the beach if it buys a burger and an apartment with six friends over an internship at IBM that smacks of the corporate America sellout.

Generation Y children of the baby boomers, typically born between 1979 and 1984. Also known as the Millennial or Internet Generation.

Table 1.4
Four Generations in Recent History

Generation	Characteristics
Great Generation	Years of the Great Depression, World War II (veterans and civilians). Culture steeped in traditional values.
Baby Boomers	Children of World War II's Great Generation, born between 1946 and 1964. Questioning of traditional values.
Gen X (Generation X)	Children of boomers born between early the 1960s to early 1980s. Generation of change, MTV, AIDS, diversity.
Gen Y (Generation Y)	Also known as Millennial Generation, persons born from early 1980s to early 2000s. Loyalty to corporations gone, frequent job changes.

Choices Are Influenced by the Stage in the Family Life Cycle The choices a person makes tend to be individualistic or familistic, depending on the stage of the family life cycle that the person is in. Before marriage, individualism characterizes the thinking and choices of most individuals. Individuals need only be concerned with their own needs. Most people delay marriage in favor of completing school, becoming established in a career, and enjoying the freedom of singlehood.

Once married, and particularly after having children, the person's familistic values and choices ensue as the needs of a spouse and children begin to influence. Evidence of familistic choices is reflected in the fact that spouses with children are less likely to divorce than spouses without children.

Making Wise Choices Is Facilitated by Learning Decision-Making Skills Choices occur at the individual, couple, and family levels. Deciding to transfer to another school or take a job out of state may involve all three levels, whereas the decision to lose weight is more likely to be an individual decision. Regardless of the level, the steps in decision making include setting aside enough time to evaluate the issues involved in making a choice, identifying alternative courses of action, carefully weighing

Courtesy of Michelle Butterfield

A major focus of Generation Yers is to have fun.

Table 1.5

"Best" and "Worst" Choices Identified by University Students

Best Choice	Worst Choice
Waiting to have sex until I was older and involved.	Cheating on my partner.
Ending a relationship with someone I did not love.	Getting involved with someone on the rebound.
Insisting on using a condom with a new partner.	Making decisions about sex when drunk.
Ending a relationship with an abusive partner.	Staying in an abusive relationship.
Forgiving my partner and getting over cheating.	Changing schools to be near my partner.
Getting out of a relationship with an alcoholic.	Not going after someone I really wanted.

Reprinted by permission of Dr. Lee Axelson.

the consequences for each choice, and being attentive to your own inner voice ("Listen to your senses"). The goal of most people is to make relationship choices that result in the most positive and least negative consequences. Students in my marriage and family class identified their "best" and "worst" relationship choices (see Table 1.5).

1-6 Research: Process and Evaluation

Research is valuable since it helps to provide evidence for or against a hypothesis. For example, there is a stigma associated with persons who have tattoos and it is often assumed that students who have tattoos make lower grades than those who do not have tattoos. But Martin and Dula (2010) compared the GPA of persons who had tattoos and those who did not and found no significant differences. Researchers follow a standard sequence when conducting a research project. We discuss these in the following section and emphasize certain caveats to be aware of when reading any research finding.

1-6a Steps in the Research Process

Several steps are used in conducting research.

1. **Identify the topic or focus of research.** Select a subject about which you are passionate. For example, are you interested in studying cohabitation of college students? Give your project a title in the form of a question—"Do People Who Cohabit Before Marriage Have Happier Marriages Than Those Who Do Not Cohabit?"

2. **Review the literature.** Go online to the various databases of your college or university and read research that has already been published on cohabitation. Not only will this prevent you from "reinventing the wheel" (you might find that a research study has already been conducted on exactly what you want to study), but it will also give you ideas for your study.

3. **Develop hypotheses.** A **hypothesis** is a suggested explanation for a phenomenon. For example, you might hypothesize that cohabitation results in greater marital happiness and less divorce because the partners have a chance to "test-drive" each other and their relationship.

4. **Decide on type of study and method of data collection.** The type of study may be **cross-sectional**, which means studying the whole population at one time—in this case, finding out from persons now living together about their experience—or **longitudinal**, which means studying the same group across time—in this case, collecting data from the same couple each of their four years of living together during college. The method of data collection could be archival (secondary sources such as journals), survey (questionnaire), interview (one or both partners), or case study (focus on one couple). A basic differentiation in method is quantitative (surveys, archival) or qualitative (interviews/case study).

5. **Get IRB approval.** To ensure the protection of people who agree to be interviewed or who complete questionnaires, researchers must obtain

hypothesis a suggested explanation for a phenomenon.

cross-sectional study means studying the whole population at one time (e.g, finding out from persons now living together about their experience).

longitudinal study means studying the same group across time (e.g., follow several couples who are living together at one-year intervals for 10 years).

IRB approval by submitting a summary of their proposed research to the Institutional Review Board (IRB) of their institution. The IRB reviews the research plan to ensure that the project is consistent with research ethics and poses no undue harm to participants. When collecting data from individuals, it is important that they are told that their participation is completely voluntary, that the study maintains their anonymity, and that the results are confidential. Respondents under age 18 need the consent of their parents. Community members were aghast when some parents of students at Memorial Middle School in Massachusetts reported that they never received consent forms for their seventh graders to take a survey on youth risk behavior that included items on oral sex and drug use (Starnes, 2011).

6. **Collect and analyze data.** Various statistical packages are available to analyze data to discover if your hypotheses are true or false.

7. **Write up and publish results.** Writing up and submitting your findings for publication are important so that your study becomes part of the academic literature.

1-6b Evaluating Research in Marriage and the Family

"New Research Study" is a frequent headline in popular magazines such as *Cosmopolitan* promising accurate information about "hooking up," "sexting," "what women/men want," or other sexual, communication, and gender issues. As you read such articles, as well as the research in such texts as this, be alert to their potential flaws.

Sample Some of the research on marriage and the family is based on random samples. In a **random sample**, each individual in the population has an equal chance of being included in the sample. Random sampling involves randomly selecting individuals from an identified population. That population often does not include the homeless or people living on military bases. Studies that use random samples are based on the assumption that the individuals studied are similar to and therefore representative of the population that the researcher is interested in.

Because of the trouble and expense of obtaining random samples, most researchers study subjects to whom they have convenient access. This often means students in the researchers' classes. The result is an overabundance of research on "convenience" samples consisting of white, Protestant, middle-class college students. Because college students cannot be assumed to be similar in their attitudes, feelings, and behaviors to their noncollege peers or older adults, research based on college students cannot be generalized beyond the base population.

In addition to having a random sample, having a large sample is important. The American Council on Education and the University of California (Eagan et al., 2013) collected a national sample of 165,743 first-semester undergraduates at 234 colleges and universities throughout the United States; the sample was designed to reflect the responses of over a million first-time, full-time students entering four-year colleges and universities. If only 50 college students had been in the sample, the results would have been less valuable in terms of generalizing beyond that sample.

Be alert to the sample size of the research you read. Most studies are based on small samples. In addition, considerable research has been conducted on college students, leaving us to wonder about the attitudes, values, and behaviors of noncollege students.

Control Groups In an example of a study that concludes that an abortion (or any independent variable) is associated with negative outcomes (or any dependent variable), the study must necessarily include two groups: (1) women who have had an abortion, and (2) women who have not had an abortion. The latter would serve as a **control group**—the group not exposed to the independent variable you are studying (**experimental group**). Hence, if you find that women in both groups in your study develop negative attitudes toward sex, you know that abortion cannot be the cause. Be alert to the existence of a control group, which is usually not included in research studies.

IRB approval
Institutional Review Board approval is the OK by one's college, university, or institution that the proposed research is consistent with research ethics standards and poses no undo harm to participants.

random sample
sample in which each person in the population being studied has an equal chance of being included in the sample.

control group group used to compare with the experimental group that is not exposed to the independent variable being studied.

experimental group the group exposed to the independent variable.

Age and Cohort Effects In some research designs, different cohorts or age groups are observed and/or tested at one point in time. One problem that plagues such research is the difficulty—even impossibility—of discerning whether observed differences between the subjects studied are due to the research variable of interest, cohort differences, or some variable associated with the passage of time (e.g., biological aging).

A good illustration of this problem is found in research on changes in marital satisfaction over the course of the family life cycle. In such studies, researchers may compare the levels of marital happiness reported by couples that have been married for different lengths of time. For example, a researcher may compare the marital happiness of two groups of people—those who have been married for 50 years and those who have been married for 5 years. However, differences between these two groups may be due to (1) differences in age (age effect), (2) differences in the historical time period that the two groups have lived through (cohort effect), or (3) differences in the lengths of time the couples have been married (research variable).

Terminology In addition to being alert to potential shortcomings in sampling and control groups, you should consider how the phenomenon being researched is defined. For example, if you are conducting research on living together, how would you define this term? How many people, of what sex, spending what amount of time, in what place, engaging in what behaviors will constitute your definition? Indeed, researchers have used more than 20 definitions of what constitutes living together.

What about other terms? Considerable research has been conducted on marital success, but how is the term to be defined? What is meant by marital satisfaction, commitment, interpersonal violence, and sexual fulfillment? Before reading too far in a research study, be alert to the definitions of the terms being used. Exactly what is the researcher trying to measure?

Researcher Bias Although one of the goals of scientific studies is to gather data objectively, it may be impossible for researchers to be totally objective. Researchers

are human and have values, attitudes, and beliefs that may influence their research methods and findings. It may be important to know what the researcher's bias is in order to evaluate the findings. For example, a researcher who does not support abortion rights may conduct research that focuses only on the negative effects of abortion. Similarly, researchers funded by corporations who have a vested interest in having their products endorsed are also suspect.

Time Lag Typically, a two-year lag exists between the time a study is completed and the study's appearance in a professional journal. Because textbook production takes even longer than getting an article printed in a professional journal, they do not always present the most current research, especially on topics in flux. In addition, even though a study may have been published recently, the data on which the study was based may be old. Be aware that the research you read in this or any other text may not reflect the most current, cutting-edge research.

Distortion and Deception In some cases there is outright deception by professionals. In a study of improvement of student scores in the Atlanta public school system, an investigation revealed that cheating occurred at 44 schools involving at least 178 teachers and principals (almost half of whom confessed) (Severson, 2011). In another example, Dr. Anil Potti (Duke University) changed data on research reports and provided fraudulent results (Darnton, 2012) to ensure the continuous flow of funds for his research projects.

Distortion and deception, deliberate or not, also exist in marriage and family research. Marriage is a very private relationship that happens behind closed doors; individual interviewees and respondents to questionnaires have been socialized not to reveal the intimate details of their lives to strangers. Hence, they are prone to distort, omit, or exaggerate information, perhaps unconsciously, to cover up what they may feel is no one else's business. Thus, researchers sometimes obtain inaccurate information.

Marriage and family researchers know more about what people say they do than about what they actually do. An unintentional and probably more common form of distortion is inaccurate recall. Sometimes researchers ask respondents to recall details of their relationships that occurred years ago. Time tends to blur some memories, and respondents may relate not what actually happened, but rather what they remember to have happened, or, worse, what they wish had happened.

Other Research Problems Nonresponse on surveys and the discrepancy between attitudes and behaviors are other research problems. With regard to nonresponse, not all individuals who complete questionnaires or agree to participate in an interview are willing to provide information about such personal issues as date rape and partner abuse. Such individuals leave the questionnaire blank or tell the interviewer they would rather not respond. Others respond but give only socially desirable answers. The implications for research are that data gatherers do not know the nature or extent to which something may be a problem because people are reluctant to provide accurate information.

It is sometimes assumed that if people have a certain attitude (e.g., a belief that extramarital sex is wrong), then their behavior will be consistent with that attitude (avoidance of extramarital sex). However, this assumption is not always accurate. People do indeed say one thing and do another. This potential discrepancy should be kept in mind when reading research on various attitudes.

Finally, most research reflects information that volunteers provide. However, volunteers may not represent nonvolunteers when they are completing surveys.

Table 1.6
Potential Research Problems in Marriage and Family

Weaknesses	Consequences	Examples
Sample not random	Cannot generalize findings	Opinions of college students do not reflect opinions of other adults.
No control group	Inaccurate conclusions	Study on the effect of divorce on children needs control group of children whose parents are still together.
Age differences between groups of respondents	Inaccurate conclusions	Effect may be due to passage of time or to cohort differences.
Unclear terminology	Inability to measure what is not clearly defined	What is definition of cohabitation, marital happiness, sexual fulfillment, good communication, and quality time?
Researcher bias	Slanted conclusions	A researcher studying the value of a product (e.g., Atkins Diet) should not be funded by the organization being studied.
Time lag	Outdated conclusions	Often-quoted Kinsey sex research is over 65 years old.
Distortion	Invalid conclusions	Research subjects exaggerate, omit information, and/ or recall facts or events inaccurately. Respondents may remember what they wish had happened.
Deception	Public misled	Dr. Anil Potti (Duke University) changed data on research reports and provided fraudulent results (Darnton, 2012).

In view of the research cautions identified here, you might ask, "Why bother to report the findings?" The quality of some family science research is excellent. For example, articles published in *Journal of Marriage and the Family* (among other journals) reflect the high level of methodologically sound articles that are being published. There are even some less sophisticated journals that provide useful information on marital, family, and other relationship data. Table 1.6 summarizes potential inadequacies of any research study.

Danny Smythe/Shutterstock.com

a high school but not a four-year college degree), marriage remains the dominant choice for most Americans, particularly for college-educated individuals with a good income (National Marriage Project, 2012). Though these individuals will increasingly delay getting married until their late twenties or early thirties (to complete their educations, launch their careers, and/or become economically independent), there is no evidence that marriage will cease to be a life goal. Indeed 6 in 10 never-married adults say they want to get married.

With the Supreme Court legalizing same-sex marriage and federal tax codes permitting the filing of joint returns by same-sex couples, an increasing number of states will legalize same-sex marriage. Marriage education programs traditionally designed for heterosexual couples will then be modified to include same-sex partners (Whitton & Buzzellla, 2012).

1-7 Trends in Marriage and Family

While there is a decline of marriage among middle Americans (defined as the nearly 60% of Americans aged 25 to 60 who have

---

STUDY TOOLS 1

Ready to study? In the book, you can:

- Rip out the chapter review card at the back of the book for a handy summary of the chapter and key terms.

- Assess your attitudes toward marriage with the Self-Assessment card at the back of the book.

Online at CENGAGEBRAIN.COM you can:

- Prepare for tests with quizzes.

- Review the key terms with Flash Cards.

- Play games to master concepts.

Singlehood, Hooking Up, Cohabitation, and Living Apart Together

"For I shall never give up the state of **living alone**, which has manifested itself as an **indescribable blessing.**"

—ALBERT EINSTEIN

What do Sheryl Crow, Tyra Banks, Diane Keaton, Hugh Grant, Ophra Winfrey, and Bill Maher have in common? Each has never married. Indeed, in 2012, 23% of men and 17% of women age 25 and older had never married (Wang & Parker, 2014). While their never having stood at the altar puts them in the minority, increasingly, more individuals are not only delaying marriage, cohabiting and rearing children outside of marriage, but may be foregoing the marriage experience permanently.

Though 53% of never-married adults report wanting to eventually get married (Wang & Parker, 2014), youth today are in no hurry. They enjoy the freedom of singlehood and put off marriage until their late twenties and beyond (the same is true of youth in France, Germany, and Italy). In the meantime, the process of courtship includes hanging out (small groups of individuals interacting), hooking up (the new term for "one-night stand"), and "being in a relationship" (seeing each other exclusively and announcing one's relationship status on Facebook), which may include cohabitation and rearing children as a prelude to marriage. We begin with an examination of singlehood.

2-1 Singlehood

In this section, we discuss the reasons for remaining single, how social movements have increased the acceptance of singlehood, the various categories of single people, and the choice to be permanently unmarried.

2-1a **Reasons for Remaining Single**

Muraco and Curran (2012) identified 13 reasons individuals delay marriage. These included financial stability, ability to pay for a wedding, doubts about self as a potential spouse, doubts about partner as spouse, quality of relationship, doubts about self as parent, doubts about partner as parent, capability of being economic provider, partner capability of being economic provider, fear of divorce, infidelity, in-laws, and bringing children from own and partner's previous relationships together.

Table 2.1 lists some other standard reasons people give for remaining single. The primary advantage of remaining single is freedom and control over one's life. Others do not set out to be single but drift into singlehood longer than they anticipated— discover that they like it, and remain single.

Still other individuals remain single out of fear. Their parents, siblings, and friends are divorced so they feel their chances for happiness are better remaining single. " I know I can be happy single," said one of our students. "It's marriage that scares me."

There is also a scarcity of men. In her article on "The End of Men" Rosin (2010) noted that women now hold the majority of the nation's jobs. The result is that men are being marginalized and bring less to the table economically. Rosin also states that modern industrial society no longer values men's size and strength and that "social intelligence, open communication, the ability to sit still and focus are, at a minimum, not predominantly male"—yet these are precisely the qualities now demanded in the global economy. Black women, particularly, report that there are few Black men who have the education and job stability they prefer in a mate. Only 26% of Black women compared to 51% of White women are married (National Survey of Family Growth, 2014).

Regardless of the reason, will those who delay getting married eventually marry? Ninety-five percent of 4,689 undergraduates (slightly higher percent of females than males) at two universities agreed that "Someday, I want to marry" (Hall & Knox, 2015).

Courtesy of Michelle North

Seventy-five percent of all young adults today will eventually marry (Wang & Parker, 2014). And the older a person, the more likely the person is to have married. By age 65, 96% of Americans have married at least once (*Proquest Statistical Abstract of the United States 2014*, Table 35). In the meantime, the existence and enjoyment of singlehood is facilitated by three social movements.

2-1b Social Movements and the Acceptance of Singlehood

The acceptance of singlehood as a lifestyle can be attributed to social movements—the sexual revolution, the women's movement, and the gay liberation movement. The sexual revolution involved openness about sexuality and permitted intercourse outside the context of marriage. No longer did people feel compelled to wait until marriage for involvement in a sexual relationship. Hence, the sequence changed from dating, love, maybe intercourse with a future spouse, and then marriage and parenthood to "hanging out," "hooking up" with numerous partners, maybe living together (in one or more relationships), marriage, and children. (For some, living together is a required relationship stage.)

The women's movement emphasized equality in education, employment, and income for women. As a result, rather than get married and depend on a husband for income, women earned higher degrees, sought career opportunities, and earned their own income. This economic independence brought with it independence of choice. Women could afford to remain single or to leave an unfulfilling or abusive relationship. Commanding respect and an egalitarian relationship have become normative to today's relationships.

Table 2.1
Reasons to Remain Single

Benefits of Singlehood	Limitations of Marriage
Freedom to do as one wishes	Restricted by spouse or children
Variety of lovers	One sexual partner
Spontaneous lifestyle	Routine, predictable lifestyle
Close friends of both sexes	Pressure to avoid close other-sex friendships
Responsible for one person only	Responsible for spouse and children
Spend money as one wishes	Expenditures influenced by needs of spouse and children
Freedom to move as career dictates	Restrictions on career mobility
Avoid being controlled by spouse	Potential to be controlled by spouse
Avoid emotional and financial stress of divorce	Possibility of divorce

Gay relationships are becoming more normative. Over 30 states now recognize same sex marriages.

The gay liberation movement, with its push for recognition of same-sex marriage (more than 30 states have legalized same-sex marriage), has increased the visibility of gay individuals and relationships. The Supreme Court ruling for gay marriage, openly gay politicians (e.g., Steve Gallardo), NFL athletes (e.g., Michael Sam), and celebrities (e.g., Ellen DeGeneres) infuse new norms into our society. Though some gay people still marry heterosexuals to provide a traditional social front, the gay liberation movement has provided support for a lifestyle consistent with one's sexual orientation.

In effect, there is a new wave of youth who feel that their commitment is to themselves in early adulthood and to marriage in their late twenties and thirties, if at all. The increased acceptance of singlehood translates into staying in school or getting a job, establishing oneself in a career, and becoming economically and emotionally independent from one's parents. The old pattern was to leap from high school into marriage. The new pattern of these Generation Yers (discussed in Chapter 1) is to wait until after college, become established in a career, and enjoy themselves.

2-1c Unmarried Equality Organization

According to the mission statement identified on the website of the Unmarried Equality (http://www.unmarried.org), the emphasis of the organization is to advocate

…for equality and fairness for unmarried people, including people who are single, choose not to marry, cannot marry, or live together before marriage. We provide support and information for this fast-growing constituency, fight discrimination on the basis of marital status, and educate the public and policymakers about relevant social and economic issues. We believe that marriage is only one of many acceptable family forms, and that society should recognize and support healthy relationships in all their diversity.

The Unmarried Equality organization exists, in part, because a stigma toward remaining single and never marrying remains. See the stereotypes, positives and negatives, identified by one of the speakers in the author's marriage and family class.

2-1d Legal Blurring of the Married and Unmarried

The legal distinction between married and unmarried couples is blurring. Whether it is called the deregulation of marriage or the deinstitutionalization of marriage, the result is the same—more of the privileges previously reserved for the married are now available to unmarried and/or same-sex couples. Domestic partnerships, recognized in some cities and by some corporations, convey rights and privileges (e.g., health benefits for a partner) previously available only to married people.

"In many ways, **single women** are under **constant social surveillance**. They are constantly being questioned: So what's new? Are you seeing anyone? What are you waiting for?! They are constantly being **warned** that they are liable to **miss their train or die alone**."

—KINNERET LAHAD, SOCIOLOGIST

Personal View: A Never-Married Single Woman's View of Singlehood

A never-married woman, 40 years of age, revealed her experience of being a single woman. The following is from the outline she used in making this presentation to the author's marriage and family class.

STEREOTYPES ABOUT NEVER-MARRIED WOMEN

Various assumptions are made about the never-married woman and why she is single. These include the following:

Unattractive. She's either overweight or homely, or else she would have a man.

Lesbian. She has no real interest in men and marriage because she is homosexual.

Workaholic. She's career-driven and doesn't make time for relationships.

Poor interpersonal skills. She has no social skills, and she embarrasses men.

History of abuse. She has been turned off to men by the sexual abuse of, for example, her father, a relative, or a date.

Negative previous relationships. She's been rejected again and again and can't hold a man.

Man-hater. Deep down, she hates men.

Frigid. She hates sex and avoids men and intimacy.

Promiscuous. She is indiscriminate in her sexuality so that no man respects or wants her.

Too picky. She always finds something wrong with each partner and is never satisfied.

Too weird. She would win the Miss Weird contest, and no man wants her.

POSITIVE ASPECTS OF BEING SINGLE

1. Freedom to define self in reference to own accomplishments, not in terms of attachments (e.g., spouse).
2. Freedom to pursue own personal and career goals and advance without the time restrictions posed by a spouse and children.
3. Freedom to come and go as you please and to do what you want, when you want.
4. Freedom to establish relationships with members of both sexes at desired level of intensity.
5. Freedom to travel and explore new cultures, ideas, and values.

NEGATIVE ASPECTS OF BEING SINGLE

1. **Increased extended-family responsibilities.** The unmarried sibling is assumed to have the time to care for elderly parents.
2. **Increased job expectations.** The single employee does not have marital or family obligations and consequently can be expected to work at night, on weekends, and holidays.
3. **Isolation.** Too much time alone does not allow others to give feedback such as "Are you drinking too much?" "Have you had a checkup lately?" or "Are you working too much?"
4. **Decreased privacy.** Others assume the single person is always at home and always available. They may call late at night or drop in whenever they feel like it. They tend to ask personal questions freely.
5. **Less safety.** A single woman living alone is more vulnerable than a married woman with a man in the house.
6. **Feeling different.** Many work-related events are for couples, husbands, and wives. A single woman sticks out.
7. **Lower income.** Single women have much lower incomes than married couples.

8. **Less psychological intimacy.** The single woman does not have an emotionally intimate partner at the end of the day.

9. **Negotiation skills lie dormant.** Because single people do not negotiate issues with someone on a regular basis, they may become deficient in compromise and negotiation skills.

10. **Patterns become entrenched.** Because no other person is around to express preferences, the single person may establish a very repetitive lifestyle.

MAXIMIZING ONE'S LIFE AS A SINGLE PERSON

1. **Frank discussion.** Talk with parents about your commitment to and enjoyment of the single lifestyle and request that they drop marriage references. Talk with siblings about joint responsibility for aging parents and your willingness to do your part. Talk with employers about spreading workload among all workers, not just those who are unmarried and childfree.

2. **Relationships.** Develop and nurture close relationships with parents, siblings, extended family, and friends to have a strong and continuing support system.

3. **Participate in social activities.** Go to social events with or without a friend. Avoid becoming a social isolate.

4. **Be cautious.** Be selective in sharing personal information such as your name, address, and phone number.

5. **Money.** Pursue education to maximize income; set up a retirement plan.

6. **Health.** Exercise, have regular checkups, and eat healthy food. Take care of yourself.

© s_bukley/Shutterstock.com

Diane Keaton is an example of a never-married woman who has maximized her life as a single person.

2-1e Categories of Singles

The term **singlehood** is most often associated with young unmarried individuals. However, there are three categories of single people: the never-married, the divorced, and the widowed. Combined, there are 102 million single people in the United States (http://www.census.gov/hhes/families/data/cps2012.html).

Never-Married Singles There are 64 million never-married adults ages 18 years and older in the United States. What are the characteristics of those who never marry? Men are likely to be less educated (Murray, 2012) and to have lower incomes (Ashwin & Isupova, 2014) than married men. Never-married women tend to be poor, to have mental/physical health issues, to use drugs, and to have children with multiple partners (Manning et al., 2010). Those who are never married (both women and men) also tend to be Black (Wang & Parker, 2014), obese (Sobal & Hansen, 2011) and to have been reared in single-parent families (Valle & Tillman, 2014). Of course, never-married individuals may share none of these characteristics just as some spouses may be less educated. A disproportionate

> **Cross-Cultural Data**
> In a review of worldwide singlehood, of over 150 countries reporting, all but 16 of the countries revealed that 20% had never married by age 49 (United Nations, 2011).

singlehood state of being unmarried.

> ## "When I'm single, I don't focus. *I focus on a guy if he's a boyfriend, but I don't focus on finding a boyfriend. They're never around when you want them.*"
>
> —SCARLETT JOHANSSON, ACTRESS

number of unmarried individuals live in large cities—New York, DC, Los Angeles, Atlanta, and Boston. These individuals are often young adults seeking jobs/careers, adventure, and relationships.

Are singles less happy than spouses? Yes. While over three-quarters of 5,500 single individuals reported that they were happy with their personal life (Walsh, 2013), when compared to those who are married, they are less happy. Wienke and Hill (2009) compared single people with married people and cohabitants (both heterosexual and homosexual) and found that single people were less happy regardless of sexual orientation. Rauer (2013) also compared the unmarried (she included divorced and widowed) with the married and found that the latter were happier, healthier, etc., and that these benefits were particularly pronounced in the elderly—that "the negative effects of being unmarried are disproportionately felt by older adults."

Increasingly, individuals are choosing to live alone: 28% of all households were single households in 2011 compared to 9% in 1950. And "why not?" asks sociologist Klinenberg (2012), "living alone helps to pursue sacred modern values such as individual freedom, personal control and self-realization." His book *Going Solo* reflects interviews with 300 who live alone and, for the most part, enjoy doing so. Eck (2013) interviewed single individuals and found most content in being unmarried.

Divorced Singles Another category of those who are single is the divorced. There were 14.2 million divorced females and 10.7 million divorced males in the United States in 2012 (*Proquest Statistical Abstract of the United States, 2014*, Table 59). While some divorced singles may have ended the marriage, others were terminated by their spouse. Hence, some are voluntarily single again while others are forced into being single again.

The divorced have a higher suicide risk. Denney (2010) examined the living arrangements of over 800,000 adults and found that being married or living with children decreases one's risk of suicide. Spouses are more likely to be "connected" to intimates; this "connection" seems to insulate a person from suicide. Of course, intimate connections can occur outside of marriage but marriage tends to ensure these connections over time.

Spouses look out for the health of each other. Spouses often prod each other to "go to the doctor," "have that rash on your skin looked at," and "remember to take your pills." Single people often have no one in their life to nudge them toward regular health maintenance.

Widowed Singles Although divorced people often choose to leave their spouses and be single again, the widowed are forced into singlehood. There are over 14 million widowed singles in the United States—11.2 million widowed females and 2.9 widowed males in the United States (*Proquest Statistical Abstract of the United States, 2014*, Table 59). Approximates a 10:1 ratio of females to males. The stereotype of the widow and widower is utter loneliness, even though there are compensations (e.g., escape from an unhappy marriage, social security). Kamiya et al. (2013) found that widowhood for men was associated with depressive symptoms. But this association was mitigated by income.

2-2 Ways of Finding a Partner

Being connected to someone is a goal for many singles. In a sample of undergraduates at two universities, 44% of 3,570 women and 37% of 1,148 men reported that finding a partner with whom to have a happy marriage was a top value (Hall & Knox, 2015). In this section, we review the various ways of finding a partner to hang out with or to marry.

"HEY BIG BOY!": WOMEN WHO INITIATE RELATIONSHIPS WITH MEN

Mae West, film actress of the 1930s and 1940s, is remembered for being very forward with men. Two classic phrases of hers are "Is that a pistol in your pocket or are you glad to see me?" and "Why don't you come up and see me sometime, I'm home every night?" As a woman who went after what she wanted, West was not alone—then or now. There have always been women not bound by traditional gender role restrictions.

Blend Images/Alamy

Ross et al. (2008) conducted a study on women, like Mae West, who have ventured beyond the traditional gender role expectations of the passive female. Data for the study consisted of 692 undergraduate women who answered "yes" or "no" on a questionnaire to the question, "I have asked a new guy to go out with me"—a nontraditional gender role behavior.

Of the 692 women surveyed, 39.1% reported that they had asked a new guy out on a date; 60.9% had not done so. Analysis of the data revealed seven statistically significant findings in regard to the characteristics of those women who had initiated a relationship with a man and those who had not done so.

1. **Nonbeliever in "one true love."** Of the women who asked men out, 42.3% did not believe in one true love in contrast to 31.7% who believed in one true love—a statistically significant difference (p < .01). These assertive women felt they had a menu of men from which to choose.

2. **Experienced "love at first sight."** Of the women who had asked a man out, 45.7% had experienced falling in love at first sight in contrast to 28.2% who had not had this experience. Hence, women who let a man know they were interested in him were likely to have already had a "sighting" of a man with whom they fell in love.

3. **Sought partner on the Internet.** Only a small number (59 of 692; 8.5%) of women reported that they had searched for a partner using the Internet. However, 54.2% of those who had done so (in contrast to only 37.5% who had not used the Internet to search for a partner) reported that they had asked a man to go out (p < .02). Because both seeking a partner on the Internet and asking a partner to go out verbally are reflective of nontraditional gender role behavior, these women were clearly in charge of their lives and moved the relationship forward rather than waiting for the man to make the first move.

4. **Nonreligious.** Respondents who were not religious were more likely to ask a guy out than those who were religious (74.6% versus 64.2%; p < .001). These women may have felt inclined to go after their man rather than have him sent from heaven.

5. **Nontraditional sexual values.** Consistent with the idea that women who had asked a guy out were also nonreligious (a nontraditional value) is the finding that these same women tended to have nontraditional sexual values. Of those women who had initiated a relationship with a guy, 44.4% reported having a hedonistic sexual value ("If it feels good, do it.") compared to 25.7% who regarded themselves as having absolutist sexual values ("Wait until marriage to have intercourse."). Hence, women with nontraditional sexual values were much more likely to be aggressive in initiating a new relationship with a man.

6. **Open to cohabitation.** Of the women who reported that they had asked a man to go out, 44.6% reported that they would cohabit with a man compared to 26.0% who would not

cohabit. This finding is consistent with the liberal female who goes after her man and who does not require marriage.

7. **White.** Over 40% (41.4%) of the White women, compared to 28.2% of Black women, in the sample reported that they had asked a guy out. This finding may be more reflection of context than race. Since most of the men available to these women were White, Black women would be crossing racial lines to ask a White male to go out. If this study were conducted at a predominately Black university, we anticipate that there would be a higher percentage of Black females who would ask out a Black male.

In summary, the study revealed that 39.1% of the undergraduate women at a large southeastern university had asked a guy to go out (a nontraditional gender role behavior). There are three implications of this finding for both women and men. Women who feel uncomfortable asking a man out, who fear rejection for doing so, or who lack the social skills to do so ("Hey big boy! Wanna hang out?") may be less likely to find the man who will be whisked away by women who have such comfort and who make their interest in a partner known.

The implication of this study for men is not to be surprised when a woman makes a direct request to hang out—relationship norms are changing. For some men, this comes as a welcome trend in that they feel burdened that they must always be the first one to indicate interest in a partner and to move the relationship forward. Men might also reevaluate their negative stereotypes of women who initiate relationships ("they are loose") and be reminded that the women in this study who had asked men out were *more* likely to have been faithful in previous relationships than those who had not.

2-2a Hanging Out

Hanging out, also referred to as getting together, refers to going out in groups where the agenda is to meet others and have fun. The individuals may watch television, rent a DVD, go to a club or party, and/or eat out. Of 4,712 undergraduates, 91% reported that "hanging out for me is basically about meeting people and having fun" (Hall & Knox, 2015). Hanging out may occur in group settings such as at a bar, a sorority or fraternity party, or a small gathering of friends that keeps expanding. Friends may introduce individuals, or they may meet someone "cold," as in initiating a conversation. Hanging out is basically about screening and interviewing a number of potential partners in one setting. At a party, one can drift over to a potential who "looks good" and start talking. If there is chemistry in the banter and perceived interest from the person, the interaction will continue and can include the exchange of phone numbers for subsequent texting. If there is no chemistry, the individual can move on to the next person without having invested any significant time. Both partners are in the process of assessing the other. There is usually no agenda beyond meeting and having fun. Of the respondents noted above, only 6% said that hanging out was about beginning a relationship that may lead to marriage (Hall & Knox, 2015).

2-2b Hooking Up

Hooking up is a sexual encounter that occurs between individuals who have no relationship commitment. Lewis et al. (2013) defined hooking up in their survey as: "event where you were physically intimate (any of the following: kissing, touching, oral sex, vaginal sex, anal sex) with someone whom you were not dating or in a romantic relationship with at the time and in which you understood there was no mutual expectation of a romantic commitment." Their sample of 1,468 revealed that, while definitions vary, most define hooking up as involving some type of sex (vaginal, oral, anal), not just kissing.

For those who hook up there is generally no expectation of seeing one another again and alcohol is often involved. Chang et al. (2012) identified the unspoken rules of hooking up—hooking up is not dating, hooking up is not a romantic relationship, hooking up is physical, hooking up is secret, one who hooks up is to expect no subsequent phone calls from their hooking up partner, and condom/protection should always occur (though only 57% of their sample reported condom use on hookups). Aubrey and Smith (2013) also noted that

hanging out going out in groups where the agenda is to meet others and have fun.

hooking up having a one-time sexual encounter in which there are generally no expectations of seeing one another again.

Pressmaster/Shutterstock.com

there is a set of cultural beliefs about hooking up. These beliefs include that hooking up is shameless/fun, will enhance one's status in one's peer group, and reflects one's sexual freedom/control over one's sexuality.

Data on the frequency of college students having experienced a hookup varies. Barriger and Velez-Blasini (2013) in their review of literature found that between 77% and 85% of undergraduates reported having hooked up within the previous year. Reporting on more than 17,000 students at 20 colleges and universities (from Stanford University's Social Life Survey), sociologist Paula England revealed that 72% of both women and men reported having hooked up (40% intercourse, 35% kiss, touch; 12% hand genital; 12% oral sex). Men reported having ten hookups, women seven (Jayson, 2011).

LaBrie et al. (2014) noted the effect of drinking alcohol on hooking up. Of 828 college students, over half (55%) reported hooking up in the last year. Of those who hooked up 31% of the females and 28% of the males reported that they would not have hooked up had they not been drinking. A similar percent reported they would not have gone as far physically had they not been drinking. Females who had been drinking and hooked up were more likely to feel discontent with their hookup decisions

Olmstead et al. (2012) surveyed 158 first-semester men to assess those factors predictive of hooking up their first semester on campus. Those who had hooked up before coming to college (77% had done so), those who had a pattern of binge drinking, and those who had casual sexual attitudes were more likely to hook up their first semester on campus.

Chang et al. (2012) surveyed 369 undergraduates (69% reported having hooked up) and found that those with profeminist attitudes were not more likely to hook up. However, women who hooked up were more likely than men to agree with the unspoken rules of hooking up—no commitment, no emotional intimacy, and no future obligation to each other. They were clear that hooking up was about sex with no future. These data reflect that hooking up is becoming the norm on college and university campuses. Not only do female students outnumber men students (60 to 40) (which means women are less able to bargain sex for commitment since men have a lot of options), individuals want to remain free for summer internships, study abroad, and marriage later (Uecker & Regnerus, 2011). Kalish and Kimmel (2011) suggested that hooking up is a way of confirming one's heterosexuality.

The hooking up experience is also variable. Bradshaw et al. (2010) compared the experiences of women and men who hooked up. Men benefited more since they were able to have casual sex with a willing partner and no commitment. Women were more at risk for feeling regret/guilt, becoming depressed, and defining the experience negatively. Women in hookup contexts were less likely to experience cunnilingus but often expected to provide fellatio (Blackstrom et al., 2012). Both women and men experienced the hookup in a context of deception (neither being open about their relationship goals) and may have exposed themselves to sexually transmitted infections (STI). While few long-term relationships begin with a "hook up," there is a belief that it is not unusual. When 4,720 undergraduates were presented with the statement "People who 'hook up' and have sex the first night don't end up in a stable relationship," 51% agreed (Hall & Knox, 2015).

2-2c The Internet: Meeting Online

Increasingly, individuals are using the Internet (and attendant technology) to find partners for fun, companionship, and marriage. Based on a sample of 2,252 adults, almost 40% (38%) of Americans who are currently single and actively looking for a partner have used online dating. Of these almost two-thirds (66%) have been out on a date with someone they met through a dating site or app. Almost a quarter have met their spouse or a long-term partner through these sites (Smith & Duggan, 2013). Individuals also use Facebook to find a partner. Hall (2014) compared spouses who had met on social networking sites (e.g., Facebook) with those who met through other online means—dating sites, online communities, and one on one communication. Individuals who met through social networking sites were younger and more likely to be African American.

"In the past 15 years, the rise of the Internet has partly displaced not only family and school, but also neighborhood, friends, and the workplace as venues for meeting partners. The Internet increasingly allows individuals to meet and form relationships with perfect strangers" (Rosenfeld & Thomas, 2012). Sociologist Rosenfeld of Stanford University followed 926 unmarried couples over a three-year period—those who met online were twice as likely to marry as those offline, 13% to 6%. Twenty percent of all new relationships begin online (Jayson, 2013).

Courtesy of Rachel Calisto

Internet Partners: The Upside
In regard to advantages, online dating services have become clear in their mission—to provide a place where people go to "shop" for potential romantic partners and to "sell" themselves in hopes of creating a successful romantic relationship. In interviews with 34 persons who had used online dating services, the respondents revealed that they used economic metaphors to describe the experience—"supermarket," "catalog," etc. (Heino et al., 2010). Indeed there were five themes: assessing the market worth of the various people online, determining one's own market worth, shopping for perfect parts, maximizing inventory, and calibrating selectivity. The latter referred to assessing one's own market worth by the number of emails received and changing their presentation of self to elicit more interest, for example, changing photos or putting up more photos (Heino et al., 2010). Such "calibration" also involved comparing what one had to offer with what one could ask for. One online dater said that since she had put on weight she had to be willing to accept older partners who might be divorced with kids.

A primary attraction of meeting someone online is efficiency. It takes time and effort to meet someone at a coffee shop for an hour, only to discover that the person has habits (e.g., does or does not smoke) or values (e.g., religious or agnostic) that would eliminate them as a potential partner. On the Internet, one can spend a short period of time and literally scan hundreds of profiles of potential partners. For noncollege people who are busy in their job or career, the Internet offers the chance to meet someone outside their immediate social circle. "There are only six guys in my office," noted one female Internet user. "Four are married and the other two are alcoholics. I don't go to church and don't like bars so the Internet has become my guy store."

Another advantage of looking for a partner online is that it removes emotion/chemistry/first meeting magic from the mating equation so that individuals can focus on finding someone with common interests, background, values, and goals. In real life, you can "fall in love at first sight" and have zero in common (Heino et al., 2010). Some websites exist to target persons with specific interests such as Black singles (BlackPlanet .com), Jewish singles (Jdate.com), and gay people (Gay.com).

These spouses, each of whom had been married before, met on the Internet and married after two years.

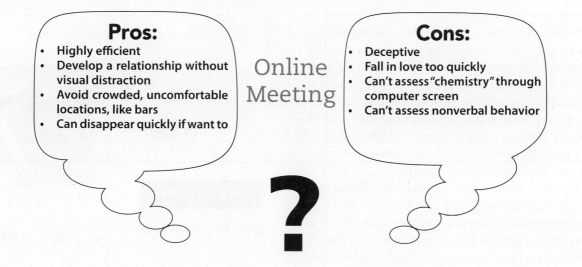

Pros:
- Highly efficient
- Develop a relationship without visual distraction
- Avoid crowded, uncomfortable locations, like bars
- Can disappear quickly if want to

Online Meeting

Cons:
- Deceptive
- Fall in love too quickly
- Can't assess "chemistry" through computer screen
- Can't assess nonverbal behavior

?

Internet Partners: The Downside There are also downsides to meeting on the Internet. Lying occurs in Internet dating (as it does in non-Internet dating). Ellison and Hancock (2013) noted that persons using the Internet to find a partner fear that others are lying in their profiles, that "fudging" (or small deceptions) is common but that "big lies" are relatively rare. However, "serious misrepresentation" was reported by half of the Pew Research Center respondents who had experience with online dating (Smith & Duggan, 2013). Hall et al. (2010) analyzed data from 5,020 individuals who posted profiles on the Internet in search of a date who revealed seven categories of misrepresentation. These included personal assets ("I own a house at the beach"), relationship goals ("I want to get married'), personal interests ("I love to exercise"), personal attributes ("I am religious"), past relationships ("I have only been married once"), weight, and age. Men were most likely to misrepresent personal assets, relationship goals, and personal interests whereas women were more likely to misrepresent weight. Heino et al. (2010) interviewed 34 online dating users and found that there is the assumption of exaggeration and a compensation for such exaggeration. The female respondents noted that men exaggerate how tall they are, so the women downplay their height. If a man said he was 5'11", the woman would assume he was 5'9". Lo et al. (2013) noted that deception is motivated by the level of attractiveness of the target person—higher deception if the target person is particularly attractive. In addition, women were more deceptive than men.

Some online users also lie about being single. They are married, older, and divorced more times than they reveal. But to suggest that the Internet is the only place where deceivers lurk is to turn a blind eye to those people met through traditional channels. Be suspicious of everyone until you know otherwise.

It is important to be cautious of meeting someone online. Although the Internet is a good place to meet new people, it also allows someone you rejected or an old lover to monitor your online behavior. Most sites note when you have been online last, so if you reject someone online by saying, "I'm really not ready for a relationship," that same person can log on and see that you are still looking. Some individuals become obsessed with a person they meet online and turn into a cyber stalker when rejected. A quarter of the respondents in the Pew Research Center study said they were harassed or made to feel uncomfortable by someone they had met online (Smith & Duggan, 2013). Some people also use the Internet to try on new identities. For example, a person who feels he or she is attracted to same-sex individuals may present a gay identity online.

Other disadvantages of online meeting include the potential to fall in love too quickly as a result of intense mutual disclosure; not being able to assess "chemistry" or how a person interacts with your friends or family; the tendency to move too quickly (from texting to phone to meeting to first date) to marriage, without spending much time to get to know each other and not being able to observe nonverbal behavior. In regard to the nonverbal issue, Kotlyar and Ariely (2013) emphasized the importance of using Skype (which allows one to see the partner/assess nonverbal cues) as soon as possible and as a prelude to meeting in person to provide more information about the person behind the profile.

Another disadvantage of using the Internet to find a partner is that having an unlimited number of options sometimes results in not looking carefully at the options one has. Wu and Chiou (2009) studied undergraduates looking for romantic partners on the Internet who had 30, 60, and 90 people to review and found that the more options the person had, the less time the undergraduate spent carefully considering each profile. The researchers concluded that it was better to examine a small number of potential online partners carefully than to be distracted by a large pool of applicants, which does not permit the time for close scrutiny.

It is also important to use Internet dating sites safely, including not giving out home or business phone numbers or addresses, always meeting the person in one's own town with a friend, and not posting photos that are "too revealing," as these can be copied and posted elsewhere. Take it slow—after connecting in an email through the dating site, move to instant messages, texting, phone calls, Skyping, then meet in a public place with friends near. Also, be clear about what you want (e.g., "If you are looking for a hookup, keep moving. If you are looking for a lifetime partner, 'I'm your gal'"). The Internet site www.wildxangel.com provides horror stories of meeting someone on the Internet and emphasizes the importance of being cautious.

The Internet may also be used to find out information about a partner. Beenverified.com provides a way to confirm that the person is who he or she says via public records, Argali.com can be used to find out where the Internet mystery person lives, Zabasearch.com for how long the person has lived there, and Zoominfo.com for where the person works. The person's birth date can be found at Birthdatabase.com.

LuLu is specifically only for females and is a website (www.onlulu.com) which rates males. Women download the app to compare notes on different aspects of a guy's "datability." It is done anonymously and linked through one's Facebook friends. First it gives statistics about the guy and a picture, asks the female's relationship to the guy, asks for an overall rating (multiple choice list), asks which positives and negatives apply, and provides an overall score out of 100%. Here's a sample review:

John Doe
Age: 33

College: Coast Guard
Last Seen: Los Angeles
Average Score: 8.4, Reviewers: 1
Best qualities: Panty Dropper, Loves His Family, Great Listener
Worst qualities: Questionable Search History, Self Absorbed, Sketchy Call Log, Wandering Eye
Appearance: 9.0
Humor: 10
Manners: 5.0
Sex: 8.0
First Kiss: 9.0
Ambition: 9.0
Commitment: 7.0

Apps Online dating is moving from websites to apps on mobile devices. Seven percent of smartphone users say they have used a dating app on their phone (Smith & Duggan, 2013). Tinder.com (on the basis of a photo) allows one to identify and connect with someone (who also selected their photo) in the area. Some users of Tinder.com refer to it as "the newest hookup device."

2-2d Speed-Dating: The Eight-Minute Date

Dating innovations that involve the concept of speed include the eight-minute date. The website http://www.8minutedating.com/identifies

these "Eight-Minute Dating Events" throughout the country, where a person has eight one-on-one "dates" at a bar that last eight minutes each. If both parties are interested in seeing each other again, the organizer provides contact information so that the individuals can set up another date. Speed-dating saves time because it allows daters to meet face to face without burning up a whole evening.

2-2e **International Dating**

Go to Google.com and type in "international brides," and you will see an array of sites dedicated to finding foreign women for Americans. American males often seek women from Asian countries who are thought to be more traditional. Women from other countries seek American males as a conduit for entry into U.S. citizenship.

> ## "In **true love** the **smallest distance** is too **great,** and the **greatest distance** can be **bridged.**"
>
> —HANS NOUWENS, WRITER

Six Functions of Involvement with a Partner

1. Confirmation of a social self
2. Recreation
3. Companionship, intimacy, and sex
4. Anticipatory socialization
5. Status achievement
6. Mate selection

2-3 Long-Distance Relationships

One outcome of online dating is that you may meet someone who does not live close to you so you end up in a long-distance relationship. Alternatively, you may already be in a relationship and the separation occurs.

Courtesy of Drew Marino

This couple is separated by a 12-hour drive—he in North Carolina and she in Florida. They Skype frequently and see each other every three months. The distance is manageable but a challenge.

Long-distance relationships must be considered with great caution. In a sample of 5,500 never-married undergraduates, about half (51% of the men and 47% of the women) reported that a long-distance relationship was "out of the question" (Walsh, 2013).

The primary advantages of involvement in a **long-distance dating relationship (LDDRs)** (lovers separated by a distance, usually 500 miles, which prevents weekly face-to-face contact)) include: positive labeling ("even though we are separated, we care about each other enough to maintain our relationship"), keeping the relationship "high" since constant togetherness may result in the partners being less attentive to each other, having time to devote to school or a career, and having a lot of one's own personal time and space. Pistole (2010) in a review of

long-distance dating relationship (LDDR) lovers are separated by a distance, usually 500 miles, which prevents weekly face-to-face contact

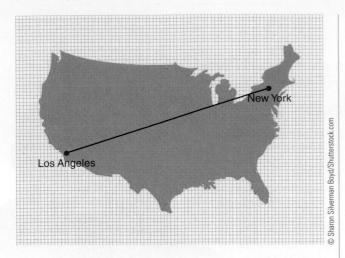

New York

Los Angeles

© Sharon Silverman Boyd/Shutterstock.com

the literature on long-distance relationships noted that they are as satisfying and as stable as geographically close relationships.

People suited for LDDRs have developed their own autonomy or independence for the times they are apart, have a focus for their time such as school or a job, have developed open communication with their partner to talk about the difficulty of being separated, and have learned to trust each other because they spend a lot of time away from each other. Another advantage is that the partner may actually look better from afar than up close. One respondent noted that he and his partner could not wait to live together after they had been separated—but "when we did, I found out I liked her better when she wasn't there."

The primary disadvantages of LDDRs include being frustrated over not being able to be with the partner and loneliness. Pistole et al. (2010) noted that being attached to someone is the strongest of social behaviors and that there is a "separation protest" when the two are separated. This longing for each other may be stronger than either anticipated. When the long-distance lovers are reunited, they spend higher order quality time together.

Other disadvantages of involvement in an LDDR are missing out on other activities and relationships, less physical intimacy, spending a lot of money on phone calls/ travel, and not discussing important relationship topics. Regarding the latter, Stafford (2010) compared the communication topics of individuals in LDDRs with those in geographically close-distance relationships (GCDRs) and found that the former avoided sensitive topics such as what household roles men and women should fulfill, the importance of marriage, if and how many children were desired, the importance of religion,

etc. "Accentuating intimacy and positive affect in their talk, and avoiding discussion of potentially problematic or taboo topics could allow geographically separated couples to maintain a positive outlook on their relationship" (p. 292).

For couples who have the goal of maintaining their LDDRs and reducing the chance that the distance will result in their breaking up, some specific behaviors to engage in include:

1. **Maintain daily contact via text messaging.** Texting allows individuals to stay in touch throughout the day. A husband we interviewed who must travel four to five days a week said, "Texting allows us to stay connected with what is going on in each other's lives throughout the day so when I get home on weekends we've been 'together' as much as possible during the week."

2. **Enjoy/ use the time when apart.** While separated, it is important to remain busy with study, friends, work, sports, and personal projects. Doing so will make the time pass faster.

3. **Avoid arguing during phone conversations.** Talking on the phone should involve the typical sharing of events. When the need to discuss a difficult topic arises (e.g., trust), the phone is not the best place for such a discussion. Rather, it may be wiser to wait and have the discussion face to face. If you decide to settle a disagreement over the phone, stick to it until you have a solution acceptable to both of you.

4. **Stay monogamous.** Agreeing not to be open to other relationships is crucial to maintaining a long-distance relationship (LDR). Individuals who say, "Let's date others to see if we are really meant to be together," often discover that they are capable of being attracted to and becoming involved with numerous "others." Such other involvements usually predict the end of an LDR. Lydon et al. (1997) studied 69 undergraduates who were involved in LDRs and found that "moral commitment" predicted the survival of the relationships. Individuals committed to maintaining their relationships are not open to becoming involved with others.

5. **Skype.** Skyping allows the partners to see and hear each other. Frequent Skype encounters

allow the partners to "date" even though they cannot touch each other. Some partners also become sexual by stripping for each other, which is a way of being intimate with each other while physically separated.

6. **Be creative.** Some partners in long-distance relationships watch Netflix's movies together—they each pull up the movie on their computer and talk on the phone while they watch it. Others send video links, photos, etc., throughout the day. One coed says "I wear his shirts" and "he has my pillow." Some find that keeping a journal of their relationship/feelings during the LDDR adventure is helpful. Be creative.

2-4 Cohabitation

Cohabitation, also known as living together, involves two adults, unrelated by blood or by law, involved in an emotional and sexual relationship who sleep in the same residence at least four nights a week for three months. About 60% of spouses report having lived together with one or more partners before their wedding. Not all cohabitants are college students. Indeed, only 18% of all cohabitants are under the age of 25. The largest percentage (36%) are between the ages of 25 and 34 (Jayson, 2012). Most cohabitants are other sex (1% same sex) and White (87%) (*Statistical Abstract of the United States, 2012–2013*).

Reasons for the increase in cohabitation include career or educational commitments; increased tolerance of society, parents, and peers; improved birth control technology; desire for a stable emotional and sexual relationship without legal ties; avoiding loneliness (Kasearu, 2010); and greater disregard for convention. Two-thirds of 122 cohabiters reported concerns about divorce (Miller et al., 2011).

2-4a Nine Types of Cohabitation Relationships

There are various types of cohabitation:

1. **Here and now.** These new partners have a fun relationship and are focused on the here and now, not the future of the relationship.

They want to be together more often and living together is one way to do so.

2. **Testers.** These couples are involved in a relationship and want to assess whether they have a future together. Sassler and Miller (2011) indentified such a couple:

> We're trying to see how it is going to work. So if our relationship's going to continue to grow and prosper, this would be one way for us to kind of gauge, you know, maybe we will get married. And if we can't live together, we're not going to get married. So this will help us make future decisions and stuff like that.

Women and men cohabitants often value the relationship differently. Rhoades et al. (2012) studied 120 cohabiting couples and found women more committed than men to the relationship in about half the couples (46%). Such discrepancy was associated with a lower quality relationship.

3. **Engaged.** These cohabiting couples are in love and are planning to marry. Among those who report having lived together, about two-thirds (64%) say they thought of their living arrangement as a step toward marriage (Pew Research Center, 2010). Engaged cohabitant couples who have an agreed upon future report the highest level of satisfaction, the lowest level of conflict, and, in general, have a higher quality relationship than other types of cohabitants (Willoughby et al., 2012). After three years, 40% of first premarital cohabitants end up getting married, 32% are still cohabiting, and 27% have broken up.

4. **Money savers.** These couples live together primarily out of economic convenience. They are open to the possibility of a future together but regard such a possibility as unlikely. Sassler and Miller (2011) noted that working class individuals tend to transition more quickly than middle class individuals to cohabitation out of economic necessity or to meet a housing need. Dew (2011) noted that financial

cohabitation (living together)
two unrelated adults (by blood or by law) involved in an emotional and sexual relationship who sleep in the same residence at least four nights a week for three months.

This 60-plus-year-old male has never been married. He and his partner are living together.

disagreements were an important factor in predicting cohabitation couples breaking up.

5. **Pension partners.** These cohabitation partners are older, have been married before, still derive benefits from their previous relationships, and are living with someone new. Getting married would mean giving up their pension benefits from the previous marriage. An example is a widow from the war in Afghanistan who was given military benefits due to a spouse's death. If remarried, the widow would forfeit both health and pension benefits, but now lives with a new partner and continues to get benefits from the previous marriage.

6. **Alimony maintenance.** Related to widows who cohabit are the divorced who are collecting alimony which they would forfeit should they remarry. They live with a new partner instead of marrying to maintain the benefits. The example is a divorced woman receiving a hefty alimony check from her ex-husband, a successful attorney. She was involved in a new relationship with a partner who wanted to marry her. She did not so in order to keep the alimony flowing. In effect, her ex-husband was paying the bills for his former wife and her new lover who had moved in.

7. **Security blanket cohabiters.** Some of the individuals in these cohabitation relationships are drawn to each other out of a need for security rather than mutual attraction. Being alone is not an option. They want somebody, anybody, in the house.

8. **Rebellious cohabiters.** Some couples use cohabitation as a way of making a statement to their parents that they are independent and can make their own choices. Their cohabitation is more about rebelling from parents than being drawn to each other.

9. **Marriage never (cohabitants forever).** Ten percent of cohabitants view their living together not as a prelude to marriage but as a way of life (Sommers et al., 2013).

What motives do individuals have for living together as a permanent lifestyle? Ortyl. (2013) interviewed 48 long-term heterosexual cohabiters who identified their motives. Six themes emerged:

1. **Marriage free.** The largest percent (38%) of Ortyl's respondents believed that marriage was unnecessary to their happiness (they used the term *marriage free* much like one would use the word *childfree*). Such couples may feel a moral commitment to each other and to their relationship yet have no interest in marriage (Pope & Cashwell, 2013).

2. **Risk aversion.** The cohabitants had parents or siblings who had divorced or who were in disastrous marriages and wanted to avoid the same fate.

3. **Marriage boycott.** The cohabitants rejected the government defining marriage as heterosexual, thus supporting gays who are denied same-sex marriage in all states.

4. **Sexism dissent.** Cohabitants rejected the patriarchal history of marriage which controlled women.

5. **American dreamer.** Some cohabitants saw the day when their school debts would be paid off and their career established, as the day it would be OK to get married.

6. **Economic disincentives.** Cohabitants knew that marriage involves being responsible for the debts of the spouse and wanted to avoid (by living together) being economically

burdened by a partner with an unstable job and money history.

Some couples who view their living together as "permanent" seek to have it defined as a **domestic partnership, a relationship** involving two adults who have chosen to share each other's lives in an intimate and committed relationship of mutual caring

2-4b **Does Cohabitation Result in Marriages That Last?**

Individuals who live together before getting married assume that doing so will increase their chances of having a happy and durable marriage relationship. But will it? The answer is, "It depends." For individuals (particularly women, Manning & Cohen, 2012) who have only one cohabitation experience with the person they eventually marry, there is no increased risk of divorce (Jose et al., 2010). The period of time while these engaged couples are cohabiting is superior to the time spent by couples who are not committed to the future (Willoughby et al., 2012).

Because people commonly have more than one cohabitation experience, the term **cohabitation effect** applies. This means that those who have multiple cohabitation experiences prior to marriage are more likely to end up in marriages characterized by lower levels of happiness and higher levels of divorce (Booth et al., 2008). Liat and Havusha-Morgenstern (2011) compared the marital adjustment among women who cohabited with their spouses before marriage versus those who did not. They found that cohabiting women reported lower levels of adjustment of spousal cohesion and display of affection than did women who did not cohabit.

What is it about serial cohabitation relationships that predict negatively for future marital happiness and durability? One explanation is that cohabitants tend to be people who are willing to violate social norms by living together before marriage. Once they marry, they may be more willing to break another social norm and divorce if they are unhappy than are unhappily married people who tend to conform to social norms and have no history of unconventional behavior. Another reason is the restraints inherent in cohabitation. Rhoades et al. (2012) studied 120 cohabiting couples and found that restraints often keep a couple together. These include signing a lease, having a joint bank account, and having a pet. In some cases couples may move forward toward marriage for reasons of constraint rather than emotional desire.

Whatever the reason, cohabitants should not assume that cohabitation will make them happier spouses or insulate them from divorce.

Not all researchers have found negative effects of cohabitation on relationships. Reinhold (2010) found that among more recent cohabitant cohorts, the negative association between living together and marital instability is weakening. Indeed, Kuperberg (2014) provided data to confirm that previous research linking cohabitation with divorce did not account for the age at which coresidence began. She suggested that it is the age at which individuals begin their lives together (coresidence) which impacts divorce, not cohabitation per se. She suggested that individuals delay their marriage into their mid-twenties "when they are older and more established in the lives, goals and careers, whether married or not at the time of co-residence rather than avoiding premarital cohabitation altogether" (p. 368)

2-5 Living Apart Together

The definition of **living apart together (LAT)** is a long-term committed couple who does not live in the same dwelling. Some couples (including spouses) find that living apart together is preferable to their living in the same place (Hess, 2012). Actress Teri Garr noted that before their daughter Molly was born, she and her former husband (John O'Neil) lived in separate houses. "We got along much better with a little distance between us." They are not alone. In a study of 68 adults (93% married), 7% reported that they preferred a LAT arrangement with their spouse. Forty-six percent said that living apart from your spouse enhances your relationship (Jacinto & Ahrend, 2012). The Census Bureau estimates that 1.7 married couples are living in this arrangement (Gottman, 2013).

domestic partnership a relationship in which individuals who live together are emotionally and financially interdependent and are given some kind of official recognition by a city or corporation so as to receive partner benefits.

cohabitation effect multiple cohabitation experiences before marriage has negative effect on a subsequent marriage such as lower levels of happiness/ higher divorce.

living apart together (LAT) a long-term committed couple who does not live in the same dwelling.

Adam Vilimek/Shutterstock.com

Three criteria must be met for a couple to be defined as a living apart together couple: (1) they must define themselves as a committed couple; (2) others must define the partners as a couple; and (3) they must live in separate domiciles. The lifestyle of living apart together involves partners in loving and committed relationships (married or unmarried) identifying their independent needs in terms of the degree to which they want time and space away from each other. People living apart together exist on a continuum from partners who have separate bedrooms and baths in the same house to those who live in a separate place (apartment, condo, house) in the same or different cities. LAT couples are not those couples who are forced by their career or military assignment to live separately. Rather, LAT partners choose to live in separate domiciles and feel that their relationship benefits from the LAT structure.

The living apart together lifestyle or family form is not unique to couples in the United States (e.g., the phenomenon is more prevalent in European countries such as France, Sweden, and Norway). Couples choose this pattern for a number of reasons, including the desire to maintain some level of independence, to enjoy their time alone, and to keep their relationship exciting.

2-5a Advantages of LAT

The benefits of LAT relationships include the following:

1. **Space and privacy.** Having two places enables each partner to have a separate space to read, watch TV, talk on the phone, or whatever. This not only provides a measure of privacy for the individuals, but also as a couple. When the couple has overnight guests, the guests can stay in one place while the partners stay in the other place. This arrangement gives guests ample space and the couple private space and time apart from the guests.

2. **Career or work space.** Some individuals work at home and need a controlled quiet space to work on projects, talk on the phone, and focus on their work without the presence of someone else. The LAT arrangement is particularly appealing to musicians for practicing, artists to spread out their materials, and authors for quiet (Hemingway built a separate building where he would do his writing in Key West).

3. **Variable sleep needs.** Although some partners enjoy going to bed at the same time and sleeping in the same bed, others like to go to bed at radically different times and to sleep in separate beds or rooms. A frequent comment from LAT partners is, "My partner thrashes throughout the night and kicks me, not to speak of the wheezing and teeth grinding, so to get a good night's sleep, I need to sleep somewhere else."

4. **Allergies.** Individuals who have cat or dog allergies may need to live in a separate antiseptic environment from their partner who loves animals and would not live without them. "He likes his dog on the bed" said one woman.

5. **Variable social needs.** Partners differ in terms of their need for social contact with friends, siblings, and parents. The LAT arrangement allows for the partner who enjoys frequent time with others to satisfy that need without subjecting the other to the presence of a lot of people in one's life space. One wife from a family of seven children enjoyed frequent contact with both her siblings and parents. The LAT arrangement allowed her to continue to enjoy her family at no expense to her husband who was upstairs in a separate condo.

6. **Blended family needs.** LAT works particularly well with a blended family in which remarried spouses live in separate places with their children from previous relationships. An example is a remarried couple who bought a duplex with each spouse living with his/her own children on the respective sides of the duplex. The parents could maintain a private living space with their children, the spouses could be next door to each other, and the stepsiblings were not required to share living quarters—a structural answer to major stepfamily blending problems.

7. **Keeping the relationship exciting.** Zen Buddhists remind us of the necessity to be in touch with polarities, to have a perspective where we can see and appreciate the larger picture—without the darkness, we cannot fully appreciate the light. The two are inextricably part of a whole. This is the same with relationships; time apart from our beloved can make time together more precious. The term **satiation** is a well-established psychological principle. The term means that a stimulus loses its value with repeated exposure. Just as we tire of eating the same food, listening to the same music, or watching the same movie, so satiation is relevant to relationships. Indeed, couples who are in a long-distance dating relationship know the joy of "missing" each other and the excitement of being with each other again. Similarly, individuals in a LAT relationship help ensure that they will not "satiate" on each other but maintain some of the excitement in seeing or being with each other.

8. **Self-expression and comfort.** Partners often have very different tastes in furniture, home décor, music, and temperature. With two separate places, each can arrange and furnish their respective homes according to their individual preferences. The respective partners can also set the heat or air conditioning according to their own preferences, and play whatever music they like.

9. **Cleanliness or orderliness.** Separate residences allow each partner to maintain the desired level of cleanliness and orderliness without arguing about it. Some individuals like their living space to be as ordered as a cockpit. Others simply do not care.

10. **Elder care.** One partner may be taking care of an elderly parent in his or her own house. The partner may prefer not to live with an elderly parent. A LAT relationship allows for the partner taking care of the elderly parent to do so and a place for the couple to be alone.

11. **Maintaining one's lifetime residence.** Some retirees, widows, and widowers meet, fall in love, and want to enjoy each other's companionship. However, they don't want to move out of their own home. The LAT arrangement does not require that the partners move.

12. **Leaving inheritances to children from previous marriages.** Having separate residences allows the respective partners to leave their family home or residential property to their biological children from an earlier relationship without displacing their surviving spouse.

2-5b Disadvantages of LAT

There are also disadvantages of the LAT lifestyle.

1. **Stigma or disapproval.** Because the norm that married couples move in together is firmly entrenched, couples who do not do so are suspect. "People who love each other should want to live in their own house together . . . those who live apart aren't really committed" is the traditional perception of people in a committed relationship.

2. **Cost.** The cost of two separate living places can be more expensive than two people living in one domicile. But there are ways LAT couples afford their lifestyle. Some live in two condominiums that are cheaper than owning one larger house. Others buy housing out of high-priced

satiation a stimulus loses its value with repeated exposure or people get tired of each other if they spend relentless amounts of time with each other.

real estate areas. One partner said, "We bought a duplex 10 miles out of town where the price of housing is 50% cheaper than in town. We have our LAT context and it didn't cost us a fortune."

3. **Inconvenience.** Unless the partners live in a duplex or two units in the same condominium, going between the two places to share meals or be together can be inconvenient.

4. **Lack of shared history.** Because the adults are living in separate quarters, a lot of what goes on in each house does not become a part of the life history of the couple. For example, children in one place don't benefit as much from the other adult who lives in another domicile most of the time.

5. **No legal protection.** The legal nature of the LAT relationship is ambiguous. Currently there are no legal protections in the United States for LAT partners as there are for spouses. Lyssens-Danneboom et al. (2013) noted that LAT partners in Belgium "believe they should be granted the same family-based benefits as those enjoyed by their cohabiting or married counterparts."

2-6 Trends in Singlehood

Singlehood will (in the cultural spirit of diversity) lose some of its stigma; more young adults will choose this option; and those who remain single will, increasingly, find satisfaction in this lifestyle.

Individuals will continue to be in no hurry to get married. Completing their education, becoming established in their career, and enjoying hanging out and hooking up will continue to delay serious consideration of marriage. The median age for women getting married is 26; for men, 28. This trend will continue as individuals keep their options open in America's individualistic society.

Cohabitation will become the typical first union for young adults (Guzzo, 2014). The percent of cohabitants will increase not just for those who live together before marriage (now about two-thirds) but also in the prevalence of serial cohabitation (Vespa, 2014). Previously, only risk takers and people willing to abandon traditional norms lived together before marriage. In the future, mainstream individuals will cohabit. And fewer cohabitation relationships will transition into marriage, even among those who are engaged (Guzzo, 2014).

STUDY TOOLS 2

Ready to study? In the book, you can:

⮌ Rip out the chapter review card at the back of the book for a handy summary of the chapter and key terms.

⮌ See if you might be right for a "living apart together relationship" with the Self-Assessment card at the back of the book.

Online at CENGAGEBRAIN.COM you can:

⮌ Prepare for tests with quizzes.

⮌ Review the key terms with Flash Cards.

⮌ Play games to master concepts.

4LTR Press solutions are designed for today's learners through the continuous feedback of students like you. Tell us what you think about **M&F** and help us improve the learning experience for future students.

YOUR FEEDBACK MATTERS.

Gender in Relationships

"The woman **most in need of liberation** is the woman in every man and the man in every woman."

—WILLIAM SLOAN COFFIN, CLERGYMAN AND POLITICAL ACTIVIST

Golf is the game which reveals that gender discrimination continues to be evident. It took the Augusta National Golf Club until 2012 to invite its first two female members (Condoleezza Rice and Darla Moore) for membership. It is not unusual for clubs around the country to have women absent on their rolls, including Burning Tree of Maryland where House Speaker John Boehner golfs. Commenting on this gender exclusion, Brennan (2014) writes, "It sends the message that men don't want women around them when they are playing golf. Discriminatory male-only clubs put out a big stop sign out that has a chilling effect of girl's and women's participation in golf." (p. 2A).

Sociologists note that one of the defining moments in an individual's life is when the sex of a fetus (in the case of an ultrasound) or an infant (in the case of a birth) is announced. "It's a boy" or "It's a girl" immediately summons an onslaught of cultural programming affecting the color of the nursery (blue for a boy and pink for a girl), name of the baby (there are few gender-free names such as Chris) and sport participation alternatives (e.g., volleyball but not golf). The social script for women and men is radically different so that being identified as either is to put one's life on a path quite different than if the person had been reared as the other gender. In this chapter, we examine variations in gender roles and how they impact relationships. We begin by looking at the terms used to discuss gender issues.

Courtesy of Brittany Bolen

3-1 Terminology of Gender Roles

In common usage, the terms *sex* and *gender* are often interchangeable, but sociologists, family or consumer science educators, human development specialists, and health educators do not find these terms synonymous. After clarifying the distinction between *sex* and *gender*, we discuss other relevant terminology, including *gender identity*, *gender role*, and *gender role ideology*.

3-1a Sex

Sex refers to the biological distinction between females and males. Hence, to be assigned as a female or male, several

sex the biological distinction between being female and being male.

factors are used to determine the biological sex of an individual:

- **Chromosomes.** XX for females; XY for males

- **Gonads.** Ovaries for females; testes for males

- **Hormones.** Greater proportion of estrogen and progesterone than testosterone in females; greater proportion of testosterone than estrogen and progesterone in males

- **Internal sex organs.** Fallopian tubes, uterus, and vagina for females; epididymis, vas deferens, and seminal vesicles for males

- **External genitals.** Vulva for females; penis and scrotum for males

Even though we commonly think of biological sex as consisting of two dichotomous categories (female and male), biological sex exists on a continuum. Sometimes not all of the five bulleted items just listed are found neatly in one person (who would be labeled as a female or a male). Rather, items typically associated with females or males might be found together in one person, resulting in mixed or ambiguous genitals; such persons are called **intersexed individuals**. Indeed, the genitals in these intersexed (or middlesexed) individuals (about 2% of all births) are not clearly male or female. **Intersex development** refers to congenital variations in the reproductive system, sometimes resulting in ambiguous genitals. The self-concept of these individuals is variable. Some may view themselves as one sex or the other or as a mix. Since our culture does not know how to relate to mixed sex

Drpixel/Shutterstock.com

individuals, the individual is typically reared as a woman or as a man. However, an intersexed person who is reared as and presents as a woman may have no ovaries and will not be able to have children.

3-1b Gender

The term **gender** is a social construct and refers to the social and psychological characteristics associated with being female or male. Women are often thought of as soft, passive, and cooperative; men as rough, aggressive, and forceful. In popular usage, gender is dichotomized as an either-or concept (feminine or masculine). Each gender has some characteristics of the other. However, gender may also be viewed as existing along a continuum of femininity and masculinity.

Gender differences are a consequence of biological (e.g., chromosomes and hormones) and social factors (e.g., male/female models such as parents, siblings, peers). The biological provides a profound foundation for gender role development. Evidence for this biological influence is the experience of the late John Money, psychologist and former director of the now-defunct Gender Identity Clinic at Johns Hopkins University School of Medicine, who encouraged the parents of a boy (Bruce Reimer) to rear him as a girl (Brenda) because of a botched circumcision that left the infant without a penis. Money argued that social mirrors dictate one's gender identity, and thus, if the parents treated the child as a girl (e.g., name, clothing, toys), the child would adopt the role of a girl and later that of a woman. The child was castrated and sex reassignment began.

However, the experiment failed miserably; the child as an adult (now calling himself David) reported that he never felt comfortable in the role of a girl and had always

---

intersexed individuals people with mixed or ambiguous genitals.

intersex development refers to congenital variations in the reproductive system, sometimes resulting in ambiguous genitals.

gender the social and psychological behaviors associated with being female or male.

Gelpi JM/Shutterstock.com

> **"A man is lucky if he is the first love of a woman.** *A woman is lucky if she is the last love of a man.* **"**
>
> —CHARLES DICKENS, AUTHOR

viewed himself as a boy. He later married and adopted his wife's three children.

In the past, David's situation was used as a textbook example of how nurture is the more important influence in gender identity, if a reassignment is done early enough. Today, his case makes the point that one's biological wiring dictates gender outcome. Indeed, David Reimer noted in a television interview, "I was scammed," referring to the absurdity of trying to rear him as a girl. Distraught with the ordeal of his upbringing and beset with financial difficulties, he committed suicide in 2004 via a gunshot to the head.

The story of David Reimer emphasizes the power of biology in determining gender identity. Other research supports the critical role of biology. Nevertheless, **socialization** (the process through which we learn attitudes, values, beliefs, and behaviors appropriate to the social positions we occupy) does impact gender role behaviors, and social scientists tend to emphasize the role of social influences in gender differences.

3-1c Gender Identity

Gender identity is the psychological state of viewing oneself as a girl or a boy, and later as a woman or a man. Such identity is largely learned and is a reflection of society's conceptions of femininity and masculinity. Some individuals experience **gender dysphoria**, a condition in which one's

socialization the process through which we learn attitudes, values, beliefs, and behaviors appropriate to the social positions we occupy.

gender identity the psychological state of viewing oneself as a girl or a boy and, later, as a woman or a man.

gender dysphoria the condition in which one's gender identity does not match one's biological sex.

Diversity in Gender Role Socialization

Although her research is controversial, Margaret Mead (1935) focused on the role of social learning in the development of gender roles in her study of three cultures. She visited three New Guinea tribes in the early 1930s, and observed that the Arapesh socialized both men and women to be feminine, by Western standards. The Arapesh people were taught to be cooperative and responsive to the needs of others. In contrast, the Tchambuli were known for dominant women and submissive men—just the opposite of our society. Both of these societies were unlike the Mundugumor, which socialized only ruthless, aggressive, "masculine" personalities. The inescapable conclusion of this cross-cultural study is that human beings are products of their social and cultural environments and that gender roles are learned.

gender identity does not match one's biological sex. An example of gender dysphoria is transgender or transsexualism.

The word **transgender** is a generic term for a person of one biological sex who displays characteristics of the other sex. A **crossdresser** is a person of one biological sex who enjoys dressing in the clothes of the other sex—for example, a biological man who enjoys dressing up as a woman. **Transsexuals** are individuals with the biological and anatomical sex of one gender (e.g., female) but the self-concept of the other sex (e.g., male). "I am a female trapped in a man's body" reflects the feelings of the male-to-female transsexual (MtF), who may take hormones to develop breasts and reduce facial hair and may have surgery to artificially construct a vagina. Such a person lives full-time as a woman.

The female-to-male transsexual (FtM) is a biological and anatomical female who feels "I am a man trapped in a female's body." This person may take male hormones to grow facial hair and deepen her voice and may have surgery to create an artificial penis. This person lives full-time as a man. Individuals need not take hormones or have surgery to be regarded as transsexuals. The distinguishing variable is living full-time in the role of the gender opposite one's biological sex. A man or woman who presents full-time as the opposite gender is a transsexual by definition.

This female-to-male transsexual (FtM, on the right) is in a romantic relationship with this self-identified lesbian (on the left). The FtM transsexual loves her as a woman; the lesbian reports that she "fell in love with him when he was a woman and still loves him as a man."

transgender
a generic term for a person of one biological sex who displays characteristics of the opposite sex.

cross-dresser
a generic term for individuals who may dress or present themselves in the gender of the opposite sex.

transsexual an individual who has the anatomical and genetic characteristics of one sex but the self-concept of the other.

gender roles
behaviors assigned to women and men in a society.

sex roles behaviors defined by biological constraints.

3-1d Gender Roles

Gender roles are social norms which specify the socially appropriate behavior for females and males in a society. All societies have expectations of how boys and girls, men and women "should" behave. Gender roles influence women and men in virtually every sphere of life, including dating. Of 2,467 adults between the ages of 18–59, 69% of the men and 55% of the women reported that the man should pay for the date (Jayson, 2014). Regarding family roles, even with transgender male-to-female individuals, women end up doing more domestic work. Pfeffer (2010) interviewed 50 women partners of transgender and transsexual men to assess the division of labor. Women often spoke of a nonegalitarian division of labor, but rationalized the reasons for this division. Ani stated: "I do the dishes; but I'm so neurotic about having a clean house and he is not.... I definitely do more than he does but, again, I'm the one that happens to be a neat freak." Women also devote more time to child rearing and child care. Barclay (2013) noted that even in countries which promote equality in division of labor, women end up taking care of children more often.

Gender impacts one's life experiences. The What's New? section provides data on how sexual and relationship experiences are different for women and for men.

The term *sex roles* is often confused with and used interchangeably with the term *gender roles*. However, whereas gender roles are socially defined and can be enacted by either women or men, **sex roles** are defined by biological constraints and can be enacted by members of one biological sex only—for example, wet nurse, sperm donor, child bearer.

What's New?

DIFFERENT GENDER WORLDS OF SEX, BETRAYAL, AND LOVE

Women and men report significantly different life experiences. The table below reflects some of these differences in a large nonrandom sample of 4,570 undergraduates.

Differences in female masturbation reflect anatomy, biology, and socialization. Unlike the penis, which is an external appendage that lends itself to being rubbed, touched, and handled, the clitoris is hidden and embedded in the woman's vaginal lips. In addition to lower anatomical availability, lower testosterone levels in females may result in a lower biological drive to reduce the sex drive. In addition, the socialization of women includes more caveats about sex. Women are taught negatives about "down there" and to regard menstruation as something to deal with. Finally, women are encouraged to experience sex in a relationship context, unlike men who are socialized to experience sexuality independent of relationship factors. The result of these influences not only reflect an overall lifetime lower frequency of masturbation for females but also an annual lower frequency. Based on a survey of 5,865 respondents in the United States of adults ages 20–29, 68% of women compared to 84% of men reported having masturbated alone (rather than with a partner) in the past 12 months (Reece et al., 2012).

The lower frequency of females hooking up with someone they just met reflects female socialization to have sex in the context of a relationship. Later in the chapter we review sociobiological theory which suggests that females are wired to be more selective about their sexual behavior due to the potential consequence of becoming pregnant and being socialized for the role of caretaker of the infant/child. Men can drop their sperm and move on. Women, in order to protect themselves and their offspring, must be more selective. The higher regret at first intercourse among women is often over the lack of insistence on a meaningful relationship as the context.

Greater experience in being betrayed is reflected in other data which point out that men are more likely to cheat. About a quarter of husbands and 20% of wives report having had sex with someone outside the marriage at some time during the marriage (Russell et al., 2013). Having a partner who is unfaithful often results in the woman learning to develop feelings of mistrust in their new relationship.

Finally, women reporting a lower frequency of experiencing love at first sight than men speaks to the fact that men may be more attentive to visual cues in partner selection. Gervais et al. (2013) confirmed that compared to women, men showed an increased tendency to exhibit "the objectifying gaze" toward women who represented the cultural ideal (i.e., hourglass shaped women with large breasts and small waist-to-hip ratios).

Source: Based on original data from Hall, S., & D. Knox. (2015). Relationship and sexual behaviors of a sample of 4,736 university students. Department of Family and Consumer Sciences, Ball State University and Department of Sociology, East Carolina University.

Percent Agreement on Sex, Betrayal, and Love,

Item	Female N = 3,581	Male N = 1,155	Sig.
Sexual experiences			
I have masturbated.	65.7%	96.7%	.000
I could hook up and have sex with someone I liked.	23.7%	60.9%	.000
I have hooked up/had sex with a person I just met.	22.1%	33.1%	.000
I regret my choice for sexual intercourse the first time.	26.1%	15.8%	.000
Betrayed			
I have been involved with someone who cheated on me.	50.2%	40.0%	.000
Love at first sight			
I have experienced love at first sight.	24.1%	35.8%	.000

3-1e Gender Role Ideology

Gender role ideology refers to beliefs about the proper role relationships between women and men in a society. Traditionally, men initiated relationships, called women for dates, and were expected to be the ones who proposed. New norms include that women may be first to initiate an interaction, text men for time together, and ask men when they can marry. Gender role ideology is also operative in cross-sex friendships. Women are more likely to regard cross-sex friendships as devoid of romantic attraction and sexual possibility, particularly if they are involved in a relationship. Men, on the other hand, are more likely to experience attraction in cross-sex friendships whether or not they are involved in a relationship. Hence when men and women friends are hanging out, he may be thinking of a romantic/sex potential while she typically does not share his view (Bleske-Rechek et al., 2012).

Traditional American gender role ideology has perpetuated and reflected patriarchal male dominance and male bias in almost every sphere of life. Even our language reflects this male bias. For example, the words *man* and *mankind* have traditionally been used to refer to all humans. There has been a growing trend away from using male-biased language. Dictionaries have begun to replace *chairman* with *chairperson* and *mankind* with *humankind*.

gender role ideology the proper role relationships between women and men in a society.

biosocial theory (sociobiology) emphasizes the interaction of one's biological or genetic inheritance with one's social environment to explain and predict human behavior.

parental investment any investment by a parent that increases the chance that the offspring will survive and thrive.

sociobiology emphasizes the interaction of one's biological or genetic inheritance with one's social environment to explain and predict human behavior.

Courtesy of Rachel Calisto

While females are often socialized into domestic roles, they often reject such roles.

3-2 Theories of Gender Role Development

Various theories attempt to explain why women and men exhibit different characteristics and behaviors.

"The same time **women** came up with **PMS**, men came up with **ESPN**."

—BLAKE CLARKE, COMEDIAN

3-2a Biosocial/Biopsychosocial

Biosocial theory emphasizes that social behaviors (e.g., gender roles) are biologically based and have an evolutionary survival function. For example, women tend to select and mate with men whom they deem will provide the maximum parental investment in their offspring. The term **parental investment** refers to any venture by a parent that increases the offspring's chance of surviving and thus increases reproductive success of the adult. Parental investments require time and energy. Women have a great deal of parental investment in their offspring (including nine months of gestation), and they tend to mate with men who have high status, economic resources, and a willingness to share those economic resources. As we will see in Chapter 12 on parenting, economic resources are not inconsequential as the average cost today of rearing a child from birth to age 18 (does not include college) is almost $250,000.

The biosocial explanation (also referred to as **sociobiology**) for mate selection is extremely controversial. Critics argue that women may show concern for the earning capacity of a potential mate because they have been systematically denied access to similar economic resources, and selecting a mate with these resources is one of their remaining options. In addition, it is argued that both women and men, when selecting a mate, think more about their partners as companions to have fun with than as future parents of their offspring. Related to the biosocial view of gender roles is the biopsychosocial view which includes psychological aspects such as level of self-control, emotions, and thinking as they impact gender role expression. How the individual perceives the social aspects of socioeconomic status, culture, poverty, technology, and religion will impact gender role expression. For example, a transgender person brings a personality to the culture that reflects various levels of acceptance or disapproval.

3-2b Bioecological Model

The bioecological model, proposed by Urie Bronfenbrenner, emphasizes the importance of understanding bidirectional influences between an individual's development and his or her surrounding environmental contexts. The focus is on the combined interactive influences so that the predispositions of the individual interact with the environment/culture/society, resulting in various gender expressions. For example, the individual will read what gender role behavior his or her society will tolerate and adapt accordingly.

3-2c Social Learning

Derived from the school of behavioral psychology, the social learning theory emphasizes the roles of reward and punishment in explaining how a child learns gender role behavior. This is in contrast to the biological explanation for gender roles. For example, consider the real-life example of two young brothers who enjoyed playing "lady"; each of them put on a dress, wore high-heeled shoes, and carried a pocketbook. Their father came home early one day and angrily demanded, "Take those clothes off and never put them on again. Those things are for women." The boys were punished for "playing lady" but rewarded with their father's approval for boxing and playing football (both of which involved hurting others).

This female is being encouraged to enjoy football, traditionally an all-male sport.

Reward and punishment alone are not sufficient to account for the way in which children learn gender roles. They also learn gender roles when parents or peers offer direct instruction (e.g., "girls wear dresses" or "a man stands up and shakes hands"). In addition, many of society's gender rules are learned through modeling. In modeling, children observe and imitate another's behavior. Gender role models include parents, peers, siblings, and characters portrayed in the media.

The impact of modeling on the development of gender role behavior is controversial. For example, a modeling perspective implies that children will tend to imitate the parent of the same sex, but children in all cultures are usually reared mainly by women. Yet this persistent female model does not seem to interfere with the male's development of the behavior that is considered appropriate for his gender. One explanation suggests that boys learn early that our society generally grants boys and men more status and privileges than it does girls and women. Therefore, boys devalue the feminine and emphasize the masculine aspects of themselves.

Regardless of the source, Witt and Wood (2010) found that expectations for following gender role standards are so powerful that one's well-being is related to the degree that one's behavior reflects these standards. Persons who accept the standards but who do not live up to them feel an impaired sense of well-being.

3-2d Identification

Although researchers do not agree on the merits of Freud's theories (and students question its relevance), Freud was one of the first theorists to study gender role acquisition. He suggested that children acquire the characteristics and behaviors of their same-sex parent through a process of identification. Boys identify with their fathers, and girls identify with their mothers.

3-2e Cognitive-Developmental Theory

The cognitive-developmental theory of gender role development reflects a blend of the biological and social learning views. According to this theory, the biological readiness of the child, in terms of cognitive development, influences how the child responds to gender cues in the environment (Kohlberg, 1966). For example, gender discrimination (the ability to identify social and psychological characteristics associated with being female or male) begins at about age 30 months. However, at this age, children do not view gender as a permanent characteristic. Thus, even though young children may define people who wear long hair as girls and those who never wear dresses as boys, they also believe they can change their gender by altering their hair or changing clothes.

Not until age 6 or 7 do children view gender as permanent (Kohlberg, 1966; 1969). In Kohlberg's view, this cognitive understanding involves the development of a specific mental ability to grasp the idea that certain basic characteristics of people do not change. Once children learn the concept of gender permanence, they seek to become competent and proper members of their gender group. For example, a child standing on the edge of a school playground may observe one group of children jumping rope while another group is playing football. That child's gender identity as either a girl or a boy connects with the observed gender-typed behavior, and the child joins one of the two groups. Once in the group, the child seeks to develop behaviors that are socially defined as gender appropriate.

3-3 Agents of Socialization

Three of the four theories discussed in the preceding section emphasize that gender roles are learned through interaction with the environment. Indeed, though biology may provide a basis for one's gender identity, cultural influences in the form of various socialization agents (parents, peers, religion, and the media) shape the individual toward various gender roles. These powerful influences, in large part, dictate what people think, feel, and do in their roles as men or women. In the next section, we look at the different sources influencing gender socialization.

3-3a Family

The family is a gendered institution in that female and male roles are highly structured by gender. The names parents assign to their children, the clothes they dress them in, and the toys they buy them all reflect gender. Parents (particularly African American mothers) may also be stricter on female children—determining the age they are allowed to leave the house at night, the time of curfew, and using directives such as "text me when you get to the party." Female children are also assigned more chores (Mandara et al., 2010).

> "We've begun to **raise daughters** more **like sons** ... but few have the **courage** to **raise** our **sons** more **like** our **daughters**."
>
> —GLORIA STEINEM, JOURNALIST AND FEMINIST

Parents who expose their daughters to Barbie dolls may inadvertently be socializing them to a more limited set of career possibilities. Sherman and Zurbriggen (2014) created a context where 37 girls aged 4–7 played with Barbie dolls for five minutes and were then asked how many of 10 different occupations they themselves could do in the future and how many of those occupations a boy could do. Girls reported that boys could do significantly more occupations than they could themselves, especially when considering male-dominated careers.

This little girl is looking to her big sister to discover how to behave.

Siblings also influence gender role learning. Growing up in a family of all sisters or all brothers intensifies social learning experiences toward femininity or masculinity. A male reared with five sisters and a single-parent mother is likely to reflect more feminine characteristics than a male reared in a home with six brothers and a stay-at-home dad.

3-3b Peers

Though parents are usually the first socializing agents that influence a child's gender role development, peers become increasingly important during the school years. In regard to the degree to which peers influence the use of alcohol, 371 adolescents ages 11 to 13 years old participated in a study (Trucco et al., 2011). The researchers found that having peers who used alcohol and who approved of alcohol use by the participant predicted initiation of alcohol use. In another study, having one close friend who engaged in a high frequency of deviant behavior (e.g., smoking, drinking, driving recklessly) resulted in a greater influence than having several friends, only one of whom engaged in deviant behavior (Rees & Pogarsky, 2011).

3-3c Religion

"Wives, submit to your husbands, as is fitting in the Lord."

—COLOSSIANS 3:18

An example of how religion impacts gender roles involves the Roman Catholic Church, which does not have female clergy. Men dominate the 19 top positions in the U.S. dioceses. In addition, the

Cultural Diversity in Gender Roles

Samoan society/culture provides a unique example of gender role socialization via the family. The **Fafafini** (commonly called Fafa) are males reared as females. There are about 3,000 Fafafini in Samoa. The practice arose when there was a lack of women to perform domestic chores and the family had no female children. Thus, effeminate boys were identified and socialized/reared as females. Fafafini represent a third gender, neither female nor male; they are unique and valued, not stigmatized. Most Samoan families have at least one Fafafini child who takes on the role of a woman, including having sex with men (Abboud, 2013).

Fafafini in Samoan society, these are effeminate males socialized and reared as females due to the lack of women to perform domestic chores.

In the Roman Catholic Church, which does not allow females to be priests, men dominate the top positions.

Mormon Church is dominated by men and does not provide positions of leadership for women. Maltby et al. (2010) also observed that the stronger the religiosity for men, the more traditional and sexist their view of women. This association was not found for women.

3-3d Education

The educational institution is another socialization agent for gender role ideology. However, such an effect must be considered in the context of the society or culture in which the "school" exists and of the school itself. Schools are basic cultures of transmission in that they make deliberate efforts to reproduce the culture from one generation to the next.

3-3e Economy

The economy of the society influences the roles of the individuals in the society. The economic institution is a very gendered institution. **Occupational sex segregation** is the concentration of men and women in different occupations which has grown out of traditional gender roles. Men dominate as airline pilots, architects, and auto mechanics; women dominate as elementary school teachers, florists, and hair stylists (Weisgram et al., 2010). Only

occupational sex segregation the concentration of women in certain occupations and men in other occupations.

recently have women become NASCAR drivers (e.g., Danica Patrick).

Female-dominated occupations tend to require less education and have lower status. England (2010) noted that gains of women have been uneven because women have had a strong incentive to enter traditionally male jobs, but men have had little incentive to take on female jobs or activities. The salaries in these occupations (e. g., elementary education) have remained relatively low.

Increasingly, occupations are becoming less segregated on the basis of gender, and social acceptance of nontraditional career choices has increased. In 1960, 98% of persons entering veterinary medicine were male; today only 49% are male (Lincoln, 2010).

3-3f Mass Media

Mass media, such as movies, television, magazines, newspapers, books, music, computer games, and music television videos, both reflect and shape gender roles. Media images of women and men typically conform to traditional gender stereotypes. Gerding and Signorielli (2014) analyzed content of gender role portrayals in 49 episodes of 40 U.S. tween television programs of two genres: teen scene (geared towards girls) and action adventure (geared toward boys). Results show that females were underrepresented in the action-adventure genre. Also, compared to males, females were more attractive, more concerned about their appearance, and received comments about their "looks."

As for music, Ter Gogt et al. (2010) studied 410 13- to 16-year-old students and found that a preference for hip-hop music was associated with gender stereotypes (e.g., men are sex driven and tough, women are sex objects). In regard to music television videos, Wallis (2011) conducted a content analysis of 34 music videos and found that significant gender displays reinforced stereotypical notions of women as sexual objects, females as subordinate, and males as aggressive.

The cumulative effects of family, peers, religion, education, the economic institution, and mass media perpetuate gender stereotypes. Each agent of socialization reinforces gender roles that are learned from other agents of socialization, thereby creating a gender role system that is deeply embedded in our culture. All of these influences affect relationship choices (see Table 3.1).

Table 3.1

Effects of Gender Role Socialization on Relationship Choices

Women	Men
1. A woman who is not socialized to pursue advanced education (which often translates into less income) may feel pressure to stay in an unhappy relationship with someone on whom she is economically dependent.	1. Men who are socialized to define themselves in terms of their occupational success and income may find their self-esteem and masculinity vulnerable if they become unemployed, retired, or work in a low-income job.
2. Women who are socialized to play a passive role and not initiate relationships are limiting interactions that might develop into valued relationships.	2. Men who are socialized to restrict their experience and expression of emotions are denied the opportunity to discover the rewards of emotional interpersonal involvement.
3. Women who are socialized to accept that they are less valuable and less important than men are less likely to seek, achieve, or require egalitarian relationships with men.	3. Men who are socialized to believe it is not their role to participate in domestic activities (child rearing, food preparation, house cleaning) will not develop competencies in these life skills. Potential partners often view domestic skills as desirable.
4. Women who internalize society's standards of beauty and view their worth in terms of their age and appearance are likely to feel bad about themselves as they age. Their negative self-concept, more than their age or appearance, may interfere with their relationships.	4. Heterosexual men who focus on cultural definitions of female beauty overlook potential partners who might not fit the cultural beauty ideal but who would be wonderful life companions.
5. Women who are socialized to accept that they are solely responsible for taking care of their parents, children, and husband are likely to experience role overload. In this regard, some women may feel angry and resentful, which may have a negative impact on their relationships.	5. Men who are socialized to have a negative view of women who initiate relationships will be restricted in their relationship opportunities.
6. Women who are socialized to emphasize the importance of relationships in their lives will continue to seek relationships that are emotionally satisfying.	6. Men who are socialized to be in control of relationship encounters may alienate their partners, who may desire equality.

3-4 Consequences of Traditional Gender Role Socialization

This section discusses positive and negative consequences of traditional female and male socialization in the United States.

3-4a Traditional Female Role Socialization

In this section, we summarize some of the negative and positive consequences of being socialized as a woman in U.S. society. Each consequence may or may not be true for a specific woman. For example, although women in general have less education and income, a particular woman may have more education and a higher income than a particular man.

Negative Consequences of Traditional Female Role Socialization There are several negative consequences of being socialized as a woman in our society.

1. **Less income.** Although women earn more college degrees than men (Wells et al., 2011) and earn 47% of Ph.D.s (National Science Foundation, 2012), they have lower academic rank and earn less money. The lower academic rank is because women give priority to the care of their children and family over their advancement in the workplace. Women still earn about

© iStockphoto.com/Ivanastar

Table 3.2
Women's and Men's Median Income with Similar Education

	Bachelor's	Master's	Doctoral Degree
Men	$56,404	$71,537	$82,376
Women	$36,812	$48,738	$63,913

Source: *Proquest Statistical Abstract of the United States, 2014.* Bethesda, MD, Table 732.

three-fourths of what men earn, even when their level of educational achievement is identical (see Table 3.2). Their visibility in the ranks of high corporate America is also still low. Less than 5% of CEOs at Fortune 500 companies are women (Sandberg, 2013). An exception is Virginia Rometty, CEO of IBM—the first female head of the company in 100 years.

However, the value women place on a high-income career job is changing. The Pew Research Center (2012) reports a reversal of traditional gender roles with young women surpassing young men in saying that achieving success in a high-paying career or profession is important in their lives. Two-thirds (66%) of young women ages 18 to 34 rate career success high on their list of life priorities, compared with 59% of young men. In 1997, 56% of young women and 58% of young men felt the same way. Today's young men are also supportive of women having careers equal to their own. Fifty-six percent of 1,141 undergraduate men agreed that it was important to them that their wives had a career of equal status to their own (Hall & Knox, 2015).

2. **Feminization of poverty.** Another reason many women are relegated to a lower income status is the **feminization of poverty**. This term refers to the disproportionate percentage of poverty experienced by women living alone or with their children. Single mothers are particularly associated with poverty.

When head-of-household women are compared with married-couple households, the median income is $33,637 versus $74,130 (*Proquest Statistical Abstract of the United States, 2014,* Table 722). The process is cyclical—poverty contributes to teenage pregnancy because teens have

feminization of poverty the idea that women disproportionately experience poverty.

limited supervision and few alternatives to parenthood (the median income for head-of-household men is $48,084). Such early childbearing interferes with educational advancement and restricts women's earning capacity, which keeps them in poverty. Their offspring are born into poverty, and the cycle begins anew.

Low pay for women is also related to the fact that they tend to work in occupations that pay relatively low incomes. Indeed, women's lack of economic power stems from the relative dispensability (it is easy to replace) of women's labor and how work is organized (men occupy and control positions of power). Women also live longer than men, and poverty is associated with being elderly.

When women move into certain occupations, such as teaching, there is a tendency in the marketplace to segregate these occupations from men's, and the result is a concentration of women in lower-paid occupations. The salaries of women in these occupational roles increase at slower rates. For example, salaries in the elementary and secondary teaching profession, which is predominately female, have not kept pace with inflation.

Conflict theorists assert that men are in more powerful roles than women and use this power to dictate incomes and salaries of women and "female professions." Functionalists also note that keeping salaries low for women keeps women dependent and in child-care roles so as to keep equilibrium in the family. Hence, for both conflict and structural reasons, poverty is primarily a feminine issue. One of the consequences of being a woman is to have an increased probability of feeling economic strain throughout life.

3. **Higher risk for sexually transmitted infections.** Due to the female anatomy, women are more vulnerable to sexually transmitted infections and HIV (they receive more bodily fluids from men). Some women also feel limited in their ability to influence their partners to wear condoms (East et al., 2011).

4. **Negative body image.** Just as young girls tend to have less positive self-concepts than boys (Yu and Zi, 2010), they also feel more negatively about their bodies due to the cultural emphasis on being thin and trim (Grogan, 2010). Darlow and Lobel (2010) noted that overweight women who endorse cultural values

of thinness have lower self-esteem. There are more than 3,800 beauty pageants annually in the United States. The effect for many women who do not match the cultural ideal is to have a negative body image. Hollander (2010) reported on a study whereby teenage girls who viewed themselves as overweight had an increased risk of having their first sexual experience by age 13.

American women also live in a society that devalues them in a larger sense. Their lives and experiences are not taken as seriously as men's. **Sexism** is an attitude, action, or institutional structure that subordinates or discriminates against individuals or groups because of their sex. Sexism against women reflects the tradition of male dominance and presumed male superiority in U.S. society.

Benevolent sexism (reviewed by Maltby et al., 2010) is a related term and reflects the belief that women are innocent creatures who should be protected and supported. While such a view has positive aspects, it assumes that women are best suited for domestic roles and need to be taken care of by a man since they are not capable of taking care of themselves.

5. **Less personal/marital satisfaction.** Demaris et al. (2012) analyzed data on 707 marriages and found that couples characterized by more traditional attitudes toward gender roles were significantly less satisfied than others. Wives in traditional marriages are particularly likely to report lower marital satisfaction (Bulanda, 2011). Such lower marital satisfaction of wives is attributed to power differentials in the marriage. Traditional husbands expect to be dominant and expect their wives to take care of the house and children. Women have been socialized to believe that they can have both a family and a career—that they can have it all. In reality, they have discovered that they "can have two things half way" and that men are not much help (Haag, 2011, p. 47). In large nonclinical samples comparing wives and husbands on marital satisfaction (which include egalitarian marriages), wives do not report lower marital satisfaction.

Poorer mental health (Read & Grundy, 2011) and higher levels of anger (Simon & Lively, 2010) among women are also associated with traditional gender roles. One interpretation of these associations is related to the sense

Young girls tend to feel more negatively about their bodies than boys due to the cultural emphasis on being thin and trim.

of powerlessness that women feel in the United States due to inequitable division of labor, their likelihood of holding lower status/lower wage jobs, less power in relationships, etc.

Positive Consequences of Traditional Female Role Socialization
We have discussed the negative consequences of being born and socialized as a woman. However, there are also decided benefits.

1. **Longer life expectancy.** Women have a longer life expectancy than men. It is not clear if their greater longevity is related to biological or to social factors. Females born in the year 2015 are expected to live to the age of 81.8, in contrast to men, who are expected

sexism an attitude, action, or institutional structure that subordinates or discriminates against an individual or group because of their sex.

benevolent sexism the belief that women are innocent creatures who should be protected and supported.

to live to the age of 77.1 (*Proquest Statistical Abstract of the United States, 2014*, Table 112).

2. **Stronger relationship focus.** Women prioritize family over work and do more child care than men (Craig & Mullan, 2010). Mothers provide more "emotion work," helping children with whatever they are struggling with (Minnottea et al., 2010).

3. **Keeps relationships on track.** Women are more likely to initiate "the relationship talk" to ensure that the relationship is moving forward (Nelms et al., 2012). In addition, when there is a problem in the relationship, it is the woman who moves the couple toward help (Eubanks Fleming & Córdova, 2012).

4. **Bonding with children.** Another advantage of being socialized as a woman is the potential to have a closer bond with children. In general, women tend to be more emotionally bonded with their children than men. Although the new cultural image of the father is of one who engages emotionally with his children, many fathers continue to be content for their wives to take care of their children, with the result that mothers, not fathers, become more emotionally bonded with their children. Table 3.3 summarizes the consequences of traditional female role socialization.

3-4b Consequences of Traditional Male Role Socialization

Male socialization in U.S. society is associated with its own set of consequences. As with women,

Courtesy of Michelle North

each consequence may or may not be true for a specific man.

Negative Consequences of Traditional Male Role Socialization There are several negative consequences associated with being socialized as a man in U.S. society.

1. **Identity synonymous with occupation.** Ask men who they are, and many will tell you what they do. Society tends to equate a man's identity with his occupational role. The recession of 2008 was difficult for men that 70% of the jobs lost were to men. Michniewicz et al. (2014) analyzed the perceived effect of threatened unemployment on men and found that men estimated a lower appraisal of themselves following an imagined job loss. That men are focused on their work role may also translate into fewer friendships and relationships. While friendships are important to both men and women (Meliksah et al., 2013), men have fewer sustained friendships than women across time. This is due not only to their work focus but also to the cultural values of independence which support men being the loner (Way, 2013).

Table 3.3
Consequences of Traditional Female Role Socialization

Negative Consequences	Positive Consequences
Less income (more dependent)	Longer life
Feminization of poverty	Stronger relationship focus
Higher STD/HIV infection risk	Keep relationships on track
Negative body image	Bonding with children
Less personal/marital satisfaction	Identity not tied to job

> "The equivalent of a **woman** being treated as a **sex object** is a **man** being treated as a **success object**."
>
> —WARREN FARRELL, MEN'S ADVOCATE

2. **Limited expression of emotions.** Most men not only cry less (Barnes et al., 2012) but are also pressured to disavow any expression that could be interpreted as feminine (e.g., emotional). Lease et al. (2010) confirmed that the socialization of men (particularly white men) involves men "being more emotionally isolated and competitive in relationships and less competent at providing support to others, disclosing their own feelings, or managing conflict effectively."

3. **Fear of intimacy.** Garfield (2010) reviewed men's difficulty with emotional intimacy and noted that their emotional detachment stems from the provider role which requires them to stay in control. Being emotional is seen as weakness. Men's groups where men learn to access their feelings and express them have been helpful in increasing men's emotionality. Murphy et al. (2010) emphasized that men profit from becoming involved in the emotional labor of maintaining a relationship—their satisfaction increases.

4. **Custody disadvantages.** Courts are sometimes biased against divorced men who want custody of their children. Because divorced fathers are typically regarded as career focused and uninvolved in child care, some are relegated to seeing their children on a limited basis, such as every other weekend or four evenings a month.

5. **Shorter life expectancy.** Men typically die five years sooner (at age 77) than women. One explanation is that the traditional male role emphasizes achievement, competition, and suppression of feelings, all of which may produce stress. Not only is stress itself harmful to physical health, but it may also lead to compensatory behaviors such as smoking, alcohol, or other drug abuse, and dangerous risk-taking behavior (e.g., driving fast, binge drinking).

> "How easy to be a **busy man**: to **sacrifice myself** on the altar of **accomplishment**; to light the incense and chant the mantra of success—not for money or glory, but **success** nonetheless."
>
> —SY SAFRANSKY, EDITOR, *THE SUN*

Benefits of Traditional Male Socialization
As a result of higher status and power in society, men tend to have a more positive self-concept and greater confidence in themselves. In a sample of 1,151 undergraduate men, 28.7% "strongly agreed" with the statement, "I have a very positive self-concept." In contrast, 19% of 3,572 undergraduate women strongly agreed with the statement (Hall & Knox, 2015). Men also enjoy higher incomes and an easier climb up the good-old-boy corporate ladder; they are stalked/followed/harassed less often than women (17.7% versus 26.2%). Other benefits are the following:

1. **Freedom of movement.** Men typically have no fear of going anywhere, anytime. Their freedom of movement is unlimited. Unlike women (who are taught to fear rape, be aware of their surroundings, walk in well-lit places, and not walk alone after dark) men are oblivious to these fears and perceptions. They can go anywhere alone and are fearless about someone harming them.

2. **Greater available pool of potential partners.** Because of the mating gradient (men marry "down" in age and education whereas women marry "up"), men tend to marry younger women so that a 35-year-old man may view women from 20 to 40 years of age as possible mates. However, a woman of age 35 is more likely to view men her same age or older as potential mates. As she ages, fewer men are available.

3. **Norm of initiating a relationship.** Men are advantaged because traditional norms allow

Table 3.4

Consequences of Traditional Male Role Socialization

Negative Consequences	Positive Consequences
Identity tied to work role	Higher income and occupational status
Limited emotionality	More positive self-concept
Fear of intimacy; more lonely	Less job discrimination
Disadvantaged in getting custody	Freedom of movement; more partners to select from; more normative to initiate relationships
Shorter life	Happier marriage

them to be aggressive in initiating relationships with women. In addition, men tend to initiate marriage proposals and the bride more often takes the last name of her husband rather than keeping her last name (Robnett & Leaper, 2013). Table 3.4 summarizes the consequences of being socialized as a male.

3-5 Changing Gender Roles

Androgyny, gender role transcendence, and gender postmodernism emphasize that gender roles are changing.

3-5a Androgyny

Androgyny is a blend of traits that are stereotypically associated with masculinity and femininity. Androgynous celebrities include Lady Gaga, David Bowie, Boy George, Patti Smith, and Annie Lennox. Two forms of androgyny are:

1. Physiological androgyny refers to intersexed individuals, discussed earlier in the chapter. The genitals are neither clearly male nor female, and there is a mixing of "female" and "male" chromosomes and hormones.

2. Behavioral androgyny refers to the blending

androgyny a blend of traits that are stereotypically associated with masculinity and femininity.

gender role transcendence abandoning gender frameworks and looking at phenomena independent of traditional gender categories

or reversal of traditional male and female behavior, so that a biological male may be very passive, gentle, and nurturing and a biological female may be very assertive, rough, and selfish.

Androgyny may also imply flexibility of traits; for example, an androgynous individual may be emotional in one situation, logical in another, and assertive in another. Gender role identity (androgyny, masculinity, femininity) was assessed in a sample of Korean and American college students with androgyny emerging as the largest proportion in the American sample and femininity in the Korean sample (Shin et al., 2010).

3-5b Gender Role Transcendence

Beyond the concept of androgyny is that of gender role transcendence. We associate many aspects of our world, including colors, foods, social or occupational roles, and personality traits, with either masculinity or femininity. The concept of **gender role transcendence** means abandoning gender frameworks and looking at phenomena independent of traditional gender categories.

Transcendence is not equal for women and men. Although females are becoming more masculine, in part because our society values whatever is masculine, men are not becoming more feminine. Indeed, adolescent boys may be described as very gender entrenched. Beyond gender role transcendence is gender postmodernism.

3-5c Gender Postmodernism

Gender postmodernism abandons the notion of gender as natural and emphasizes that gender is socially constructed. Fifteen years ago, Monro (2000) noted that people in the postmodern society would no longer be categorized as male or female but be recognized as capable of many identities— "a third sex" (p. 37). A new

photobank.ch/Shutterstock.com

Personal View: "My Life as a Cross-Dresser Husband and Dad"*

Being a cross-dresser can be very confusing and very frustrating. I told my ex, about a month before we married, that I liked wearing women's clothing. I told her that I could quit and burned all of my clothing (called **purging**). How wrong I was! After about three months of being married, I told her that I needed to get some hosiery and panties to wear. She said that this would be OK, as long they were not lace. This progressed to the point that I needed to wear shoes, and later boots. Over time, I escalated to 5-inch stiletto heels, and I began to wear women's pajamas at night because I liked the silky feel.

All of this progressed into my wearing various forms of lingerie, which I only wore with my wife's consent and presence. While I always tried to be very respectful of my wife's wishes, I soon realized that I needed to cross-dress more often and with more intensity than she was comfortable with. When my two daughters were born, I had to be very discreet as to when I would dress as a woman. I remember many a night of getting up with my oldest daughter and rocking her to sleep while wearing hose, high heels, and lingerie, in our moonlit living room.

On several occasions, my ex told me that she wanted me to get rid of my things. She "wanted my things out of her house." So, I put my clothes in a box that I kept in a utility building. But I missed my cross-dressing, continued to do it, and became very sneaky about it. When my workweek would end on Friday, I would rush home quickly, and dress for several hours before I picked up my daughters from kindergarten. Of course, I felt very guilty about my sneaking around.

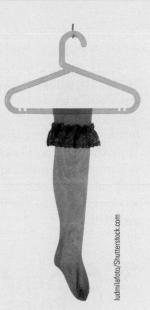

ludmilafoto/Shutterstock.com

When the girls were about 8 and 5, we bought a house. Again, my ex told me that she wanted me to get rid of my "things"; she wanted them out of her house. I only became sneakier and sneakier. For example, I started hiding my shoes and hosiery under a dresser in the bedroom, so that when the family was gone, to let's say, the grocery store, I could quickly and easily slip into some of my things for a few moments of ecstatic relaxing bliss. Several times, I would underestimate when they would return from the store and they would almost catch me while dressed. . . .

The older my girls got, the less time I had to dress. It finally got to the point that the only time that I could dress would be when the family would go out of town (to grandmother's house). I loved for them to be gone and was irritated by their return. Indeed, my daughters asked their mom if I was having an affair with another woman while they were gone. In essence, there was another woman, but this woman is the one who resided within me.

It is hard to explain the frustration of not being able to dress; it would lead to anger and anxiety. This anger reached its greatest point about two years before my wife and I ended our marriage. What caused this was that my ex demanded that I get rid of my feminine clothing. She actually insisted on one occasion that she accompany me to get rid of them. I finally gave the clothes to Goodwill so that I could get a tax break. I also went to see a psychiatrist who helped me realize that this was an important need, that cross-dressing was something that I was probably going to continue. My ex totally rejected this advice and we divorced.

*This heterosexual, a formerly married father of two, prefers to remain anonymous.

purging for cross-dressers, the term means to destroy one's clothes of the other sex as a means of distancing one's self from cross-dressing. The practice is rarely effective as the cross-dresser acquires new clothes.

conceptualization of "trans" people calls for new social structures, "based on the principles of equality, diversity and the right to self-determination" (p. 42). No longer would our society telegraph transphobia but would instead embrace pluralization "as an indication of social evolution, allowing greater choice and means of self-expression concerning gender" (p. 42).

Our society is becoming increasingly tolerant of variations in gender expression. But stigma remains, both inside and outside the family. The Personal View section emphasizes the life of a cross-dressing husband and father.

3-6 Trends in Gender Roles

Imagine a society in which women and men each develop characteristics, lifestyles, and values that are independent of gender role stereotypes. Characteristics such as strength, independence, logical thinking, and aggressiveness are no longer associated with maleness, just as passivity, dependence, emotions, intuitiveness, and nurturance are no longer associated with femaleness. Both sexes are considered equal, and women and men may pursue the same occupational, political, and domestic roles. These changes are occurring…slowly. Lucier-Greer and Adler-Baeder (2010) provided data which showed that, compared to 2000, gender role attitudes of today are becoming more egalitarian. Fisher (2010) emphasized that peer marriage or marriage between equals is the most profound change in marriage in recent years. McGeorge et al. (2012) noted that couple therapists are being trained to be sensitive to and encourage relationship equality in terms of power and division of labor in relationships.

Another change in gender roles is the independence and ascendency of women. Women will less often require marriage for fulfillment, will increasingly take care of themselves economically, and will opt for having children via adoption or donor sperm rather than foregoing motherhood. That women are slowly outstretching men in terms of education will provide the impetus for these changes.

STUDY TOOLS ➲ **3**

Ready to study? In the book, you can:

➲ Rip out the chapter review card at the back of the book for a handy summary of the chapter and key terms.

➲ Assess your attitudes toward men and women with the Self-Assessment card at the back of the book.

Online at CENGAGEBRAIN.COM you can:

➲ Prepare for tests with quizzes.

➲ Review the key terms with Flash Cards.

➲ Play games to master concepts.

Love and Relationship Development

"We're all a little **weird**. And life is a little weird. And when we find someone whose weirdness is **compatible** with ours, we join up with them and fall into mutually satisfying weirdness—and call it love—**true love**."

—ROBERT FULGHUM

Love is an experience we are taught to seek, cherish, and be sad about when it ends. Eighty-five percent of 5,500 single individuals (no gender difference) reported that they had been in love (Walsh, 2013). For most, it is an important motivation for and prerequisite for marriage. Ninety-three percent of 1,306 spouses and 83% of 1,385 unmarried individuals said that "love" was an important reason for getting married (Cohn, 2013). In this chapter, we review the various views and origins of love, how it develops, and the various factors involved in finding and maintaining love with a specific person.

4-1 Ways of Viewing Love

A common class exercise among professors who teach marriage and the family is to randomly ask class members to identify one word they most closely associate with love. Invariably, students identify different words (e. g., commitment, feeling, trust, altruism), suggesting great variability in the way we think about love (Berscheid, 2010). Passionate love is also universal—it has existed through history and is not unique to Western society (p. 154).

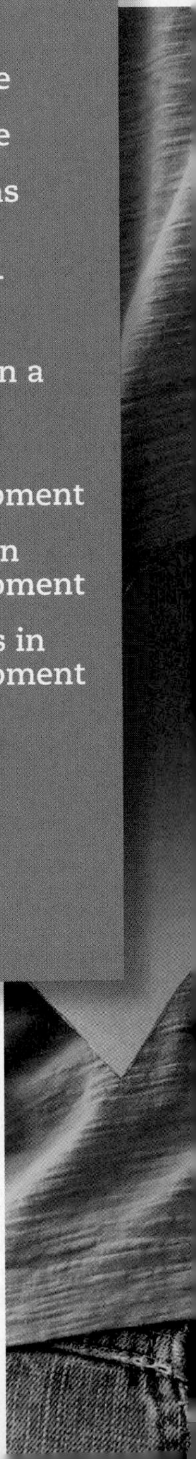

Courtesy of Rachel Calisto

Love is often confused with lust and infatuation. Love (and attachment) is about deep, abiding feelings with a focus on the long term (Langeslag et al., 2013; Foster 2010); **lust** is about sexual desire and the present. The word **infatuation** comes from the same root word as *fatuous*, meaning "silly" or "foolish," and refers to a state of passion or attraction that is not based on reason. Infatuation is characterized by euphoria (Langeslag et al., 2013) and by the tendency to idealize the love partner. People who are infatuated magnify their lovers' positive qualities ("My partner is always happy") and overlook or minimize their negative qualities ("My partner doesn't have a problem with alcohol; he just likes to have a good time").

In the following section, we look at the various ways of conceptualizing love.

lust sexual desire.

infatuation emotional feelings based on little actual exposure to the love object.

> "I was **born** when you **kissed me**. I **died** when you **left me**. I **lived**
> a few weeks **while you loved me**."
>
> —HUMPHREY BOGART TO GLORIA GRAHAME IN THE MOVIE, *IN A LONELY PLACE*

4-1a **Romantic versus Realistic Love**

Love may also be described as being on a continuum from romanticism to realism. For some people, love is romantic; for others, it is realistic. **Romantic love** is characterized in modern America by such beliefs as "love at first sight," "one true love" and "love conquers all."

Regarding love at first sight, 36% of 1,144 undergraduate males and 25% of 3,554 undergraduate females reported that they had experienced love at first sight (Hall & Knox, 2015). One explanation for men falling in love more quickly than women is that (from a biological/evolutionary perspective) men must be visually attracted to young, healthy females to inseminate them. This biologically based reproductive attraction is interpreted as a love attraction so that the male feels immediately drawn to the female, but what he may actually see is an egg needing fertilization.

An openness to falling in love and developing an intimate relationship is Erik Erikson's sixth stage of psychosocial development. He noted that between the ages of 19 and 40, most individuals move from "isolation to intimacy" wherein they establish committed loving relationships. Failure to do so is to leave one vulnerable to loneliness and depression.

In contrast to romantic love is realistic love. Realistic love is also known as conjugal love. **Conjugal love** is the love between married people characterized by companionship, calmness, comfort, and security. Conjugal love is in contrast to romantic love, which is characterized by excitement and passion. Stanik et al. (2013) interviewed 146 African American couples who had been married from 3 to 25 years and confirmed a decrease in the intensity of love feelings across time. However, couples with a traditional division of labor (compared to egalitarian couples) showed a greater decline.

Both romantics and realists can be happy individuals and successful relationship partners. Love also conveys enormous benefits, including positive mental health. Plant et al. (2010) emphasized that these benefits are so pronounced for the individual that he or she will protect these benefits by diverting themselves when they feel attracted to an alternative.

romantic love an intense love whereby the lover believes in love at first sight, only one true love, and love conquers all.

conjugal (married) love the love between married people characterized by companionship, calmness, comfort, and security.

Diversity in Love throughout the World

The theme of U.S. culture is individualism, which translates into personal fulfillment, emotional intimacy, and love as the reason for marriage. In Asian cultures (e.g., China) the theme is collectivism, which focuses on "family, comradeship, obligations to others, and altruism" with love as secondary (Riela et al., 2010). While arranged marriages are and have been the norm in Eastern societies, love marriages are becoming more frequent (Allendorf, 2013).

© teolin/Shutterstock.com

4-1b Love Styles

Theorist John Lee (1973, 1988) identified various styles of love that describe the way individuals view love and relate to each other.

1. **Ludic.** The **ludic love style** views love as a game in which the player has no intention of getting seriously involved. The ludic lover refuses to become dependent on any one person and does not encourage another's intimacy. Two essential skills of the ludic lover are to juggle several partners at the same time and to manage each relationship so that no one partner is seen too often.

   These strategies help ensure that the relationship does not deepen into an all-consuming love. Don Juan represented the classic ludic lover, embodying the motto of "Love 'em and leave 'em." Tzeng et al. (2003) found that whereas men were more likely than women to be ludic lovers, ludic love characterized the love style of college students the least.

2. **Pragma.** The **pragma love style** is the love of the pragmatic—that which is logical and rational. Pragma lovers assess their partners on the basis of assets and liabilities. One undergraduate female hung out with a guy because he had a car and could drive her home on weekends to see her boyfriend. An undergraduate male dated his partner because she had an apartment and would cook for him. Pragma lovers do not become involved in interracial, long-distance, or age-discrepant relationships because logic argues against doing so. The Personal View section emphasizes using one's heart or head in making decisions.

3. **Eros.** Just the opposite of the pragmatic love style, the **eros love style** (also known as romantic love) is imbued with passion and sexual desire. Eros is the most common love style of college women and men (Tzeng et al., 2003).

4. **Mania.** The **mania love style** is the out-of-control love whereby the person "must have" the love object. Jealousy, possessiveness, dependency, and controlling are symptoms of manic love. Zamora et al. (2013) identified the love styles of 72 gay men who revealed that a mania love style was the most frequent. These gay men were anxious for reassurance (mania) from their partner.

5. **Storge.** The **storge love style**, also known as companionate love, is a calm, soothing, nonsexual love devoid of intense passion. Respect, friendship, commitment, and familiarity are characteristics that help define the storge (pronounced STOR-jay) love relationship. The partners care deeply about each other but not in a romantic or lustful sense. Their love is also more likely to endure than fleeting romance.

> "We want things not because we have reasons for them. *We have reasons for them because we want them.*"
>
> —SCHOPENHAUER, GERMAN PHILOSOPHER

> "Therefore **love** moderately, long **love** doth so."
>
> —FRIAR LAURENCE IN SHAKESPEARE'S, *ROMEO AND JULIET*

ludic love style love style that views love as a game in which the love interest is one of several partners, is never seen too often, and is kept at an emotional distance.

pragma love style love style that is logical and rational; the love partner is evaluated in terms of assets and liabilities.

eros love style love style characterized by passion and romance.

mania love style an out-of-control love whereby the person "must have" the love object; obsessive jealousy and controlling behavior are symptoms of manic love.

storge love style a love consisting of friendship that is calm and nonsexual.

Personal View: "Using One's Heart or Head to Make Relationship Decisions?"

This couple represent high school sweethearts in their first year of college together. In general, the younger the individuals the more romantic is their view of love and the more they tend to make decisions with their heart.

Lovers are frequently confronted with the need to make decisions about their relationships, but they are divided on whether to let their heart or their head rule in such decisions. Some evidence suggests that the heart rules. Fifty-nine percent of 3,563 undergraduate females and 54% of 1,145 undergraduate males agreed with the statement, "I make relationship decisions more with my heart than my head" (Hall & Knox, 2015), suggesting that the heart rules more often in relationship matters of undergraduates. We asked students in our classes to fill in the details about deciding with their heart or their head. Some of their answers follow.

Heart

Those who relied on their hearts for making decisions felt that emotions were more important than logic and that listening to their heart made them happier. One woman said:

In deciding on a mate, my heart would rule because my heart has reasons to cry and my head doesn't. My heart knows what I want, what would make me most happy. My head tells me what is best for me. But I would rather have something that makes me happy than something that is good for me.

Some men also agreed that the heart should rule. One said:

I went with my heart in a situation, and I'm glad I did. I had been hanging out with a girl for two years when I decided she was not the one I wanted and that my present girlfriend was. My heart was saying to go for the one I loved, but my head was telling me not to because if I broke up with the first girl, it would hurt her, her parents, and my parents. But I decided I had to make myself happy and went with the feelings in my heart and started dating the girl who is now my fiancée.

Relying on one's emotions does not always have a positive outcome, as the following experience illustrates:

Last semester, I was dating a guy I felt more for than he did for me. Despite that, I wanted to spend any opportunity I could with him when he asked me to go somewhere with him. One day he had no classes, and he asked me to go to the park by the river for a picnic. I had four classes that day and exams in two of them. I let my heart rule and went with him. He ended up breaking up with me on the picnic.

Head

Some undergraduates make relationship choices based on their head as some of the following comments show:

In deciding on a mate, I feel my head should rule because you have to choose someone that you can get along with after the new wears off. If you follow your heart solely, you may not look deep enough into a person to see what it is that you really like. Is it just a pretty face or a nice body? Or is it deeper than that, such as common interests and values? The "heart" sometimes can fog up this picture of the true person and distort reality into a fairy tale.

Another student said:

Love is blind and can play tricks on you. Two years ago, I fell in love with a man whom I later found out was married. Although my heart had learned to love this man, my mind knew the consequences and told me to stop seeing him. My heart said, "Maybe he'll leave her for me," but my mind said, "If he cheated on her, he'll cheat on you." I broke up with him and am glad that I listened to my head.

Some individuals feel that both the head and the heart should rule when making relationship decisions.

When you really love someone, your heart rules in most of the situations. But if you don't keep your head in some matters, then you risk losing the love that you feel in your heart. I think that we should find a way to let our heads and hearts work together.

There is an adage, "Don't wait until you find the person you can live with; wait and find the person that you can't live without!" In your own decisions you might consider the relative merits of listening to your heart or head and moving forward recognizing there is not one "right" answer for all individuals on all issues.

One's grandparents who have been married 50 years are likely to have a storge type of love. Neto (2012) compared love perceptions by age group and found that the older the individual, the more important love became and the less important sex became.

6. **Agape.** **Agape love style** is characterized by a focus on the well-being of the person who is loved, with little regard for reciprocation.

Older couples married many years are likely to have a storge type of love.

Key qualities of agape love are not responding to a partner's negativity and not expecting an exchange for positives but believing that the other means well and will respond kindly in time. The love parents have for their children is often described as agape love.

7. **Compassionate love.** While not one of the six love styles identified by John Lee, **compassionate love** (emotional feelings toward another that generate behaviors to promote the partner's well-being) is a unique and important style of love (Fehr et al., 2014). Specific behaviors associated with compassionate love include: concern for the other's well-being (doing things to help the partner achieve his or her goals/experience satisfaction), understanding/acceptance (adopting the other person's point of view), respect and admiration (value the person), and openness (being receptive to the other's preferences and opinions) (Reis et al., 2014).

agape love style love style characterized by a focus on the well-being of the love object, with little regard for reciprocation; the love of parents for their children is agape love.

compassionate love emotional feelings toward another that generate behaviors to promote the partner's well-being.

While eros/romantic love style is hormonally driven, compassionate love involves self-sacrifice and support for the partner over time. The genders do not differ in their capacity for compassionate love (which is associated with relationship quality in terms of closeness, satisfaction, commitment) (Fehr et al., 2014).

4-1c Triangular View of Love

Sternberg (1986) developed the "triangular" view of love, which consists of three basic elements: intimacy, passion, and commitment. The presence or absence of these three elements creates various types of love experienced between individuals, regardless of their sexual orientation. These various types include the following:

Intimacy

Commitment

Passion

1. **Nonlove.** The absence of intimacy, passion, and commitment. Two strangers looking at each other from afar are experiencing nonlove.

2. **Liking.** Intimacy without passion or commitment. A new friendship may be described in these terms of the partners liking each other.

3. **Infatuation.** Passion without intimacy or commitment. Two people flirting with each other in a bar may be infatuated with each other.

4. **Romantic love.** Intimacy and passion without commitment. Love at first sight reflects this type of love.

5. **Conjugal love** (also known as married love). Intimacy and commitment without passion. A couple married for 50 years are said to illustrate conjugal love.

6. **Fatuous love.** Passion and commitment without intimacy. Couples who are passionately wild about each other and talk of the future but do not have an intimate connection with each other have a fatuous love.

7. **Empty love.** Commitment without passion or intimacy. A couple who stay together for social (e.g., children) and legal reasons but who have no spark or emotional sharing between them have an empty love.

8. **Consummate love.** Combination of intimacy, passion, and commitment; Sternberg's view of the ultimate, all-consuming love.

Individuals bring different combinations of the elements of intimacy, passion, and commitment (the triangle) to the table of love. One lover may bring a predominance of passion, with some intimacy but no commitment (romantic love), whereas the other person brings commitment but no passion or intimacy (empty love). The triangular theory of love allows lovers to see the degree to which they are matched in terms of passion, intimacy, and commitment in their relationship.

4-1d Love Languages

Gary Chapman's (2010) **five love languages** have become part of U.S. love culture. These five languages

Flowers and candy are understood in relationships to imply feelings of love and affection.

are gifts, quality time, words of affirmation, acts of service, and physical touch. Chapman encourages individuals to use the language of love most desired by the partner rather than the one preferred by the individual providing the love. Chapman has published nine books (e.g., for couples, singles, men, military) to illustrate the application of five love languages in different contexts. No empirical studies have been conducted on the five languages.

4-1e Polyamory

Poly (many) and amory (love) make up the word **polyamory**—open emotional and sexual involvement with three or more people (honesty and respect are norms common upon polyamorists). An online survey of 34 self-identified polyamorists revealed that they tended to be Democratic, liberal, nonreligious but "spiritual," and accepting of abortion and gay marriage (Jenks, 2014). Being in their thirties and forties is common with as few as 1.2 million practicing polyamory (Scheff, 2014). Polyamorists are different from **swinging** (persons who exchange partners for the purpose of sex) in that the latter are more focused on sex. Swingers also tend to be older, more educated, and have higher incomes.

4-2 Social Control of Love

Though we think of love as an individual experience, the society in which we live exercises considerable control over our love object or choice. The ultimate social control of love is **arranged marriage**—mate selection pattern whereby parents select the spouse of their offspring. Parents arrange 80% of marriages in China, India, and Indonesia (three countries representing 40% of the world's population). In most Eastern countries marriage is regarded as the linking of two families; the love feelings of the respective partners are irrelevant. Love is expected to follow marriage, not precede it. Arranged marriages not only help guarantee that cultural traditions will be carried on and passed to the new generation, but they also link two family systems together for mutual support of the couple.

Parents may know a family who has a son or daughter whom they would regard as a suitable partner for their offspring. If not, they may put an advertisement in a newspaper identifying the qualities they are seeking. The prospective mate is then interviewed by the family to confirm his or her suitability. Or a third person—a matchmaker—may be hired to do the screening and introducing.

Selecting a spouse for a daughter may begin early in the child's life. In some countries (e.g., Nepal and Afghanistan), child marriage occurs whereby young females (ages 8 to 12) are required to marry an older man selected by their parents. Suicide is the only alternative "choice" for these children.

While parents in Eastern societies may exercise direct control by selecting the partner for their son or daughter, American parents influence mate choice by moving to certain neighborhoods, joining certain churches, and enrolling their children in certain schools. Doing so increases the chance that their offspring will "hang out" with, fall in love with, and marry people who are similar in race, religion, education, and social class. Parents want their offspring to meet someone who will "fit in" and with whom they will feel comfortable.

Peers also exert an influence on homogenous mating by approving or disapproving of certain partners. Their motive is similar to that of parents—they want to feel comfortable around the people their peers bring with them to social encounters. Both parents and peers are influential, as most offspring and friends end up falling in love with and marrying people of the same race, education, and social class. Social approval of one's partner is normally important for a love relationship to proceed. If your

polyamory open emotional and sexual involvement with three or more people.

swinging persons who exchange partners for the purpose of sex.

arranged marriage mate selection pattern whereby parents select the spouse of their offspring.

parents and peers disapprove of the person you are involved with, it is difficult for you to continue the relationship long term.

Diamond (2003) emphasized that individuals are biologically wired and capable of falling in love and establishing intense emotional bonds with members of their own or opposite sex. Discovering that one's offspring is in love with and wants to marry someone of the same sex is a challenge for many parents.

"Love—it's everything I understand and all the things I never will."

—MARY CHAPIN CARPENTER, SINGER

4-3 Love Theories: Origins of Love

Various theories have been suggested with regard to the origins of love.

4-3a Evolutionary Theory

The **evolutionary theory of love** is that individuals are motivated to emotionally bond with a partner to ensure a stable relationship for producing and rearing children. In effect, love is a bonding mechanism between the parents during the time their offspring are dependent infants. Love's strongest bonding lasts about four years after birth, the time when children are most dependent and when two parents are most beneficial to the developing infant. "If a woman was carrying the equivalent of a twelve-pound bowling ball in one arm and a pile of sticks in the other, it was ecologically critical to pair up with a mate to rear the young," observed anthropologist Helen Fisher (Toufexis, 1993). The "four-year itch" is Fisher's term for the time at which parents with one child are most likely to divorce—the time when the

evolutionary theory of love theory that individuals are motivated to emotionally bond with a partner to ensure a stable relationship for producing and rearing children.

woman can more easily survive without parenting help from the male. If the couple has a second child, doing so resets the clock, and the "seven-year itch" is the next most vulnerable time.

4-3b Learning Theory

Unlike evolutionary theory, which views the experience of love as innate, learning theory emphasizes that love feelings develop in response to certain behaviors engaged in by the partner. Individuals in a new relationship who look at each other, smile at each other, compliment each other, touch each other endearingly, do things for each other, and do enjoyable things together are engaging in behaviors that encourage the development of love feelings. In effect, love can be viewed as a feeling that results from a high frequency of positive behavior and a low frequency of negative behavior. One high-frequency behavior is positive labeling whereby the partners flood each other with positive statements. We asked one of our students who reported that she was deliriously in love to identify the positive statements her partner had said to her. She kept a list of these statements for a week, which included:

Angel, Sweetie, Cinderella, Sleeping Beauty.
You understand me so well.
Precious jewel, Sugar bear, Snow White, Honey.
You are my best friend.
I never thought you were out there.
You know me better than anyone in my whole life.
You're always safe in my arms.
I would sell my guitar for you if you needed money.

People who "fall out of love" may note the high frequency of negatives on the part of their partner and the low frequency of positives. People who say, "this is not the person I married," are saying the ratio of positives to negatives has changed dramatically.

4-3c Sociological Theory

Sixty years ago, Ira Reiss (1960) suggested the wheel model as an explanation for how love develops. Basically, the wheel has four stages—rapport, self-revelation, mutual dependency, and fulfillment of personality needs. In the rapport stage, each partner has the feeling of having known the partner before, feels comfortable with the partner, and wants to deepen the relationship.

Such desire leads to self-revelation or self-disclosure, whereby each reveals intimate thoughts to the other about oneself, the partner, and the relationship. Such revelations deepen the relationship because it is assumed that the confidences are shared only with special people, and each partner feels special when listening to the revelations of the other. Indeed, Shelon et al. (2010) confirmed that intimacy develops when individuals disclose personal information about themselves and perceive that the listener understands/validates and cares about the disclosure. As the level of self-disclosure becomes more intimate, a feeling of mutual dependency develops. Each partner is happiest in the presence of the other and begins to depend on the other for creating the context of these euphoric feelings. "I am happiest when I am with you" is the theme of this stage.

4-3d Psychosexual Theory

According to psychosexual theory, love results from blocked biological sexual desires. In the sexually repressive mood of his time, Sigmund Freud (1905/1938) referred to love as "aim-inhibited sex." Love was viewed as a function of the sexual desire a person was not allowed to express because of social restraints. In Freud's era, people would meet, fall in love, get married, and have sex. Freud felt that the socially required delay from first meeting to having sex resulted in the development of "love feelings." By extrapolation, Freud's theory of love suggests that love dies with marriage (access to one's sexual partner).

4-3e Biochemical Theory

"Love is deeply biological" wrote Carter and Porges (2013), who reviewed the biochemistry involved in the development and maintenance of love. They noted that oxytocin and vasopressin are hormones involved in the development and maintenance of social bonding. The hormones are active in forging emotional connections between adults and infants and between adults. They are also necessary for our social and physiological survival.

Oxytocin is released from the pituitary gland during the expulsive stage of labor that has been associated with the onset of maternal behavior in lower animals (but oxytocin may be manufactured in both women and men when an infant or another person is present—hence it is not dependent on the birth process)

The love these individuals feel for each other is driven by the biochemical interaction of hormones (oxytocin and vasopressin) when they kiss.

Courtesy of Brittany Bolen

(Carter & Porges, 2013). Oxytocin has been referred to as the "cuddle chemical" because of its significance in bonding. Later in life, oxytocin seems operative in the development of love feelings between lovers during sexual arousal. Oxytocin may be responsible for the fact that more women than men prefer to continue cuddling after intercourse.

Phenylethylamine (PEA) is a natural, amphetamine-like substance that makes lovers feel euphoric and energized. The high that they report feeling just by being with each other is from PEA that the brain releases into their bloodstream. The natural chemical high associated with love may explain why the intensity of passionate love decreases over time. As with any amphetamine, the body builds up a tolerance to PEA, and it takes more and more to produce the special kick. Hence, lovers develop a tolerance for each other. "Love junkies" are those who go from one love affair to the next to maintain the high. Alternatively, some lovers break up and get back together frequently as a way of making the relationship new again and keeping the high going.

4-3f Attachment Theory

The attachment theory of love emphasizes that a primary motivation in life is to be emotionally connected with other people. Children abandoned by their parents and placed in foster care (400,000 are in foster care

oxytocin a hormone released from the pituitary gland during the expulsive stage of labor that has been associated with the onset of maternal behavior in lower animals.

Table 4.1
Love Theories and Criticisms

Theory	Criticism
Evolutionary. Love is the social glue that bonds parents with dependent children and spouses with each other to care for offspring.	The assumption that women and children need men for economic/emotional survival is not true today. Women can have and rear children without male partners.
Learning. Positive experiences create love feelings.	The theory does not account for (1) why some people share positive experiences but do not fall in love, and (2) why some people stay in love despite negative behavior by their partner.
Psychosexual. Love results from blocked biological drive.	The theory does not account for couples who report intense love feelings and have sex regularly.
Sociological. The wheel theory whereby love develops from rapport, self-revelation, mutual dependency, and personality need fulfillment.	Not all people are capable of rapport, revealing oneself, and so on.
Biochemical. Love is chemical. Oxytocin is an amphetamine-like chemical that bonds mother to child and produces a giddy high in young lovers.	The theory does not specify how much of what chemicals result in the feeling of love. Chemicals alone cannot create the state of love; cognitions are also important.
Attachment. Primary motivation in life is to be connected to others. Children bond with parents and spouses to each other.	Not all people feel the need to be emotionally attached to others. Some prefer to be detached.

in the United States) are vulnerable to having their early emotional attachment to their parents disrupted and developing "reactive attachment disorder" (Stinehart et al., 2012). This disorder involves a child who is anxious and insecure since he or she does not feel to be in a safe and protected environment. Such children find it difficult to connect emotionally, and this deficit continues into adulthood. Conversely, a secure emotional attachment with loving adults as a child is associated with the capacity for later involvement in satisfying, loving, communicative adult relationships.

Each of the theories of love presented in this section has critics (see Table 4.1).

4-4 Love as a Context for Problems

"The sweetest joy, the wildest woe is love."

—PEARL BAILEY, SINGER

"Love" answers Hester when asked by her husband Bill, "What has happened to you?"

unrequited love a one-sided love where one's love is not returned.

She has had an affair with Freddie, fallen hopelessly in love, with divorce and attempted suicide in the wake. *The Deep Blue Sea*, a 2012 movie, reveals up close and personal, how love can become a context for problems. In this section we review seven such problems.

4-4a Unrequited/ Unreciprocated Love

Unrequited love is a one-sided love where one's love is not returned. In the 16th century, it was called "lovesickness"; today it is known as erotomania or erotic melancholy (Kem, 2010). An example is from the short story "Winter Dreams" by F. Scott Fitzgerald. Dexter Green is in love with Judy Jones. "He loved her, and he would love her until the day he was too old for loving—but he could not have her."

Blomquist and Giuliano (2012) assessed the reactions of a sample of adults and college students to a partner telling them "I love you." The predominant response by both men and women was "I'm just not there yet." Both genders acknowledged that while this response was honest, it would hurt the individual who was in love.

4-4b Making Risky, Dangerous Choices

Some research suggests that individuals in love make risky, dangerous, or questionable decisions.

Nonsmokers who become romantically involved with a smoker are more likely to begin smoking (Kennedy et al., 2011). Similarly, couples in love and in a stable relationship are less likely to use a condom (doing so is not very romantic) (Warren et al., 2012). Of 381 undergraduates, 70% reported that they had engaged in unprotected sex with their romantic partner (Elliott et al., 2012).

4-4c Ending the Relationship with One's Parents

Some parents disapprove of the partner their son or daughter is involved with (e.g., race, religion, age) to the point that their offspring will end the relationship with their parents. "They told me I couldn't come home if I kept dating this guy, so I stopped going home" said one of our students who was involved with a partner of a different race. Choosing to end a relationship with one's parents is a definite downside of love.

4-4d Simultaneous Loves

Sometimes an individual is in love with two or more people at the same time. While this is acceptable in polyamorous relationships where the partners agree on multiple relationships, simultaneous loves become a serious problem.

4-4e Abusive Relationships

Twenty-two percent of 1,148 undergraduate males and 36% of 3,574 undergraduate females reported that they had been involved in an emotionally abusive relationship with a partner. As for physical abuse, 3% of the males and 11% of the females reported such previous involvement. (Hall & Knox, 2015). The primary reason individuals report that they remain in an abusive relationship is love for the partner.

"Love is a grave mental illness."

—PLATO, GREEK PHILOSOPHER

Valua Vitaly/Shutterstock.com

4-4f Profound Sadness/ Depression When a Love Relationship Ends

Fisher et al. (2010) noted that "romantic rejection causes a profound sense of loss and negative affect. It can induce clinical depression and in extreme cases lead to suicide and/or homicide." The researchers studied brain changes via magnetic resonance imaging of 10 women and five men who had recently been rejected by a partner but reported they were still intensely "in love." Participants alternately viewed a photograph of their rejecting beloved and a photograph of a familiar individual interspersed with a distraction-attention task. Their responses while looking at the photo of the person who rejected them included feelings of love, despair, good, and bad memories, and wondering about why this happened. Brain image reactions to being rejected by a lover are similar to withdrawal from cocaine.

There is also physical pain. "The pain one experiences in response to an unwanted breakup is identical to the pain one experiences when physically hurt" noted Dr. Steven Richeimer (2011) of the University of Southern California Pain Center. He was speaking in reference to a University of Michigan study where researchers asked people who recently had an unwanted romantic breakup to look at a picture of their ex or hold a hot cup of coffee. The pain reaction in the brain was exactly the same.

While there is no recognized definition or diagnostic criteria for "love addiction," some similarities to substance dependence include: euphoria and unrestrained desire in the presence of the love object or associated stimuli (drug intoxication); negative mood and sleep disturbance when separated from the love object (drug withdrawal); intrusive thoughts about the love object; and problems associated with love which may lead to clinically significant impairment or distress (Reynaud et al., 2011).

Problems associated with love begin early. Starr et al. (2012) conducted a study of 83 seventh and eighth grade girls (mean age 13.8) and found an association between those who reported involvement in romantic activities (e.g., flirting, dating, kissing) and depression, anxiety, and eating disorders. The researchers pointed out that it was unclear if romance led to these outcomes or if the girls who were depressed, anxious, etc., sought romance to help cope with their depressive symptoms.

4-5 How Love Develops in a New Relationship

Various social, physical, psychological, physiological, and cognitive conditions affect the development of love relationships.

4-5a Social Conditions for Love

Our society promotes love through popular music, movies, and novels. These media convey the message that love is an experience to pursue, enjoy, and maintain. People who fall out of love are encouraged to try again: "Love is lovelier the second time you fall." Unlike people reared in Eastern cultures, Americans grow up in a cultural context which encourages them to turn on their radar for love.

4-5b Psychological Conditions for Love

Five psychological conditions associated with falling in love are perception of reciprocal liking, personality, high self-esteem, self-disclosure, and gratitude.

Perception of Reciprocal Liking Riela et al. (2010) conducted two studies on falling in love using

It is important to feel good about oneself if one is to fall in love.

both American and Chinese samples. The researchers found that one of the most important psychological factors associated with falling in love was the perception of reciprocal liking. When one perceives that he or she is desired by someone else, this perception has the effect of increasing the attraction toward that person. Such an increase is particularly strong if the person is very physically attractive (Greitemeyer, 2010).

Personality Qualities The personality of the love object has an important effect on falling in love (Riela et al., 2010). Viewing the partner as intelligent or having a sense of humor is an example of a quality that makes the lover want to be with the beloved.

Self-Esteem High self-esteem is important for falling in love because it enables individuals to feel worthy of being loved. Feeling good about yourself allows you to believe that others are capable of loving you. Individuals with low self-esteem doubt that someone else can love and accept them.

Self-Disclosure Disclosing oneself is necessary if one is to fall in love—to feel invested in another. Ross (2006) identified eight dimensions of self-disclosure: (1) background and history, (2) feelings toward the partner, (3) feelings toward self, (4) feelings about one's body, (5) attitudes toward social issues, (6) tastes and interests, (7) money and work, and (8) feelings about friends. Disclosed feelings about the partner included "how much I like the partner," "my feelings about our sexual relationship," "how much I trust my partner," "things I dislike about my partner," and "my thoughts about the future of our relationship"—all of which are associated with relationship satisfaction. Of

Diversity in Body Preference throughout the World

The U.S. preference for a thin and trim female is not universal. Two researchers compared body-mass preferences among 300 cultures throughout the world and found that 81% of cultures preferred a female body size that would be described as "plump" (Brown & Sweeney, 2009).

LEARNING THE BASICS OF INTERACTION VIA SURROGATE THERAPY

Not everyone has the basic skills of being able to talk with, touch, and kiss a potential love partner. One example is a 29-year-old virgin male who sought therapy to enable him to learn the basic skills of talking with, touching, and kissing a woman. To assist the client in achieving his goal, the therapist enlisted the aid of three female relationship surrogates. A surrogate is person who works with clients in therapy as part of a three-way therapeutic team, consisting of the therapist, the surrogate partner, and the client.

The first stage of therapy involved the use of two female surrogates who met with the client in a social context (e.g., Starbucks) and taught him the basic communication skills of how to talk with a woman (e.g., use of open-ended questions, reflective statements, eye contact, etc.). He met with the first surrogate on five occasions and the second surrogate on four occasions to develop skills in talking with different women. These appointments were scheduled through the therapist, who paid the surrogates ($25 a session—money provided by the client) after they gave feedback about their meeting with the client. No independent contact was ever made between the client and the surrogates. The surrogates were part of the therapeutic team, not potential girlfriends. After six weeks the client was able to talk to and touch (e.g., hold hands with) each of the surrogates. His self-esteem soared.

Having developed comfort (anxiety 3 or below) in talking with and touching the two female surrogates, the client wanted to overcome his fear of kissing. To assist the client in achieving his goal, a third relationship surrogate (a 42-year-old woman) was identified and socialized as a member of the therapeutic team. She agreed (with the prior permission of her husband) to assist the client in overcoming his anxiety about kissing a woman. The client and the third surrogate met for a total of four sessions over a period of two months. By the fourth session with the third surrogate, the client was able to kiss her with low anxiety.

The client was also proud of his accomplishment and his self-esteem soared. Subsequently, he was able to make dates with nonsurrogate women in his social network, talk with them, touch them, and kiss them (and more, he reported). This case history illustrates that learning the basics is sometimes necessary to create a context where love can develop.

Source: Adapted from Zentner, M., & D. Knox. (2013). Surrogates in relationship therapy: A case study in learning how to talk, touch, and kiss. *Psychology Journal, 10*: 63–68.

interest in Ross's findings is that disclosing one's tastes and interests was negatively associated with relationship satisfaction. By telling a partner too much detail about what he or she liked, partners discovered something that turned them off and lowered their relationship satisfaction.

It is not easy for some people to let others know who they are, what they feel, or what they think. **Alexithymia** is a personality trait which describes a person with little affect. The term means "lack of words for emotions," which suggests that the person does not experience or convey emotion. Persons with alexithymia are not capable of experiencing psychological intimacy. Frye-Cox (2012) studied 155 couples who had been married an average of 18.6 years and found that an alexithymic spouse reported lower marital quality, as did his or her spouse. Alexithymia also tends to repel individuals in mate selection in that persons who seek an emotional relationship are not reinforced by alexithymics.

4-5c What Makes Love Last

Caryl Rusbult's investment model of commitment has been used to identify why relationships last. While love is important, there are other factors involved:

People become dependent on their relationships because they (a) are satisfied with the relationship—it gratifies important needs,

alexithymia a personality trait which describes a person with little affect.

This fence is next to the harbor in Portland, Maine. Lovers put their names and date on a lock, secure it to the fence, and throw the key in the harbor.

4-6 Cultural Factors in Relationship Development

Individuals are not free to become involved with/ marry whomever they want. Rather, their culture and society radically restrict and influence their choice. The best example of mate choice being culturally and socially controlled is the fact that *less than* 1% of the over 63 million marriages in the United States consist of a Black spouse and a White spouse (*Statistical Abstract of the United States, 2012–2013*, Table 60). Indeed, up until the late sixties, such marriages in most states were a felony and mixed spouses were put in jail. In 1967, the Supreme Court ruled that mixed marriages (e.g., between Whites and Blacks) were legal. Only recently have homosexuals received federal recognition for same-sex marriage.

Endogamy and exogamy are also two forms of cultural pressure operative in mate selection.

4-6a Endogamy

Endogamy is the cultural expectation to select a marriage partner within one's own social group, such as in the same race, religion, and social class. Endogamous pressures involve social approval from parents for selecting a partner within one's own group and disapproval for selecting someone outside one's own group. As noted above, the pressure toward an endogamous mate choice is especially strong when race is involved. Love may be blind, but it knows the color of one's partner as the overwhelming majority of individuals end up selecting someone of the same race to marry.

4-6b Exogamy

Exogamy is the cultural pressure to marry outside the family group (e.g., you cannot marry your siblings). Woody Allen experienced enormous social disapproval because he fell in love with and married in 1997 his long-time partner's adopted daughter Soon-Yi Previn. (They remain married.)

Incest taboos are universal; in addition, children are not permitted to marry the parent of the other sex in any society. In the United States, siblings and (in some states) first cousins are also prohibited from marrying each other. The reason for such restrictions

including companionship, intimacy, and sexuality; (b) believe their alternatives are poor—their most important needs could not be gratified independent of the particular relationship (e.g., in an alternative relationship, by friends and kin); and (c) have invested many important resources in the relationship (e.g., time, effort, shared friendship network, and material possessions). The model thereby includes not only internal factors (i.e., satisfaction) to explain why partners stick with each other but also external, structural factors to capture individuals in their interpersonal context. (*Finkenauer, 2010, p. 162*)

Hence, people stay in a relationship because their needs are being met, they have no place to go, and they have made considerable investment in getting where they are and do not want to give it all up. Lambert et al. (2012) also identified a behavior associated with weakening one's love/commitment to a partner—pornography. In a study of 240 undergraduates, higher use of pornography was associated with more flirting and infidelity in the form of hooking up. "Our research suggests that there is a relationship cost associated with pornography" (p. 432).

endogamy the cultural expectation to select a marriage partner within one's social group.

exogamy the cultural expectation that one will marry outside the family group.

is fear of genetic defects in children whose parents are too closely related.

Once cultural factors have determined the general **pool of eligibles** (the population from which a person may select a mate), individual mate choice becomes more operative. However, even when individuals feel that they are making their own choices, social influences are still operative (e.g., approval from one's peers).

4-7 Sociological Factors in Relationship Development

Numerous sociological factors are at work in bringing two people together who eventually marry.

4-7a Homogamy

Although endogamy refers to cultural pressure to select within one's own group, **homogamy** ("like selects like") refers to the tendency for the individual to seek a mate with similar characteristics (e.g., age, race, education). In general, the more couples have in common, the higher the reported relationship satisfaction and the more durable the relationship (Amato et al., 2007). Mcintosh et al. (2011) studied older adults who posted online ads and found that they were concerned about homogamous factors in terms of age, race, religion, etc., and they were willing to travel great distances to find such a partner. A small contingent of Japanese and Koreans believe that similarity of blood type is an important factor in selecting one's partner. For them, "what is your blood type?" is an important and appropriate question to ask a person they consider as a potential mate.

Race Race refers to physical characteristics that are given social significance. Yoo et al. (2010) noted the difference between blatant and subtle racism. Blatant racism involves outright name calling and discrimination. Tiger Woods was once denied the right to play golf on a Georgia golf course because he is Black. Subtle racism involves omissions, inactions, or failure to help, rather than a conscious desire to hurt. Not sitting at a table where persons of another race are sitting or failing to stop to help a person because of his or her race is subtle racism. The roots or racism run deep, the horror of which was depicted in the academy award winning movie, *12 Years a Slave*.

Increasingly, American married couples involve someone whose partner is of a different race/ethnicity. Blacks, younger individuals, and the politically liberal are more willing to cross racial lines for dating and for marriage than are Whites, older individuals, and conservatives (Tsunokai & McGrath, 2011). These findings emerged from a study of 1,335 respondents on Match.com. Being from the South and identifying with religion (Catholic, Protestant, Jewish) was associated with less openness toward interracial relationships. The greatest resistance to interracial relationships of Whites was to African Americans, then Hispanics, then Asians. Gays also tended to select partners of the same racial and ethnic background. When an African American male marries a white female, he more often has higher educational credentials than she does (Hou & Myles, 2013).

In general, as one moves from dating to living together to marriage, the willingness to marry interracially decreases. Of 1,144 undergraduate males, 34% reported that they had dated interracially; 39% of 3,541 undergraduate females had done so (Hall & Knox, 2015). But only about 15% actually marry interracially. Some individuals prefer and seek interracial/interethnic relationships (Yodanis et al., 2012). But these are often not Black–White mergers. Black/White marriages, like that of Grace Hightower and Robert De Niro (as well as the marriage of Chaz Bono (a Black woman) to the late Roger Ebert), are less than 1% of all marriages.

We have been discussing undergraduate attitudes toward interracial dating. But what about their parents? Ozay et al. (2012) analyzed data on 251 parents of undergraduates and found that that over a third (35%) disapproved of their child's involvement in an interracial relationship. Parental disapproval (which did not vary by gender of the parent) increased with the seriousness of relationship involvement (e.g., dating or marriage) and parents were more disapproving of their daughters' than of their sons' interracial involvement.

In regard to parental approval of their son's or daughter's interracial involvement, Black mothers and White fathers have different roles in their respective Black and White communities, in terms of setting the

pool of eligibles the population from which a person selects an appropriate mate.

homogamy tendency to select someone with similar characteristics.

norms of interracial relationships. Hence, the Black mother who approves of her son's or daughter's interracial relationship may be less likely to be overruled by the father than the White mother. The White husband may be more disapproving and have more power over his White wife than the Black father/husband over his partner.

Age Most individuals select a partner who is relatively close in age. Men tend to select women three to five years younger than themselves. The result is the **marriage squeeze**, which is the imbalance of the ratio of marriageable-aged men to marriageable-aged women. In effect, women have fewer partners to select from because men choose from not only their same age group but also from those younger than themselves. One 40-year-old recently divorced woman said, "What chance do I have with all these guys looking at all these younger women?"

Intelligence Dijkstra et al. (2012) studied the mate selection preferences of the intellectually gifted and found that intelligence was one of the primary qualities sought in a potential partner. While both genders valued an intelligent partner, women gave intelligence a higher priority.

Education Educational homogamy also operates in mate selection. Not only does college provide an opportunity to meet, hang out with and cohabit with potential partners, but it also increases one's chance that only a college-educated partner becomes acceptable as a potential cohabitant or spouse. Education becomes an important criterion to find in a partner.

Open-Mindedness People vary in the degree to which they are **open-minded** (receptive to understanding alternative points of view, values, and behaviors). Homogamous pairings in regard to open-mindedness reflect that partners seek like-minded individuals. Such open-mindedness translates into tolerance of various religions, political philosophies, and lifestyles.

marriage squeeze the imbalance of the ratio of marriageable-age men to marriageable-age women.

open-minded being open to understanding alternative points of view, values, and behaviors.

mating gradient the tendency for husbands to be more advanced than their wives with regard to age, education, and occupational success.

> "If you would be **married fitly, wed** your **equal**."
>
> —OVID, PHILOSOPHER

Social Class Brown (2010) interviewed undergraduate students to assess their awareness of social class differences. Having a large number of designer clothes and owning one's own new car were viewed as characteristics of middle- or upper-class students in contrast to students in the working or lower class. Also, the degree to which an item was sought because it was extravagant or functional was an indicator of class. Upper-class students have expensive, fancy computers; middle-class students have computers that are functional—they work, but are not the latest model. Having discretionary income to frequently replace items such as cars, iPhones, or laptops was also seen as an indicator of one's social class.

Social class reflects your parents' occupations, incomes, and educations as well as your residence, language, and values. If you were brought up by parents who were physicians, you probably lived in a large house in a nice residential area—summer vacations and a college education were givens. Alternatively, if your parents dropped out of high school and worked blue-collar jobs, your home was likely smaller and in a less expensive part of town, and your opportunities (such as education) were more limited. Social class affects one's comfort in interacting with others. We tend to select as friends and mates those with whom we feel most comfortable. One undergraduate from an upper-middle-class home said, "When he pulled out coupons at Subway for our first date, I knew this was going nowhere."

The **mating gradient** refers to the tendency for husbands to be more advanced than their wives with regard to age, education, and occupational success. Indeed, husbands are typically older than their wives, have more advanced education, and earn higher incomes (*Statistical Abstract of the United States, 2012–2013*, Table 702).

Physical Appearance Homogamy is operative in regard to physical appearance in that people tend to become involved with those who are similar in degree of physical attractiveness. However, a partner's attractiveness may be a more important consideration for men than for women. Meltzer and McNulty (2014) emphasized that "body valuation by a committed male partner is positively associated with women's relationship satisfaction when that partner also values them for their nonphysical qualities, but negatively associated with women's relationship satisfaction when that partner is not committed or does not value them for their nonphysical qualities" (p. 68).

Career Individuals tend to meet and select others who are in the same career. In 2012, country western singers Blake Shelton and Miranda Lambert won the Male and Female Vocalist of the Year awards, respectively (she won it again in 2014). They are married. Michelle and Barack Obama are both attorneys. Angelina Jolie and Brad Pitt are both movie celebrities. Danica Patrick and Ricky Stenhouse Jr. are both NASCAR drivers.

Marital Status Never-married people tend to select other never-married people as marriage partners; divorced people tend to select other divorced people; and widowed people tend to select other widowed people.

Religion/Spirituality/Politics Religion may be defined as a set of beliefs in reference to a supreme being which involves practices or rituals (e.g., communion) generally agreed upon by a group of people. Of course, some individuals view themselves as "not religious" but "spiritual," with spirituality defined as belief in the spirit as the seat of the moral or religious nature that guides one's decisions and behavior. Religious/spiritual homogamy is operative in that people of a similar religion or spiritual philosophy tend to seek out each other. Over a third (33%) of 1,144 undergraduate males and 44% of 3,559 undergraduate females agreed that "It is important that I marry someone of my same religion" (Hall & Knox, 2015). Ellison et al. (2010) observed that couples who have shared religious beliefs and who practice in-home family devotional activities report higher relationship quality than couples without common religious values/no ritual sharing.

Alford et al. (2011) emphasized that homogamy is operative in regard to politics. An analysis of national data on thousands of spousal pairs in the United States revealed that homogamous political attitudes were the strongest of all social, physical, and personality traits. Further, this similarity derived from initial mate choice "rather than persuasion and accommodation over the life of the relationship."

Personality Gonzaga et al. (2010) asked both partners of 417 eHarmony couples who married to identify the degree to which various personality trait terms reflected who they, as individuals, were. Examples of the terms included warm, clever, dominant, outgoing, quarrelsome, stable, energetic, affectionate, intelligent, witty, content, generous, wise, bossy, kind, calm, outspoken, shy, and trusting. Results revealed that those who were more similar to each other in personality characteristics were also more satisfied with their relationship.

Circadian Preference Circadian preference refers to an individual's preference for morningness–eveningness in regard to intellectual and physical activities. In effect, some prefer the morning while others prefer late afternoon or evening hours. In a study of 84 couples, Randler and Kretz (2011) found that partners in a romantic relationship tended to have similar circadian preferences. The couples also tended to have similar preferences as to when they went to bed in the evening and when they rose in the morning.

Traditional Roles Partners who have similar views of what their marital roles will be are also attracted to each other. Abowitz et al. (2011) found that about a third of the 692 undergraduate women wanted to marry a "traditional" man—one who viewed his primary role as that of provider and who would be supportive of his wife staying home to rear the children. In a related study of 1,027 undergraduate men, 31% reported that they wanted a traditional wife—one who viewed her primary role as wife and mother, staying home and rearing the children (not having a career) (Knox & Zusman, 2007).

Geographic Background Haandrikman (2011) studied Dutch cohabitants and found that the romantic partners tended to have grown up within six kilometers of each other (**spatial homogamy**). While the university context draws people from different regions of the country, the demographics of state universities tend to reflect a preponderance of those from within the same state.

Economic Values, Money Management, and Debt Individuals vary in the degree to which they have money, spend money, and save money. Some are deeply in debt and carry

circadian preference refers to an individual's preference for morningness-eveningness in regard to intellectual and physical activities.

spatial homogamy the tendency for individuals to marry who grew up in close physical proximity.

significant educational debt. The average debt for those with a bachelor's degree is $23,000. Almost half (48%) of those who borrowed money for their undergraduate education report that they are burdened by this debt (Pew Research Center, 2011). Of 1,142 undergraduate males, 4.5% (and 3.1% of 3,549 undergraduate females) owed more than a thousand dollars on a credit card (Hall & Knox, 2015). Money becomes an issue in mate selection in that different economic values predict conflict. One undergraduate male noted, "There is no way I would get involved with this woman—she jokes about maxing out her credit cards. She is going to be someone else's nightmare."

4-8 Psychological Factors in Relationship Development

Psychologists have focused on complementary needs, exchanges, parental characteristics, and personality types with regard to mate selection.

4-8a Complementary-Needs Theory

Complementary-needs theory (also known as "opposites attract") states that we tend to select mates whose needs are opposite yet complementary to our own. For example, some partners may be drawn to each other on the basis of nurturance versus receptivity. These complementary needs suggest that one person likes to give and take care of another, whereas the other likes to be the benefactor of such care. Other examples of complementary needs are responsibility versus irresponsibility, peacemaker versus troublemaker, and disorder versus order. Former *Tonight Show* host Jay Leno revealed in his autobiography that he and his wife of over 30 years are very different:

complementary-needs theory tendency to select mates whose needs are opposite and complementary to one's own needs.

exchange theory theory that emphasizes that relations are formed and maintained between individuals offering the greatest rewards and least costs to each other.

We were, and are, opposites in most every way. Which I love. There's a better balance…I don't consider myself to have much of a spiritual side, but Mavis has almost a sixth sense about people and situations. She has deep focus, and I fly off in 20 directions at once. She reads 15 books a week, mostly classic literature. I collect classic car and motorcycle books. She loves European travel; I don't want to go anywhere people won't understand my jokes. (*Leno, 1996, pp. 214–215*)

4-8b Exchange Theory

Exchange theory in mate selection is focused on finding the partner who offers the greatest rewards at the lowest cost. The following five concepts help explain the exchange process in mate selection:

1. **Rewards.** Rewards are the behaviors (your partner looking at you with the eyes of love), words (saying "I love you"), resources (being beautiful or handsome, having a car, condo at the beach, and money), and services (cooking for you, typing for you) your partner provides that you value and that influence you to continue the relationship. Similarly, you provide behaviors, words, resources, and services for your partner that he or she values. Relationships in which the exchange is equal are happiest. Increasingly, men are interested in women who offer "financial independence" and women are interested in men who "cook and do the dishes."

2. **Costs.** Costs are the unpleasant aspects of a relationship. A woman identified the costs associated with being involved with her partner: "He abuses drugs, doesn't have a job, and lives nine hours away." The costs her partner associated with being involved with this woman included "she nags me," "she doesn't like sex," and "she insists that we live in the same town they live in if we marry."

3. **Profit.** Profit occurs when the rewards exceed the costs. Unless the couple described in the preceding paragraph derive a profit from staying together, they are likely to end their relationship and seek someone else with whom there is a higher profit margin. Biographer Thomas Maier (2009) said of Virginia Johnson (of the famous Masters and Johnson

team) that she was motivated by money to marry Dr. William Masters. "I never wanted to be with him, but when you are making $200,000 a year, you don't walk," said Johnson (p. 235).

Profit may also refer to nonmonetary phenomenon. Moore et al. (2013) found that being altruistic (giving selflessly) was a characteristic of a potential mate valued by both women and men. In effect, such a person provides a higher profit margin for any relationship.

4. **Loss.** Loss occurs when the costs exceed the rewards. Partners who feel that they are taking a loss in their relationship are vulnerable to looking for another partner who offers a higher profit.

5. **Alternative.** Is another person currently available who offers a higher profit? Individuals on the marriage market have an understanding of what they are worth and whom they can attract (Bredow et al., 2011). Oberbeek et al. (2012) found that facially unattractive individuals with high body mass indexes were less selective in a speed dating context. In effect, they had less to offer so adjusted their expectations in terms of what they felt they could get. You will stay in a relationship where you have a high profit at a low cost. You will leave a relationship where your costs are high and you have an alternative partner who offers you a relationship with high rewards.

4-8c Role Theory

Freud suggested that the choice of a love object in adulthood represents a shift in libidinal energy from the first love objects, the parents. **Role theory of mate selection** emphasizes that a son or daughter models after the parent of the same sex by selecting a partner similar to the one the parent selected as a mate. This means that a man looks for a wife who has similar characteristics to those of his mother and that a woman looks for a husband who is very similar to her father.

4-8d Attachment Theory

The **attachment theory of mate selection** emphasizes the drive toward psychological intimacy and

Courtesy of Chelsea Curry

The drive for intimate attachment is fueled by the comfort one experiences when in a reciprocal love relationship.

a social and emotional connection (Sassler, 2010). One's earliest experience as a child is to be emotionally bonded to one's parents (usually the mother) in the family context. The emotional need to connect remains as an adult and expresses itself in relationships with others, most notably the romantic love relationship. Children diagnosed with oppositional-defiant disorder (ODD) or post-traumatic stress disorder (PTSD) have had disruptions in their early bonding and frequently display attachment problems, possibly due to early abuse, neglect, or trauma. Children reared in Russian orphanages in the fifties where one caretaker was assigned to multiple children learned "no one cares about me" and "the world is not safe." Reversing early negative or absent bonding is difficult.

McClure and Lydon (2014) noted that anxiety attachment may express itself in meeting a new partner in the form of not knowing what to say and feeling socially awkward. The result is that would-be partners may be put off by this anxiety, disengage, and prevent a relationship from developing, the very context which could assist the anxious person in becoming more comfortable.

role theory of mate selection emphasizes that a son or daughter models after the parent of the same sex by selecting a partner similar to the one the parent selected as a mate.

attachment theory of mate selection emphasizes the drive toward psychological intimacy and a social and emotional connection.

4-8e Undesirable Personality Characteristics of a Potential Mate

Researchers have identified several personality factors predictive of relationships which end in divorce or are unhappy (Burr et al., 2011; Foster, 2008). Potential partners who are observed to consistently display these characteristics might be avoided.

1. **Controlling.** The behavior that 60% of a national sample of adult single women reported as the most serious fault of a man was his being "too controlling" (Edwards, 2000).

2. **Narcissistic.** Individuals who are narcissistic view relationships in terms of what they get out of them. When satisfactions wane and alternatives are present, narcissists are the first to go. Because all relationships have difficult times, a narcissist is a high risk for a durable marriage relationship.

3. **Poor impulse control.** Lack of impulse control is problematic in relationships because such individuals are less likely to consider the consequences of their actions. Having an affair is an example of failure to control one's behavior to insure one's fidelity. Such people do as they please and worry about the consequences later.

4. **Hypersensitive.** Hypersensitivity to perceived criticism involves getting hurt easily. Any negative statement or criticism is received with a greater impact than a partner intended. The disadvantage of such hypersensitivity is that a partner may learn not to give feedback for fear of hurting the hypersensitive partner. Such lack of feedback to the hypersensitive partner blocks information about what the person does that upsets the other and what could be done to make things better. Hence, the hypersensitive one has no way of learning that something is wrong, and the partner has no way of alerting the hypersensitive partner. The result is a relationship in which the partners cannot talk about what is wrong, so the potential for change is limited.

5. **Inflated ego.** An exaggerated sense of oneself is another way of saying a person has a big ego and always wants things to be his or her way. A person with an inflated sense of self may be less likely to consider the other person's opinion in negotiating a conflict and prefer to dictate an outcome. Such disrespect for the partner can be damaging to the relationship.

6. **Perfectionist.** Individuals who are perfectionists may require perfection of themselves and others. They are rarely satisfied and always find something wrong with their partner or relationship. Living with a perfectionist will be a challenge since whatever one does will not be good enough.

7. **Insecure.** Feelings of insecurity also make relationships difficult. The insecure person has low self-esteem, constantly feels that something is wrong, and feels disapproved of by the partner. The partner must constantly reassure him or her that all is well—a taxing expectation over time.

8. **Controlled.** Individuals who are controlled by their parents, former partner, child, or anyone else compromise the marriage relationship because their allegiance is external to the couple's relationship. Unless the person is able to break free of such control, the ability to make independent decisions will be thwarted, which will both frustrate the spouse and challenge the marriage. An example is a wife whose father dictated that she spend all holidays with him. The husband ultimately divorced her since he felt she was more of a daughter than a wife and that she had never emotionally left home. The late film critic Roger Ebert did not marry until he was 50—after his mother was dead—since he said she approved of none of the women he wanted to marry and he could not break free of her disapproval.

Individuals may also be controlled by culture. The example is a Muslim man who fell in love with a woman and she with him. They dated for four years and developed an intense emotional and sexual relationship. But he was not capable of breaking from his socialization to marry a woman of another faith.

9. **Substance abuser.** Heavy drug use does not predict well for relationship quality or stability. Blair (2010) studied a national sample of adolescent females and noted a negative relationship between substance abuse (alcohol and marijuana) and lower likelihood of marriage. The researcher reasoned that being labeled as a "drunk" or a "stoner" as a female alienated her from her peers and reduced her chance of being selected as a romantic partner. Wiersma et al. (2011) also noted negative relationship outcomes for the female cohabitant who drinks significantly more than her partner.

10. **Unhappy.** Personal happiness is associated with relationship happiness. Stanley et al. (2012) noted that being unhappy in one's personal life is predictive of having an unhappy relationship once married. Conversely, having high life satisfaction before marriage is predictive of high relationship satisfaction once married. Hence, selecting an upbeat, optimistic, happy individual predicts well for a future marriage relationship with this person.

In addition to undesirable personality characteristics, Table 4.2 reflects some particularly troublesome personality disorders and how they may impact a relationship.

4-8f Female Attraction to "Bad Boys"

Carter et al. (2014) reviewed the **dark triad personality** in some men and confirmed that some women are attracted to these "bad boys." The dark triad is a term for intercorrelated traits of narcissism (sense of entitlement and grandiose self-view), Machiavellianism (deceptive and insincere), and psychopathy (callous and no empathy). These men are socially skilled and manipulative, have a high number of sex partners,

dark triad personality term identifying traits of "bad boys" including narcissistic, deceptive, and no empathy.

Table 4.2

Personality Disorders Problematic in a Potential Partner

Disorder	Characteristics	Impact on Partner
Paranoid	Suspicious, distrustful, thin-skinned, defensive	Partners may be accused of everything.
Schizoid	Cold, aloof, solitary, reclusive	Partners may feel that they can never connect and that the person is not capable of returning love.
Borderline	Moody, unstable, volatile, unreliable, suicidal, impulsive	Partners will never know what their Jekyll-and-Hyde partner will be like, which could be dangerous.
Antisocial	Deceptive, untrustworthy, conscienceless, remorseless	Such a partner could cheat on, lie to, or steal from a partner and not feel guilty.
Narcissistic	Egotistical, demanding, greedy, selfish	Such a person views partners only in terms of their value. Don't expect such a partner to see anything from your point of view; expect such a person to bail in tough times.
Dependent	Helpless, weak, clingy, insecure	Such a person will demand a partner's full time and attention, and other interests will incite jealousy.
Obsessive-compulsive	Rigid, inflexible	Such a person has rigid ideas about how a partner should think and behave and may try to impose them on the partner.
Neurotic	Worries, obsesses about negative outcomes	This individual will impose negative scenarios on the partners and couple.

Courtesy of Rachel Calisto

and engage in mate poaching. The researchers analyzed data from 128 British undergraduate females and found that they were attracted to these bad boys. Explanations for such attraction was the self-confidence of the bad boys as well as their skill in manipulating the females. And there is the challenge. "I always knew I wasn't the only one, but I wanted to be the girl that changed him," said one undergraduate who dated a bad boy.

4-9 Engagement

Being identified as a couple occurs after certain events have happened. Chaney and Marsh (2009) interviewed 62 married and 60 cohabiting couples to find out when they first identified themselves as a couple. There were four markers: relationship events, affection/sex, having or rearing children, and time and money.

1. Relationship events included a specific event such as visiting the parents of one's partner, becoming engaged, or moving in together.

2. Affection/sexual events such as the first time the couple had sex. Losing one's virginity was a salient event.

3. Children—becoming pregnant, having a child together, or the first time the partner assumed a parenting role.

4. Time/money—spending a lot of time together, sharing funds, or exchanging financial support.

engagement
period of time during which committed, monogamous partners focus on wedding preparations and systematically examine their relationship.

Some engagements happen very fast. Paul Jobs (father of Steve Jobs) and his wife, Clara, were engaged 10 days after they met (the marriage lasted over 40 years and ended by death). Mariah Carey and Nick Cannon married within two months of their first date.

Engagement moves the relationship of a couple from a private love-focused experience to a public, parent-involved experience. Unlike casual dating, **engagement** is a time in which the romantic partners are sexually monogamous, committed to marry, and focused on wedding preparations. The engagement period molds the intimate relationship of the couple by means of the social support and expectations of family and friends. It is the last opportunity before marriage to systematically examine the relationship—to become confident in one's decision to marry a particular person. Johnson and Anderson (2011) studied 610 newly married couples and found that those spouses who were more confident in their decision to marry ended up investing more in their relationship with greater marital satisfaction.

4-9a Premarital Counseling

Jackson (2011) developed and implemented a unique evidence-informed treatment protocol for premarital counseling including six private sessions and two postmarital booster sessions. Some clergy require one or more sessions of premarital counseling before they agree to marry a couple. Premarital counseling is a process of discovery. Because partners might have hesitated to reveal information that they feel may have met with disapproval during casual dating, the sessions provide the context to be open with the other about their thoughts, feelings, values, goals, and expectations.

There is no shortage of advice for those about to marry. Allgood and Gordon (2010) studied 56 couples who were given advice from 412 individuals. The most frequent advice givers were friends (32%), followed by parents (30%) and religious leaders (20%). Results revealed that the closer the individual was to the advice giver, the more likely the person was to use the advice and to regard it as useful. Both men and women regarded their close friends (who happened to be of the same sex) as sources of advice they listened to and found most helpful.

4-9b Visiting Your Partner's Parents

Fisher and Salmon (2013) identified the reasons individuals take their potential partners home to meet their parents—to seek parental approval and feedback and to confirm to their partner that they are serious about the relationship. They also want to meet their partner's parents—to see how their potential mate will look when older, their future health, and potential familial resources that will be available.

Seize the opportunity to discover the family environment in which your partner was reared, and consider the implications for your subsequent marriage. When visiting your partner's parents, observe their standard of living and the way they interact (e.g., level of affection, verbal and nonverbal behavior, marital roles) with one another. How does their standard of living compare with that of your own family? How does the emotional closeness (or distance) of your partner's family compare with that of your family? Such comparisons are significant because both you and your partner will reflect your respective family or origins. "This is the way we did it in my family" is a phrase you will hear your partner say from time to time.

If you want to know how your partner is likely to treat you in the future, observe the way your partner's parent of the same sex treats and interacts with his or her spouse. If you want to know what your partner may be like in the future, look at your partner's parent of the same sex. There is a tendency for a man to become like his father and a woman to become like her mother. A partner's parent of the same sex and the parents' marital relationship are the models of a spouse and a marriage relationship that the person is likely to duplicate.

4-10 Delay or Call Off the Wedding If...

"No matter how far you have gone on the wrong road, turn back" is a Turkish proverb. Behavioral psychologist B. F. Skinner noted that one should not defend a course of action that does not feel right, but stop and reverse directions.

If your engagement is characterized by the following factors, consider delaying your wedding at least until the most distressing issues have been resolved. Alternatively, break the engagement (which happens in 30% of formal engagements), which will have fewer negative consequences and involve less stigma than ending a marriage.

4-10a Age 18 or Younger

The strongest predictor of getting divorced is getting married during the teen years. Individuals who marry at age 18 or younger have three times the risk of divorce than those who delay marriage into their late twenties or early thirties. Teenagers may be more at risk for marrying to escape an unhappy home and may be more likely to engage in impulsive decision making and behavior. Early marriage is also associated with an end to one's education, social isolation from close friends, early pregnancy or parenting, and locking oneself into a low income. Increasingly, individuals are delaying when they marry. As noted earlier, the median age at first marriage in the United States is 29 for men and 27 for women.

4-10b Known Partner Less Than Two Years

Thirty-one percent of 1,142 undergraduate males and 31% of 3,552 undergraduate females agreed, "If I were really in love, I would marry someone I had known for only a short time" (Hall & Knox, 2015). Impulsive marriages in which the partners have known each other for less than a month are associated with a higher-than-average divorce rate. Indeed, partners who date each other for at least two years (25 months to be exact) before getting married report the highest level of marital satisfaction and are less likely to divorce (Huston et al., 2001). A short courtship does not allow partners enough time to learn about each other's background, values, and goals and does not permit opportunity to observe and scrutinize each other's behavior in a variety of settings (e.g., with one's close friends and family).

To increase the knowledge you and your partner have about each other, find out each other's answers to the questions in the Involved Couple's Inventory, take a five-day "primitive" camping trip, take a 15-mile hike together, wallpaper a small room together, or spend several days together when one partner is sick. If the couple plans to have children, they may want to offer to babysit a 6-month-old of their friends for a weekend.

4-10c Abusive Relationship

Abusive lovers become abusive spouses, with predictable negative outcomes. Though extricating oneself from an abusive relationship is difficult before the wedding, it becomes even more difficult after marriage, particularly when children are involved. Abuse is a serious red flag of impending relationship doom that should not be overlooked, and one should seek the exit ramp as soon as possible (see Chapter 10 on the details of leaving an abusive relationship).

4-10d High Frequency of Negative Comments/Low Frequency of Positive Comments

Markman et al. (2010) studied couples across the first five years of marriage and found that more negative and less positive communication before marriage tended to be associated with subsequent divorce. In addition, the researchers emphasized that "negatives tend to erode positives over time." Individuals who criticize each other end up damaging their relationship in a way which does not make it easy for positives to erase.

4-10e Numerous Significant Differences

Relentless conflict often arises from numerous significant differences. Though all spouses are different from each other in some ways, those who have numerous differences in key areas such as race, religion, social class, education, values, and goals are less likely to report being happy and to divorce. Amato et al. (2007) also found that the

cyclical relationships when couples break up and get back together several times.

less couples had in common, the more their marital distress.

4-10f On-and-Off Relationship

A roller-coaster premarital relationship is predictive of a marital relationship that will follow the same pattern. Partners in **cyclical relationships** (break up and get back together several times) have developed a pattern in which the dissatisfactions in the relationship become so frustrating that separation becomes the antidote for relief. Dailey et al. (2013) identified five types of on-and-off relationships (the percent who represent each are from Dailey et al., 2013):

1. **Controlling partner (26%).** One partner more persistent to keep the relationship together.

2. **Capitalized on transitions (22%).** Partners improve the relationship after a breakup.

3. **Mismatched (19%).** Differences in personalities and desires or geographic distance.

4. **Habitual (14%).** Partners break up but fall back together with little or no negotiation since it was convenient to resume the relationship.

5. **Gradual separators (11%).** Partners recognized relationship was over and ended it (7% could not be categorized).

Vennum (2011) studied individuals in cyclical relationships and found that they reported lower-quality relationships compared to those in relationships

© nick vangopoulos/Shutterstock.com

which were uninterrupted. Partners in cyclical relationships tended to be African Americans in a long-distance dating relationship who expressed uncertainty about the future of the relationship and had less constructive communication.

4-10g Dramatic Parental Disapproval

Parents usually have an opinion of their son's or daughter's mate choice. Mothers disapprove of the mate choice of their daughters if they predict that he will be a lousy father, and fathers disapprove if they predict the suitor will be a poor provider (Dubbs & Buunk, 2010). When student and parent ratings of mate selection traits were compared, parents ranked religion higher than did their offspring, whereas offspring ranked physical attractiveness higher than did parents (Perilloux et al., 2011). Parents were also more focused on earning capacity and education (e.g., college graduate) in their daughter's mate selection than in their son's.

4-10h Low Sexual Satisfaction

Sexual satisfaction is linked to relationship satisfaction, love, and commitment. Sprecher (2002) followed 101 dating couples across time and found that low sexual satisfaction for both women and men was related to reporting low relationship quality, less love, lower commitment, and breaking up. Hence, couples who are dissatisfied with their sexual relationship might explore ways of improving it (alone or through counseling) or consider the impact of such dissatisfaction on the future of their relationship.

4-10i Limited Relationship Knowledge

Individuals and couples are most likely to have a positive future together if they have relationship knowledge. Bradford et al. (2012) validated the Relationship Knowledge Questionnaire as a way of assessing relationship knowledge. Such knowledge included knowing how to listen effectively, settle disagreements/solve problems/reach compromise, deepen a loving relationship, develop a strong friendship, and spend time together.

4-10j Wrong Reasons for Getting Married

Some reasons for getting married are more questionable than others. These reasons include the following:

Rebound A rebound marriage results when you marry someone immediately after another person has ended a relationship with you. It is a frantic attempt on your part to reestablish your desirability in your own eyes and in the eyes of the partner who dropped you. To marry on the rebound is usually a bad decision because the marriage is made in reference to the previous partner and not to the current partner. In effect, you are using the person you intend to marry to establish yourself as the "winner" in the previous relationship.

Barber and Cooper (2014) used a longitudinal, online diary method to examine trajectories of psychological recovery and sexual experience following a romantic relationship breakup among 170 undergraduate students. Consistent with stereotypes about individuals on the rebound, those respondents who had been "dumped" used sex to cope with feelings of distress, anger, and diminished self-esteem. And those who had sex for these reasons were more likely (not initially, but over time) to continue having sex with different new partners. Caution about becoming involved with someone on the rebound may be warranted. One answer to the question, "How fast should you run from a person on the rebound?" may be "as fast as you can." Waiting until the partner has 12 to 18 months distance from the previous relationships provides for a more stable context for the new relationship.

Escape A person might marry to escape an unhappy home situation in which the parents are oppressive, overbearing, conflictual, alcoholic and/or abusive. Marriage for escape is a bad idea. It is far better to continue the relationship with the partner until mutual love and respect become the dominant forces propelling you toward marriage, rather than the desire to escape an unhappy situation. In this way you can evaluate the marital relationship in terms of its own potential and not as an alternative to unhappiness.

Unplanned Pregnancy Getting married because a partner becomes pregnant should be considered carefully. Indeed, the decision of whether to marry should be kept separate from decisions about a pregnancy. Adoption, abortion, single parenthood, and unmarried parenthood (the couple can remain together as an unmarried couple and have the baby) are all alternatives to simply deciding to marry if a partner becomes pregnant. Avoiding feelings of being trapped or later feeling that the marriage might not have happened without the pregnancy are two reasons for not rushing into marriage because of pregnancy. Couples who marry when the woman becomes pregnant have an increased chance of divorce.

Psychological Blackmail Some individuals get married because their partner takes the position that "I will commit suicide if you leave me." Because the person fears that the partner may commit suicide, he or she may agree to the wedding. The problem with such a marriage is that one partner has been reinforced for threatening the other to get what he or she wants. Use of such power often creates resentment in the other partner, who feels trapped in the marriage. Escaping from the marriage becomes even more difficult. One way of coping with a psychological blackmail situation is to encourage the person to go with you to a therapist to discuss the relationship. Once inside the therapist's office, you can tell the counselor that you feel pressured to get married because of the suicide threat. Counselors are trained to respond to such a situation. Alternatively, another response to a partner who threatens suicide is to call the police and say, "Name, address, and phone number has made a serious threat on his or her own life." They police will dispatch a car to have the person picked up and evaluated.

Insurance Benefits In a poll conducted by the Kaiser Family Foundation, 7% of adults said someone in their household had married in the past year to gain access to insurance. In effect, marital decisions had been made to gain access to health benefits. "For today's couples, 'in sickness and in health' may seem less a lover's troth than an actuarial contract. They marry for better or worse, for richer or poorer, for co-pays and deductibles" (Sack, 2008). While selecting a partner who has resources (which may include health insurance) is not unusual, to select a partner solely because he or she has health

benefits is dubious. Both parties might be cautious if the alliance is more about "benefits" than the relationship.

Pity Some partners marry because they feel guilty about terminating a relationship with someone whom they pity. The fiancée of an Afghanistan soldier reported that "when he came back with his legs blown off I just changed inside and didn't want to stay in the relationship. I felt guilty for breaking up with him...." Regardless of the reason, if one partner becomes brain-damaged or fails in the pursuit of a major goal, it is important to keep the issue of pity separate from the advisability of the marriage. The decision to marry should be based on factors other than pity for the partner.

Filling a Void A former student in the authors' classes noted that her father died of cancer. She acknowledged that his death created a vacuum, which she felt driven to fill immediately by getting married so that she would have a man in her life. Because she was focused on filling the void, she had paid little attention to the personality characteristics of the man who had asked to marry her. She discovered on her wedding night that her new husband had several other girlfriends whom he had no intention of giving up. The marriage was annulled.

In deciding whether to continue or terminate a relationship, listen to what your senses tell you ("Does it feel right?"), listen to your heart ("Do you love this person or do you question whether you love this person?"), and evaluate your similarities ("Are we similar in terms of core values, goals, view of life?"). Also, be realistic. Indeed, most people exhibit some negative and some positive indicators before they marry.

4-11 Trends in Love Relationships

Love will continue to be one of the most treasured experiences in life. Love will be sought, treasured, and, when lost or ended, will be met with despair and sadness. After a period of recovery, a new search will begin. As our society becomes more diverse, the range of potential love partners will widen to include those with demographic characteristics different from our own. Romantic love will continue and love will maintain its innocence as those getting remarried love just as deeply and invest in the power of love all over again.

The development of a new love relationship will involve the same cultural constraints and sociological and psychological factors identified in the chapter. Individuals are not "free" to select their partner but do so from the menu presented by their culture. Once at the relationship buffet, factors of homogamy and exchange come into play. These variables will continue to be operative. Becoming involved with a partner with similar characteristics as one's own will continue to be associated with a happy and durable relationship.

STUDY TOOLS ⮕ **4**

Ready to study? In the book, you can:

⮕ Rip out the chapter review card at the back of the book for a handy summary of the chapter and key terms.

⮕ Find out how romantic or realistic you are in your attitudes toward love with the Self-Assessment card at the back of the book.

Online at CENGAGEBRAIN.COM you can:

⮕ Prepare for tests with quizzes.

⮕ Review the key terms with Flash Cards.

⮕ Play games to master concepts.

Communication and Technology in Relationships

"I should just **change my voicemail** greeting to: 'Please hang up and **text me**, thanks!'"

—UNIVERSITY STUDENT

Communication is the barometer of a good and bad relationship. "We talked all night" confirms the joy and quality of one's relationship just as "we have nothing to say to each other" confirms that the relationship is over. You can make choices about how you communicate with your partner and these choices will impact your relationship. Two researchers followed 136 newlyweds over four years and revealed that those who ended up getting divorced did not make communication choices to enhance their relationship—they expressed anger, they were critical, and they chose not to make positive remarks to and about each other (Lavner & Bradbury, 2012). Those who stayed together chose to speak respectfully to each other and made positive, affirming statements to each other. The communication pattern you choose to have with your partner will influence whether you live in misery or enjoy life with your partner. In this chapter, we review the basics of relationship communication and emphasize how technology (texting, Tinder, sexting, Skype, Facebook) impacts how relationships are initiated, maintained, and terminated. We begin by defining the nature of interpersonal communication.

"You say it **best when** you **say nothing at all**."

—ALISON KRAUSS, SINGER

5-1 Communication: Verbal and Nonverbal

Communication can be defined as the process of exchanging information and feelings between two or more people. Wickrama et al. (2010) studied 540 couples and found that spouses mirror each other in terms of how they relate to each other. When one spouse insults, swears, or shouts, the other is likely to engage in the same behavior.

Communication is both verbal and nonverbal. Although most communication is focused on verbal content, most (estimated

communication
the process of exchanging information and feelings between two or more people.

Nonverbal behavior is a powerful communicator. In this photo, each partner is communicating, "I am mad at you and don't want to talk to you."

to be as high as 80%) interpersonal communication is nonverbal. **Nonverbal communication** is the "message about the message," the gestures, eye contact, body posture, tone, and rapidity of speech. Even though a person says, "I love you and am faithful to you," crossed arms and lack of eye contact will convey a different meaning from the same words accompanied by a tender embrace and sustained eye-to-eye contact. The greater the congruence between verbal and nonverbal communication, the better. One aspect of nonverbal behavior is the volume of speech. The volume one uses when interacting with a partner has relevance for the relationship. When one uses a high frequency (yells) there are physiological (e.g., high blood pressure, higher heart rate, higher cortisol levels) and therapeutic (e.g., more limited gains in counseling—higher divorce rate) outcomes.

Flirting is a good example of both nonverbal and verbal behavior. One researcher defined **flirting** as showing another person romantic interest without serious intent (Moore, 2010). Examples of how interest is shown include preening, such as stroking one's hair or adjusting one's clothing, and positional cues, such as leaning toward or away

nonverbal communication
the "message about the message," using gestures, eye contact, body posture, tone, volume, and rapidity of speech.

flirting showing another person romantic interest without serious intent.

from the target person. Individuals go through a series of steps on their way to sexual intercourse, and the female is responsible for signaling the male that she is interested so that he moves the interaction forward. At each stage, she must signal readiness to move to the next stage. For example, if the male holds the hand of the female, she must squeeze his hand before he can entwine their fingers. Females generally serve as sexual gatekeepers and control the speed of the interaction toward sex.

5-1a Words versus Action

A great deal of social discourse depends on saying things that sound good but that have no meaning in terms of behavioral impact. "Let's get together" or "let's hang out" sounds good since it implies an interest in spending time with each other. But the phrase has no specific plan so that the intent is likely never to happen—just the opposite. So let's hang out really means "we won't be spending any time together." Similarly, come over anytime means "never come."

Where there is behavioral intent, a phrase with meaning is "let's meet Thursday night at seven for dinner" (rather than "let's hang out") and "can you come over Sunday afternoon at four to play video games?" (rather than "come over any time"). With one's partner, "let's go camping sometime" means we won't ever go camping. "Let's go camping at the river this coming weekend" means you are serious about camping.

5-2 Technology-Mediated Communication in Romantic Relationships

> "People have **entire relationships** via **text messages** now."
>
> —DANIELLE STEEL

The following scenario from one of our students reflects that technology has an enormous effect on romantic relationships.

"I saw his photo on Tinder.com"

"I typed his name into Facebook and friend requested him."

"He accepted, messaged me and we began to text each other."

"We were long distance so we began to have long talks on Skype."

"To keep his interest I would sex text him ('What would you like me to do to you?')"

"After we moved in together, I snooped, checked his cell phone—discovered other women."

"I sent him a text message that it was over, and moved out."

Taylor et al. (2013) surveyed 1,003 emerging adults (18–25) about the use of technology in romantic relationships. Both men and women reported that texting is appropriate with a potential romantic partner; however, neither agree with announcing a relationship on Facebook before having the "relationship talk" and becoming a committed couple.

Rappleyea et al. (2014) analyzed data on 1,003 young adults in regard to technology and relationship formation and found that the respondents believe that "talking," "hanging out," and "sharing intimate details" are more important when compared with using communication technologies to establish a relationship" (p. 269).

Text messages have become a primary means for flirting (73% of 18–25-year-olds—Gibbs, 2012) and the initiation, escalation, and maintenance of romantic relationships (Bergdall et al., 2012). Some (to the disapproval of most) end a relationship with a text message (Gershon, 2010). Text messages and communication technology are also used by divorced parents in the coparenting of their children (Ganong et al., 2011).

5-2a Texting and Interpersonal Communication

Personal, portable, wirelessly networked technologies in the form of iPhones, Droids, iPads, etc., have become commonplace in the lives of individuals (Gibbs, 2012; Looi et al., 2010) as a way of staying connected with offline friends and partners (Reich et al., 2012). Bauerlein (2010) noted that today's youth are being socialized in a hyper-digital age where traditional modes of communication will be replaced by gadgets and texting will become the primary mode of communication. This shift to greater use of technology affects relationships in both positive and negative ways. On the positive side, it allows for instant and unabated connection—individuals can text each other throughout the day so that they are "in effect, together all the time." One of our students noted that on a regular day, she and her boyfriend will exchange 50 to 60 text messages. On the negative side is **nomophobia**, where the individual is dependent on virtual environments to the point of having a social phobia (King et al., 2013). He or she finds personal interaction difficult.

Cell phone use now consumes the life space of most Americans. In a Bank of America Trends in Mobility Survey Report regarding cell phone use in a typical day, 51% reported once or more every hour, 26% a few times a day, 13% hardly ever, 8% morning and evening, and 2% don't know (Carey & Bravo, 2014).

nomophobia the individual is dependent on virtual environments to the point of having a social phobia.

National and International Data

Based on a survey of 4,700 respondents in the United States and seven other countries (China, the United Kingdom, India, South Korea, South Africa, Indonesia, and Brazil), 9 in 10 carry a cell phone (1 in 4 check it every 30 minutes; 1 in 5 check it every 10 minutes) (Gibbs, 2012).

Coyne et al. (2011) examined the use of technology by 1,039 individuals in sending messages to their romantic partner. The respondents were more likely to use their cell phones to send text messages than any other technology. The most common reasons were to express affection (75%), discuss serious issues (25%), and apologize (12%). There were no significant differences in use by gender, ethnicity, or religion. Pettigrew (2009) emphasized that **texting**, or text messaging (short typewritten messages—maximum of 160 characters sent via cell phone) is used to "commence, advance, maintain" interpersonal relationships and is viewed as more constant and private than talking on a cell phone. Women text more than men. In a Nielsen State of the Media Survey conducted in the third quarter of 2010, it was reported that women spent 818 minutes texting 640 messages; men, 716 minutes texting 555 messages (Carey & Salazar, 2011). On the negative side, these devices encourage the continued interruption of face-to-face communication between individuals and encourage the intrusion of one's work/job into the emotional intimacy of a couple and the family.

Huang and Leung (2010) studied instant messaging and identified four characteristics of "addiction" in teenagers: preoccupation with IM, loss of relationships due to overuse, loss of control, and escape. Results also showed that shyness and alienation from family, peers, and school were significantly and positively associated with levels of IM addiction. As expected, both the level of IM use and level of IM addiction were significantly linked to teenagers' poorer academic performance.

The extent to which technology and relationships has entered U.S. culture is reflected in the 2013 movie Her, which features lonely and heartbroken Theodore Twombly's involvement in a relationship with a woman who exists only as a computer operating system. Samantha asks him how his day was, she flirts with him, she understands him. . . . they fall in love with each other. He comes to depend on her being there for him. Alas, she is only there for him as an operating system which eventually closes down, making the point that technology has its limits.

Technology is also used to convey both good and bad news to a romantic partner (see the What's New? section).

> "You have **nothing** if you're **texting a guy** in a relationship. We can **text six women a minute**."
>
> —STEVE HARVEY, COMEDIAN

5-2b When Texting and Facebook Become a Relationship Problem

Schade et al. (2013) studied the effects of technology on romantic relationships. They analyzed data from 276 adults ages 18–25 in committed relationships and found that male texting frequency was negatively associated with relationship satisfaction and stability scores for both partners while female texting frequency was positively associated with their own relationship stability scores. Hence, females thrived on texting, which had a positive relationship effect. Males tolerated it, which had the opposite/negative effect.

Cell phones/text messaging may be a source of conflict (e.g., partner text messages while the lover is talking to him or her). Over 60% of Chinese respondents and a quarter of U.S. respondents noted that a mobile device had come between them and their spouse (Gibbs, 2012). But these devices may also help reduce conflict. Perry and Werner-Wilson (2011) assessed the use of technology in conflict resolution in romantic relationships and found that text messaging

Heyyyy!
1:47PM

Heyy girl! What's up??
1:48PM

Not much! Want to go get a drink tonight?
1:49PM

I'm down.
1:50PM

Wanna grab dinner first?
1:52PM

Enter message

Courtesy of Chelsea Curry

texting text messaging (short typewritten messages—maximum of 160 characters sent via cell phone).

What's New?

THE GOOD, THE BAD, AND TECHNOLOGY IN ROMANTIC RELATIONSHIPS

Three-hundred-and-fifty-four undergraduates at a large southeastern university completed a 42-item Internet survey designed to reveal the degree to which electronic delivery (e.g., texting/email versus face-to-face communication) was used to deliver good (e.g., "I love you," "Let's get married," "Are you down for sex tonight?") and bad (e.g., "I think we should break up," "I cheated," "I have an STI," "I got into graduate school and will be moving") news to a romantic partner.

Findings

1. **Frequency.** Nearly half of the respondents had experience receiving bad news via an electronic method— 46% had been broken up with via text, email, or another electronic method, and 44% had been told that their partner had been unfaithful. Respondents not only received but also gave bad

Courtesy of Rachel Calisto

Technology can interfere with face-to-face interaction.

news—39% reported that they had ended a relationship and 25% had communicated an infidelity to a partner via electronic means.

The respondents also revealed that technology-mediated communication (TMC) was used to convey good news to romantic partners. Over half (51%) reported that their partner communicated "I love you" for the first time via electronic methods such as a text or an email. In addition, approximately 50% of the respondents noted that their partners had used text or email to broach the subject of readiness to have sex. Thirty-five percent reported informing partners of good news about a job promotion, and 6% reported receiving a marriage proposal by text or email.

2. **Gender.** In regard to disclosing that one had been unfaithful, 15% more males than females used TMC. However, a similar percent of males and females responded that they had ended a relationship with a romantic partner electronically (35% and 39%, respectively).

3. **Media ideologies and emotional response.** While technology was used to convey both negative and positive content, it was not identified as preferable to face-to-face delivery. Almost 90% (88%) of the respondents listed face to face as their preferred method of receiving bad news, and 71% listed face to face as their preferred method of delivering bad news. Similarly, face to face was the preferred method for delivering and receiving good news with both receiving a preference over 90%.

Source: Abridged and adapted from Faircloth, M., D. Knox, & J. Brinkley. (2012, March 21–24). The good, the bad and technology mediated communication in romantic relationships. Paper presented at the Southern Sociological Society, New Orleans, LA.

allowed for de-escalation of conflict and provided time to construct ideas—to think about what they were going to say.

Norton and Baptist (2012) identified how social networking sites (e.g., Facebook, with over a billion users and 140 billion friendship connections) are problematic for couples—the sites are intrusive (e.g.,

partner surfs while lover is talking), and they encourage compulsive use (e.g., partner is always sending/receiving messages) and infidelity (e.g., flirting/cheating online). They studied how 205 married individuals mitigated the impact of technology on their relationship. Three strategies included openness (e.g., each spouse knew the passwords and online friends and had access to each other's online social networking accounts, email, etc.), fidelity (e.g., flirting and online relationships were off limits), and appropriate people (e.g., knowing the friends of the partner and no former partners allowed).

5-2c Sexting

Another way in which technology affects communication, particularly in romantic relationships, is **sexting** (sending erotic text and photo images via a cell phone). Sexting begins in high school. Strassberg et al. (2013) surveyed 606 high school students and found that almost 20% reported having *sent* a sexually explicit image of themselves via cell phone while almost twice as many reported that they had *received* a sexually explicit picture via cell phone (of these, over 25% indicated that they had *forwarded* this picture to others). Of those who sent a sexually explicit cell phone picture, over a third did so despite believing that there could be serious legal consequences attached to the behavior.

Burke-Winkelman et al. (2014) reported that 65% of 1,652 undergraduates reported sending sexually suggestive texts or photos to a current or potential partner (69% reported receiving). Almost a third (31%) reported sending the text messages to a third party. In regard to how they felt about sending nude photos, less than half were positive, and females were more likely to feel pressure to send nude photos.

Parker et al. (2011) analyzed data from 483 undergraduates at a large southeastern university who completed a 25-item Internet questionnaire. They found that about two-thirds (64%) reported sending a sex text message; 43%, a sex photo. There were no gender differences, but, as also occurred in the Burke-Winkelman research, African Americans reported higher frequencies of sending sex content to a romantic partner.

Dir et al. (2013) surveyed 278 undergraduates in regard to their receiving and sending sex text messages and sex photos. Gender and relationship status were significant predictors of specific sexting behaviors. Males reported sending more sex photos. In addition, those involved in a relationship (dating, serious relationship, cohabiting) sent more sexts than uninvolved singles. Males reported more positive outcomes of receiving sexts (e.g., sexual excitement); women were more likely to feel uncomfortable (e.g., embarrassed). Weisskirch and Delevi (2011) examined the role of attachment anxiety in sexting and found attachment anxiety predicted positive attitudes toward sexting such as accepting it as normal, believing that it would enhance the relationship, and stating that partners expected it. Temple et al. (2012) studied sexting in over 900 public high school students and found that doing so was associated with the likelihood of having sex, having multiple sex partners, and using alcohol before sex.

While undergraduates are not at risk as long as the parties are age 18 or older, sending erotic photos of individuals younger than 18 can be problematic. Sexting is considered by many countries as child pornography and laws related to child pornography have been applied in cases of sexting. Six high school students in Greensburg, Pennsylvania, were charged with child pornography after three teenage girls allegedly took nude or seminude photos of themselves and shared them with male classmates via their cell phones.

Some undergraduate females are under age 18. Having or sending nude images of underage individuals is a felony which can result in fines, imprisonment, and a record. We noted above that Strassberg et al. (2013) reported that a third of their high school respondents reported that they had forwarded a sexually explicit photo that could get them in serious legal trouble.

5-2d Video-Mediated Communication

Communication via computer between separated lovers, spouses, and family members is becoming more common. Furukawa and Driessnack (2013) assessed the use of **video-mediated communication (VMC)** in a sample of 341 online participants (ages 18 to 70-plus). Ninety-six percent reported that VMC was the most common method they used to communicate with their family, and 60% reported doing so at least once a week. VMC allows the person to see and hear what is going on; for example, while the grandparents can't be present Christmas morning they can see the excitement of their grandchildren opening their gifts.

sexting sending erotic text and photo images via a cell phone.

video-mediated communication (VMC) communication via computer between separated lovers, spouses, and family members.

5-3 Principles of Effective Communication

"Polite conversation is rarely either."

—FRANCES ANN LEBOWITZ, AUTHOR

The following identify the ways we can ensure effective communication in our relationships:

1. **Prioritize communication.** Communicating effectively implies making communication an important priority in a couple's relationship. When communication is a priority, partners make time for it to occur in a setting without interruptions: they are alone; they are not texting or surfing the Internet; they do not answer the phone; and they turn the television off. Making communication a priority results in the exchange of more information between partners, which increases the knowledge each partner has about the other.

2. **Avoid negative and make positive statements to your partner.** Because intimate partners are capable of hurting each other so intensely, it is important to avoid brutal statements to the partner. Such negativity is associated with vulnerability to divorce (Woszidlo & Segrin, 2013). Indeed, be very careful how you give negative feedback or communicate disapproval to your partner. Markman et al. (2010a) noted that couples in marriage counseling often will report "it was a bad week" based on one negative comment made by the partner.

   Markman et al. (2010a) also emphasized the need for partners to make positive comments to each other and that doing so was associated with more stable relationships. People like to hear others say positive things about them. These positive statements may be in the form of compliments ("You look terrific!") or appreciation ("Thanks for putting gas in the car"). Maatta and Uusiautti (2013) emphasized that compliments to the partner were associated with being involved in a romantic love relationship.

Partners who look at each other when they are talking not only communicate an interest in each other but also are able to gain information about the partner's feelings and responses to what is being said.

3. **Establish and maintain eye contact.** Shakespeare noted that a person's eyes are the "mirrors to the soul." Partners who look at each other when they are talking not only communicate an interest in each other but also are able to gain information about the partner's feelings and responses to what is being said. Not looking at your partner may be interpreted as lack of interest and prevents you from observing nonverbal cues.

4. **Establish empathy.** **Empathy** is the ability to emotionally experience and cognitively understand another person and his or her experiences. To the degree that partners in a relationship have dyadic empathy (empathy with each other) they report satisfaction in their relationship (Kimmes et al., 2014).

> **empathy** the ability to emotionally experience and cognitively understand another person and his or her experiences.

5. **Ask open-ended questions.** When your goal is to find out your partner's thoughts and feelings about an issue, using **open-ended questions** is best. Such questions (e.g., "How do you feel about me?") encourage your partner to give an answer that contains a lot of information. **Closed-ended questions** (e.g., "Do you love me?"), which elicit a one-word answer such as "yes" or "no," do not provide the opportunity for the partner to express a range of thoughts and feelings.

6. **Use reflective listening.** Effective communication requires being a good listener. One of the skills of a good listener is the ability to use the technique of **reflective listening**, which involves paraphrasing or restating what the person has said to you while being sensitive to what the partner is feeling. For example, suppose you ask your partner, "How was your day?" and your partner responds, "I felt exploited today at work because I went in early and stayed late and a memo from my new boss said that future bonuses would be eliminated because of a company takeover." Listening to what your partner is both saying and feeling, you might respond, "You feel frustrated because you really worked hard and felt unappreciated."

Reflective listening serves the following functions: (1) it creates the feeling for speakers that they are being listened to

and are being understood; and (2) it increases the accuracy of the listener's understanding of what the speaker is saying. If a reflective statement does not accurately reflect what a speaker has just said, the speaker can correct the inaccuracy by saying it again.

An important quality of reflective statements is that they are nonjudgmental. For example, suppose two lovers are arguing about spending time with their respective friends and one says, "I'd like to spend one night each week with my friends and not feel guilty about it." The partner may respond by making a statement that is judgmental (critical or evaluative), such as those exemplified in Table 5.1. Judgmental responses serve to punish or criticize people for what they think, feel, or want and often result in an argument.

Table 5.1 also provides several examples of nonjudgmental reflective statements.

7. **Use "I" statements. "I" statements** focus on the feelings and thoughts of the communicator *without* making a judgment on others. Because "I" statements are a clear and nonthreatening way of expressing what you want and how you feel, they are likely to result in a positive change in the listener's behavior. Making "I" statements reflects being authentic. Impett et al. (2010) emphasized the need to be authentic when communicating. Being **authentic** means speaking and acting in a manner according to what one feels. Being authentic in a relationship means being open with the partner about one's preferences and feelings about the partner's behavior. Being authentic has positive consequences for the relationship in that one's thoughts and feelings are out in the open (in contrast to being withdrawn and resentful).

"You" statements blame or criticize the listener and often result in increasing negative feelings and behavior in the relationship. For example, suppose you are angry with your partner for being late. Rather than say, "You are always late and irresponsible" (a "you" statement), you might respond with, "I get upset when you are late and ask that you call me when you will be delayed." The latter focuses on your feelings and a desirable future behavior rather than blaming the partner for being late.

open-ended questions questions that encourage answers that contain a great deal of information.

closed-ended questions questions that allow for a one-word answer and do not elicit much information.

reflective listening paraphrasing or restating what a person has said to indicate that the listener understands.

"I" statements statements that focus on the feelings and thoughts of the communicator without making a judgment on others.

authentic speaking and acting in a manner according to what one feels.

"you" statements statements that blame or criticize the listener and often result in increasing negative feelings and behavior in the relationship.

Table 5.1

Judgmental and Nonjudgmental Responses to a Partner's Saying, "I'd Like to Go Out with My Friends One Night a Week"

Nonjudgmental, Reflective Statements	Judgmental Statements
You value your friends and want to maintain good relationships with them.	You only think about what you want.
You think it is healthy for us to be with our friends some of the time.	Your friends are more important to you than I am.
You really enjoy your friends and want to spend some time with them.	You just want a night out so that you can meet someone new.
You think it is important that we not abandon our friends just because we are involved.	You just want to get away so you can drink.
You think that our being apart one night each week will make us even closer.	You are selfish.

8. **Touch.** Hertenstein et al. (2007) identified the various meanings of touch such as conveying emotion, attachment, bonding, compliance, power, and intimacy. The researchers also emphasized the importance of using touch as a mechanism of nonverbal communication to emphasize one's point or meaning.

> "An **argument** usually consists of **two people** each **trying to get** in the **last word**—first!"
>
> —LAWRENCE PETER, HUMORIST

9. **Use soft emotions/take responsibility.** Sanford and Grace (2011) identified "hard" emotions (e.g., angry or outraged) or "soft" emotions (sad or hurt) displayed during conflict. Examples of flat emotions are being bored or indifferent. The use of hard emotions resulted in an escalation of negative communication, whereas the display of soft emotions resulted in more benign

communication and an increased feeling regarding the importance of resolving interpersonal conflict.

10. **Identify specific new behavior you want.** Focus on what you want your partner to do rather than on what you do not want. Rather than say, "You spend too much time with your Xbox playing Halo, Call of Duty, and Gears of War," an alternative might be "When I come home, please help me with dinner, ask me about my day and turn on your Xbox after 10 P.M.—after you have turned me on." Rather than say, "You never call me when you are going to be late," say, "Please call me when you are going to be late." Notice that you are asking your partner for what you want, not demanding. Demanding is associated with a demand/withdraw pattern in which one partner demands and the other withdraws. Couples with this pattern are more likely to be dissatisfied with their relationship, and it is more likely for one of the partners to be having an affair (Balderrama-Durbin et al., 2012).

> Couples who focus on the issue to get it resolved handle conflict with minimal marital fallout.

11. **Stay focused on the issue.** Wheeler et al. (2010) studied conflict resolution strategies and found that couples who have a solution-oriented style in contrast to a confrontational (attack) or nonconfrontational (ignore) style report higher levels of marital satisfaction. Hence, couples who focus on the issue to get it resolved handle conflict with minimal marital fallout.

   Branching refers to going out on different limbs of an issue rather than staying focused on the issue. If you are discussing the overdrawn checkbook, stay focused on the checkbook. To remind your partner that he or she is equally irresponsible when it comes to getting things repaired or doing house work is to go off target. Stay focused.

12. **Make specific resolutions to disagreements.** To prevent the same issues or problems from recurring, agreeing on what each partner will do in similar circumstances in the future is important. For example, if going to a party together results in one partner's drinking too much and drifting off with someone else, what needs to be done in the future to ensure an enjoyable evening together? In this example, a specific resolution would be to decide how many drinks the partner will have within a given time period and make an agreement to stay together and dance only with each other.

13. **Give congruent messages. Congruent messages** are those in which the verbal and nonverbal behaviors match. A person who says, "Okay, you're right" and smiles while embracing the partner is communicating a congruent message. In contrast, the same words accompanied by leaving the room and slamming the door communicate a very different message.

branching in communication, going out on different limbs of an issue rather than staying focused on the issue.

congruent message one in which verbal and nonverbal behaviors match.

power the ability to impose one's will on one's partner and to avoid being influenced by the partner.

principle of least interest principle stating that the person who has the least interest in a relationship controls the relationship.

14. **Share power. Power** is the ability to impose one's will on the partner and to avoid being influenced by the partner. One of the greatest sources of dissatisfaction in a relationship is a power imbalance and conflict over power (Kurdek, 1994). Over half (63%) of 4,577 undergraduates from two universities reported that they had the same amount of power in the relationship as their partner. Twenty-four percent of males reported that they have more power (21% of women). Sixteen percent of males and females reported that they had less power (Hall & Knox, 2015). Perhaps males are reluctant to admit that they have less power!

One way to assess power is to identify who has the least interest in the relationship. Waller and Hill (1951) observed that the person who has the least interest in continuing the relationship is in control of the relationship. This **principle of least interest** is illustrated by the woman who said, "He wants us to stay together more than I do so I am in control and when we disagree about something he gives in to me." Expressions of power in a relationship are numerous and include the following:

Withdrawal (not speaking to the partner)
Guilt induction ("How could you ask me to do this?")
Being pleasant ("Kiss me and help me move the sofa.")
Negotiation ("We can go to the movie if we study for a couple of hours before we go.")
Deception (running up credit card debts of which the partner is unaware)
Blackmail ("I'll find someone else if you won't have sex with me.")
Physical abuse or verbal threats ("You will be sorry if you try to leave me.")
Criticism ("You are stupid and fat.")
Dominance ("I make more money than you so I will decide where we go.")

Power may also take the form of love and sex. The person in the relationship who loves less and who needs sex less has enormous power over the partner who is very much in love and who is dependent on the partner for sex. This pattern reflects the principle of least interest we discussed earlier in the text.

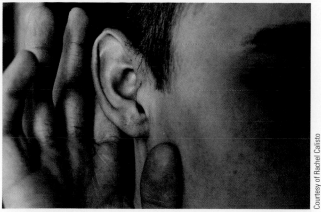

Courtesy of Rachel Calisto

15. **Keep the process of communication going.** Communication includes both content (verbal and nonverbal information) and process (interaction). It is important not to allow difficult content to shut down the communication process. To ensure that the process continues, the partners should focus on the fact that talking is important and reinforce each other for keeping the communication process alive. For example, if your partner tells you something that you do that bothers him or her, it is important to thank your partner for telling you rather than becoming defensive. In this way, your partner's feelings about you stay out in the open rather than being hidden behind a wall of resentment. If you punish such disclosure because you do not like the content, disclosure will stop. For example, a wife told her husband that she felt his lunches with a woman at work were becoming too frequent and wondered if it were a good idea. Rather than the husband becoming defensive and saying he could have lunch with whomever he wanted he might say, "I appreciate your telling me how you feel about this…you're right…maybe it would be best to cut back."

5-4 Gender, Culture, and Communication

How individuals communicate with each other depends on which gender is talking/listening and the society/culture in which they were socialized and live.

5-4a Gender Differences in Communication

Numerous jokes address the differences between how women and men communicate. One anonymous quote on the Internet follows:

> When a woman says, "Sure . . . go ahead," what she means is "I don't want you to." When a woman says, "I'm sorry," what she means is "You'll be sorry." When a woman says, "I'll be ready in a minute," what she means is "Kick off your shoes and start watching a football game on TV."

Women and men differ in their approach to and patterns of communication. Women are more communicative about relationship issues, view a situation emotionally, and initiate discussions about relationship problems. Deborah Tannen (1990; 2006) is a specialist in communication. She observed that, to women, conversations are negotiations for closeness in which they try "to seek and give confirmations and support, and to reach consensus" (1990, p. 25). To men, conversations are about winning and achieving the upper hand.

The genders differ in regard to emotionality. Garfield (2010) reviewed men's difficulty with emotional intimacy. He noted that their emotional detachment stems from the provider role which requires them to stay in control. Being emotional is seen as weakness. Men's groups where men learn to access and express their feelings have been helpful in increasing men's emotionality.

In contrast, women tend to approach situations emotionally. For example, if a child is seriously ill, wives will want their husbands to be emotional, to cry, to show that they really care that their child is sick. But a husband might react to a seriously ill child by putting pressure on the wife to be "mature" about the situation and by encouraging stoicism.

Women disclose more in their relationships than men do (Gallmeier et al., 1997). In a study of 360 undergraduates, women were more likely to disclose information about previous love relationships, previous sexual relationships, their love feelings for the partner, and what they wanted for the future of the relationship. They also wanted their partners to reciprocate this level of disclosure, but such disclosure from their partners was, generally, not forthcoming.

5-4b Cultural Differences in Communication

The meaning of a particular word varies by the country in which it is used. Xu (2013) emphasized the importance of being aware of cultural differences in communication. An American woman was dating a man from Iceland. When she asked him, "Would you like to go out to dinner?" he responded, "Yes, maybe." She felt confused by this response and was uncertain whether he wanted to eat out. It was not until she visited his home in Iceland and asked his mother, "Would you like me to set the table?"—to which his mother replied, "Yes, maybe"—that she discovered "Yes, maybe" means "Yes, definitely."

Individuals reared in France, Germany, Italy, or Greece regard arguing as a sign of closeness—to be blunt and argumentative is to keep the interaction alive and dynamic; to have a tone of agreement is boring. Asian cultures (e.g., Japanese, Chinese, Thai, and Pilipino) place a high value on avoiding open expression of disagreement and emphasizing harmony. Deborah Tannen (1998) observed the different perceptions of a Japanese woman married to a Frenchman:

> He frequently started arguments with her, which she found so upsetting that she did her best to agree and be conciliatory. This only led him to seek another point on which to argue. Finally she lost her self-control and began to yell back. Rather than being angered, he was overjoyed. Provoking arguments was his way of showing interest in her, letting her know how much he respected her intelligence. To him, being able to engage in spirited disagreement was a sign of a good relationship. (p. 211)

5-5 Self-Disclosure and Secrets

Shakespeare noted in *Macbeth* that "the false face must hide what the false heart doth know," suggesting that withholding information and being dishonest may affect the way one feels about oneself and relationships with others. All of us make decisions about the degree to which we disclose, are honest, and/or keep secrets.

5-5a Self-Disclosure in Intimate Relationships

One aspect of intimacy in relationships is self-disclosure, which involves revealing personal information and feelings about oneself to another person.

Relationships become more stable when individuals disclose themselves to their partners (Tan et al., 2012). Areas of disclosure include one's formative years, previous relationships (positive and negative), experiences of elation and sadness, and goals (achieved and thwarted). We noted in the discussion of love in Chapter 4 that self-disclosure is a psychological condition necessary for the development of love. To the degree that you disclose yourself to another, you invest yourself in and feel closer to that person. People who disclose nothing are investing nothing and remain aloof. One way to encourage disclosure in one's partner is to make disclosures about one's own life and then ask about the partner's life.

> "A **wonderful fact** to reflect upon, that **every human** creature is constituted to be that profound **secret and mystery to every other.**"
>
> —CHARLES DICKENS

5-5b Secrets in Romantic Relationships

Most lovers keep a secret or two from their partners. Oprah Winfrey's biographer revealed a secret that

Personal View: How Much Do You Tell Your Partner about Your Past?

The most frequent lie partners tell each other is the number of past sexual partners.

Because of the fear of HIV infection and other sexually transmitted infections (STIs), some partners want to know the details of each other's previous sex life, including how many partners they have had sex with and in what contexts (e.g., hookups or stable relationships). Those who are asked will need to decide whether to disclose the requested information, which may include one's sexual orientation, present or past STIs, and any sexual preferences the partner might find bizarre (e.g., sadism). Ample evidence suggests that individuals are sometimes dishonest with regard to the sexual information about their past they provide to their partners. The "number of previous sexual partners" is the most frequent lie undergraduates report telling each other. One female undergraduate who has had more partners than she wants to reveal said that when she is asked about her number, she smiles and says, "A lady never kisses and tells." Another said, "Whatever number the person tells me, I double it."

In deciding whether or not to talk honestly about your past to your partner, you may want to consider the following questions: How important is it to your partner to know about your past? Do you want your partner to tell you (honestly) about her or his past? What impact on your relationship will open disclosure have? What impact will withholding such information have on the level of intimacy you have with your partner?

Oprah kept from her long-term boyfriend Stedman Graham. When the couple was vacationing at a resort and Stedman left for a round of golf, Oprah promptly called room service and ordered two whole pecan pies. She called room service back a short time later to come and remove the empty tin plates so that her partner would not know of her food binge (Kelley, 2010).

College students also keep secrets from their partners. In a study of 431 undergraduates, Easterling et al. (2012) found the following:

1. **Most kept secrets.** Over 60% of the respondents reported having kept a secret from a romantic partner, and over one-quarter of respondents reported currently doing so.

2. **Females kept more secrets.** Sensitivity to the partner's reaction, desire to avoid hurting the partner, and desire to avoid damaging the relationship were the primary reasons why females

were more likely than males to keep a secret from a romantic partner.

3. **Spouses kept more secrets.** Spouses have a great deal to lose if there is an indiscretion or if one partner does something the other will disapprove of (e.g., hook up). Partners who are dating or "seeing each other" have less to lose and are less likely to keep secrets.

4. **Blacks kept more secrets.** Blacks are a minority who live in a racist society and are still victimized by the White majority. One way to avoid such victimization is to keep one's thoughts to oneself—to keep a secret. This skill of deception may generalize to one's romantic relationships.

clok/Shutterstock.com

Edyta Pawlowska/Shutterstock.com

5. **Homosexuals kept more secrets.** Indeed, the phrase "in the closet" means "keeping a secret." Transgendered individuals in Europe are required to reveal their secret before marriage (Sharpe, 2012).

Respondents were asked why they kept a personal secret from a romantic partner. "To avoid hurting the partner" was the top reason reported by 39% of the respondents. "It would alter our relationship" and "I feel so ashamed for what I did" were reported by 18% and 11% of the respondents, respectively.

Some secrets are embedded in technology—text messages, emails, and cell phone calls. Being deceptive with one's emails and cell phone is disapproved of by both women and men. In a study of 5,500 never-married individuals, 76% of the women and 53% of the men reported that being secretive with emails was a behavior they would not tolerate. Similarly, 69% of the women (47% of the men) said they would not put up with a partner who answered cell phone calls discretely (Walsh, 2013).

5-5c Family Secrets

Just as romantic partners have secrets, so do families. The family secret that takes the cake is the one Bernie Madoff kept from his wife and two adult sons—that he had defrauded almost 5,000 clients of over $50 billion over a number of years. Madoff's wife and sons were adamant that their husband and father had acted alone and without their knowledge. Sadly, Madoff's elder son, Mark, committed suicide in the wake of the intense public scrutiny.

Oprah Winfrey also had a family secret. At age 15 she gave birth to a son in her seventh month of pregnancy. The baby died a month later. When Oprah ran for Miss Black Tennessee she completed an application on which she stated that she had never had a child (Kelley, 2010).

5-6 Dishonesty, Lying, and Cheating

Relationships are compromised by dishonesty, lying, and cheating.

5-6a Dishonesty

Dishonesty and deception take various forms. One is a direct lie—saying something that is not true (e.g., telling your partner that you have had 6 previous sexual partners when, in fact, you have had 13). Not correcting an assumption is another form of dishonesty (e.g., your partner assumes you are heterosexual but you are bisexual).

5-6b Lying in American Society

Lying, a deliberate attempt to mislead, is pervasive. Lance Armstrong was stripped of his seven Tour de France

One aspect of intimacy in relationships is self-disclosure, which involves revealing personal information and feelings about oneself to another person.

titles for lying about doping. Journalist Mike Daisey admitted to fabricating stories about oppressed workers at an Apple contractor's factory in China. He said of his deception, "It's not journalism. It's theater." Sixty-two percent of 125 Harvard students admitted to cheating on either tests or papers (Webley, 2012). Politicians routinely lie to citizens ("Lobbyists can't buy my vote"), and citizens lie to the government (via cheating on taxes). Teachers lie to students ("The test will be easy"), and students lie to teachers ("I studied all night"). Parents lie to their children ("It won't hurt"), and children lie to their parents ("I was at my friend's house"). Dating partners lie to each other ("I've had a couple of previous sex partners"), women lie to men ("I had an orgasm"), and men lie to women ("I'll call"). The price of lying is high—distrust and alienation. A student in class wrote:

> At this moment in my life I do not have any love relationship. I find hanging out with guys to be very hard. They lie to you about anything and you wouldn't know the truth. I find that college dating is mostly about sex and having a good time before you really have to get serious. That is fine, but that is just not what I am all about.

Catfishing refers to a process whereby a person makes up an online identity and an entire social facade to trick a person into becoming involved in an emotional relationship. The catfish is the lonely person on the Internet who is susceptible to being seduced into this fake relationship. University of Notre Dame football player Manti Te'o reported that he was a victim of an online hoax, fooled into a relationship by someone pretending to be a woman named Lennay Kekua. The creator of the pretend Lennay Kekua then conspired with others to lead Te'o to believe that Kekua had died of leukemia.

newphotoservice/Shutterstock.com

5-6c Lying and Cheating in Romantic Relationships

Lying is epidemic in college student romantic relationships. In response to the statement, "I have lied to a person I was involved with," 69% of 4,690 undergraduates (men more than women) reported "yes." Nineteen percent of men and 14% of women reported

having lied to a partner about their previous number of sexual partners (Hall & Knox, 2015).

Cheating may be defined as having sex with someone else while involved in a relationship with a romantic partner. When 4,690 undergraduates were asked if they had cheated on a partner they were involved with, almost a quarter (23%) reported that they had done so (men more than women) (Hall & Knox, 2015). Even in monogamous relationships, there is considerable cheating. Vail-Smith et al. (2010) found that of 1,341 undergraduates, 27.2% of the males and 19.8% of the females reported having oral, vaginal, or anal sex outside of a relationship that their partner considered monogamous.

People most likely to cheat in these monogamous relationships were men over the age of 20, those who were binge drinkers, members of a fraternity, male NCAA athletes, and those who reported that they were nonreligious. White et al. (2010) also studied 217 couples where both partners reported on their own risk behaviors and their perceptions of their partner's behavior; 3% of women and 14% of men were unaware that their partner had recently had a concurrent partner. Eleven percent and 12%, respectively, were unaware that their partner had ever injected drugs; 10% and 12% were unaware that their partner had recently received an STI diagnosis; and 2% and 4% were unaware that their partner was HIV-positive. These data suggest a need for people in committed relationships to reconsider their risk of STI and to protect themselves via condom usage. In addition, one of the ways in which college students deceive their partners is by failing to disclose that they have an STI. Approximately 25% of college students will contract an STI while they are in college (Purkett, 2014).

Strickler and Hans (2010) conceptualized infidelity (cheating) as both sexual and nonsexual. Sexual cheating was intercourse, oral sex, and kissing. Nonsexual cheating could be interpersonal (secret time together, flirting), electronic (text messaging, emailing), or solitary (sexual fantasies, pornography, masturbation). Of 400 undergraduates, 74% of the males and 67% of the females in a committed relationship reported that they had

catfishing
process whereby a person makes up an online identity and an entire social facade to trick a person into becoming involved in an emotional relationship.

cheated according to their own criteria. Hence, in the survey, they identified a specific behavior as cheating and later reported whether they had engaged in that behavior.

5-7 Theories of Relationship Communication

Symbolic interactionism and social exchange are theories that help explain the communication process.

5-7a Symbolic Interactionism

Interactionists examine the process of communication between two actors in terms of the meanings each attaches to the actions of the other. Definition of the situation, the looking-glass self, and taking the role of the other (discussed in Chapter 1) are all relevant to understanding how partners communicate. With regard to resolving a conflict over how to spend the semester break (e.g., vacation alone or go to see parents), the respective partners must negotiate their definitions of the situation (is it about their time together as a couple or their loyalty to their parents?). The looking-glass self involves looking at each other and seeing the reflected image of someone who is loved and cared for and someone with whom a productive resolution is sought. Taking the role of the other involves each partner's understanding the other's logic and feelings about how to spend the break.

5-7b Social Exchange

Exchange theorists suggest that the partners' communication can be described as a ratio of rewards to costs. Rewards are positive exchanges, such as compliments, compromises, and agreements. Costs refer to negative exchanges, such as critical remarks, complaints, and attacks. When the rewards are high and the costs are low, the outcome is likely to be positive for both partners (profit). When the costs are high and the rewards low, neither may be satisfied with the outcome (loss).

When discussing how to spend the semester break, the partners are continually in the process of exchange—not only in the words they use but also in the way they use them. If the communication is to continue, both partners need to feel acknowledged for their points of view and to feel a sense of legitimacy and respect. Communication in abusive relationships is characterized by the parties criticizing and denigrating each other, which usually results in a shutdown of the communication process and a drift toward ending the relationship.

5-8 Fighting Fair: Steps in Conflict Resolution

Resolving stress via conflict resolution is important for one's health. Lund et al. (2014) analyzed data from 9,875 Danish men and women aged 36–52 years to assess associations between stressful social relations with partner and with children and mortality. Results revealed that frequent worries/demands from partner or children were associated with 50%–100% increased mortality risk.

When a disagreement begins, it is important to establish rules for fighting that will leave the partners and their relationship undamaged. Indeed, Lavner and Bradbury (2010) studied 464 newlyweds over a four-year period, noticed the precariousness of relationships (even those reporting considerable satisfaction divorced), and recommended that couples "impose and regularly maintain ground rules for safe and nonthreatening communication." Such guidelines for fair fighting/effective communication include not calling each other names, not bringing up past misdeeds, and not attacking each other.

Gottman (1994) identified destructive communication patterns to avoid which he labeled as "the four horsemen of the apocalypse"—criticism, defensiveness, contempt (the most damaging), and stonewalling. He also noted that being positive about the partner is essential—partners who said positive things to each

Fighting fairly also involves keeping the interaction focused and respectful, and moving toward a win–win outcome.

other at a ratio of 5:1 (positives to negatives) were more likely to stay together. We have noted that "avoiding giving your partner a zinger" is also essential to maintaining a good relationship.

Fighting fairly also involves keeping the interaction focused and respectful, and moving toward a win–win outcome. If recurring issues are not discussed and resolved, conflict may create tension and distance in the relationship, with the result that the partners stop talking, stop spending time together, and stop being intimate. Developing and using skills for fair fighting and conflict resolution are critical for the maintenance of a good relationship.

Howard Markman is head of the Center for Marital and Family Studies at the University of Denver. He and his colleagues have been studying 150 couples at yearly intervals (beginning before marriage) to determine those factors most responsible for marital success. They have found that a set of communication skills that reflect the ability to handle conflict, which they call "constructive arguing," is the single biggest predictor of marital success over time (Marano, 1992). According to Markman, "Many people believe that the causes of marital problems are the differences between people and problem areas such as money, sex, and children. However, our findings indicate it is not the differences that are important, but how these differences and problems are handled, particularly early in marriage" (Marano, 1992, p. 53). Markman et al. (2010b) provide details for constructive communication in their book *Fighting for Your Marriage*. The following sections identify standard steps for resolving interpersonal conflict.

5-8a Address Recurring, Disturbing Issues

Addressing issues in a relationship is important. But whether partners do so is related to their level of commitment to the relationship. Partners who are committed to each other and to their relationship invest more time and energy to resolving problems. Those who feel stuck in relationships with barriers to getting out (e.g., children, economic dependence), avoid problem resolution (Frye, 2011). The committed are intent on removing relationship problems.

5-8b Identify New Desired Behaviors

Dealing with conflict is more likely to result in resolution if the partners focus on what they *want* rather than what they *don't want*. Tell your partner specifically what you want him or her to do. For example, if your partner routinely drives the car but never puts gas in it, rather than say, "Stop driving the gas out of the car," you might ask him or her to "always keep at least a fourth tank of gas in the car."

5-8c Identify Perceptions to Change

Rather than change behavior, changing one's perception of a behavior may be easier and quicker. Rather than expect one's partner to always be on time, it may be easier to drop this expectation and to stop being mad about something that doesn't matter. South et al. (2010) emphasized the importance of perception of behavior in regard to marital satisfaction.

5-8d Summarize Your Partner's Perspective

We often assume that we know what our partner thinks and why he or she does things. Sometimes we are wrong. Rather than assume how our partner thinks and feels about a particular issue, we might ask open-ended questions in an effort to learn our partner's thoughts and feelings about a particular situation. The answer to "How do you feel about me taking an internship abroad next semester?" will give you valuable information.

5-8e Generate Alternative Win–Win Solutions

Looking for win–win solutions to conflicts is imperative. Solutions in which one person wins means that the other person is not getting needs met. As a result, the person who loses may develop feelings of resentment, anger, hurt, and hostility toward the winner and may even look for ways to get even. In this way, the winner is also a loser. In intimate relationships, one winner really means two losers.

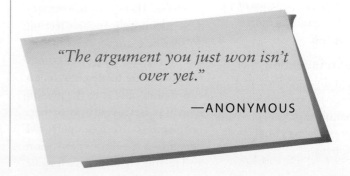

"The argument you just won isn't over yet."

—ANONYMOUS

Generating win–win solutions to interpersonal conflict often requires **brainstorming**. The technique of brainstorming involves suggesting as many alternatives as possible without evaluating them. Brainstorming is crucial to conflict resolution because it shifts the partners' focus from criticizing each other's perspective to working together to develop alternative solutions.

Kurdek (1995) emphasized that conflict-resolution styles that stress agreement, compromise, and humor are associated with marital satisfaction, whereas conflict engagement, withdrawal, and defensiveness styles are associated with lower marital satisfaction. In his own study of 155 married couples, the style in which the wife engaged the husband in conflict and the husband withdrew was particularly associated with low marital satisfaction for both spouses.

Communicating effectively and creating a context of win–win in one's relationship contributes to a high-quality marital relationship, which is good for one's health.

Courtesy of Rachel Calisto

5-8f **Forgive**

Too little emphasis is placed on forgiveness as an emotional behavior that can move a couple from a deadlock to resolution. Merolla and Zhang (2011) noted that offender remorse positively predicted forgiveness and that such forgiveness was associated with helping resolve the damage. Hill (2010) studied forgiveness and emphasized that it is less helpful to try to "will" oneself to forgive the transgressions of another than to engage a process of self-reflection—that one has also made mistakes, hurt others, and is guilty—and to empathize with the fact that we are all fallible and need forgiveness. In addition, forgiveness ultimately means letting go of one's anger, resentment, and hurt, and its power comes from offering forgiveness as an expression of love to the person who has betrayed him or her. Forgiveness also has a personal benefit—it reduces hypertension and feelings of stress. To forgive is to restore the relationship—to pump life back into it. Of course, forgiveness given too quickly may be foolish. A person who has deliberately hurt his or her partner without remorse may not deserve forgiveness.

It takes more energy to hold on to resentment than to move beyond it. One reason some people do not forgive a partner for a transgression is that one can use the fault to control the relationship. "I wasn't going to let him forget," said one woman of her husband's infidelity.

A related concept to forgiveness is **amae**. Marshall et al. (2011) studied the concept of amae in Japanese romantic relationships. The term means expecting a close other's indulgence when one behaves inappropriately. Thirty Japanese undergraduate romantic couples kept a diary for two weeks that assessed their amae behavior (requesting, receiving, providing amae). Results revealed that amae behavior was associated with greater relationship quality and less conflict. "Cutting one some slack" may be another way of expressing amae.

5-8g **Avoid Defense Mechanisms**

Effective conflict resolution is sometimes blocked by **defense mechanisms**—techniques that function without awareness to protect individuals from anxiety and to minimize emotional hurt. The following paragraphs discuss some common defense mechanisms.

Escapism is the simultaneous denial of and withdrawal from a problem. The usual form of escape is avoidance. The spouse becomes "busy" and "doesn't have time" to think about or deal with the problem, or the partner may escape into recreation, sleep, alcohol, marijuana, or work. Denying and withdrawing from problems in relationships offer no possibility for confronting and resolving the problems.

Rationalization is the cognitive justification for one's own behavior that unconsciously conceals one's

---

brainstorming suggesting as many alternatives as possible without evaluating them.

amae expecting a close other's indulgence when one behaves inappropriately.

defense mechanisms unconscious techniques that function to protect individuals from anxiety and minimize emotional hurt.

escapism the simultaneous denial of and withdrawal from a problem.

rationalization the cognitive justification for one's own behavior that unconsciously conceals one's true motives.

true motives. For example, one wife complained that her husband spent too much time at the health club in the evenings. The underlying reason for the husband's going to the health club was to escape an unsatisfying home life. However, the idea that he was in a dead marriage was too painful and difficult for the husband to face, so he rationalized to himself and his wife that he spent so much time at the health club because he made a lot of important business contacts there. Thus, the husband concealed his own true motives from himself (and his wife).

Projection occurs when one spouse unconsciously attributes individual feelings, attitudes, or desires to the partner. For example, the wife who desires to have an affair may accuse her husband of being unfaithful to her. Projection may be seen in such statements as "You spend too much money" (projection for "I spend too much money") and "You want to break up" (projection for "I want to break up"). Projection interferes with conflict resolution by creating a mood of hostility and defensiveness in both partners. The issues to be resolved in the relationship remain unchanged and become more difficult to discuss.

Displacement involves shifting your feelings, thoughts, or behaviors from the person who evokes them onto someone else. The wife who is turned down for a promotion and the husband who is driven to exhaustion by his boss may direct their hostilities (displace them) onto each other rather than toward their respective employers. Similarly, spouses who are angry at each other may displace this anger onto someone else, such as the children.

By knowing about defense mechanisms and their negative impact on resolving conflict, you can be alert to them in your own relationships.

5-9 Trends in Communication and Technology

The future of communication will increasingly involve technology in the form of texting, smartphones, Facebook, etc. Such technology will be used to initiate, enhance, and maintain relationships. Indeed, intimates today may text each other 60 times a day. Over 2,000 messages a month are not unusual. Parental communication with children will also be altered. Aponte and Pessagno (2010) noted that technology may have positive and negative effects on the family. A positive effect is that parents will be able to use technology to monitor content as their children surf the Internet, send text messages, and send/receive photos on their cell phone. Parents may also use technology to know where their children are by global tracking systems embedded in their cell phones. The downside is that children can use this same technology to establish relationships external to the family which may be nefarious (e.g., child predators).

projection
attributing one's own feelings, attitudes, or desires to one's partner while avoiding recognition that these are one's own thoughts, feelings, and desires.

displacement
shifting one's feelings, thoughts, or behaviors from the person who evokes them onto someone else.

STUDY TOOLS 5

Ready to study? In the book, you can:

- ⮑ Rip out the chapter review card at the back of the book for a handy summary of the chapter and key terms.

- ⮑ Find out how supportive the communication is in your relationship with the Self-Assessment card at the back of the book.

Online at CENGAGEBRAIN.COM you can:

- ⮑ Prepare for tests with quizzes.

- ⮑ Review the key terms with Flash Cards.

- ⮑ Play games to master concepts.

Sexuality in Relationships

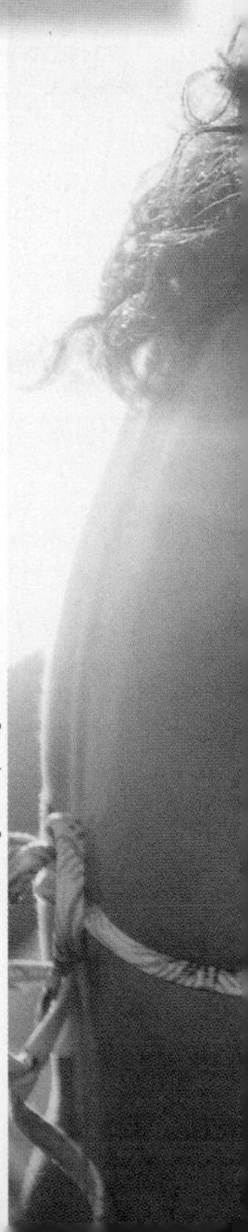

"**Sex isn't good unless it means something.** It doesn't necessarily need to mean "love" and it doesn't necessarily need to happen in a relationship, but it does need to mean **intimacy and connection.**...There exists a very fine line between being sexually liberated and being sexually used...."

—LAURA SESSIONS STEPP, UNHOOKED

Think about the following situations:

Two people are at a party, drinking and flirting. Although they met only two hours ago, they feel a strong attraction to each other. Each is wondering what level of sexual behavior will occur when they go back to one of their rooms later that evening. How much sexual involvement is appropriate in a first-time encounter?

***

Two students have decided to live together, but they know their respective parents would disapprove. If they tell their parents, the parents may cut off their money and both students will be forced to drop out of school. Should they tell?

***

While Maria was away for a weekend visiting her grandmother, her live-in partner hooked up with his ex-girlfriend. He regretted his behavior, asked for forgiveness, and promised never to be unfaithful again. Should Maria take him back?

***

Hero Images/Getty Images

A woman is married to a man whose career requires that he be away from home for extended periods. Although she loves her husband, she is lonely, bored, and sexually frustrated in his absence. She has been asked out by a colleague at work whose wife also travels. He, too, is in love with his wife but is lonely for emotional and sexual companionship. They are ambivalent about whether to hook up occasionally while their spouses are away. What advice would you give to this couple?

Each of these scenarios involves **sexual values**—moral guidelines for sexual behavior in relationships. Values sometimes predict sexual behavior. One's sexual values may be identical to one's sexual behavior. For example, a person who values abstinence until marriage may not have sexual intercourse until marriage. But one's sexual behavior does not always correspond with one's sexual values. One explanation for the discrepancy between values and behavior is that a person may engage in a sexual behavior, then decide the behavior was wrong,

sexual values
moral guidelines for sexual behavior in relationships.

and adopt a sexual value against it. We begin this chapter with an overview of various sexual values.

6-1 Alternative Sexual Values

At least three sexual values guide choices in sexual behavior: absolutism, relativism, and hedonism. See Table 6.1 for the respective sexual values of over 4,500 undergraduates. Individuals may have different sexual values at different stages of the family life cycle. For example, young and elderly individuals are more likely to be absolutist, whereas those in the middle years are more likely to be relativistic. Young unmarried adults are more likely to be hedonistic.

6-1a **Absolutism**

Absolutism is a sexual value system which is based on unconditional allegiance to tradition or religion (i.e., waiting until marriage to have sexual intercourse). People who are guided by absolutism in their sexual choices have a clear notion of what is right and wrong.

The official creeds of fundamentalist Christian and Islamic religions encourage absolutist sexual values. Intercourse is solely for procreation, and any sexual acts that do not lead to procreation (masturbation, oral sex, homosexuality) are immoral and regarded as sins against God, Allah, self, and community. Waiting until marriage to have intercourse is also an absolutist sexual value. This value is often promoted in the public schools.

Individuals conceptualize their virginity in one of three

absolutism belief system based on unconditional allegiance to the authority of religion, law, or tradition.

This coed is an absolutist. She has never had oral, vaginal, or anal sex and plans to only be sexual with her spouse when she is married.

ways—as a process, a gift, or a stigma. The process view regards first intercourse as a mechanism of learning about one's self and one's partner and sexuality. The gift view regards being a virgin as a valuable positive status wherein it is important to find the right person since sharing the gift is special. The stigma view considers virginity as something to be ashamed of, to hide, and to rid oneself of. When 215 undergraduates were asked their view, 54% classified themselves as process oriented, 38% as gift oriented, and 8.4% as stigma oriented at the time of first coitus (Humphreys, 2013).

"You've got to **stand for something** or you'll fall for anything."

—AARON TIPPIN, SINGER

"True Love Waits" is an international campaign designed to challenge teenagers and college students to remain sexually abstinent until marriage. Under this program, created and sponsored by the Baptist Sunday School Board, young people are asked to agree to the absolutist position and sign a commitment to the following: "Believing that true love waits, I make a commitment to God, myself, my family, my friends, my future mate, and my future children to sexual purity including abstinence from this day until the day I enter a biblical marriage relationship" (True Love Waits, 2014) How effective

Table 6.1
Sexual Values of 4,720 Undergraduates

Respondents	Absolutism, %	Relativism, %	Hedonism , %
Male students (N = 1,149)	12	58	30
Female students (N = 3,571)	16	66	18

Source: Hall, S., & D. Knox. (2015). Relationship and sexual behaviors of a sample of 4,720 university students. Unpublished data collected for this text. Department of Family and Consumer Sciences, Ball State University and Department of Sociology, East Carolina University.

are the "True Love Waits" and "virginity pledge" programs in delaying sexual behavior until marriage? Data from the National Longitudinal Study of Adolescent Health revealed that, although youth who took the pledge were more likely than other youth to experience a later "sexual debut," have fewer partners, and marry earlier, most eventually engaged in premarital sex, were less likely to use a condom when they first had intercourse, and were more likely to substitute oral and/or anal sex in the place of vaginal sex. There was no significant difference in the occurrence of sexually transmitted infections (STIs) between "pledgers" and "nonpledgers" (Brucker & Bearman, 2005). The researchers speculated that the emphasis on virginity may have encouraged the pledgers to engage in noncoital (nonintercourse) sexual activities (e.g., oral sex), which still exposed them to STIs and to be less likely to seek testing and treatment for STIs. Landor and Simons (2010) studied 1,215 undergraduates and found that male pledgers (in contrast to female pledgers) were less likely to remain virgins.

Some individuals still define themselves as virgins even though they have engaged in oral sex. Of 1,141 undergraduate males, 75% (70% of 3,530 undergraduate females) agreed with the statement "If you have oral sex, you are still a virgin." Hence, according to these undergraduates, having oral sex with someone is not really having sex (Hall & Knox, 2015). Persons most likely to agree that "oral sex is not sex" are freshmen/sophomores, those self-identifying as religious (Dotson-Blake et al., 2012). Individuals may engage in oral sex rather than sexual intercourse to avoid getting pregnant, to avoid getting an STI, to keep their partner interested, to avoid a bad reputation, and to avoid feeling guilty over having sexual intercourse (Vazonyi & Jenkins, 2010).

Indeed, rather than a dichotomous "one is or is not a virgin" concept that gets muddled by one's view of oral sex being "sex," a three-part view of virginity might be adopted—oral sex, vaginal sex, and anal sex. No longer might the term "virgin" be used to reveal sexual behaviors in these three areas. Rather, whether

kamomeen/Shutterstock.com

one has engaged in each of the three behaviors must be identified. Hence, an individual would not say "I am a virgin" but "I am an oral virgin" (or intercourse virgin or anal virgin as the case may be).

Crisma/Getty Images

Virginity loss for heterosexuals typically refers to vaginal sex (though some would say they are no longer virgins if there has been oral or anal sex). Lesbian and gay males typically refer to virginity loss if there has been oral or anal sex.

Mega Pixel/Shutterstock.com

A subcategory of absolutism is **asceticism**. Ascetics believe that giving in to carnal lust is unnecessary and attempt to rise above the pursuit of sensual pleasure into a life of self-discipline and self-denial. Accordingly, spiritual life is viewed as the highest good, and self-denial helps one to achieve it. Catholic priests, monks, nuns, and some other celibate people have adopted the sexual value of asceticism.

6-1b Relativism

Fifty-eight percent of 1,149 undergraduate males and 66% of 3,571 undergraduate females identified relativism as their sexual value (Hall & Knox, 2015). **Relativism** is a value system emphasizing that sexual decisions should be made

asceticism the belief that giving in to carnal lusts is wrong and that one must rise above the pursuit of sensual pleasure to a life of self-discipline and self-denial.

relativism sexual decisions are made in reference to the emotional, security, and commitment aspects of the relationship.

in reference to the emotional, security, and commitment aspects of the relationship. Although absolutists might feel that having intercourse is wrong for unmarried people, relativists might feel that the moral correctness of sex outside marriage depends on the particular situation. For example, a relativist might say that marital sex between two spouses who are emotionally and physically abusive to each other is not to be preferred over intercourse between two unmarried individuals who love each other, are kind to each other, and are committed to the well-being of each other.

Relativists apparently do value having a good sex partner. About half of never-married individuals want to stay in a relationship only if the sex is good. In a study of 5,500 never- married people, half of the women and 44% of the men said that "bad sex would be a deal-breaker" (Walsh, 2013).

A disadvantage of relativism as a sexual value is the difficulty of making sexual decisions on a relativistic case-by-case basis. The statement "I don't know what's right anymore" reflects the uncertainty of a relativistic view. Once a person decides that mutual love is the context justifying intercourse, how often and how soon is it appropriate for the person to fall in love? Can love develop after some alcohol and two hours of conversation? How does one know that love feelings are genuine? The freedom that relativism brings to sexual decision making requires responsibility, maturity, and judgment. In some cases, individuals may convince themselves that they are in love so that they will not feel guilty about having intercourse.

Personal View:
A Script for Delaying Intercourse in a Relationship

It is sometimes the case where one partner wants to de-escalate or slow down the sex in a relationship. Communicating this desire to the partner is tricky—one wants to communicate interest in sex but not allow the relationship to become sex focused. The script to a partner follows:

I need to talk about sex in our relationship. I am anxious talking about this and not sure exactly what I want to say so I have written it down to try and help me get the words right. I like you, enjoy the time we spend together, and want us to continue seeing each other. The sex is something I need to feel good about to make it good for you, for me, and for us.

I need for us to slow down sexually. I need to feel an emotional connection that goes both ways—we both have very strong emotional feelings for each other. I also need to feel secure that we are going somewhere—that we have a future and that we are committed to each other. We aren't there yet so I need to wait till we get there to be sexual (have intercourse) with you.

How long will this take? I don't know—the general answer is "longer than now." This may not be what you have in mind and you may be ready for us to increase the sex now. I'm glad that you want us to be sexual and I want this too but I need to feel right about it. So, for now, let me drive the bus sexually…when I feel the emotional connection and that we are going somewhere, I'll be the best sex partner you ever had. But let me move us forward. If this is too slow for you or not what you have in mind, maybe I'm not the girl for you. It is certainly OK for you to tell me you need more and move on. Otherwise, we can still continue to see each other and see where the relationship goes.

So…tell me how you feel and what you want…don't feel like you need to tell me you love me and want us to get married…ha! I'm not asking that…I'm just asking for time….

4. **Transition out.** The couple were romantic, the relationship ended, but the sexual relationship continued. Eleven percent reported this type of FWBR.

5. **Successful transition.** Intentional use of friends with benefits relationship to transition into a romantic relationship. Eight percent reported this type of FWBR.

6. **Unintentional transition.** Sexual relationship morphs into romantic relationship without intent. Relationship results from regular sex, hanging out together, etc. Eight percent reported this type of FWBR.

7. **Failed transition.** One partner becomes involved while the other does not. The relationship stalls. The lowest percentage (7%) of the 258 respondents reported this type of FWBR.

Friends with Benefits Relationship Friends with benefits is often part of the relational sexual landscape. Mongeau et al. (2013) defined the **friends with benefits relationship (FWBR)** as platonic friends (i.e., those not involved in a romantic relationship) who engage in some degree of sexual intimacy on multiple occasions. This sexual activity could range from kissing to sexual intercourse and is a repeated part of a friendship, not just a one-night hookup. Forty-four percent of 1,148 undergraduate males reported that they had been in a FWBR (40% of 3,570 undergraduate females). These undergraduates were primarily first- and second-year students (Hall & Knox, 2015). Mongeau et al. (2013) identified seven types of friends with benefits relationships:

1. **True friends.** Close friends who have sex on multiple occasions (similar to but not labeled as romantic partner). The largest percent (26%) of the 258 respondents reported this type of FWBR.

2. **Network opportunism.** Part of same social networks who just hang out and end up going home to have sex together since there is no better option; a sexual fail-safe. Fifteen percent reported this type of FWBR.

3. **Just sex.** Focus is sex, serial hookup with same person. Do not care about person other than as sexual partner. Twelve percent reported this type of FWBR.

Advantages to having sex with one's former partner include having a "safe" sexual partner, having a predictably "good" sexual partner who knows likes/dislikes and does not increase the number of lifetime sexual partners, and "fanning sexual flames might facilitate rekindling partners' emotional connections" (Mongeau et al., 2013). Disadvantages include developing a bad reputation as someone who does not really care about emotional involvement, coping with discrepancy of becoming more or less involved than the partner, and losing the capacity to give oneself emotionally.

Lehmiller et al. (2014) analyzed data on 376 individuals in a relationship and compared those who were in a friends with benefits relationship (50.5%) with those who were romantically involved but not in a FWB relationship (49.5%). Differences included that FWB partners were

friends with benefits relationship
a relationship between nonromantic friends who also have a sexual relationship.

less likely to be sexually exclusive, were less sexually satisfied, more likely to practice safe sex, and generally communicated less about sex than romantic partners who were not in an FWB relationship.

Concurrent sexual partnerships are those in which the partners have sexual relationships with several individuals with whom they have a relationship. In a study of 783 adults ages 18 to 59, 10% reported that both they and their partners had had other sexual partners during their relationship (Paik, 2010). Men were more likely than women to have been nonmonogamous (17% versus 5%). A serious relationship was the context in which going outside the relationship for sex was least likely. However, if the partners had a casual or FWBR, the chance of nonmonogamy rose 30% and 44%, respectively. Wester and Phoenix (2013) studied relationships at the "talking" phase and noted that men were much more likely to approve of all forms of sexual behavior in multiple relationships than women.

Swinging Swinging relationships (also called **open relationships**) consist of focused recreational sex. These relationships involve married/pair-bonded individuals agreeing that they may have sexual encounters with others.

Kimberly and Hans (2012) reviewed the literature on swinging couples. There are about 3 million swinging couples in the United States with most reporting positive emotional and sexual relationships. Swinging for them is a mutually agreeable recreational behavior that has positive benefits for the relationship.

Over 20% (21%) of 1,032 undergraduate males reported that they were comfortable with their partner being emotionally and sexually involved with someone else; 11% of 3,113 undergraduate females agreed (Hall & Knox, 2015).

6-1c Hedonism

Hedonism is the belief that the ultimate value and motivation for human actions lie in the pursuit of pleasure and the avoidance of pain. Thirty percent of 1,149 undergraduate males and 18% of 3,571 undergraduate females identified hedonism as their primary sexual value (Hall & Knox, 2015). Bersamin and colleagues (2014) analyzed data on single, heterosexual college students (N = 3,907) ages 18 to 25 from 30 institutions across the United States.

A greater proportion of men (18.6%) compared to women (7.4%) reported having had casual sex in the month prior to the study. The researchers also found that casual sex was negatively associated with psychological well-being (defined in reference to self-esteem, life satisfaction, and eudamonic well-being—having found oneself). Casual sex was also positively associated with psychological distress (e.g., anxiety, depression). There were no gender differences. Sandberg-Thoma and Kamp Dush (2014) also found suicide ideation and depressive symptoms associated with casual sexual relationships (sample of 12,401 adolescents). Fielder et al. (2014) studied hookups in first-year college women and found an association with experiencing depression, sexual victimization, and STIs.

Hedonism is sometimes viewed as sexual addiction. Levine (2010) noted that the term *sexual addiction* has no professional agreement and is sometimes applied to those who watch pornography, have commercial sex, and engage in cybersex. More accurately, sex addition applies to those who have lost control over their sexual behavior which is often accompanied by spiraling psychological deterioration, that is depression.

The **sexual double standard** is the view that encourages and accepts sexual expression of men more than women. Table 6.1 revealed that men are almost two times more hedonistic than women (Hall & Knox, 2015). Acceptance of the double standard is evident in the words used to describe hedonism—hedonistic men are thought of as "studs" but hedonistic women as "sluts." Indeed, Porter (2014) emphasized the double standard in her presentation on "slut-shaming" which she defined as "the act of making one feel guilty or inferior for engaging in certain sexual behaviors that violate traditional dichotomous gender roles." She pointed out that Charlie Sheen was a national celebrity for his flagrant debauchery but Kristen Stewart was shamed for her infidelity. Porter surveyed 240 undergraduates and found that 81% of the females reported having been slut-shamed (7.3% of the males).

concurrent sexual partnerships those in which the partners have sex with several individuals concurrently.

swinging relationships (open relationships) involve married/pair-bonded individuals agreeing that they may have sexual encounters with others.

hedonism belief that the ultimate value and motivation for human actions lie in the pursuit of pleasure and the avoidance of pain.

sexual double standard the view that encourages and accepts sexual expression of men more than women.

6-2 Sources of Sexual Values

The sources of one's sexual values are numerous and include one's education, religion, and family, as well as technology, television, social movements, and the Internet. Public schools in the United States typically promote absolutist sexual values through abstinence education.

Religion is also an important influence. Thirty-six percent of 1,154 undergraduate males (44% of 3,574 undergraduate females) self-identified as being "very" or "moderately" religious (Hall & Knox, 2015). In regard to sexual behavior, researchers have found that religiously active young adults are more likely to agree that sexting and sexual intercourse are inappropriate activities to engage in before being in a committed dating relationship (Miller et al., 2011).

Parents/family are also influential. Wetherill et al. (2010) found that for individuals with parents who were aware of what their children were doing and cared about them, their behavior reported engaging in less frequent sexual behaviors. Purity Balls, which are events where fathers and daughters alike pledge purity—the fathers pledge that they will be faithful to their wives and the daughters pledge (both sign "pledge cards") that they will wait until marriage to have sexual intercourse—are held in 48 states. Purity Balls have their basis in evangelical conservative religious families. The documentary *Virgin Tales* focuses on one family in Colorado and their involvement in Purity Balls. As noted earlier, most pledgers end up having sexual intercourse before marriage and are more likely than nonpledgers to engage in oral sex. Pledgers are also less likely to use a condom than nonpledgers. You can read other examples of familial influences on ones sexuality in the What's New? section on page 124.

Reproductive technologies such as birth control pills, the morning-after pill, and condoms influence sexual values by affecting the consequences of behavior. Being able to reduce the risk of pregnancy and HIV infection with the pill and condoms allow one to consider a different value system than if these methods of protection did not exist.

Television also influences sexual values. A television advertisement shows an affectionate couple with minimal clothes on in a context where sex could occur. "Be ready for the moment" is the phrase of the announcer, and Levitra, the new quick-start Viagra, is the product for sale. The advertiser used sex to get the attention of the viewer and punch in the product to elicit buying behavior. Ballard et al. (2011) studied the sexual socialization of young adults. Some of the respondents noted the mixed messages in their socialization, e.g., the discrepancy between media messages and church messages regarding sexuality: *"They [church members] say abstain, abstain, abstain, but it's so hard when sex is being thrown at you constantly from, you know, watching TV."*

In regard to magazines, exposure outcome is more positive. Walsh and Ward (2010) assessed sexual health behaviors and magazine reading among 579 undergraduate students. They found that more frequent reading of mainstream magazines was associated with greater sexual health knowledge, safe-sex self-efficacy, and consistency of using contraception (although results varied across sex and magazine genre).

The women's movement affects sexual values by empowering women with an egalitarian view of their sexuality. This view translates into encouraging women to be more assertive about their own sexual needs and giving them the option to experience sex in a variety of contexts (e.g., without love or commitment) without self-deprecation. The net effect is a potential increase in the frequency of recreational, hedonistic sex. The gay liberation movement has also been influential in encouraging values that are accepting of sexual diversity.

Finally, the Internet has an influence on sexual values. The Internet features erotic photos, videos, and "live" sex acts/stripping by webcam sex artists. Individuals can exchange nude photos, have explicit sex dialogue, arrange to have phone sex or meet in person, or find a prostitute. Indeed, the adult section of Craigslist was shut down because it featured blatant prostitution.

Kuleczka/Shutterstock.com

Dusan Zidar/Shutterstock.com

What's New?

UNDERGRADUATES REPORT WHAT THEY LEARNED FROM FAMILY ABOUT SEX

One's family (both nuclear and extended) is a major source of learning about sexuality. A team of researchers conducted an online survey of 101 undergraduates (from a midwest, northeast, and southern campus) who identified a memory or event from their family that had an impact on their current sexual behavior.

Key Findings

The narratives reflected a balance of negative (37.6%), positive (34.7%), and neutrally framed (27.7%) familial messages about sex. The five themes included: (1) practice safe sex, (2) premarital sex is wrong, (3) wait until you are ready/for the right person, (4) sex as natural or pleasurable, and (5) sex as negative, abusive, and/or taboo. Examples follow:

Practice Safe Sex

My mother had an abortion when she was 18 years old. She has always been very open with me about her decision as well has how I should practice safe sex so I would never have to make that decision.

Premarital Sex is Wrong

From my mother and others I was always told to wait until I was married to have sex—I was actually kind of scared out of it completely...

Wait Until You Are Ready/ for the Right Person

I have decided to sustain from any sexual activity until I am in a committed relationship because of a promise I made to my aunt. My aunt Crystal and I are extremely close and my promise to her will remain sacred until I am in a committed and respectable relationship.

Sex as Natural or Pleasurable

When I was 9 years old I wanted to sleep in my mother's bed. She allowed it and after a minute I felt something weird. I had a used condom stuck to my leg. My mother called her boyfriend and they laughed hysterically. I guess I learned early that sex is fun and lighthearted.

Sex as Negative, Abusive, and/or Taboo

My father was arrested for a sexual offense when I was 10 years old. He had solicited a girl just three years older than me online, sent her explicit material, and made plans to meet. During the custody battle, my sister and I had to endure a rape kit. Although I have no recollection of sexual abuse, the tests revealed heavy sexual abuse from extraneous objects, which revealed why I had been ridden with infection as a young child. The wounds were mostly healed, and despite no memory of the indicated events, to this day it takes me a very long time and a lot of trust to take steps towards more intimacy.

ducu59us/Shutterstock.com

Source: Kauffman, L., M. P. Orbe, A. L. Johnson, & A. Cooke-Jackson. (2013, June 20). Memorable familial messages about sex: A qualitative content analysis of college student narratives. *Electronic Journal of Human Sexuality, 16*. Online at www.ejhs.org.

6-3 Sexual Behaviors

We have been discussing the various sources of sexual values. We now focus on what people report that they do sexually.

6-3a What Is Sex?

Horowitz and Spicer (2013) asked 124 emerging adults (40 male heterosexuals, 42 female heterosexuals, and 42 lesbians) to identify various sexual behaviors on a six-point scale (from "definitely" to "definitely not") as "having sex." There was agreement that vaginal and anal sex were "definitely" sex while kissing was "definitely not" sex. Ratings of heterosexual males and females did not differ significantly, but gays were more likely than heterosexuals to rate various forms of genital stimulation as "having sex."

Some individuals are **asexual**, which means there is an absence of sexual behavior with a partner and oneself (masturbation). About 4% of females and 11% of males reported being asexual in the last 12 months (DeLamater & Hasday, 2007). In contrast, most individuals report engaging in various sexual behaviors. Penhollow et al. (2010) found that, for both male and female students, participation in recreational sexual behaviors (with or without a partner) enhanced their overall sexual satisfaction. The following discussion includes kissing, masturbation, oral sex, vaginal intercourse, and anal sex. Sexual behavior is dictated by **social scripts**—the identification of the roles in a social situation, the nature of the relationship between the roles, and the expected behaviors of those roles. In regard to kissing, two individuals kiss because they are in a relationship where the expectation is such that they are expected to kiss.

> ## "Remember, we're madly **in love**, so it's **all right to kiss me** anytime you feel like it."
>
> —SUZANNE COLLINS, *THE HUNGER GAMES*

6-3b Kissing

Kissing has been the subject of literature and science. The meanings of a kiss are variable—love, approval, hello, goodbye, or as a remedy for a child's hurt knee. There is also a kiss for luck, a stolen kiss, and a kiss to seal one's marriage vows. Kisses have been used to denote hierarchy. In the Middle Ages, only peers kissed on the lips; a person of lower status kissed someone of higher status on the hand, and a person of lower status showed great differential of status by kissing on the foot. Kissing also has had a negative connotation as in the "kiss of death," which reflects the kiss Judas gave Jesus as he was about to betray him. Kissing may be considered an aggressive act as some do not want to be kissed.

Courtesy of Rachel Calisto

Vlorika Prikhodko/iStockphoto.com

Although the origin of kissing is unknown, one theory posits that kissing is associated with parents putting food into their offspring's mouth—the bird pushes food down the throat of a chick in the nest. Some adult birds also exchange food by mouth during courtship. The way a person kisses reflects the person's country, culture, and society. The French kiss each other once on each cheek or three times in the same region. Greeks tend to kiss on the mouth, regardless of the sex of the person. Anthropologists note that some cultures (e.g., Eskimos, Polynesians) promote meeting someone by rubbing noses.

6-3c Masturbation

Masturbation involves stimulating one's own body with the goal of experiencing pleasurable sexual sensations. Ninety-seven percent of 1,141 undergraduate males and 65% of 3,534 undergraduate females

asexual an absence of sexual behavior with a partner and oneself.

social scripts the identification of the roles in a social situation, the nature of the relationship between the roles, and the expected behaviors of those roles.

masturbation stimulating one's own body with the goal of experiencing pleasurable sexual sensations.

reported having masturbated at some point (Hall & Knox, 2015).

Alternative terms for masturbation include *auto-eroticism, self-pleasuring, solo sex,* and *sex without a partner*. An appreciation of the benefits of masturbation has now replaced various myths about it (e.g., it causes blindness). Most health care providers and therapists today regard masturbation as a normal and healthy sexual behavior. Masturbation is safe sex in that it involves no risk of transmitting diseases (such as HIV) or producing unintended pregnancy.

Masturbation is also associated with orgasm during intercourse. In a study by Thomsen and Chang (2000), 292 university undergraduates reported whether they had ever masturbated and whether they had an orgasm during their first intercourse experience. The researchers found that the strongest single predictor of orgasm and emotional satisfaction with first intercourse was previous masturbation.

6-3d Oral Sex

In a sample of 1,142 undergraduate males, 72% reported that they had given oral sex; 75% of 3,539 undergraduate women had done so (Hall & Knox, 2015). Fellatio is oral stimulation of the man's genitals by his partner. In many states, legal statutes regard fellatio as a "crime against nature." "Nature" in this case refers to reproduction, and the "crime" is sex that does not produce babies. Nevertheless, most men have experienced fellatio. Cunnilingus is oral stimulation of the woman's genitals by her partner.

Malacad and Hess (2010) observed among Canadian adolescent females that oral sex has become an increasingly common and casual activity. In a study of 181 women aged 18–25 years, approximately three-quarters reported having engaged in oral sex, a prevalence rate almost identical to that for vaginal intercourse. The respondents were an average age of 17 (for both oral sex and intercourse) at their first experience. Most reported that their most recent oral sex experience was positive and in a committed relationship. Adolescent females who reported oral sex with a partner they were not in love with reported the most negative experience.

There is the mistaken belief that only intercourse carries the risk of contracting an STI. However, STIs as well as HIV can be contracted orally. Use of a condom or dental dam, a flat latex device that is held over the vaginal area, is recommended.

6-3e Vaginal Intercourse

Vaginal intercourse, or **coitus**, refers to the sexual union of a man and woman by insertion of the penis into the vagina. In a study of 4,698 undergraduates, 71% reported that they had had sexual intercourse (men 67.4%; women 71.4%) (Hall & Knox, 2015). In regard to university students, each academic year, there are fewer virgins.

Vasilenko et al. (2012) reported on the consequences of vaginal sex for 209 first-year undergraduates at a large northeastern university. Students kept a diary for 28 days on their sexual behavior and the personal and interpersonal consequences. The most commonly reported (81%) positive consequence of having vaginal sex was feeling physically satisfied; the most commonly reported negative consequence was worry about pregnancy (17%). In regard to interpersonal consequences, the most common consequence was feeling closer to the partner (89%); the most common negative consequence was worrying if the partner wanted more commitment. More positive outcomes resulted from dating rather than casual partners.

6-3f First Intercourse

A study of first intercourse experience of 475 young Canadian adults revealed orgasm for 6% of females and 62% of males with alcohol/drugs associated with fewer positive experiences and higher sexual regret (Reissing et al., 2012). Having first intercourse "early" in adolescence (age 14) was remembered as a positive experience when there was an emotional relationship with the partner and when the experience was physically pleasurable. Early negative experiences were associated with no relationship involvement, physical pain/discomfort, and little preparation. It was not uncommon for the early debut adolescents to report that they were drunk. These findings are based on quantitative data ($N = 705$) and in-depth interviews (24 young people ages 16–18 years) (Symons et al., 2014).

Sprecher (2014) examined data from 5,769 respondents over a 23-year period in regard to gender differences in pleasure, anxiety, and guilt in response to first intercourse. Men reported more pleasure and anxiety than women, and women reported more guilt

coitus sexual union of a man and woman by insertion of the penis into the vagina.

than men. Anxiety decreased over the three decades for men; pleasure increased and guilt decreased for women. The result is that "although gender differences in emotional reactions to first intercourse have decreased over time, the first intercourse experience continues to be a more positive experience for men than for women."

Hawes et al. (2010) emphasized that **sexual readiness** is helpful in determining when one is ready for first intercourse. Such readiness can be determined in reference to contraception, autonomy of decision (not influenced by alcohol or peer pressure), consensuality (both partners equally willing), and the absence of regret (the right time for me). Using these criteria, the negative consequences of first intercourse are minimized.

6-3g Anal Sex

In a study of 4,689 undergraduates (primarily heterosexuals), 23% reported that they had had anal sex (men 25%; women 22%) (Hall & Knox, 2015). Younger individuals, those with higher numbers of partners, and those with STIs are more likely to report having participated in anal sex. Motivations associated with anal sex include intimacy ("anal sex is more intimate than regular sex"), enjoyment in variety, domination by male, breaking taboos, and pain–pleasure enjoyment (McBride & Fortenberry, 2010).

The greatest danger of anal sex is that the rectum might tear, in which case blood contact can occur; STIs (including HIV infection) may then be transmitted. Partners who use a condom during anal intercourse reduce their risk of not only HIV infection but also other STIs. Pain (physical as well as psychological) may occur for anyone of any sexual identity involved in receiving anal sex. Such pain is called **anodyspareunia**—frequent, severe pain during receptive anal sex. Of 505 women who experienced anal intercourse, 9% reported severe pain during every penetration (Stulhofer & Ajdukovic, 2011).

6-3h Cybersex

Cybersex is any consensual sexual experience mediated by a computer that involves at least two people. In this context, sexual experience includes sending text messages or photographic images that are sexual. Individuals typically send sex text and photos with the goal of arousal or looking at each other naked or

Courtesy of Chelsea Curry

This couple agree on the respective roles that they play for their enjoyment.

masturbating when viewing each other on a webcam.

6-3i Kink

Individuals enjoy a range of sex play including what is referred to as a **kink**. Examples include sadism (a partner enjoys giving pain to another) and masochism (a partner enjoys receiving pain), which are sometimes viewed as pathological. However, Dr. Julie Fennell (2014) noticed that such behaviors are often pathologized by the psychiatric community (e.g., DSM-5).

In much the same way that gays and lesbians feel that their sexual desires were unfairly pathologized by the medical establishment prior to the removal of homosexuality as a mental disorder from the DSM, people who participate in BDSM (Bondage & Discipline/Dominance & Submission/Sadism & Masochism) feel that they have been unfairly pathologized for their sexual desires. People who engage in BDSM apply principles of consent to their practices—meaning that there are no victims or abusers, only "tops" and "bottoms." As explained by the National Coalition for Sexual Freedom, when BDSM is practiced correctly, only people who want to get hurt get hurt. People who

sexual readiness determining when one is ready for first intercourse in reference to contraception, autonomy of decision, consensuality, and absence of regret.

anodyspareunia frequent, severe pain during receptive anal sex.

cybersex any consensual sexual experience mediated by a computer that involves at least two people.

kink typically refers to BDSM (bondage and discipline/ dominance and submission/sadism and masochism).

have medical disorders associated with sadism and masochism are not engaging in consensual behaviors, and most people in the BDSM community view the diagnoses of "fetishism" and "transvestism" as obsolete and heteronormative, respectively.

6-4 Sexuality in Relationships

Sexuality occurs in a social context that influences its frequency and perceived quality.

6-4a Sexual Relationships among Never-Married Individuals

Never-married individuals and those not living together report more sexual partners than those who are married or living together. The never married also report the lowest level of sexual satisfaction. One-third of a national sample of people who were not married and not living with anyone reported that they were emotionally satisfied with their sexual relationships. In contrast, 85% of the married and pair-bonded individuals reported emotional satisfaction in their sexual relationships. Hence, although never-married individuals have more sexual partners, they are less emotionally satisfied (Michael et al., 1994).

6-4b Sexual Relationships among Married Individuals

Marital sex is distinctive for its social legitimacy, declining frequency, and satisfaction (both physical and emotional).

1. **Social legitimacy.** In our society, marital intercourse is the most legitimate form of sexual behavior. Those who engage in homosexual, premarital, and extramarital intercourse do not experience as high a level of social approval as do those who engage in marital sex. It is not only OK to have intercourse when married, it is expected. People assume that married couples make love and that something is wrong if they do not.

2. **Declining frequency.** Sexual intercourse between spouses occurs about six times a month, which declines in frequency as spouses age. Pregnancy also decreases the frequency of sexual intercourse (Lee et al., 2010). In addition to biological changes due to aging and pregnancy, satiation also contributes to the declining frequency of intercourse between spouses and partners in long-term relationships. Psychologists use the term **satiation** to mean that repeated exposure to a stimulus results in the loss of its ability to reinforce. The thousandth time that a person has sex with the same partner is not as new and exciting as the first few times.

3. **Satisfaction (emotional and physical).** Despite declining frequency and less satisfaction over time, marital sex remains a richly satisfying experience. As noted above, in a national sample, 88% of married people said they received great physical pleasure from their sexual lives, and almost 85% said they received great emotional satisfaction (Michael et al., 1994).

6-4c Sexual Relationships among Divorced Individuals

Of the almost 2 million people getting divorced, most will have intercourse within one year of being separated from their spouses. The meanings of intercourse for separated or divorced individuals vary. For many, intercourse is a way to reestablish—indeed, repair—their crippled self-esteem. Questions such as, "What did I do wrong?" "Am I a failure?" and "Is there anybody out there who will love me again?" loom in the minds of divorced people. One way to feel loved, at least temporarily, is through sex. Being held by another and being told that it feels good give people some evidence that they are desirable. Because divorced people may be particularly vulnerable, they may reach for sexual encounters as if for a lifeboat. "I felt that, as long as someone was having sex with me, I wasn't dead and I did matter," said one recently divorced person.

Because divorced individuals are usually in their mid-thirties or older, they may not be as sensitized to the danger of contracting HIV as are people in their twenties. Divorced individuals should always use a condom to lessen the risk of an STI, including HIV infection and autoimmune deficiency syndrome (AIDS).

satiation the state in which a stimulus loses its value with repeated exposure.

6-4d Sexual Problems: General

Sexual relationships are not without sexual problems. Hendrickx and Enzlin (2014) analyzed data on an Internet survey of 35,132 heterosexual Flemish men and women (aged 16 to 74 years). In men, the most common sexual difficulties (impairment in sexual function regardless of level of distress) were hyperactive sexual desire (frequent sexual urges or activity) (27.7%), premature ejaculation (12.2%), and erectile difficulty (8.3%). For women, the most common sexual difficulties were absent or delayed orgasm (20.1%), hypoactive sexual desire (19.3%), and lack of responsive desire (13.7%).

6-4e Sexual Problems: Pornography

While not typically regarded as a "sexual problem," the discrepancy in pornography use by men and women creates a context for conflict. All research reports higher pornography use by men than women. Brown et al. (2013) analyzed national data and found that the discrepancy drops as men move into more committed relationships. For example, among casually dating individuals, over half of men reported regular pornography use (weekly or more frequent) while only 1% of women report similar patterns—45% of men and 43% of women reported conflict over pornography use. Among married couples, the pornography gap was less with 20% for men and 3% for women reporting regular pornography use—21% of men and 27% of women reported conflict over pornography use. Whether reduction in conflict about porn use as relationship seriousness increased was due to lower frequency of use as a result of the partner's surveillance or the partner was more secure as relationship commitment increased remains an open question.

Courtesy of Rachel Calisto

6-5 Sexual Fulfillment: Some Prerequisites

There are several prerequisites for having a good sexual relationship.

6-5a Self-Knowledge, Body Image, and Health

Sexual fulfillment involves knowledge about yourself and your body. Such information not only makes it easier for you to experience pleasure but also allows you to give accurate information to a partner about pleasing you. It is not possible to teach a partner what you do not know about yourself.

Sexual fulfillment also implies having a positive body image. To the degree that you have positive feelings about your body, you will regard yourself as a person someone else would enjoy touching, being close to, and making love with. If you do not like yourself or your body, you might wonder why anyone else would. Van den Brink et al. (2013) analyzed body image data on 319 Dutch undergraduates. Most reported neutral or mildly positive body evaluations, and in 30% of the sample these evaluations were clearly positive. Comparisons between women who reported positive versus neutral body evaluations showed that the body-satisfied women had lower body mass indexes (BMIs) and reported less body image investment, less overweight preoccupation, and less body self-consciousness during sexual activity. These women also reported higher sexual self-esteem and better sexual functioning. Woertman and Van den Brink (2012) found sexual arousal, initiating sex, sexual satisfaction, and orgasm related to a positive body image in women.

Effective sexual functioning also requires good physical and mental health. This means regular exercise, good nutrition, lack of disease, and lack of fatigue. Performance in all areas of life does not have to diminish with age—particularly if people take care of themselves physically.

Good health also implies being aware that some drugs may interfere with sexual performance. Alcohol is the drug most frequently used by American adults (including college students). Although a moderate amount of alcohol can help a person become aroused through a lowering of inhibitions, too much alcohol can slow the physiological processes and deaden the senses. Shakespeare may have said it best: "It [alcohol] provokes the desire, but it takes away the performance" (*Macbeth*, act 2, scene 3). The result of an excessive intake of alcohol for women is a reduced chance of orgasm; for men, overindulgence results in a reduced chance of attaining an erection.

The reactions to marijuana are less predictable than the reactions to alcohol. Though some individuals report a short-term enhancement effect, others say that marijuana just makes them sleepy. In men, chronic use may decrease sex drive because marijuana may lower testosterone levels.

"Sex is hardly ever just about sex."

—SHIRLEY MACLAINE, ACTRESS

6-5b A Committed Loving Relationship

A guideline among therapists who work with couples who have sexual problems is to treat the relationship before focusing on the sexual issue. The sexual relationship is part of the larger relationship between the partners, and what happens outside the bedroom in day-to-day interaction has a tremendous influence on what happens inside the bedroom. Indeed, relationship satisfaction is associated with sexual satisfaction (Stephenson et al., 2013). The statement "I can't fight with you all day and want to have sex with you at night" illustrates the social context of the sexual experience. Partners in committed relationships reported the highest sexual satisfaction (Galinsky & Sonenstein, 2013).

In the chapter on love, we reviewed the concept of alexithymia or the inability to experience and express emotion. Scimeca et al. (2013) studied a sample of 300 university students who revealed that higher alexithymia scores were associated with lower levels of sexual satisfaction and higher levels of sexual detachment for females, and with sexual shyness and sexual nervousness for both genders. Conversely, being able to experience and express emotion has positive outcomes for one's sexual relationship.

6-5c An Equal Relationship

Sanchez et al. (2012) emphasized that traditional gender roles interfere with and inhibit a sexually fulfilling relationship. These roles dictate that the woman not initiate sex, be submissive, disregard her own pleasure, and not give accurate feedback to the male. By disavowing these roles, adopting an egalitarian perspective, and engaging in new behavior (initiating sex, taking the dominant role, insisting on her own sexual pleasure, and informing her partner about what she needs), the couple are on the path to a new and more fulfilling sexual relationship. Of course, such a change in the woman requires that the male give up that he must always initiate sex and belief in the double standard (e.g., the belief that women who love sex are sluts), and delight in not having to drive the sexual bus all the time.

6-5d Condom Assertiveness

Wright et al. (2012) reviewed the concept of **condom assertiveness**—the unambiguous messaging that sex without a condom is unacceptable—and identified the characteristics of undergraduate women who are more likely to insist on condom use. Compared with less condom-assertive females, more condom-assertive females

condom assertiveness the unambiguous messaging that sex without a condom is unacceptable.

Good Health = Good Sex

have more faith in the effectiveness of condoms, believe more in their own condom communication skills, perceive that they are more susceptible to STIs, believe there are more relational benefits to being condom assertive, believe their peers are more condom assertive, and intend to be more condom assertive.

Condom assertiveness is important not only for sexual and anal intercourse but also for oral sex. Not to use a condom or dental dam is to increase the risk of contracting a STI. Indeed, individuals think "I am on the pill and won't get pregnant" or "No way I am getting pregnant by having oral sex" only to discover Human papillomavirus (HPV) or another STI in their mouth or throat.

Also known as sexually transmitted disease, or STD, **STI** refers to the general category of sexually transmitted infections such as chlamydia, genital herpes, gonorrhea, and syphilis. The most lethal of all STIs is that caused by the human immunodeficiency virus (HIV), which attacks the immune system and can lead to AIDS.

6-5e Open Sexual Communication (Sexual Self-Disclosure) and Feedback

Sexually fulfilled partners are comfortable expressing what they enjoy and do not enjoy in the sexual experience. Unless both partners communicate their needs, preferences, and expectations to each other, neither is ever sure what the other wants. In essence, the Golden Rule ("Do unto others as you would have them do unto you") is *not* helpful, because what you like may not be the same as what your partner wants.

Sexually fulfilled partners take the guesswork out of their relationship by communicating preferences and giving feedback. Ali (2011) compared Australian and Malaysian couples and found that the former were

sexually self-disclosing with their partners (telling the partner what he or she wanted sexually) while the latter were not. She noted that the norms of Malaysian parental socialization do not allow for talking about sex since doing so is thought to destroy childhood innocence. Even adult married couples do not discuss sex as it is not regarded as proper.

6-5f Frequent Initiation of Sexual Behavior

Simms and Byers (2013) assessed the sexual initiation behaviors of 151 individuals (33% men and 66% women) who were 18–25 years of age, had been in an exclusive, heterosexual, noncohabitating dating relationship between 3 and 18 months, had seen their dating partners at least 3–4 days per week over the previous month, and had engaged in genitally focused sexual activities. The researchers found that both men and women who reported initiating sex more frequently and who perceived their partner as initiating more frequently reported greater sexual satisfaction.

6-5g Having Realistic Expectations

To achieve sexual fulfillment, expectations must be realistic. A couple's sexual needs, preferences, and expectations may not coincide. It is unrealistic to assume that your partner will want to have sex with the same frequency and in the same way that you do on all occasions. It may also be unrealistic to expect the level of sexual interest and frequency of sexual interaction to remain consistently high in long-term relationships.

Sexual fulfillment means not asking things of the sexual relationship that it cannot deliver. Failure to develop realistic expectations will result in frustration and resentment. One's age, health (both mental and physical), sexual dysfunctions of self and partner, and previous sexual

STI sexually transmitted infection.

The wedding night is certainly a night of high sexual expectations.

compliance comprising 17% of all sexual activity recorded over a three-week period. Indeed, sexual compliance was a mechanism these individuals used in their committed relationships to resolve the issue of different levels of sexual desire that is likely to happen over time in a stable couple's relationship. Others felt guilty they did not desire sex, and still others did it because their partner provided sex when the partner was not in the mood.

There were no gender differences in differences of sexual desire and no gender difference in providing sexual compliant behavior. The majority of participants reported enjoying the sexual activity despite not wanting to engage in it at first.

6-5i Job Satisfaction

Stulhofer et al. (2013) analyzed data on job satisfaction and sexual health among a sample of over 2,000 males and found that negative mood resulting from job stress/unhappiness at work was associated with sexual difficulties. Having low job stress, high income, emotional intimacy with one's partner, and having children were associated with sexual health.

6-5j Avoiding Spectatoring

One of the obstacles to sexual functioning is **spectatoring**, which involves mentally observing your sexual performance and that of your partner. When the researchers in one extensive study observed how individuals actually behave during sexual intercourse, they reported a tendency for sexually dysfunctional partners to act as spectators by mentally observing their own and their partners' sexual performance. For example, the man would focus on whether he was having an erection, how complete it was, and whether it would last. He might also watch to see whether his partner was having an orgasm (Masters & Johnson, 1970). Just focusing on one's own body can have an effect. Van den Brink et al. (2013) confirmed that body image self-consciousness was negatively associated with sexual functioning and frequency of sexual activity with a partner. Spectatoring, as Masters and Johnson conceived it, interferes with each partner's sexual enjoyment because it creates anxiety about performance, and anxiety blocks performance. A man who worries about getting an erection reduces his chance of doing so. A woman

experiences will have an effect on one's sexuality and one's sexual relationship and sexual fulfillment (McCabe & Goldhammer, 2012).

Each partner brings to a sexual encounter, sometimes unconsciously, a motive (pleasure, reconciliation, procreation, duty), a psychological state (love, hostility, boredom, excitement), and a physical state (tense, exhausted, relaxed, turned on). The combination of these factors will change from one encounter to another. Tonight one partner may feel aroused and loving and seek pleasure, but the other partner may feel exhausted and hostile and have sex only out of a sense of duty. Tomorrow night, both partners may feel relaxed and have sex as a means of expressing their love for each other.

6-5h Sexual Compliance

Given that partners may differ in sexual interest and desire, Vannier and O'Sullivan (2010) identified the concept of **sexual compliance** whereby an individual willingly agrees to participate in sexual behavior without having the desire to do so. The researchers studied 164 heterosexual young (18–24) adult couples in committed relationships to assess the level of sexual compliance. Almost half (46%) of the respondents reported at least one occasion of sexual compliance with sexual

sexual compliance an individual willingly agrees to participate in sexual behavior without having the desire to do so.

spectatoring mentally observing one's own and one's partner's sexual performance.

who is anxious about achieving an orgasm probably will not. Montesi et al. (2013) confirmed the negative effect of anxiety on a couple's sexual relationship. The desirable alternative to spectatoring is to relax, focus on and enjoy your own pleasure, and permit yourself to be sexually responsive.

6-5k Female Vibrator Use, Orgasm, and Partner Comfort

It is commonly known that vibrators (also known as sex toys and novelties) are beneficial for increasing the probability of orgasmic behavior in women. During intercourse women typically report experiencing a climax 30% of the time; vibrator use increases orgasmic reports to over 90%. Herbenick et al. (2010) studied women's use of vibrators within sexual partnerships. They analyzed data from 2,056 women aged 18–60 years in the United States. Partnered vibrator use was common among heterosexual-, lesbian-, and bisexual-identified women. Most vibrator users indicated comfort using them with a partner and related using them to positive sexual function. In addition, partner knowledge and perceived liking of vibrator use was a significant predictor of sexual satisfaction for heterosexual women.

That men are accepting of using a vibrator with their female partners was confirmed by Reece et al. (2010), who surveyed a nationally representative sample of heterosexual men in the United States. Forty-three percent reported having used a vibrator. Of those who had done so, most vibrator use had occurred within the context of sexual interaction with

Eillen/Shutterstock.com

a female partner. Indeed, 94% of male vibrator users reported that they had used a vibrator during sexual play with a partner, and 82% reported that they had used a vibrator during sexual intercourse. In another study which used a national sample, women and men held positive attitudes about female vibrator use (Herbenick et al., 2011).These data support recommendations of therapists and educators who often suggest the incorporation of vibrators into partnered relationships.

6-6 Trends in Sexuality in Relationships

The future of sexual relationships will involve continued individualism as the driving force in sexual relationships. Numerous casual partners (hooking up) with predictable negative outcomes—higher frequencies of STIs, unexpected pregnancies, sexual regret, and relationships going nowhere—will continue to characterize individuals in late adolescence and early twenties. As these persons reach their late twenties, the goal of sexuality begins to transition to seeking a partner not just to hook up and have fun with but to settle down with. This new goal is accompanied by new sexual behaviors such as delayed first intercourse in the new relationship, exclusivity, and movement toward marriage. The monogamous move toward the marriage context creates a transitioning of sexual values from hedonism to relativism to absolutism, where strict morality rules become operative in the relationship (expected fidelity).

STUDY TOOLS ⮕ **6**

Ready to study? In the book, you can:

- ⮑ Rip out the chapter review card at the back of the book for a handy summary of the chapter and key terms.

- ⮑ Assess how conservative or liberal you are in your attitudes toward sex with the Self-Assessment card at the back of the book.

Online at CENGAGEBRAIN.COM you can:

- ⮑ Prepare for tests with quizzes.

- ⮑ Review the key terms with Flash Cards.

- ⮑ Play games to master concepts.

GLBTQ Relationships

"The **only queer** people are those who don't love anybody."

—RITA MAE BROWN, AMERICAN WRITER

Featured on the cover of *Time Magazine* in 2014, Jason Collins as a center for the Brooklyn Nets is the first openly gay National Basketball Association player. His coming out reflects the increased visibility of gay individuals propelled by a revolution of cultural change—the Supreme Court legalizing same-sex marriage, more than 30 states legalizing same-sex marriage, mainstream television programs featuring gay characters, and the Boy Scouts of America lifting their ban on admitting gays into their ranks.

In this chapter, we discuss gay, lesbian, bisexual, and transgender individuals and couples as well as those who are fluid in their thinking and behavior leading to a questioning of one's sexual orientation and sexuality. We begin by defining some terms.

sexual orientation (sexual identity) a classification of individuals as heterosexual, bisexual, or homosexual, based on their emotional, cognitive, and sexual attractions and self-identity.

heterosexuality the predominance of emotional and sexual attraction to individuals of the other sex.

7-1 Language and Identification

Sexual orientation (also known as **sexual identity**) is a classification of individuals as heterosexual, bisexual, or gay, based on their emotional, cognitive, and sexual attractions and self-identity. **Heterosexuality** refers to the predominance of cognitive, emotional, and sexual attraction to individuals of the other sex. **Homosexuality** refers to the predominance of cognitive, emotional, and sexual attraction to individuals of the same sex, and **bisexuality** is cognitive, emotional, and sexual attraction to members of both sexes. The term **lesbian** refers to women who prefer same-sex partners; **gay** can refer to either women or men who prefer same-sex partners. Of U.S. adults, about 1% of females self-identify as lesbian, 2% of males self-identify as gay, and 1.5% of adults self-identify as bisexual. Hence, about 3.5% or about ten million individuals in the United States are LGB (Mock & Eibach, 2012).

The word **transgender** is a generic term for a person of one biological sex who displays characteristics of the other sex. Kuper et al. (2012) identified 292 transgendered individuals online. Most self-identified as gender queer (their gender identity was neither male nor female) and pansexual/queer (they were attracted to men,

homosexuality
predominance of emotional and sexual attraction to individuals of the same sex.

bisexuality
emotional and sexual attraction to members of both sexes.

lesbian homosexual woman.

gay homosexual women or men.

women, bisexuals) as their sexual orientation. An estimated 0.3% adults or about 700,000 individuals in the United States self-identify as transgender (Gates, 2011).

Transsexuals are individuals with the biological and anatomical sex of one gender (e.g., male) but the self-concept of the other sex (e.g., female). In the Kuper et al. (2012) sample of 292 transgendered individuals, most did not desire to (or were unsure of their desire to) take hormones or undergo sexual reassignment surgery.

"I am a woman trapped in a man's body" reflects the feelings of the male-to-female transsexual (MtF), who may take hormones to develop breasts and reduce facial hair and may have surgery to artificially construct a vagina. Such a person lives full time as a woman. The female-to-male transsexual (FtM) is one who is a biological and anatomical female but feels "I am a man trapped in a woman's body." This person may take male hormones to grow facial hair and deepen her voice and may have surgery to create an artificial penis. This person lives full time as a man. Individuals need not take hormones or have surgery to be regarded as transsexuals. The distinguishing variable is living full time in the role of the other biological sex. A man or woman who presents full time as the other gender is a transsexual by definition.

Johnson et al. (2014) noted the complexity of gender, sexuality, and sexual orientation issues as experienced by transgender, queer, and questioning individuals. One of the participants in their study who identified themselves as TQQ, explained:

I would consider myself to be bi-gendered or gender fluid. Which is probably like the most complicated thing or decision that I have ever made . . . because there aren't very many people that understand it. Being bi-gendered makes a lot more sense for me just because like my sexuality in general is just really fluid and it's really hard to identify myself in one particular box for very long at all 'cause it's always changing.

Cross-dresser is a broad term for individuals who may dress or present themselves in the gender of the other sex. Some cross-dressers are heterosexual adult males who enjoy dressing and presenting themselves as women. Cross-dressers may also be women who dress as men and present themselves as men. Cross-dressers may be heterosexual, homosexual, or bisexual.

The term **queer** is typically used by a male (but it could be used by a female) as a self-identifier to indicate that the person has a sexual orientation other than heterosexual (**genderqueer** means that the person does not identify as either male or female since he or she does not feel sufficiently like one or the other). Traditionally, the term *queer* was used to denote a gay person, and the connotation was negative. More recently, individuals have begun using the term *queer* with pride, much the same way African Americans called themselves Black during the 1960s civil rights era as part of building ethnic pride and identity. Hence, LGBT people took the term *queer*, which was used to demean them, and started to use it with pride. The term also has shock value, which some people who identify strongly with being queer seem to savor when they introduce themselves as being queer.

Queer theory refers to a movement or theory dating from the early 1990s. Queer theorists advocate for less labeling of sexual orientation and a stronger "anyone can be anything he or she wants" attitude. The theory is that society should support a more fluid range of sexual orientations and that individuals can move through the range as they become more self-aware. Queer theory may be seen as a very specific subset of gender and human sexuality studies.

7-1a Kinsey Scale

Early research on sexual behavior by Kinsey and his colleagues (1948; 1953) found that, although 37% of men and 13% of women had had at least one same-sex sexual experience since adolescence, few of the individuals reported exclusive homosexual behavior. Kinsey suggested that heterosexuality and homosexuality represent two ends of a sexual-orientation continuum and that most individuals are neither entirely homosexual nor entirely heterosexual, but fall somewhere along this continuum. The Heterosexual–Homosexual Rating Scale that Kinsey et al. (1953) developed allows individuals to identify their sexual orientation

transgender individuals who express their masculinity and femininity in nontraditional ways consistent with their biological sex.

transsexual individual with the biological and anatomical sex of one gender but the self-concept of the other sex.

cross-dresser broad term for individuals who may dress or present themselves in the gender of the other sex.

queer a self-identifier to indicate that the person has a sexual orientation other than heterosexual.

genderqueer the person does not identify as either male or female since he or she does not feel sufficiently like one or the other.

on a continuum. Very few individuals are exclusively a 0 (exclusively heterosexual) or 6 (exclusively homosexual), prompting Kinsey to believe that most individuals are bisexual.

7-1b Complications in Identifying Sexual Orientation

Lyons et al. (2014) investigated the accuracy of one's "gaydar" or the ability to identify sexual orientation by looking at a person. In a study of heterosexual (N = 80) and homosexual (N = 71) women who rated the faces of heterosexual/homosexual men and women, detection accuracy was better than chance but male targets were more likely to be falsely labeled as homosexual than were female targets.

Not all people who are sexually attracted to or have had sexual relations with individuals of the same sex view themselves as homosexual or bisexual. For example, Priebe and Svedin (2013) surveyed 3,432 Swedish high school seniors who completed an anonymous survey about sexuality. Prevalence rates of sexual minority orientation varied between 4.3% for sexual behavior (males 2.9%, females 5.6%) and 29.4% for emotional or sexual attraction (males 17.7%, females 39.5%). Hence, five times as many reported same-sex emotional/sexual attraction as actually engaged in same-sex sexual behavior.

Some researchers have also emphasized that sexual orientation of women is sometimes fluid and noted that social context impacts self-identify development as a lesbian. Davis-Delano (2014) interviewed 56 women to explore the range of activities which influenced the development of their same-sex attractions and relationships. These included activities described as involving lesbians, were primarily composed of women, affirmed women, facilitated bonding, featured a climate of acceptance of lesbians/gays/bisexuals, and did not emphasize heteronormativity.

7-2 Sexual Orientation

Same-sex behavior has existed throughout human history. Much of the biomedical and psychological research on sexual orientation attempts to identify one or more "causes" of sexual-orientation diversity. The driving question behind this research is this: "Is sexual orientation inborn or is it learned or acquired

Most gay males believe that their sexual orientation is a result of biological wiring.

sam100/Shutte-stock.com

from environmental influences?" Although a number of factors have been correlated with sexual orientation, including genetics, gender role behavior in childhood, fraternal birth order, and child sex abuse (Roberts et al., 2013), no single theory can explain diversity in sexual orientation.

7-2a Beliefs about What "Causes" Homosexuality

Aside from what "causes" homosexuality, most gay people believe that homosexuality is an inherited, inborn trait. In a national study of homosexual men, 90% reported that they believed that they were born with their homosexual orientation; only 4% believed that environmental factors were the sole cause (Lever, 1994). Overby (2014) analyzed Internet data of over 20,000 respondents (primarily heterosexual) and found that roughly half (52%) thought of homosexuality as being based primarily in "biological make-up" compared to 32% who saw sexual orientation as more of a lifestyle choice.

The terms *sexual preference* and *sexual orientation* are often used interchangeably. However, those who believe that sexual orientation is inborn more often use the term *sexual orientation*, and those who think that individuals choose their sexual orientation tend to use the term *sexual preference*. The term *sexual identity* is, increasingly, being used since it connotes more about the whole person than does sexuality.

7-2b Can Homosexuals Change Their Sexual Orientation?

Individuals who believe that homosexual people choose their sexual orientation tend to think that homosexuals can and should change their sexual orientation. While about 2% of U.S. adults report that there has been a change in their sexual orientation (Mock & Eibach, 2012), various forms of **conversion therapy (reparative therapy)** are focused on changing homosexuals' sexual orientation. Some religious organizations sponsor "ex-gay ministries," which claim to "cure" homosexuals and transform them into heterosexuals by encouraging them to ask for "forgiveness for their sinful lifestyle" through prayer and other forms of "therapy." Serovich et al. (2008) reviewed 28 empirically based, peer-reviewed articles and found them methodologically problematic, which threatens the validity of interpreting available data on this topic.

While most researchers confirm that changing one's sexual orientation is infrequent, some research brings into the question that reparative therapies "never work." Karten and Wade (2010) reported that the majority of 117 men who were dissatisfied with their sexual orientation and who sought sexual orientation change efforts (SOCE) were able to reduce their homosexual feelings and behaviors and increase their heterosexual feelings and behaviors. The primary motivations for their seeking change were religion ("homosexuality is wrong") and emotional dissatisfaction with the homosexual lifestyle. Being married, feeling disconnected from other men prior to seeking help, and feeling able to express nonsexual affection toward other men were the factors predictive of greatest change. Developing nonsexual relationships with same-sex peers was also identified as helpful in change. Jones and Yarhouse (2011) also provided data which suggest that religiously mediated sexual orientation change is possible. Better understanding the causes of homosexuality and one's homosexuality and one's emotional needs and issues as well as developing nonsexual relationships with same-sex peers were also identified as most helpful in change.

conversion therapy (reparative therapy) therapy designed to change a person's homosexual orientation to a heterosexual orientation.

heterosexism the denigration and stigmatization of any behavior, person, or relationship that is not heterosexual.

7-3 Heterosexism, Homonegativity, Etc.

The United States, along with many other countries throughout the world, is predominantly heterosexist. **Heterosexism** refers to "the institutional and societal reinforcement of heterosexuality as the privileged and powerful norm." Heterosexism is based on the belief that heterosexuality is superior to homosexuality. Of 4,711 undergraduates, 25% agreed (men more than women) with the statement "It is better to be heterosexual than homosexual" (Hall & Knox, 2015). Heterosexism results in prejudice and discrimination against homosexual and bisexual people. The word *prejudice* refers to negative attitudes, whereas *discrimination* refers to behavior that denies equality of treatment for individuals or groups.

7-3a Attitudes Toward Homosexuality: Homonegativity and Homophobia

Adolfsen et al. (2010) noted that there are five dimensions of attitudes about homosexuality:

1. **General attitude.** Is homosexuality considered to be normal or abnormal? Do people think that homosexuals should be allowed to live their lives just as freely as heterosexuals?

2. **Equal rights.** Should homosexuals be granted the same rights as heterosexuals in regard to marriage and adoption?

3. **Close quarters.** Feelings in regard to having a gay neighbor or a lesbian colleague.

NEW AMERICAN FAMILY? DO GAYS CONSTITUTE A "REAL" FAMILY?

To what degree are gay individuals, couples , and those with children regarded by mainstream adults in the United States as having "real" families? The Pew Research Center surveyed attitudes toward gay and lesbian couples raising one or more children via landline and cell phone interviews with 2,691 adults (Becker & Todd, 2013).

Almost two-thirds of the respondents (63%) agreed that a gay or lesbian couple living together raising one or more children constitute a family. In contrast, other household arrangements with children received broader public support: (1) an unmarried man and woman who live together with one or more children (80% said this is a family), (2) a single parent living with one or more children (86% said this is a family), and (3) a husband and wife with one or more children (99% say this is a family). And only 45% of respondents indicate that they define "a gay or lesbian couple living together with no children" as a family. Hence, the presence of children in households led by gay and lesbian couples has an influence on attitudes.

4. **Public display.** Reactions to a gay couple kissing in public.

5. **Modern homonegativity.** Feeling that homosexuality is accepted in society and that all kinds of special attention are unnecessary.

The term **homophobia** is commonly used to refer to negative attitudes and emotions toward homosexuality and those who engage in homosexual behavior. Homophobia is not necessarily a clinical phobia (i.e., one involving a compelling desire to avoid the feared object despite recognizing that the fear is unreasonable). Other terms that refer to negative attitudes and emotions toward homosexuality include **homonegativity** (attaching negative connotations to being gay) and antigay bias.

Homophobia and homonegativity are sometimes expressed as **hate crimes** or violence against homosexuals. Such crimes include verbal harassment (the most frequent form of hate crime experienced by victims), vandalism, sexual assault and rape, physical assault, and murder. Hate crimes also target transsexuals (more so than gay individuals).

The Sex Information and Education Council of the United States (SIECUS) notes that "individuals have the right to accept, acknowledge, and live in accordance with their sexual orientation, be they bisexual, heterosexual, gay or lesbian. The legal system should guarantee the civil rights and protection of all people, regardless of sexual orientation" (SIECUS, 2014).

Characteristics associated with positive attitudes toward homosexuals and gay rights include younger age, advanced education, no religious affiliation, liberal political party affiliation, and personal contact with homosexual individuals (Lee & Hicks, 2011). Women also tend to have more positive attitudes toward homosexuality (Wright & Bae, 2013).

Negative social meanings associated with homosexuality can affect the self-concepts of LGBT individuals. **Internalized homophobia** is a sense of personal failure and self-hatred among lesbians and gay men resulting from social rejection and the stigmatization of being gay and has been linked to increased risk for depression, substance abuse and addiction, anxiety, and suicidal thoughts (Oswalt & Wyatt, 2011; Ricks, 2012; Rubinstein, 2010). Being gay and growing up in a religious context that is antigay may be particularly difficult (Lalicha & McLaren, 2010). Todd et al. (2013) conducted focus groups of LGBTQ individuals who confirmed that religion had a negative impact on them. One group member said, "Religion made me feel that something was wrong with me . . . it made me feel worse about myself and become depressed." In contrast, McLaren et al. (2013) confirmed in a study of lesbian and gay men that involvement in the gay and lesbian community was related to fewer depressive symptoms.

homophobia negative (almost phobic) attitudes toward homosexuality.

homonegativity a construct that refers to antigay responses such as negative feelings (fear, disgust, anger), thoughts ("homosexuals are HIV carriers"), and behavior ("homosexuals deserve a beating").

hate crime instances of violence against homosexuals.

internalized homophobia a sense of personal failure and self-hatred among lesbians and gay men resulting from the acceptance of negative social attitudes and feelings toward homosexuals.

7-3b Biphobia and Transphobia

Just as the term *homophobia* is used to refer to negative attitudes toward homosexuality, gay men, and lesbians, **biphobia** (also referred to as **binegativity**) refers to a parallel set of negative attitudes toward bisexuality/those identified as bisexual. Just as bisexuals are often rejected by heterosexuals (men are more rejecting than women), they are also rejected by many homosexual individuals (Yost & Thomas, 2012). Thus, bisexuals experience "double discrimination."

Lesbians are a major source of negative views toward bisexuals. In a study of 346 self-identified lesbians, Rust (1993) found that lesbians view bisexuals as less committed to other women than are lesbians, perceive bisexuals as disloyal to lesbians, and resent bisexual women who have close relationships with the "enemy" of lesbians—men. Some negative attitudes toward bisexual individuals "are based on the belief that bisexual individuals are really lesbian or gay individuals who are in transition or in denial about their true sexual orientation" (Israel & Mohr, 2004, p. 121).

Transsexuals are targets of **transphobia**, a set of negative attitudes toward transsexuality or those who self-identify as transsexual. Singh et al. (2011) interviewed 21 transsexuals and found various strategies of coping with the discrimination/being resilient—a positive definition of self ("I am proud of who I am"), connecting with a supportive trans community, and cultivating hope for the future. In regard to the latter, one transsexual noted, "I have a right to be here and hope for a better future" (p. 24).

7-3c Effects of Antigay/Trans Bias and Discrimination on Heterosexuals

biphobia (binegativity) refers to a parallel set of negative attitudes toward bisexuality and those identified as bisexual.

transphobia negative attitudes toward transsexuality or those who self-identify as transsexual.

The antigay and heterosexist social climate of our society is often viewed in terms of how it victimizes the gay population. However, heterosexuals are also victimized by heterosexism and antigay prejudice and discrimination. Some of these effects follow:

1. **Heterosexual victims of hate crimes.** Because hate crimes are crimes of perception, victims may not be homosexual; they may just be perceived as being homosexual. The National Coalition of Anti-Violence Programs (2012) reported that, in 2011, heterosexual individuals in the United States were victims of antigay hate crimes, representing 15% of all antigay hate crime victims. An example of heterosexuals being harmed by a hate crime directed toward gays is that heterosexuals may be enjoying an evening at a gay bar with friends and the establishment is bombed by those who hate gays.

2. **Concern, fear, and grief over well-being of gay or lesbian family members and friends.** Many heterosexual family members and friends of homosexual people experience concern, fear, and grief over the mistreatment of their gay or lesbian friends and/or family members. In 2011, there were 30 murders of LGBTQH (lesbian, gay, bisexual, transgender, queer, HIV-infected individuals) (National Coalition of Anti-Violence Programs, 2012). Heterosexual parents who have a gay or lesbian teenager often worry about how the harassment, ridicule, rejection, and violence experienced at school might affect their gay or lesbian child. Will their child be traumatized/make bad grades or drop out of school to escape the harassment, violence, and alienation they endure there? Will the gay or lesbian child respond to the antigay victimization by turning to drugs or alcohol? Newcomb et al. (2014) compared national samples of gay and straight youth and found higher drug use among sexual minorities. Peter and Taylor (2014) analyzed data from of 1,205 university students. Findings showed, compared to non-LGBTQ respondents, sexual minority youth were are at a greater risk for serious suicidal ideation (8.9% versus 23%, respectively) and suicide attempt (3.5% versus 26.2%). Van Bergen et al. (2013) also found higher suicide rates among gays than straights. Higher rates of anxiety, depression, and panic attacks are associated with being gay (Oswalt & Wyatt, 2011). According to the Human Rights Campaign website (HRC.org), only 29 states have laws banning discrimination based on sexual orientation (Human Rights Campaign, 2014).

3. **Restriction of intimacy and self-expression.** Because of the antigay social climate,

heterosexual individuals, especially males, are hindered in their own self-expression and intimacy in same-sex relationships. Males must be careful about how they hug each other so as not to appear gay. Homophobic scripts also frighten youth who do not conform to gender role expectations, leading some youth to avoid activities—such as arts for boys, athletics for girls—and professions such as elementary education for males.

4. **Early sexual behavior.** Homonegativity also encourages early sexual activity among adolescent men. Adolescent male virgins are often teased by their male peers, who say things like "You mean you don't do it with girls yet? What are you, a fag or something?" Not wanting to be labeled and stigmatized as a "fag," some adolescent boys "prove" their heterosexuality by having sex with girls.

5. **School shootings.** Antigay harassment has also been a factor in many of the school shootings in recent years. For example, in May of 2013, 15-year-old Charles Andrew Williams fired more than 30 rounds in a San Diego, California, suburban high school, killing 2 and injuring 13 others. A woman who knew Williams reported that the students had teased him and called him gay.

Smit/Shutterstock.com

7-4 Coming Out

"But my point is, **homosexual is genetic**. 'Gay' is a choice. **'Gay'** is the admitting to yourself that you are not only homosexual, but you **embrace it**. You find joy in it. You're thankful for it!"

—B.G. THOMAS, *THE BOY WHO CAME IN FROM THE COLD*

"Coming out" is a major decision with which GLBT individuals struggle. Svab and Kuhar (2014) studied the coming out process of 443 gay men and lesbians. Over three-fourths (77%) came out first to their friends in comparison to 7% who first came out to their mother, 5% to their brother/sister, and 3% to their father. Seventy-four percent of the surveyed gay men and lesbians reported experiencing a positive and supportive reaction to their first coming out, 18% reported neutral, and 4% had negative reactions. However, some respondents interpreted silence and nonreaction as a form of a positive or at least a neutral reaction because they had expected a much worse response.

Parents of gay children often discover very positive outcomes from their child's coming out. Gonzalez et al. (2013) interviewed 142 parents of LGBTQ children and identified five primary themes: personal growth (open mindedness, new perspectives, awareness of discrimination, and compassion), positive emotions (pride and unconditional love), activism, social connection, and closer relationships (closer to child and family closeness).

7-4a Risks and Benefits of Coming Out

In a society where heterosexuality is expected and considered the norm, heterosexuals do not have to choose whether or not to tell others that they are heterosexual. However, decisions about **coming out**, or being open and honest about one's sexual orientation and identity (particularly to one's parents), are agonizing for LGBT individuals. As noted above, friends, mothers, siblings, and fathers, in that

coming out being open and honest about one's sexual orientation and identity.

Personal View: A Coming Out Letter to One's Parents and Parental Reactions

Courtesy of Rachel Calisto

Below is a letter one of our former students wrote to her parents disclosing her being gay.

Dear Mom and Dad:

I love you very much and would give you the world if I could. You've always given me the best. You've always been there when I needed you. The lessons of love, strength, and wisdom you've taught me are invaluable. Most importantly, you've taught me to take pride and stand up for what I believe. In the past six months I've done some very serious thinking about my goals and outlook on life. It's a tough and unyielding world that caters only to those who do for themselves. Having your help and support seems to make it easier to handle. But I've made a decision that may test your love and support. Mom and Dad, I've decided to live a gay lifestyle.

I'm sure your heads are spinning with confusion and disbelief right now but please try to let me finish. As I said before, I've given this decision much thought. First and foremost, I'd like for you to know that I'm happy. All my life as a sexual being has been spent frustrated in a role I could not fulfill. Emotionally and mentally, I'm relieved. Believe me, this was not an easy decision to make. (What made it easier was having the strength to accept myself). Who I choose to sleep with does not make me more or less of a human being. My need for love and affection is the same as anyone else's; it's just that I fulfill this need in a different way. I'm still the same person I've always been.

It's funny I say I'm still the same person, yet society seems to think I've changed. They seem to think that I don't deserve to be treated as a respectable citizen. Instead, they think I should be treated as a deranged maniac needing constant supervision to prevent me from molesting innocent children. I have the courage and strength to face up to this opposition, but I can't do it alone. Oh, what I would give to have your support!

I realize I've thrown everything at you rather quickly. Please take your time. I don't expect a response. And please don't blame yourselves, for there is no one to blame. Please remember that I'm happy and I felt the need to share my happiness with two people I love with all my heart.

Your loving daughter,
Maria

Below are examples of the reactions of parents to their child's coming out (Grafsky, 2014).

He came downstairs and said he had to talk to me. I'm like, okay, and popped off the TV. He just sat down and said that he had something to tell me and he didn't want me to worry about him. Then he told me he was bisexual. And I said, "Okay. What does that mean?" So then we talked a lot of that kind of stuff over.... I asked him pretty personal questions. (Hannah, age 55)

***

Everything that I had envisioned, you know parents have this like little ball of fantasy in their head for their children, the white picket fence, the dog, the kids, the wife. [Gay son]'s popped [snaps fingers] and it was like this void…. And it's really scary 'cause you don't know how to fill it…. I've spent twenty-some years with this whole reality for him and now there's this empty void and I don't know how to fill it. (Alice, age 42)

* * *

I was afraid. Fear, fear was my biggest thing. I was petrified. I thought somebody would hurt him, I mean, all the horrors of um, people not accepting him. Is life gonna be hard? Afraid of him being harassed at school. How do I tell people? How are people going to treat him? (Eve, age 47)

It is often thought that coming out to parents is a one-time event. But data by Denes and Afini (2014) revealed that approximately one-fourth of 106 GLBQ individuals (ages 18 to 55) reported coming out a second time. The reason for doing so was to reinforce their sexual orientation (e.g., not going through a "phase"), to clarify aspects of their identity (e.g., from "bi" to "gay"), and to share more information about their GLBQ lifestyle (e.g., their partner). The more individuals perceived their parents to react to their first coming out with denial and the lower their reported relationship satisfaction with their parents after their first coming out, the more likely they were to come out again.

order, are those to whom disclosure is first made (Svab & Kuhar, 2014).

"**Openness** may not completely disarm prejudice, but it's a **good place to start.**"

—JASON COLLINS, ATHLETE

Risks of Coming Out Whether GLBT individuals come out is influenced by how tired they have become hiding their sexual orientation, how much they feel they need to be "honest" about who they are, and their prediction of how others will respond. Some of the risks involved in coming out include disapproval and rejection by parents and other family members, harassment and discrimination at school, discrimination and harassment in the workplace, and hate crime victimization.

1. **Parental and family members' reactions.** Rothman et al. (2012) studied 177 LBG individuals who reported that two-thirds of the parents to whom they first came out responded with social and emotional support. Their research is in contrast to that of Mena and Vaccaro (2013), who interviewed 24 gay and lesbian youth about their coming out experience to their parents. All reported a less than 100% affirmative "we love you"/"being gay is irrelevant" reaction which resulted in varying degrees of sadness or depression (3 became suicidal).

Stephen Orsillo/Shutterstock.com

Svab and Kuhar (2014) identified the concept of the "transparent closet" to describe a situation in which parents are informed about a child's homosexuality but do not talk about it…a form of rejection. The "family closet" refers to the wider kinship system having knowledge of a child's homosexuality but "keeping it quiet" (a form of rejection).

Padilla et al. (2010) found that parental reaction to a son or daughter coming out had a major effect on the development of their child. Acceptance had an enormous positive effect. When GLBT individuals in their study come out to their parents, parental reactions range from "I already knew you were gay and I'm glad that you feel ready to be open with me about it" to "get out of this house, you are no longer welcome here." We know of a father who responded to the disclosure of his son, "I'd rather have a dead son than a gay son." Parental rejection of GLBT individuals is related to suicide ideation and suicide attempts (Van Bergen et al., 2013).

Because Black individuals are more likely than White individuals to view homosexual relations as "always wrong," African Americans who are gay or lesbian are more likely to face disapproval from their families than are White lesbians and gays (Glass, 2014). *The Resource Guide to Coming Out for African Americans* (2011) is a useful model. Because most parents are heavily invested in their children, they find a way not to make an issue of their son or daughter being gay. "We just don't talk about it," said one parent.

Parents and other family members can learn more about homosexuality from the local chapter of Parents, Families, and Friends of Lesbians and Gays (PFLAG) and from books and online resources, such as those found at Human Rights Campaign's National Coming Out Project (http://www.hrc.org/). Mena and Vaccaro (2013) emphasized the importance of parents educating themselves about gay/lesbian issues and to know the importance of loving and accepting their son or daughter at this most difficult time.

2. **Harassment and discrimination at school.** LGBT students are more vulnerable to being bullied, harassed, and discriminated against. The negative effects are predictable including "a wide range of health and mental health concerns, including sexual health risk, substance abuse, and suicide, compared with their heterosexual peers (Russell et al., 2011). The U.S. Department of Health (2012) published new guidelines to be sensitive to the needs of and to protect GLBT youth. GLBT individuals are often targets of discrimination and bullying. Gilla et al. (2010) found evidence that gays are subjected to negative comments and name calling in contexts such as physical education classes or physical activities where they are perceived to be small or gay. Some communities offer charter schools which are "GLBT friendly" and which promote tolerance. Google "charter schools for GLBT youth" for examples of such schools offering a safe haven from the traditional bullying. About a third of the students at the Arts and College Preparatory Academy in Ohio identify with the GLBT community. Parental acceptance of a child who is LGBTQ is also important in buffering the negative effects on a child who is bullied at school due to their sexual orientation and difference (Russell, 2013).

3. **Discrimination and harassment at the workplace.** The workplace continues to be a place where the 8 million LGBT individuals experience discrimination and harassment.

Specifically, gay men are paid less than heterosexual men, LGBT individuals feel their potential for promotion is less than the heterosexual majority, and many remain closeted for fear of retribution. There is no federal law that explicitly prohibits sexual orientation and gender identity discrimination against LGBTs (Pizer et al., 2012). McIntyre et al. (2014) noted that being a racial minority and a sexual minority is associated with more stigma and psychological distress than just being a sexual minority.

4. **Hate crime victimization.** Another risk of coming out is being victimized by antigay hate crimes against individuals or their property that are based on bias against the victim because of his/her perceived sexual orientation. Such crimes include verbal threats and intimidation, vandalism, sexual assault and rape, physical assault, and murder. "Homosexuals are far more likely than any other minority group in the United States to be victimized by violent hate crime" (Potok, 2010, p. 29).

"There are **those** from religious backgrounds **who resist and oppose LGBT equality**; some very obsessively and publicly. They make bold accusations and negative statements about gay and lesbian people, their supposed 'lifestyle' and relationships. **But when** a son, daughter, brother, sister, or close friend comes out it is no longer an 'issue,' **it becomes a person**. They **realize** everything they'd said was painfully **targeted at someone they love**. Then...**everything changes**."

—ANTHONY VENN-BROWN, *A LIFE OF UNLEARNING: ONE MAN'S JOURNEY TO FIND THE TRUTH*

Benefits of Coming Out Coming out to parents is associated with decided benefits. D'Amico and Julien (2012) compared 111 gay, lesbian, and bisexual youth who disclosed their sexual orientation to their parents with 53 who had not done so. Results showed that the former reported higher levels of acceptance from their parents, lower levels of alcohol and drug consumption, and fewer identity and adjustment problems. Similarly, Rothman et al. (2012) noted that for lesbian and bi females (not males), higher levels of illicit drug use, poorer self-reported health status, and being more depressed were associated with nondisclosure to parents.

One of the strongest contexts against coming out is that of professional football. Defensive lineman Michael Sam is an example of an openly gay male in this macho context. Since U.S culture is becoming more accepting, gays in all sports, including football, will eventually (and increasingly) come out.

The basic trajectory of reacting to a partner's homosexuality is not unlike learning of the death of one's beloved. Life as one knew it is altered. The stages of this transition are shock, disbelief, numbness, and mourning for the partner/life that was, readjustment, and moving on. The last two stages may involve staying in the relationship with the gay partner (the choice of about 15% of straight spouses) or divorcing/ending the relationship (the choice of 85%).

7-5 Mixed-Orientation Inside Relationships

Research suggests that gay and lesbian couples tend to be more similar to than different from heterosexual couples (Kurdek, 2004; 2005). However, Muraco et al. (2012) studied sexual orientation and romantic relationship quality in a sample of 15,534 adolescents and found that relationship quality was influenced by gender and age with older gay women reporting higher relationship quality. In the following section we look more closely at the similarities as well as the differences between heterosexual, gay male, and lesbian relationships in regard to relationship satisfaction, conflict and intimate partner violence, and monogamy and sexuality. We also look at relationship issues involving bisexual individuals and mixed-orientation couples.

7-5a Relationship Satisfaction

In a review of literature on relationship satisfaction and sexual orientation, Kurdek (1994) concluded, "The most striking finding regarding the factors linked to relationship satisfaction is that they seem to be the same for lesbian couples, gay couples, and heterosexual couples" (p. 251). These factors include having equal power and control, being emotionally expressive, perceiving many attractions and few alternatives to the relationship, placing a high value on attachment, and sharing decision making. In a comparison of relationship quality of cohabitants over a 10-year period involving both partners from 95 lesbian, 92 gay male, and 226 heterosexual couples living without children, and both partners from 312 heterosexual couples living with children, the researcher found that lesbian couples showed the highest levels of relationship quality averaged over all assessments (Kurdek, 2008). This finding is not surprising. Females love deeply and are more focused on developing and maintaining good relationships. Issues unique to gay couples include if, when, and how to disclose their relationships to others and how to develop healthy intimate relationships in the absence of same-sex relationship models.

Individuals who identify as bisexual have the ability to form intimate relationships with both sexes. However, research has found that the majority of bisexual women and men tend toward primary relationships

Lesbian couples report the highest relationship quality of all couples.

with the other sex (McLean, 2004). Contrary to the common myth that bisexuals are, by definition, non-monogamous, some bisexuals prefer monogamous relationships (especially in light of the widespread concern about HIV). Some gay and bisexual men have monogamous relationships in which both men have agreed that any sexual activity with casual partners must happen when both members of the couple are present and involved (e.g., three-ways or group sex) (Parsons et al., 2013).

7-5b Conflict Resolution and Intimate Partner Violence

Same-sex couples and heterosexual couples tend to differ in how they resolve conflict. Gay and lesbian partners begin their discussions more positively and maintain a more positive tone throughout the discussion than do partners in heterosexual marriages (Gottman et al., 2003). One explanation is that same-sex couples value equality more and are more likely to have equal power and status in the relationship than are heterosexual couples. In addition, same-sex partners may have a higher motivation to reduce conflict because dissolution may involve losing a tight friendship network (Van Eeden-Moorefield et al., 2011).

Some same-sex relationships involve interpersonal violence. Finneran and Stephenson (2014) analyzed national Internet data from 1,575 men who had had sex with men. Nine percent reported that they had experienced some form of physical violence from a male partner in the previous 12 months.

7-5c Sexuality

Oswalt and Wyatt (2013) reported data on 25,553 heterosexual, gay, lesbian, bisexual, and unsure college students throughout the United States and found that being uncertain about one's sexual orientation was significantly associated with having more sexual partners than gay, bisexual, and heterosexual men. When compared to heterosexuals, gay, bisexual, and bisexual women have significantly more sexual partners.

The fact that gay males have a higher number of casual sexual relationships is better explained by the fact that they are male than by the fact that they are gay. In this regard, gay and straight men have a lot in common: They both tend to have fewer barriers to engaging in casual sex than do women (heterosexual or lesbian). However, Starks and Parsons (2014) found

Securely attached males report high levels of communication with their partners and fewer outside partners.

that males who reported secure adult attachment with their partners also reported higher levels of sexual communication with their partners, higher frequencies of sex with their partners, and fewer outside sexual partners.

One way that uninvolved gay men meet partners is through the Internet. Blackwell and Dziegielewski (2012) studied men who seek men for sex on the Internet and noted that these sites promote higher-risk sexual activities. "Party and play" (PNP) is one such activity and involves using crystal methamphetamine and having unprotected anal sex.

Like many heterosexual women, most gay women value stable, monogamous relationships that are emotionally as well as sexually satisfying. Gay and heterosexual women in U.S. society are taught that sexual expression should occur in the context of emotional or romantic involvement.

Nonuse of condoms results in the high rate of human immunodeficiency virus (HIV) infection and acquired immunodeficiency syndrome (AIDS). Approximately 50,000 new cases of HIV are reported annually, and male-to-male sexual contact is the most common mode of transmission in the United States (29,700 infections annually) (Centers for Disease Control and Prevention, 2012). Women who have sex exclusively with other women have a much lower rate of HIV infection than do men (both gay and straight) and women who have sex with men. Many gay men have lost a love partner to HIV infection or AIDS; some have experienced multiple losses.

7-5d Love and Sex

Rosenberger et al. (2014) analyzed data from 24,787 gay and bisexual men who were members of online websites facilitating social or sexual interactions with other men. Over half of those who completed the questionnaire (61.4 %) reported that they did not love their sexual partner during their most recent sexual encounter (28.3 % reported being in love with their most recent sexual partner). Hence, most of these men did not require love as a context to having sex. However, they were mostly accurate in identifying their own feelings and those of their partners. Over 90% (91%) were matched by presence ("I love him/he loves me"), absence ("I don't love him/he doesn't love me"), or uncertainty ("I don't know if I do/I don't know if he does") of feelings of love with their most recent sexual partner.

7-5e Division of Labor

Sutphina (2010) studied the division of labor patterns reported by 165 respondents in same-sex relationships. Partners with greater resources (e.g., income, education) performed fewer household tasks. Satisfaction with division of labor and sense of being appreciated for one's contributions to household tasks were positively correlated with overall relationship satisfaction.

Esmaila (2010) noted that lesbians struggle to maintain a sense of fairness in regard to the division of labor. This includes invoking how heterosexual couples often slip into an unequal division of labor and how gay couples try to raise the bar on this issue.

7-5f Mixed-Orientation Relationships

Mixed-orientation couples are those in which one partner is heterosexual and the other partner is lesbian, gay, or bisexual. Reinhardt (2011) studied a sample of bisexual women in heterosexual relationships and found that these women maintained a satisfactory relationship with the male partner. One half of these women were maintaining sexual relationships with other women while in primary heterosexual relationships. Their average sexual contact with another female was 1½ times per month. The majority of the women were satisfied with their sexual experiences with their male partner, and they were having sexual intercourse, on average, 3 times per week. Moss (2012) interviewed a sample of married bisexual women in a polyamorous relationship. These women recognized the non-normativeness of their relationships but were able to negotiate their needs with a supportive partner.

Gay and bisexual men are also in heterosexual relationships. Peter Marc Jacobson is gay and was married to Fran Drescher (actress in *The Nanny*) for 21 years. They report a wonderful relationship/friendship during and after the marriage. In a study of 20 gay or bisexual men who had disclosed their sexual orientation to their wives, most of the men did not intentionally mislead or deceive their future wives with regard to their sexuality. Rather, they did not fully grasp their feelings toward men, although they had a vague sense of their same-sex attraction prior to the marriage (Pearcey, 2004). The majority of the men in this study (14 of 20) attempted to stay married after disclosure of their sexual orientation to their wives, and nearly half (9 of 20) stayed married for at least three years. The Straight Spouse Network (www.straightspouse.org) provides support to heterosexual spouses or partners, current or former, of GLBT mates.

Some males are on the down low. Men on the **down low** (nongay-identifying men who have sex with men and women meet their partners out of town, not in predictable contexts—e.g., park—or on the Internet) (Schrimshaw et al., 2013). There is considerable disruption in the marriage if their down low behavior is discovered by the straight spouse.

7-5g Transgender Relationships

While a great deal is known about transgender individuals, little is known about transgender relationships. The exception is the research by Iantiffi and Bockting (2011) who reported on an Internet study of 1,229 transgender individuals over the age of 18 living in the United States. Fifty-seven percent were transwomen (MtF); 43%, transmen (FtM). The median age was 33 and most were White (80%). The average income was $32,000. Almost 70% (69.2%)

down low nongay-identifying men who have sex with men and women meet their partners out of town, not in predictable contexts, or on the Internet.

of the transwomen were living together; 57% of the transmen. While those who were living together were in a relationship, another 15% of the transwomen were in a couple relationship, but not living together; 29% of the transmen—hence around 85% of the transwomen and transmen were involved in a relationship. In regard to sexual fidelity 65% of the transwomen were monogamous (26% were not); 70% of the transmen (21% were not). Some of the partners of the transgender individuals did not know their partner was trans—8% of transwomen and 2% of transmen. Transgender individuals live in fear they will not be accepted. Twenty-five percent of transwomen and 9% of transmen reported that they were afraid to tell their partner they were trans (which is required to a future marriage partner in Europe) (Sharpe, 2012). Having sex in the dark/giving oral sex were strategies used to avoid discovery that one's genitals do not match their presentation of self (if they had not had surgery). Indeed, the whole subject of the feminine/masculine parts of one's body is a subject to be avoided (56% of transwomen; 51% of transmen). Sexual-orientation attractions were predictable. Trans-gendered individuals who were mostly lesbian, gay, or bisexual in sexual orientation were mostly attracted to women, men, and both sexes.

© koosen/Shutterstock.com

What is it like to be a spouse of a partner who is transitioning? Chase (2011) interviewed partners of transgendered individuals. One wife commented on what it was like to watch her male to female transgendered husband transition:

> He's more girly than I am! Which [chuckles] is kind of frustrating at times. You know, he'll want to wear all these bracelets and you know jewelry and stuff. Oh, I just think sometimes he kind of goes overboard on it. And he'll always ask, I mean he asks me a lot my opinion on what he wears, you know, if I think this is going to look good or not and things like that. So, he's always asking about that, does this go with this and everything. [And]

he's probably bought more makeup through the years than I have. (p. 436).

7-6 Same-Sex Marriage

With the Supreme Court ruling in 2013 (*United States v. Windsor*), the federal government recognizes same-sex marriages (e.g., the survivor of a legal same-sex marriage can collect Social Security benefits as is true of a spouse in a traditional heterosexual marriage). No longer in effect is the **Defense of Marriage Act (DOMA)** (passed in 1996), which states that marriage is a "legal union between one man and one woman" and denies federal recognition of same-sex marriage. In 2014, the Justice Department affirmed that it will extend equal benefits to spouses in same-sex marriages. For example, spouses are not required to testify against each other in court, and spouses now have visitation rights for a spouse in prison, etc. (Doaring, 2014).

While more than 30 states have legalized same-sex marriage, it remains a contested issue in many states and political candidates are careful about their position in regard to same-sex relationships for fear of losing votes. After Obama announced his support for gay marriage, a quarter (25%) of Americans reported that they felt less favorable toward him because of his approval; 19% felt more favorable toward him (Pew Research Center, 2012).

Defense of Marriage Act (DOMA) legislation passed by Congress denying federal recognition of homosexual marriage and allowing states to ignore same-sex marriages licensed by other states.

> " Gay marriage is like a bra and a bikini. *One is allowed in public and the other isn't even though they're exactly the same.* "
>
> —ANONYMOUS

7-6a Arguments in Favor of Same-Sex Marriage

A major argument for same-sex marriage is that it would promote relationship stability among gay and lesbian couples. "To the extent that marriage provides status, institutional support, and legitimacy, gay and lesbian couples, if allowed to marry, would likely experience greater relationship stability" (Amato, 2004, p. 963). Indeed, same-sex relationships, like cohabitation relationships, end at a higher rate than marriage relationships (Wagner, 2006). This higher rate is attributed to the lack of institutional support. Supreme Court recognition of same-sex marriage will likely result in an increase of more stable unions.

tony4urban/Shutterstock.com

Advocates of same-sex marriage argue that banning or refusing to recognize same-sex marriages granted in some states is a violation of civil rights that denies same-sex couples the many legal and financial benefits that are granted to heterosexual married couples. Rights and benefits accorded to married people include the following:

- The right to inherit from a spouse who dies without a will.

- The benefit of not paying inheritance taxes upon the death of a spouse.

- The right to make crucial medical decisions for a spouse and to take care of a seriously ill spouse or a parent of a spouse under current provisions in the federal Family and Medical Leave Act.

- The right to collect Social Security survivor benefits.

- The right to receive health insurance coverage under a spouse's insurance plan.

Other rights bestowed on married (or once-married) partners include assumption of a spouse's pension, bereavement leave, burial determination, domestic violence protection, reduced-rate memberships, divorce protections (such as equitable division of assets and visitation of partner's children), automatic housing lease transfer, and immunity from testifying against a spouse. All of these advantages are now available to same-sex married couples because of the Supreme Court decision.

Positive outcomes for being married as a gay couple have been documented. Ducharme and Kollar (2012) evaluated a sample of 225 lesbian married couples in Massachusetts who reported physical, psychological, and financial well-being in their relationships. The researchers noted that these data support the finding in the heterosexual marriage literature that healthy marriage is associated with distinct well-being benefits for lesbian couples. Wright et al. (2013) found that same-sex married lesbian, gay, and bisexual persons were significantly less distressed than lesbian, gay, and bisexual persons not in a legally recognized relationship. However, Ocobock (2013) studied 32 gay men who were married to same-sex partners and found that while the legitimacy of marriage often led to positive family outcomes, it also commonly had negative consequences, including new and renewed experiences of family rejection. Hence, negative attitudes toward gay marriage may continue even when the marriage is legal.

Children of same-sex parents also benefit from legal recognition of same-sex marriage. Children living in gay- and lesbian-headed households are denied a range of securities that protect children of heterosexual married couples. These include the right to get health insurance coverage and Social Security survivor benefits from a nonbiological parent (Chonody et al., 2012).

Finally, there are religion-based arguments in support of same-sex marriage. Although many religious leaders teach that homosexuality is sinful and prohibited by God, some religious groups, such as the Quakers and the United Church of Christ (UCC), accept homosexuality, and other groups have made reforms toward increased acceptance of lesbians and gays.

7-6b Arguments against Same-Sex Marriage

Although advocates of same-sex marriage argue that they will not be regarded as legitimate families by the larger society so long as same-sex couples cannot be legally married, opponents do not want to legitimize same-sex couples and families. Opponents of same-sex marriage who view homosexuality as unnatural, sick, and/or immoral do not want their children to view homosexuality as socially acceptable.

Opponents of gay marriage also suggest that gay marriage leads to declining marriage rates, increased divorce rates, and increased nonmarital births. However, data in Scandinavia reflect that these trends were occurring 10 years before Scandinavia adopted registered same-sex partnership laws, liberalized alternatives to marriage (such as cohabitation), and expanded exit options (such as no-fault divorce; Pinello, 2008).

Opponents of same-sex marriage commonly argue that such marriages would subvert the stability and integrity of the heterosexual family. However, Sullivan (1997) suggests that homosexuals are already part of heterosexual families:

> [Homosexuals] are sons and daughters, brothers and sisters, even mothers and fathers, of heterosexuals. The distinction between "families" and "homosexuals" is, to begin with, empirically false; and the stability of existing families is closely linked to how homosexuals are treated within them. (*p. 147*)

Many opponents of same-sex marriage base their opposition on their religious views (viewing homosexuality as immoral). However, churches have the right to deny marriage for gay people in their congregations. Legal marriage is a contract between the spouses and the state; marriage is a civil option that does not require religious sanctioning.

7-7 Parenting Issues

About 90,000 same-sex households (15% of the 800,000 same-sex households) have children. This is a low estimate of children who have gay or lesbian parents, as it does not count children in same-sex households who did not identify their relationship in the census, those headed by gay or lesbian single parents, or those whose gay parent does not have physical custody but is still actively involved in the child's life. Regardless of the number,

hetero-gay family a heterosexual mother and a gay father conceive and raise a child together but reside separately.

Becker and Todd (2013) noted that the children of gay and lesbian couples tend to face more challenges than children from other types of family arrangements. Anti-gay views concerning gay parenting include the belief that homosexual individuals are unfit to be parents and that children of lesbians and gays will not develop normally and/or that they will become homosexual.

7-7a Gay Families: Lesbian Mothers and Gay Fathers

Many gay and lesbian individuals and couples have children from prior heterosexual relationships or marriages. Some of these individuals married as a "cover" for their homosexuality; others discovered their interest in same-sex relationships after they married. Children with mixed-orientation parents may be raised by a gay or lesbian parent, a gay or lesbian stepparent, a heterosexual parent, and a heterosexual stepparent. A gay or lesbian individual or couple may have children through the use of assisted reproductive technology, including in vitro fertilization, surrogate mothers, and donor insemination. Other gay couples adopt or become foster parents. Less commonly, some gay fathers are part of an emergent family form known as the **hetero-gay family**. In a hetero-gay family, a heterosexual mother and a gay father conceive and raise a child together but reside separately.

While the struggle for acceptance of gay parenting has been difficult, data on lesbian mothers has revealed that they tend to have high levels of shared decision making, parenting, and family work, all reflecting an egalitarian ideology. When compared to heterosexual mothers, gay mothers tended to spend more time with

Children of lesbian parents live in "real" families.

their children and expressed more warmth and affection (Biblarz & Stacy, 2010). Data on gay fathers has also revealed that they are more likely to coparent equally and compatibly than fathers in heterosexual relationships. Gay fathers are also less likely to spank their children as a method of discipline (Biblarz & Savci, 2010). Bergman et al. (2010) also found that gay men who became parents had heightened self-esteem as a result of becoming parents and raising children. Hence, the act of becoming a father has a very positive outcome on gay men's sense of self-worth. Part of this effect may be due to the fact that some gay men think that gay fatherhood is an unattainable role. While both gay females and gay males report increases in individual happiness during the first year of having a baby/adopting a child, relationship happiness decreases (Goldberg et al., 2010). This drop in relationship satisfaction after a child arrives in the gay relationship is the same as what happens in heterosexual relationships.

7-7b Bisexual Parents

Power et al. (2013) surveyed 48 bisexuals who were parenting inside a variety of family structures (heterosexual relationships, same-sex relationships, co-parenting with ex-partners or nonpartners, sole parenting) and revealed issues relevant to all parents—discipline, combining work/parenting, etc. The dimension of bisexuality rarely surfaced. When it did, it was in the form of being closeted to help prevent their child being subjected to prejudice and dealing with prejudiced ex-partners, in-laws, and grandparents.

> **"Exposing 'innocent' children to gay relationships won't make them gay.... I was exposed to straight relationships, and yet, I'm still gay!"**
>
> —ANONYMOUS

7-7c Development and Well-Being of Children with Gay or Lesbian Parents

A growing body of research on gay and lesbian parenting supports the conclusion that children of gay and lesbian parents are just as likely to flourish as are children of heterosexual parents.

There is the belief that for children to develop totally, they need both a mother and father together. Biblarz and Stacey (2010) compared children from two-parent families with same or different sex coparents and single-mother with single-father families and found that the strengths typically associated with married mother–father families appeared to the same extent in families with two mothers and potentially in those with two fathers. Hence, children seem to benefit when there are two parents in the household (rather than a single mom or dad) but the gender of the parents is irrelevant—the structure can be a woman and a man, two women, or two men. Crowl et al. (2008) also reviewed 19 studies on the developmental outcomes and quality of parent–child relationships among children raised by gay and lesbian parents. Results confirmed previous studies that children raised by same-sex parents fare equally well to children raised by heterosexual parents. Regardless of family type, adolescents were more likely to show positive adjustment when they perceived more caring from adults and when parents described having close relationships with them. Thus, the qualities of adolescent-parent relationships rather than the sexual orientation of the parents were significantly associated with adolescent adjustment.

What do children who are reared in a home with lesbian parents say? Gartrell et al. (2012) provided data from 78 adolescents who were being reared in lesbian families. Results revealed that these 17-year-old adolescents were academically successful in supportive school environments, had active social networks, and close family bonds. Almost all considered their mothers good role models and reported their overall well-being an average of 8.1 on a 10-point scale.

In a series of interviews, daughters and sons (32) of lesbian parents reported that, in general, the reactions from their peers was largely positive. Most reported experiencing stigma at some point in their lives due to their mothers' lesbianism. A range of coping mechanisms were employed, including confrontation, secrecy, and seeking outside support (counseling) (Leddy et al., 2012). Gay parents may be particularly sensitive to potential stigmatization and seek gay-friendly neighborhoods to rear their children (Goldberg et al., 2012).

Some children of lesbian parents report appreciation for being connected to a unique group: "I've always been grateful for the queer community I was raised in, which provided me with lots of adults to be close to and lots of different models for how to be a person" (Leddy et al., 2012, p. 247). Bos and Sandfort (2010) studied children in lesbian and heterosexual families and found that children in lesbian families felt less parental pressure to conform to gender stereotypes, were less likely to experience their own gender as superior, and were more likely to be uncertain about future heterosexual romantic involvement. The last finding does not suggest that the children were more likely to feel that they were gay, but that they were living in a social context which allowed for more than the heterosexual adult model. One of our former female students who was reared by two gay moms (in a same-sex relationship with each other) said, "My experience was variable. It was a challenge since there is a taboo against having gay parents. But other times family life felt normal and I certainly feel that I was taken care of and loved just as much as if I had had heterosexual parents."

7-7d Development and Well-Being of Children in Transgender Families

Research on children in transgender families is virtually nonexistent (Biblarz & Savci, 2010). What is known focuses on the stress transgender children experience as they try to "fit in" to please parents and society at the expense of personal depression and loss of well-being. Meanwhile, children of transgender parents struggle with new definitions of who their parents are and how this affects them. Male children of male-to-female transsexuals may have a particularly difficult time adapting.

7-7e Discrimination in Child Custody, Visitation, Adoption, and Foster Care

A student in one of our classes reported that, after she divorced her husband, she became involved in a lesbian relationship. She explained that she would like to be open about her relationship to her family and friends, but she was afraid that if her ex-husband found out that she was in a lesbian relationship, he might take her to court and try to get custody of their children. Although several respected national organizations including the American Academy of Pediatrics, the Child Welfare League of America, the American Bar Association, the American Medical Association, the American Psychological Association, the American Psychiatric Association, and the National Association of Social Workers have gone on record in support of treating gays and lesbians without prejudice in parenting and adoption decisions, lesbian and gay parents are often discriminated against in child custody, visitation, adoption, and foster care.

Lehman (2010) identified the three issues the courts have used to deny homosexuals custody of their children—per se, presumption, and nexus. Per se takes the position that if the parent is homosexual, he or she is unfit to parent. The presumption position is similar to the per se position except the parent has the right to prove otherwise. The nexus approach focuses on the connection between being homosexual resulting in negative outcomes for the child. These outcomes include that the child will be ridiculed, that the parent will turn the child gay, and that homosexuality is immoral. Each of these positions is rarely taken seriously by the courts. Most adoptions by gay people are second-parent adoptions. A **second-parent adoption** (also called **coparent adoption**) is a legal procedure that allows individuals to adopt their partner's biological or adoptive child without terminating the first parent's legal status as parent. Second-parent adoption gives children in same-sex families the security of having two legal parents. Second-parent adoption potentially benefits a child by:

- Placing legal responsibility on the parent to support the child.

- Allowing the child to live with the legal parent in the event that the biological (or original adoptive) parent dies or becomes incapacitated.

- Enabling the child to inherit and receive Social Security benefits from the legal parent.

- Enabling the child to receive health insurance benefits from the parent's employer.

- Giving the legal parent standing to petition for custody or visitation in the event that the parents break up. (Clunis & Green, 2003)

Second-parent adoption is not possible when a parent in a same-sex relationship has a child from a previous heterosexual marriage or relationship, unless the former spouse or partner is willing to give up parental rights.

second-parent adoption (coparent adoption) a legal procedure that allows individuals to adopt their partner's biological or adoptive child without terminating the first parent's legal status as parent.

7-7f When LGB Relationships End: Reaction of the Children

How do children react to the ending of the relationship of their LGB parents? Goldberg and Allen (2013) interviewed 20 young adults who experienced their LGB parents' relationship dissolution and/or the formation of a new LGB stepfamily. Almost all families negotiated relational transitions informally and without legal intervention; the relationship with one's biological mother was the strongest tie from breakup to repartnering and stepfamily formation; and geographic distance from their nonbiological parents created hardships in interpersonal closeness. Overall, "young people perceived their families as strong and competent in handling familial transitions" (p. 529).

7-8 Trends

Moral acceptance and social tolerance/acceptance of gays, lesbians, bisexuals, and transsexuals as individuals, couples, and parents will come slowly. Heterosexism, homonegativity, biphobia, and transphobia are entrenched in American society. However, as more states recognize same-sex marriage, more GLBT individuals will come out, their presence will become more evident, and tolerance, acceptance, and support will increase…slowly. The media will also be influential as GLBT individuals, couples, and families will increase in visibility.

STUDY TOOLS 7

Ready to study? In the book, you can:

- ⊃ Rip out the chapter review card at the back of the book for a handy summary of the chapter and key terms.

- ⊃ Learn more about your attitude toward homosexuality with the Self-Assessment card at the back of the book.

Online at CENGAGEBRAIN.COM you can:

- ⊃ Prepare for tests with quizzes.

- ⊃ Review the key terms with Flash Cards.

- ⊃ Play games to master concepts.

Marriage Relationships

"**Love** is an **ideal** thing, **marriage** a **real** thing."

—JOHANN WOLFGANG VON GOETHE

Marriage continues to be the relationship most individuals prefer. Even though young adults assert that they are not ready to "settle down" and that marriage is for "later," most acknowledge that they will probably end up getting married and having children. The majority of all U.S. adults will do just that—get married and have children.

But marriages are like cars...there are different models, sizes, shapes, and colors. The title of this chapter, Marriage Relationships with plural "relationships," confirms that marriages are different. *Diversity* is the term that best describes relationships, marriages, and families today. No longer is there a one-size-fits-all cultural norm of what a relationship, marriage, or family should be. In this chapter, we review the diversity of relationships. We begin by looking at some of the different reasons people marry.

8-1 Motivations, Functions, and Transition to Egalitarian Marriage

In this section, we discuss both why people marry and the functions that getting married serve for society.

Red Chopsticks/Getty Images

8-1a **Motivations to Marry**

We have defined marriage in the United States as a legal contract between two adults that regulates their economic and sexual interaction. However, individuals in the United States tend to think of marriage in personal more than legal terms. The following are some of the reasons people give for getting married.

1. **Love.** Unlike individuals in Middle Eastern countries (e.g., Iran), Americans view marriage as the ultimate expression of their love for each other—the desire to spend their lives together in a secure, legal, committed relationship. In U.S. society, love is expected to precede marriage—thus, only couples in love consider marriage. Those not in love are ashamed to admit it.

"To **love** and **be loved** is to **feel** the **sun** from both sides."

—DAVID VISCOTT, PSYCHIATRIST

2. **Personal fulfillment.** Americans also marry because they anticipate a sense of personal fulfillment by doing so. As Americans, we were born into a family (family of origin) and want to create a family of our own (family of procreation). We remain optimistic that our marriage will be a good one. Even if our parents divorced or we have friends who have done so, we feel that our relationship will be different.

3. **Companionship.** Most of the earlier societal functions for marriage (protective, religious, educational, procreative) have been reevaluated due to these functions being provided in alternative ways. Police provide protection, churches provide religion, schools provide education, and single parenthood is gaining increased approval. However, a stable emotional relationship over time—companionship—remains a strong incentive for marriage.

4. **Parenthood.** Most people want to have children. In response to the statement, "Someday, I want to have children," 89% of 1,137 undergraduate males and 90% of 3,555 undergraduate females answered "yes" (Hall & Knox, 2015).

   Although some people are willing to have children outside marriage (e.g., in a cohabiting relationship or in no relationship at all), most Americans prefer to have them in a marital context. Previously, a strong norm existed in our society (particularly among Whites) that individuals should be married before they have children. This norm is becoming more relaxed, with more individuals willing to have children without being married.

5. **Economic security.** Married people report higher household incomes than do unmarried people. Indeed, almost 80% of wives work outside the home so that two incomes are available to the couple. One of the disadvantages of remaining single is that the lifestyle is associated with lower income. The median income for a single female household is $25,269; for a married couple, it is $71,830 (*Statistical Abstract of the United States, 2012–2013*, Table 692).

6. **Psychological well-being.** Regardless of sexual orientation, being married is associated with lower levels of distress and higher levels of psychological well-being than being single. This conclusion is based on analysis of data from the California Health Interview Survey of 2009 adults ages 18 to 70 (Wright et al., 2013).

8-1b Functions of Marriage

As noted in Chapter 1, important societal functions of marriage are to bind a male and a female together who will reproduce, provide physical care for their dependent young, and socialize them to be productive members of society who will replace those who die (Murdock, 1949). Marriage helps protect children by giving the state legal leverage to force parents to be responsible to their offspring whether or not they stay married. If couples did not have children, the state would have no interest in regulating marriage.

Additional functions of marriage include regulating sexual behavior (spouses are expected to be faithful, which results in less exposure to sexually transmitted infections than being single) and stabilizing adult personalities by providing a companion and "in-house" counselor. As noted in the section on motivations for marriage, in the past, marriage and family served protective, educational, and religious functions for its members. These functions have been taken over by the legal, educational, and religious institutions of our society. Only the companionship-intimacy function of marriage/family has remained virtually unchanged.

8-1c Transition to Egalitarian Marriage

The very nature of the marriage relationship has also changed from being very traditional or male-dominated to being very modern or egalitarian. A summary of these differences is presented in Table 8.1. Keep in mind that these are stereotypical marriages and that only a small percentage of today's modern marriages have all the traditional or egalitarian characteristics that are listed.

8-2 Weddings and Honeymoons

A **rite of passage** is an event that marks the transition from one social status to another. Starting school, getting a driver's license, and

rite of passage an event that marks the transition from one social status to another.

Table 8.1

Traditional versus Egalitarian Marriages

Traditional Marriage	Egalitarian Marriage
Limited expectation of husband to meet emotional needs of wife and children.	Husband is expected to meet emotional needs of his wife and children.
Wife is not expected to earn income.	Wife is expected to earn income.
Emphasis is on ritual and roles.	Emphasis is on companionship.
Couples do not live together before marriage.	Couples often live together before marriage.
Wife takes husband's last name.	Wife may keep her maiden name. In some cases, he will take her last name.
Husband is dominant; wife is submissive.	Neither spouse is dominant.
Roles for husband and wife are rigid.	Roles for spouses are flexible.
Husband initiates sex; wife complies.	Either spouse initiates sex.
Wife takes care of children.	Fathers more involved in child rearing.
Education is important for husband, not for wife.	Education is important for both spouses.
Husband's career decides family residence.	Career of either spouse may determine family residence.

The wedding is an important rite of passage for a female and marks her transition from a girl to a woman.

Courtesy of Brittany Bolen

graduating from high school or college are events that mark major transitions in status (to student, to driver, and to graduate). The wedding itself is another rite of passage that marks the transition from fiancée and fiancé to spouse. Preceding the wedding is the traditional bachelor party for the soon to be groom. Somewhat new on the cultural landscape is the bachelorette party (sometimes wilder than the bachelor party), which conveys the message that both soon to be spouses can have their version of a premarital celebration—their last hurrah!

> **"The music at a wedding procession always reminds me of the music of soldiers going into battle."**
>
> —HEINRICH HEINE

8-2a Weddings

To obtain a marriage license, three states require one or both partners to have blood tests to certify the absence of a sexually transmitted infection (STI). The document is then taken to the county courthouse, where the couple applies for a marriage license. Two-thirds of states require a waiting period between the issuance of the license and the wedding. A member of the clergy marries 80% of couples; the other 20% (primarily remarriages) go to a justice of the peace, judge, or magistrate.

The wedding is a rite of passage that is both religious and civil. To the Catholic Church, marriage is a sacrament that implies that the union is both sacred and indissoluble. According to Jewish and most Protestant faiths, marriage is a special bond between the husband and wife sanctified by God, but divorce and remarriage are permitted.

Wedding ceremonies still reflect traditional cultural definitions of women as property. For example, the father of the bride usually walks the bride down the aisle and "gives her away" to her husband. In some cultures, the bride is not even present at the time of the actual marriage. For example, in the upper-middle-class Muslim Egyptian wedding, the actual marriage contract signing occurs when the bride is in another room with her mother and sisters. The father of the bride and the new husband sign the actual marriage contract (identifying who is marrying whom, the families they come from, and the names of the two witnesses). The father will then place his hand on the hand of the groom, and the maa'zun, the presiding official, will declare that the marriage has transpired.

That marriage is a public event is emphasized by weddings in which the couple invites their family and friends to participate. The wedding is a time for the respective families to learn how to cooperate with each other for the benefit of the couple. Conflicts over the number of bridesmaids and ushers, the number of guests to invite, and the place of the wedding are not uncommon. Campbell et al. (2011a) surveyed 610 spouses and found that those who had an elaborate wedding reported less present-day satisfaction and commitment, suggesting that individuals who idealize their relationship may enact elaborate weddings, and when their high relational expectations go unmet, satisfaction and commitment decline.

Klos and Sobal (2013) discussed weight management in anticipation of one's wedding.

artifact concrete symbol that reflects the existence of a cultural belief or activity.

gualtiero boffi/Shutterstock.com

While women typically engage in losing weight to approximate the desired cultural image of the bride, men also try to lose weight. Of 163 engaged men, 39% reported they were actively trying to lose weight (about nine pounds).

Brides often wear traditional **artifacts** (concrete symbols that reflect the existence of a cultural belief or activity): something old, new, borrowed, and blue. The old wedding artifact represents the durability of the impending marriage (e.g., an heirloom gold locket). The new wedding artifact, perhaps in the form of new undergarments, emphasizes the new life to begin. The borrowed wedding artifact is something that has already been worn by a currently happy bride (a wedding veil). The blue wedding artifact represents fidelity. When the bride throws her floral bouquet, the single woman who catches it will be the next to be married; the rice thrown by the guests at the newly married couple signifies fertility.

Couples now commonly have weddings that are neither religious nor traditional. In the exchange of vows, the couple's relationship may be spelled out by the partners rather than by tradition, and neither partner may promise to obey the other. Vows often include the couple's feelings about equality, individualism, humanism, and openness to change.

The wedding marks a rite of passage from the role of lover to spouse.

In 2014, the average cost of a wedding for a couple getting married for the first time was estimated to be around $30,000 (www.theknot.com). Ways in which couples lower the cost of their wedding include marrying any day but Saturday, or marrying off season (not June) or off locale (in Mexico or on a Caribbean Island where fewer guests will attend). They may also broadcast their wedding over the Internet through websites such as webcastmywedding.net. Streaming capability means that the couple can get married in Hawaii and have their ceremony beamed back to the mainland where well-wishers can see the wedding without leaving home.

> ## "Tom and I will **always** be in our **honeymoon phase**."
>
> —KATIE HOLMES (MARRIED 6 YEARS TO TOM CRUISE, NOW DIVORCED)

8-2b Honeymoons

Traditionally, another rite of passage follows immediately after the wedding—the **honeymoon** (the time following the wedding whereby the couple isolates themselves to recover from the wedding and to solidify their new status change from lovers to spouses). The functions of the honeymoon are both personal and social.

The personal function is to provide a period of recuperation from the usually exhausting demands of preparing for and being involved in a wedding ceremony and reception. The

social function is to provide a time for the couple to be alone to solidify their new identity from that of an unmarried to a married couple. Now that they are married, their sexual expression and childbearing with each other achieves full social approval and legitimacy.

Not all couples take a honeymoon. Thirty-nine percent of Campbell's et al. (2011a) 610 spouses did not take a honeymoon. The most common reasons identified by spouses included financial (59%) and lack of time (39%). Only 6% said they and their partners (5%) were not interested in taking a honeymoon.

8-3 Changes after Marriage

After the wedding and honeymoon, the new spouses begin to experience changes in their legal, personal, and marital relationship.

> ## "**Marriage** is like a **phone call** in the night; first the ring, and then you **wake up**."
>
> —EVELYN HENDRICKSON, ACTRESS

8-3a Legal Changes

Unless the partners have signed a prenuptial agreement specifying that their earnings and property will remain separate, the wedding ceremony makes each spouse part owner of what the other earns in income and accumulates in property. Although the laws on domestic relations differ from state to state, courts typically award to each spouse half of the assets accumulated during the marriage (even though one of the partners may have contributed a smaller proportion).

For example, if a couple buys a house together, even though one spouse invested more money in the initial purchase, the other will likely be awarded half of the value of the house if they divorce. Having children complicates the distribution of assets because the house is often awarded to the custodial parent. In the case of death of the spouse, the remaining spouse is legally entitled to inherit between one-third and one-half of the partner's estate, unless a will specifies otherwise.

honeymoon the time following the wedding whereby the couple becomes isolated to recover from the wedding and to solidify their new status change from lovers to spouses.

Personal View: The Wedding Night

Ninety-five spouses (75% wives, 25% husbands) reported on their "most recent wedding night."

1. **Ratings.** When asked "On a scale of one to ten with zero being awful and ten being wonderful, what number would you select to describe your wedding night experience?," the average was 6.81, looking at responses from both men and women. Grooms reported more positive experiences than brides, 7.19 to 6.68, respectively. When second marriages were the focus, the average was 8.4.

2. **Summaries.** When asked to "summarize your wedding night experience" some of the comments were:
   a. "We only went away for one night as the next day was Christmas. I was 17, it was lots of fun, we ate out, and had a great night."
   b. "It was a fulfilling, happy occasion, knowing that I was starting a new chapter in life."
   c. "It was a disaster. He was thinking about his old girlfriend and wishing he had married her."
   d. "We had an amazing night. It was so hard to keep our hands off each other."
   e. "It was so wonderful but we were tired. It could have been better."

3. **Best part.** When asked to identify "the best part of the wedding night," 29.7% listed "just being with my new partner"; 17.8% listed "sex"; 10.5% said "nothing"; and 9.4% listed the "reception."

4. **Worst part.** When asked to identify the "worst part of the wedding night," 23.1% listed "accommodations and the partner's demeanor"; 21% said "being so tired"; 9.4% said the "end of celebration"; 8.4% said "nothing"; 6.3% listed "sex"; and 3.1% listed "pain."

5. **Change.** To the question, "If you could replay your wedding night, what would you change?" 34.7% of respondents answered "nothing"; 18.9% responded "not be tired"; 14.7% listed "different time/place"; and 9.5% listed "different person."

The data suggested that the wedding night for these respondents was predominately a positive experience. A recommendation was to plan a wedding early in the day so that the reception is over early. Also, the couple might plan to spend the first night a short drive from the reception. "Avoid leaving a reception at 11:00 P.M. and driving for hours." Another suggestion was "Avoid an early-morning flight the day after your wedding."

Shaley et al. (2013) reported the experience of 12 modern-orthodox couples on their wedding night. They had had no previous sexual experience and the transition from "this is forbidden" to "everything is OK" was difficult and challenging. Two sources of help were sought—close friends and the Internet.

Sources: This research is based on unpublished data collected for this text. Appreciation is expressed to Kelly Woody for distribution of the questionnaires and to Emily Richey for tabulating the data.

8-3b Personal/Health Changes

New spouses experience an array of personal changes in their lives. One initial consequence of getting married may be an enhanced self-concept. Parents and close friends usually arrange their schedules to participate in your wedding and give gifts to express their approval of you and your marriage. In addition, the strong evidence that your spouse approves of you and is willing to spend a lifetime with you also tells you that you are a desirable person.

Married people also begin adopting new values and behaviors consistent with the married role. Although new spouses often vow that "marriage won't change me," it does. For example, rather than stay out all night at a party, which is not uncommon for single people who may be looking for a partner, spouses (who are already paired off) tend to go home early. Their roles of spouse, employee, and parent result in their adopting more regular, alcohol- and drug-free hours. Averett et al. (2013) confirmed that being married is not only associated with lower alcohol use and improved mental health but with a higher body mass index (BMI). In effect, spouses put on weight after they say "I do." Weight gain is particularly true for married men when compared to single men (Berge et al., 2014).

8-3c Friendship Changes

Marriage affects relationships with friends of the same and the other sex. Although time with same-sex friends will continue (Hall & Adams, 2011), it will decrease because of the new role demands of the spouses. More time will be spent with other married couples who will become powerful influences on the new couple's relationship. Indeed, couples who have the same friends report increased marital satisfaction.

What spouses give up in friendships, they gain in developing an intimate relationship with each other. However, abandoning one's friends after marriage may be problematic because one's spouse cannot be expected to satisfy all of one's social needs. Because many marriages end in divorce, friendships that have been maintained throughout the marriage can become a vital source of support for a person adjusting to a divorce. "Don't forget your friends on your way up, you'll need them on your way down" reflects the sentiment of maintaining one's friends after getting married.

> "Keep your **eyes** wide **open before marriage,** half **shut afterwards.**"
>
> —BEN FRANKLIN

8-3d Relationship Changes

Totenhagen et al. (2011) studied the variability in the relationships of 328 individuals on seven variables (satisfaction, commitment, closeness, maintenance, love, conflict, and ambivalence) and found that there was greater variability for newer couples than for longer-term couples. But variability is to be expected. A couple happily married for 45 years spoke to our class and began their presentation with, "Marriage is one of life's biggest disappointments." They spoke of the difference between all the hype and the cultural ideal of what marriage is supposed to be . . . and the reality.

One effect of getting married is **disenchantment**—the transition from a state of newness and high expectation to a state of mundaneness tempered by reality. It may not happen in the first few weeks or months of marriage, but it is almost inevitable. Although courtship is the anticipation of a life together, marriage is the day-to-day reality of that life together—and reality does not always fit the dream. "Moonlight and roses become daylight and dishes" is an old adage reflecting the realities of marriage. Musick and Bumpass (2012) compared spouses, cohabitants, and singles and noted that the advantages of being married over not being married tended to dissipate over time. **Satiation**, a stimulus loses its value with repeated exposure, speeds disenchantment. Individuals tire of each other because they are no longer new. Happy long-term spouses unaffected by the principle of satiation do not require newness for excitement and contentment.

Disenchantment after marriage is also related to the partners shifting their focus away from each other to work or children; each partner usually gives and gets less attention in marriage than in courtship. College students are not oblivious to the change after marriage. Twenty-seven percent of 1,142 undergraduate males and 23% of 3,546 undergraduate females agreed that "most couples become disenchanted with marriage within five years" (Hall & Knox, 2015).

In addition to disenchantment, a couple will experience numerous changes once they marry:

1. **Loss of freedom.** Single people do as they please. They make up their own rules and answer to no one.

2. **More responsibility.** Single people are responsible for themselves only. Spouses are responsible for the needs of each other and sometimes resent it.

3. **Less alone time.** Aside from the few spouses who live apart, most live together. They wake up together, eat their evening meals together, and

disenchantment the transition from a state of newness and high expectation to a state of mundaneness tempered by reality.

satiation a stimulus loses its value with repeated exposure.

go to bed together. Each may feel too much togetherness. "This altogether, togetherness thing is something I don't like," said one spouse.

4. **Change in how money is spent.** Entertainment expenses in courtship become allocated to living expenses and setting up a household together.

5. **Sexual changes.** The frequency with which spouses have sex with each other decreases after marriage (Hall & Adams, 2011).

6. **Power changes.** The power dynamics of the relationship change after marriage with men being less patriarchal/collaborating more with their wives while women change from deferring to their husbands' authority to challenging their authority (Huyck & Gutmann, 1992). In effect, with marriage, men tend to lose power and women gain power. This shift to a more egalitarian relationship is associated with marital happiness. Wives' perception of power may be assessed by asking questions about the degree to which one's husband "tries to dominate and control," "expects me to play a dependent role," "treats me as an equal," and "makes the major decisions." LeBaron et al. (2014) examined the power perceptions of 67 wives of medical students in 1990 and again 15 years later. Results revealed an association between having an egalitarian relationship at midlife and reporting happiness in one's marriage.

8-3e Parents and In-Law Changes

Marriage affects relationships with parents. Time spent with parents and extended kin radically increases when a couple has children. Indeed, a major difference between couples with and without children is the amount of time they spend with relatives. Parents and kin rally to help with the newborn and are typically there for birthdays and family celebrations.

8-3f Financial Changes

An old joke about money in marriage is that "two can live as cheaply as one as long as one doesn't eat." The reality behind the joke is that marriage involves the need for spouses to discuss and negotiate how they are going to get and spend money in their relationship. Some spouses bring considerable debt into the marriage or amass great debt during the marriage. Such debt affects marital interaction.

Marriage is also associated with the male becoming more committed to earning money. Ashwin and Isupova (2014) noted that husbands "implicitly commit themselves to a 'responsible' version of masculine identity" (e.g., rather than hard drinking) and that wives monitor their behavior to ensure a productive outcome. One example is a wife who made it clear that she did not want her husband to quit his job regardless of how unhappy he was. Her focus on keeping the family income coming in and her desire to avoid the "my husband is unemployed" stigma kept her husband on the job.

Marriage involves the need for spouses to discuss and negotiate how they are going to get and spend money in their relationship

8-4 Diversity in Marriage

Any study of marriage relationships emphasizes the need to understand the diversity of marriage/family life. Researchers (Ballard & Taylor, 2011; Wright et al., 2012) have emphasized the various racial, ethnic, structural, geographic location, and contextual differences in marriage and family relationships. In this section, we review Hispanic, Mormon, and military families. We also look at other examples of family diversity: interracial, interreligious, international, and age discrepant.

Flashon Studio/Shutterstock.com

"Once we figured out that
we could **not change each other,**
we became **free to celebrate**
ourselves **as we are.**"

—H. DEAN RUTHERFORD (IN A LETTER TO HIS
WIFE ON THEIR 59TH WEDDING ANNIVERSARY)

8-4a Hispanic Families

The panethnic term *Hispanic* refers to both immigrants and U.S. natives with an ancestry to one of 20 Spanish-speaking countries in Latin America and the Caribbean. There are about 53 million Hispanics in the United States who represent 17% of the population (United States Census Bureau, 2014). Hispanic families vary not only by where they are from but by whether they were born in the United States. About 40% of U.S. Hispanics are foreign born and immigrated here, 32% have parents who were born in the United States, and 28% were born here of parents who were foreign born.

Great variability exists among Hispanic families. Although it is sometimes assumed that immigrant Hispanic families come from rural impoverished Mexico where family patterns are traditional and unchanging, immigrants may also come from economically developed urbanized areas in Latin America (Argentina, Uruguay, and Chile), where family patterns include later family formation, low fertility, and nuclear family forms.

Hispanics tend to have higher rates of marriage, early marriage, higher fertility, nonmarital child rearing, and prevalence of female householder. They also have two micro family factors: male power and strong familistic values.

1. **Male power.** The husband and father is the head of the family in most Hispanic families.

2. **Strong familistic values.** The family is the most valued social unit in the society—not only the parents and children but also the extended family. Hispanic families have a moral responsibility to help family members with money, health, or transportation needs. Children are taught to respect their parents as well as the elderly. Indeed, elderly parents may live with the Hispanic family where children may address their grandparents in a formal way. Spanish remains the language spoken in the home as a way of preserving family bonds.

8-4b Mormon Families

There are about 6 million Mormons in the United States (and the 14 million worldwide). Also known as the Church of Jesus Christ of Latter-day Saints, the Mormons have been associated with polygyny. But this practice was disavowed in 1890 by mainstream Mormons. Sects of the Mormon Church continuing the practice are not included in the following discussion (Dollahite & Marks, 2012).

Unique characteristics of Mormon beliefs/families include:

1. **Eternal marriage.** Mormon doctrine holds that Mormon spouses are married not only until death but throughout eternity. Mormon spouses also believe that their children become a permanent part of their family both in this life and in the afterlife. Hence, the death of a

This Mormon family enjoys spending time together, including fishing for flounder.

spouse or child is viewed as a family member who has gone to heaven only to be reunited with other family members at their earthly death.

2. **Family rituals.** Mormons are expected to pray both as spouses and as a family, to study the scripture (*Book of Mormon*), and to observe family home evening every Monday night. The latter involves the family praying, singing, having a lesson taught by a parent or older child, experiencing a fun activity (e.g., board game/charades), and enjoying refreshments (e.g., homemade cookies).

3. **Frequent prayer.** While most religions encourage prayer, the Mormon faith encourages the family to pray together three times a day—morning, at mealtimes, and at bedtime. The ritual provides an emotional bond of family members to each other.

4. **Substance prohibitions.** Mormons avoid alcohol, tobacco, coffee, and some teas (e.g., caffeinated). The health benefits include lower cancer rates and increased longevity (8 to 11 years).

5. **Extended family/intergeneration support.** Family reunions, family web pages, and ties with parents and grandparents are core to Mormon family norms. The result is a close family system of children, parents, and grandparents.

6. **Intramarriage.** Selecting another Mormon to marry is encouraged. Only devout members of the church may be married in the temple, which seals the couple for this life and for eternity.

7. **Early marriage/large families.** Mormons typically marry younger (remember, no sex before marriage so there is some incentive to marry soon) and have a higher number of children than the national average. One result of larger families is that adults stay in the role of parents for a longer period of their lives.

8. **Lower divorce rate.** It comes as no surprise that the Mormon emphasis on family rituals and values results in strong family ties and a lower divorce rate. While 40% to 50% of marriages in general in the United States end in divorce, only about 10% of Mormon marriages do so.

8-4c Military Families

Although the war in Iraq is over and U.S. troops are being withdrawn from Afghanistan, approximately 1.5 million U.S. citizens are active-duty military personnel. Over half (56%) are married (726,000) (Lacks et al., 2013). Another 819,000 are in the military reserve and National Guard (*Statistical Abstract of the United States, 2012–2013*, Table 508).

There are three main types of military marriages. In one type, an individual falls in love, gets married, and subsequently joins the military. A second type of

This couple has been married for 19 years during which the husband was deployed for 4 years.

military marriage is one in which one or both of the partners is already a member of the military before getting married. The final and least common type is known as a **military contract marriage**, in which a military person will marry a civilian to get more money and benefits from the government, such as additional housing allowance. Contract military marriages are not common but they do exist.

Some ways in which military families are unique include:

1. **Traditional sex roles.** Although both men and women are members of the military service, the military has considerably more men than women (85% versus 15%). In the typical military family, the husband is deployed (sent away to serve) and the wife is expected to understand his military obligations and to take care of the family in his absence. The wife often has to sacrifice her career to follow (or stay behind in the case of deployment) and support her husband in his fulfillment of military duties.

   There are also circumstances in which both parents are military members, and this can blur traditional sex roles because the woman has already deviated from a traditional "woman's job." Military families in which both spouses are military personnel are rare.

2. **Loss of control—deployment.** Military families have little control over their lives as the chance of deployment is ever present. Where one of the spouses will be next week and for how long are beyond the control of the spouses

Shortly after this photo was taken, this husband and father was deployed to Iraq and killed.

and parents. Easterling and Knox (2010) surveyed 259 military wives (whose husbands had been deployed) who reported feelings of loneliness, fear, and sadness (other researchers have identified stress (Lacks et al., 2013). Some women had gone for extended periods of time without communicating with their husbands and in constant worry over their well-being. Talking with other military wives who understood was the primary mechanism for coping with the husband's deployment. Getting a job, participating in military-sponsored events, and living with a family were also helpful. On the positive side, wives of deployed husbands reported feelings of independence and strength. They were the sole family member available to take care of the house and children, and they rose to the challenge.

Adjusting to the return of the deployed spouse has its own challenges. Some deployed spouses who were exposed to combat have had their brain chemistry permanently altered and are never the same again. "These spouses rarely recover completely—they need to accept that their symptoms (e. g., depression, anxiety) can be managed but not cured," noted Theron Covin, who specializes in treating PTSD among the combat deployed (Covin, 2013). Spouses have a particular challenge of adjusting to their altered spouse. "You have to learn to dance all over again" said one wife (Aducci et al., 2012). A team of researchers observed an increased incidence of spousal violence related to PTSD as a result of having been deployed (Teten et al., 2010). Foran et al. (2013) examined military marriages after the deployment of a combat-exposed spouse; the greater the combat exposure, the greater the intent to divorce.

3. **Infidelity.** Although most spouses are faithful to each other, being separated for months (sometimes years) increases the vulnerability of both spouses to infidelity. The double standard may also be operative, whereby "men are expected to have other women when they are away" and "women are expected to remain faithful and be understanding." Separated spouses

military contract marriage a military person will marry a civilian to get more money and benefits from the government

try to bridge the time they are apart with text messages, emails, Skype, and phone calls, but sometimes the loneliness becomes more difficult than anticipated.

4. **Frequent moves and separation from extended family or close friends.** Because military couples are often required to move to a new town, parents no longer have doting grandparents available to help them rear their children. And although other military families become a community of support for each other, the consistency of such support may be lacking.

5. **Lower marital satisfaction and higher divorce rates among military families.** Solomon et al. (2011) compared 264 veterans who experienced combat stress reaction (CSR) with 209 veterans who did not experience such stress. Results show that traumatized veterans reported lower levels of marital adjustment and more problems in parental functioning. Wick and Nelson Goff (2014) reaffirmed the challenge miltary marriages experience when a deployed spouse returns with PTSD. The divorce rate is also higher in military than in civilian marriages (Lundquist, 2007).

8-4d **Interracial Marriages**

About 15% of all marriages in the United States are mixed racially, with Hispanic–non-Hispanic being the most frequent. Nine percent of Whites, 16% of Blacks, and 26% of Hispanics marry someone whose race or ethnicity is different from their own (Passel et al., 2010). The least likely intermarriage (1.3%) is between Black women and White men (Qian & Lichter, 2011).

In discussing interracial marriages, a complicating factor is that one's racial identity may be mixed. Tiger Woods refers to his race as "Cablinasian," which combines Caucasian, Black, Native American (Indian), and Asian origins—he is one-quarter Chinese, one-quarter Thai, one-quarter Black, one-eighth Native American, and one-eighth Dutch. Some individuals seek partners with a different racial/ethnic heritage (Yodanis et al., 2012).

Black–White marriages are the most infrequent. In spite of the cultural visibility of the interracial marriage of Kim Kardashian and Kanye West, fewer than 1% of the over 63 million marriages in the United States are between a Black person and a White person (*Statistical Abstract of the United States, 2012–2013*, Table 60). Segregation in religion (the races worship in separate churches), housing (White and Black neighborhoods), and education (White and Black colleges), not to speak of parental and peer endogamous pressure to marry within one's own race, are factors that help to explain the low percentage of interracial Black and White marriages. Perry (2013) found that interracial friendships are associated with positive intermarriage attitudes. Living in a neighborhood or attending church with other racial groups is associated with racial intermarriage tolerance primarily when friendships develop in these contexts.

Field et al. (2013) examined interracial attitudes among 1,173 college students at five universities. Attitudes at historically Black universities were less positive than at predominantly White universities with Black students disapproving more of Black/White relationships than Whites (no gender differences). White students perceived their parents as being more disapproving of Black/White relationships than Black students did. The spouses in Black and White couples are more likely to have been married before, to be age discrepant, to live far away from their families of orientation, to have been reared in racially tolerant homes, and to have educations beyond high school. Some may also belong to religions that encourage interracial unions. The Baha'i religion, which has more than 6 million members worldwide and 84,000 in the United States, teaches that God is particularly pleased with interracial unions. Finally, interracial spouses may tend to seek contexts of diversity. "I

Courtesy of Brittany Bolen

Less than 1% of the 63 million married couples in the United States consist of a White and a Black spouse.

have been reared in a military family, been everywhere, and met people of different races and nationalities throughout my life. I seek diversity," noted one student.

Black–White interracial marriages are likely to increase—slowly. Not only has white prejudice against African Americans in general declined, but also segregation in school, at work, and in housing has decreased, permitting greater contact between the races. One-third of 1,144 undergraduate males (39% of 3,541 undergraduate females) reported that they have dated someone of another race (Hall & Knox, 2015). Most Americans say they approve of racial or ethnic intermarriage—not just in the abstract, but in their own families. More than six in ten say it "would be fine" with them if a family member told them they were going to marry someone from any of three major race/ethnic groups other than their own (Passel et al., 2010).

8-4e Interreligious Marriages

Of married couples in the United States, 37% have an interreligious marriage (Pew Research, 2008). Although religion may be a central focus of some individuals and their marriage, Americans in general have become more secular, and as a result religion has become less influential as a criterion for selecting a partner. In a survey of 1,144 undergraduate males and 3,559 undergraduate females, 33% and 43%, respectively, reported that marrying someone of the same religion was important for them (Hall & Knox, 2015).

Are people in interreligious marriages less satisfied with their marriages than those who marry someone of the same faith? The answer depends on a number of factors. First, people in marriages in which one or both spouses profess "no religion" tend to report lower levels of marital satisfaction than those in which at least one spouse has a religious tie. People with no religion are often more liberal and less bound by traditional societal norms and values; they feel less constrained to stay married for reasons of social propriety.

The impact of a mixed religious marriage may also depend more on the devoutness of the partners than on the fact that the partners are of different religions. If both spouses are devout in their respective religious beliefs, they may expect some problems in the relationship. Less problematic is the relationship in which one spouse is devout but the partner is not. If neither spouse in an interfaith marriage is devout, problems regarding religious differences may be minimal or nonexistent. In their marriage vows, one interfaith couple who married (he was Christian, she was Jewish) said that they viewed their different religions as an opportunity to strengthen their connections to their respective faiths and to each other. "Our marriage ceremony seeks to celebrate both the Jewish and Christian traditions, just as we plan to in our life together."

8-4f International Marriages

With increased globalization, international matchmaking Internet opportunities, and travel abroad programs, there is greater opportunity to meet/marry someone from another country. Levchenko and Solheim (2013) studied international marriages between Eastern European-born women and U.S.-born men. They found that these pairings reflected homogamy in race and having been previously married. However, complementarity was evident in terms of age—with women being about nine years younger. Eastern European women were also more willing to be traditional in their role relationships—a trait American men seek in their wives.

Over three-fourths of 1,145 undergraduate males (80%) and over two-thirds of 3,447 undergraduate females (69%) reported that they would be willing to marry someone from another country (Hall & Knox, 2015). The opportunity to meet that someone is increasing, as upwards of 800,000 foreign students are studying at American colleges and universities.

This couple met in graduate school in the United States. The husband is 100% Taiwanese and the wife is 50% Chinese and 50% Taiwanese. The husband's parents, radical supporters of Taiwan independence, disapprove of their offspring marrying anyone of Chinese descent.

Some people from foreign countries marry an American citizen to gain citizenship in the United States, but immigration laws now require the marriage to last two years before citizenship is granted. If the marriage ends before two years, the foreigner must prove good faith (that the marriage was not just to gain entry into the country) or he or she will be asked to leave the country.

Cultural differences may sometimes surface. An American woman fell in love with and married a man from Fiji. She was unaware of a norm from his culture that anyone from his large kinship system could visit at any time and stay as long as they liked—indeed it would be impolite to ask them to leave. On one occasion two "cousins" showed up unannounced at the home of the couple. The wife was in the last week of exams for her medical degree and could not tolerate visitors. The husband told her, "I cannot be impolite." But with the marriage threatened, the husband told his cousins they had to leave.

©rtguest/Shutterstock.com

8-4g Age-Discrepant Relationships and Marriages

Although people in most pairings are of similar age, sometimes the partners are considerably different in age. In marriage, these are referred to as ADMs (age-dissimilar marriages) and are in contrast to ASMs (age-similar marriages). ADMs are also known as **May–December marriages**. Typically, the woman is in the spring of her youth (May) whereas the man is in the later years of his life (December). There have been a number of May–December celebrity marriages, including that of Celine Dion, who is 26 years younger than René Angelil (in 2015, they were aged 47 and 73). Michael Douglas is 25 years older than his wife, Catherine Zeta-Jones, and Ellen DeGeneres is

May–December marriage age-dissimilar marriage (ADM) in which the woman is typically in the spring of her life (May) and her husband is in the later years (December).

15 years older than her spouse, Portia de Rossi.

Sociobiology suggests that men select younger women since doing so results in healthier offspring. Women benefit from obtaining sperm from older men who have more resources for their off-spring. Burrows (2013) studied age-discrepant patterns in both heterosexual and homosexuals and found the same pattern regardless of sexual orientation. He explained in his findings a strong cultural programming that reaches gay individuals to select an older person to pair-bond with (greater security).

In a study of 433 spouses where the husband was older, marital effects were less time spent together and more marital problems. Having less in common was the presumed reason for the negative effects (Wheeler et al., 2012).

Perhaps the greatest example of a May–December marriage that worked is of Oona and Charles Chaplin. She married him when she was 18 (he was 54). Their alliance was expected to last the requisite six months, but they remained together for 34 years (until his death at 88) and raised eight children, the last of whom was born when Chaplin was 73.

Although less common, some age-discrepant relationships are those in which the woman is older than

This man is 49; she is 30.

What's New?

U.S. AND ICELANDIC COLLEGE STUDENT ATTITUDES TOWARD RELATIONSHIPS

It is easy to forget that people in other countries may not believe as do American youth. Iceland became part of U.S. consciousness in 2010 when the Eyjafjallajökull volcano erupted on April 14 causing the cancellation of thousands of flights across Europe and to Iceland. Anyone in the United States trying to fly to Europe was likely affected. Iceland was on the nightly national news for a week.

It is axiomatic that societies differ in regard to norms regarding relationships and sexuality. Iceland and the United States represent two diverse societies/cultures. Iceland is one of five Nordic countries (the others are Denmark, Finland, Sweden, and Norway) representing over 40,000 square miles with a population of 320,000, mostly of Norwegian and Irish origins. In contrast, the United States is 3.79 million square miles with a population of around 330 million.

The Study

A team of American and Icelandic researchers compared the responses of 722 undergraduates from a large southeastern university in the United States on 100 items about relationships with the responses of

Johann Herbertsson was born and reared in Iceland. He is pointing to the town of Reykjavik where he lives.

368 undergraduates from the University of Iceland in Reykjavik, Iceland. The sample consisted of 813 females and 277 males. The proportion of males and females in the respective countries was very similar—74% of the U.S. sample was female compared to 76% of the Icelandic students. The racial composition of the sample differed between the two countries. The U.S. sample was 76% White, 15% Black, 3% Asian, 2% mixed, and 2% reporting "other." The Icelandic sample was 98% White and 2% Black.

To test for differences in the responses of the U.S. and Icelandic students when responses were made on a five-point response scale, the test for mean differences between independent groups was used. When two point yes/no responses were analyzed, t-tests for differences in proportions were used.

Results

Several significant differences were found between U.S. and Icelandic undergraduates:

- **Desire to marry.** Students were presented with the item "Someday I want to marry" and asked to respond on a five-point scale from "strongly agree" (1) to "strongly disagree" (5), the lower the score the more important the goal to marry and the higher the score the lower the importance of getting married. Scores for the U.S. and Icelandic students were .97 and .89 respectively (significant at the .01 level) revealing that U.S. students have a higher desire to marry than Icelandic students. Explanation of the data include that Icelandics have long held the tradition of cohabitation independent of marriage, hence, their lower desire to marry.

- **Interracial marriage.** Students were presented with the item "It is important to me that I marry of someone my race" and asked to respond on a five-point scale. Scores for the U.S. and Icelandic students were 2.78 and 4.27, respectively (significant at the .01 level), revealing that U.S. students were much more open to interracial marriage with few Icelandic students reporting a willingness to cross racial lines to marry. Explanation of the data include that the United States is a much larger country with a more diverse population compared to Iceland. Icelandic undergraduates have had limited exposure to those outside their own group.

- **Interreligious marriage.** Students were presented with the item "It is important to me that I marry someone of my same religion" and asked to respond on a five-point scale. Scores for the U.S. and Icelandic students were 2.77 and 3.80, respectively (significant at the .01 level), revealing that marrying someone of the same faith was more important to U.S. than Icelandic students. This finding is not a surprise since the U.S. sample is from the southern region of the United States where conservative religion has, traditionally, been dominant. Hence, these students are simply reflecting their heritage whereas Icelandic students are more moderate in their religious views and more open to diversity.

- **Impulsiveness about love.** Students were presented with the item "If I were really in love, I would marry someone I had known for only a short time" and asked to respond on a five-point scale. Scores for the U.S. and Icelandic students were 2.77 and 3.80, respectively (significant at the .01 level), revealing that a higher number of the U.S. students would be willing to marry someone they were madly in love with even though they had only known that person for a short time. Romantic love is a pervasive concept in the United States, which may be at the basis of marrying in a short time if one is in love (Riela et al., 2010).

- **Cheating on a partner.** Students were presented with the item "I have cheated on a partner I was involved with" and asked to respond on a five-point scale. Scores for the U.S. and Icelandic students were

3.46 and 4.18, respectively (significant at the .01 level), revealing that U.S. undergraduates were much more likely to have cheated whereas Icelandic students were more likely to be faithful. The explanation for this finding is that Icelandic students were more likely to be in cohabitation relationships (e.g., more committed relationships) than U.S. students. With less commitment/structure among U.S. relationships, greater infidelity would be expected.

Additional Results

Other findings included that, compared to Icelandics, U.S. undergraduates were less likely to have lived together, less likely to have had sex without love, and more likely to view marriage as a goal. However, there were no differences between U.S. and Icelandic students in desire to have children (both had high desire to have children), in openness to marry someone outside of their own country (both were moderately open), in commitment to end a relationship with a partner who cheated on them, in willingness to see a marriage counselor before seeing a lawyer (both were moderately willing), and in condom use (both did not use consistently).

Source: Adapted and abridged from Halligan, C., D. Knox, F. J. Freysteinsdottir, & S. Skulason. (2014, March). U.S. and Icelandic college student attitudes toward relationships. Poster presented at the annual meeting of the Southern Sociological Society, Charlotte, NC.

---

her partner. Mariah Carey was 11 years older than her former husband, Nick Cannon. Valerie Gibson (2002), the author of *Cougar: A Guide for Older Women Dating Younger Men*, notes that the current use of the term **cougars** refers to "women, usually in their thirties and forties, who are financially stable and mentally independent and looking for a younger man to have fun with." Gibson noted that one-third of women between the ages of 40 and 60 are dating younger men. Financially independent women need not select a man in reference to his breadwinning capabilities. Instead, these women are looking for men not to marry but to enjoy. The downside of such relationships comes if the man gets serious and wants to have children, which may spell the end of the relationship.

cougar a woman, usually in her thirties or forties, who is financially stable and mentally independent and looking for a younger man with whom to have fun.

8-4h College Marriages

Cottle et al. (2013) analyzed data from 429 currently and formerly married college students. The ages ranged from 18 to 62. The newlyweds reported significantly greater life satisfaction, marital satisfaction, relationships with in-laws, communication about sex, working out problems, etc., than the older college-married students. A major finding was that students who quit college were less likely to report satisfaction in these areas.

8-5 Marriage Success

A successful marriage is the goal of most couples. But what is a successful marriage and what are its characteristics?

8-5a Definition and Characteristics of Successful Marriages

Marital success refers to the quality of the marriage relationship measured in terms of stability and happiness. Stability refers to how long the spouses have been married and how permanent they view their relationship to be, whereas happiness refers to more subjective/emotional aspects of the relationship. In describing marital success, researchers have used the terms *satisfaction, quality, adjustment, lack of distress,* and *integration.* Marital success is often measured by asking spouses how happy they are, how often they spend their free time together, how often they agree about various issues, how easily they resolve conflict, how sexually satisfied they are, how equitable they feel their relationship is, and how often they have considered separation or divorce.

Ron Dale/Shutterstock.com

RossHelen/Shutterstock.com

Are wives or husbands happier? Jackson et al. (2014) reported data on 226 independent samples comprising 101,110 participants and found statistically significant yet very small gender differences in marital satisfaction between wives and husbands, with wives slightly less satisfied than husbands. However, the researchers noted that this difference was due to the inclusion of clinical samples, with wives in marital therapy 51% less likely to be satisfied. When nonclinical community-based samples were used, no significant gender differences were observed in marital satisfaction among couples in the general population.

Researchers have also identified characteristics associated with enduring happy marriages (Choi and Marks, 2013; Stanley et al., 2012; DeMaris, 2010; Amato et al., 2007). Their findings and those of other researchers include the following:

1. **Intimacy/partner attachment.** Patrick et al. (2007) found that intimacy (feeling close to spouse, showing physical affection, sharing ideas/events, sharing hobbies/leisure activities) was related to marital satisfaction. They studied both spouses in 124 marriages of employees at two major state universities in the midwest. Most were in the mid-to-late forties, married an average of 20 years, and 80% in their first marriages. Wilson and Huston (2011) confirmed that partner similarity of depth of love feelings in courtship was predictive of remaining married. Finally, Kilmann et al. (2013) found that securely emotionally attached spouses reported higher levels of relationship satisfaction.

2. **Communication/humor.** Gottman and Carrere (2000) studied the communication patterns of couples over an 11-year period and emphasized that those spouses who stayed together were five times more likely to lace their arguments with positives ("I'm sorry I hurt your feelings") and to consciously choose to say things to each other that nurture the relationship.

marital success
the quality of the marriage relationship measured in terms of stability and happiness.

Successful spouses also feel comfortable telling each other what they want and not being defensive at feedback from the partner. Duba et al. (2012) assessed the marital satisfaction of 30 couples married at least 40 years and found the area of affective communication was particularly problematic. Hence, keeping the emotional connection on one's relationship is important. Successful spouses also have a sense of humor. Indeed, a sense of humor is associated with marital satisfaction across cultures—in the United States, China, Russia, and elsewhere (Weisfeld et al., 2011).

3. **Common interests/positive self-concepts.** Spouses who have similar interests, values, and goals, as well as positive self-concepts, report higher marital success (Arnold et al., 2011).

4. **Not materialistic.** Being nonmaterialistic is characteristic of happily married couples (Carroll et al., 2011). Although a couple may live in a nice house and have expensive toys (e.g., a boat and an RV), they are not tied to material comforts. "You can have my things, but don't take away my people" is a phrase from one husband reflecting his feelings about his family.

5. **Role models.** Successfully married couples speak of having positive role models in their parents. Good marriages beget good marriages—good marriages run in families. It is said that the best gift you can give your children is a good marriage.

6. **Religiosity.** A strong religious orientation and practicing one's religion is associated with being committed to one's marriage (Jorgensen et al., 2011). Religion provides spouses with a strong common value. In addition, religion provides social, spiritual, and emotional support from church members and with moral guidance in working out problems.

marriage rituals
deliberate repeated social interactions that reflect emotional meaning to the couple.

connection rituals
habits which occur daily in which the couple share time and attention.

7. **Trust.** Trust in the partner provided a stable floor of security for the respective partners and their relationship. Neither partner feared that the other partner would leave or become involved in another relationship. "She can't take

him anywhere he doesn't want to go" is a phrase from a country-and-western song that reflects the trust that one's partner will be faithful.

8. **Personal and emotional commitment to stay married.** Divorce was not considered an option. The spouses were committed to each other for personal reasons rather than societal pressure.

9. **Sexual satisfaction.** Barzoki et al. (2012) studied wives and found that marital dissatisfaction leads to sexual dissatisfaction but that this connection can become reciprocal—sexual dissatisfaction can lead to marital dissatisfaction.

10. **Equitable relationships.** Amato et al. (2007) observed that the decline in traditional gender attitudes and the increase in egalitarian decision making were related to increased happiness in today's couples. DeMaris (2010) also found that spouses who regarded their relationships as those with equal contribution reported higher quality marriages.

11. **Marriage/connection rituals. Marriage rituals** are deliberate repeated social interactions that reflect emotional meaning to the couple. **Connection rituals** are those which occur daily in which the couple share time and attention. Campbell et al. (2011b) studied 129 unmarried individuals (involved with a partner) who identified 13 different types of rituals (average of 6). The most frequent was enjoyable activities (23%) such as having meals together and watching TV together. Intimacy expressions (19%) included "taking a shower together every morning and washing each other's hair." Communication ritual (14%) examples were sending frequent text messages and having pet names for each other.

12. **Absence of negative statements and attributions.** Not making negative remarks to the spouse is associated with higher marital quality (Woszidlo & Segrin, 2013). In addition, spouses who do not attribute negative motives to their partner's behavior reported higher levels of marital satisfaction than spouses who ruminated about negative motives. Dowd et al. (2005) studied 127 husbands and 132 wives and found that the absence of negative attributions was associated with higher marital quality.

13. **Forgiveness.** At some time in all marriages, each spouse engages in behavior that hurts or disappoints the partner. Forgiveness rather than harboring and nurturing resentment allows spouses to move forward. Spouses who do not "drop the lowest test score" (an academic metaphor) of their partner find that they inadvertently create an unhappy marital context which they endure. However, Woldarsky and Greenberg (2014) noted that it is not forgiveness but the perception that the partner experiences shame for the transgression that is restorative to the couple. Perhaps the two work in combination?

> "A **happy marriage** is the union of **two good forgivers.**"
>
> —ROBERT QUILLEN

14. **Economic security.** Although money does not buy happiness, having a stable, secure economic floor is associated with marital quality (Amato et al., 2007) and marital happiness (Mitchell, 2010). Indeed, higher incomes are associated with marital happiness (Choi & Marks, 2013).

15. **Physical health.** Increasingly, research emphasizes that the quality of family relationships affects family member health and that the health of family members influences the quality of family relationships and family functioning (Choi & Marks, 2013; Proulx & Snyder, 2009). Indeed, such an association begins early as the stress of a dysfunctional family environment can activate the physiological responses to stress, change the brain structurally, and leave children more vulnerable to negative health outcomes. High conflict spouses also experience chronic stress, high blood pressure, and depression .
The family is also the primary socialization

Robert Kneschke/Shutterstock.com

unit for physical health in reference to eating nutritious food, avoiding smoking, and getting regular exercise. The first author of this text was reared in a home where a high fat diet was routine, and his father was a chronic smoker who never exercised. He died of a coronary at age 46, illustrating the need for attending to one's health.

16. **Psychological health.** One's personal psychological health is also related to marital success. Having a positive self-concept is associated with viewing one's marriage as a success (Arnold et al., 2011). Similarly, being happy as an individual is associated with being happy in one's marital relationship. Stanley et al. (2012) found that an important predictor of later marital satisfaction is being happy in one's own life.

17. **Flexibility.** Flexibility is an important quality for long-term stability. Flexible couples make mutual decisions, accommodate as necessary to each other's schedule, and ensure that they spend time together. Hence, whatever the issue of contention, their relationship is more important.

8-5b Theoretical Views of Marital Happiness and Success

Interactionists, developmentalists, exchange theorists, and functionalists view marital happiness and success differently. Symbolic interactionists emphasize the subjective nature of marital happiness and point out that the definition of the situation is critical. Indeed, a happy marriage exists only when spouses define the verbal and nonverbal behavior of their partner as positive, and only when they label themselves as being in love and happy. Marital happiness is not defined by the existence of specific criteria (time together) but by the subjective definitions of the respective partners. Indeed, spouses who work together may spend all of their time together but define their doing so as a negative. Shakespeare's phrase "Nothing is either good or bad but thinking makes it so" reflects the importance of perception.

Family developmental theorists emphasize developmental tasks that must be accomplished to enable a couple to have a happy marriage. Wallerstein and Blakeslee (1995) identified several of these tasks, including separating emotionally from one's parents, building a sense of "we-ness," establishing an imaginative and pleasurable sex life, and making the relationship safe for expressing differences.

Exchange theorists focus on the exchange of behavior of a kind and at a rate that is mutually satisfactory to both spouses. When spouses exchange positive behaviors at a high rate (with no negatives), they are more likely to feel marital happiness than when the exchange is characterized by high-frequency negative behavior (and no positives).

Structural functionalists regard marital happiness as contributing to marital stability, which is functional for society. When two parents are in love and happy, the likelihood that they will stay together to provide for the physical care and emotional nurturing of their offspring is increased. Furthermore, when spouses take care of their own children, society is not burdened with having to pay for their care through welfare payments, paying foster parents, or paying for institutional management (group homes).

> "Being in a **long marriage** is a little bit **like** that nice **cup of coffee** every morning—I might have it **every day**, but I **still enjoy it.**"
>
> —STEPHEN GAINES, AMERICAN AUTHOR

8-5c Marital Happiness across Time

Anderson et al. (2010) analyzed longitudinal data of 706 individuals over a 20-year period. Over 90% were in their first marriage; most had two children and 14 years of education. Reported marital happiness, marriage problems, time spent together, and economic hardship were assessed. Five patterns emerged:

1. **High stable 2** (started out happy and remained so across time) = 21.5%

2. **High stable 1** (started out slightly less happy and remained so across time) = 46.1%

3. **Curvilinear** (started out happy, slowly declined, followed by recovery) = 10.6%

4. **Low stable** (started out not too happy and remained so across time) = 18.3%

5. **Low declining** (started out not too happy and declined across time) = 3.6%

The researchers found that, for couples who start out with a high level of happiness, they are capable of rebounding if there is a decline. But for those who start out at a low level, the capacity to improve is more limited.

Spencer and Amato (2011) emphasized the importance of using a number of variables such as interaction and conflict rather than just marital happiness in the examination of marital relationships over time. Based on data from couples married over 20 years they found that marital happiness shows a U-shape distribution but interaction declined across time. The researchers hypothesized that couples do not interact less because they are unhappy but because they have other interests.

Plagnol and Easterlin (2008) conceptualized happiness as the ratio of aspirations and attainments in the areas of family life and material goods. They studied 47,000 women and men and found that up to age 48, women are happier than men in both domains. After age 48, a shift causes women to become less satisfied with family life (their children are gone) and material goods (some are divorced and have fewer economic resources). In contrast, men become more satisfied with family life (the empty nest is more a time of joy) and finances (men typically have more economic resources than women, whether married or divorced). Individuals who are Black and/or with lower education report less happiness.

8-6 Trends in Marriage Relationships

Diversity will continue to characterize marriage relationships of the future. The traditional model of the husband provider, stay-at-home mom, and two children will continue to transition to other forms including more women in the workforce, single parent families, and smaller families. What will remain is the intimacy/companionship focus that spouses expect from their marriages.

Openness to interracial, interreligious, cross-national, and age-discrepant relationships will increase. The driving force behind this change will be the U.S. value of individualism which minimizes parental disapproval. An increased global awareness, international students, and study abroad programs will facilitate increased opportunities and a mindset of openness to diversity in terms of one's selection of a partner.

STUDY TOOLS ⮞ 8

Ready to study? In the book, you can:

- ⮞ Rip out the chapter review card at the back of the book for a handy summary of the chapter and key terms.

- ⮞ Assess your attitudes toward interracial dating with the Self-Assessment card at the back of the book.

Online at CENGAGEBRAIN.COM you can:

- ⮞ Prepare for tests with quizzes.

- ⮞ Review the key terms with Flash Cards.

- ⮞ Play games to master concepts.

Money, Work, and Relationships

"I **wanted** the **gold**, and I **got it**—
Came out with a fortune last fall,—
Yet somehow life's not what I thought it,
And somehow the **gold isn't** all."

—ROBERT W. SERVICE, *SPELL OF THE YUKON*

Romance does not occur in a vacuum. The context of all relationships is work which impacts one's availability for romance. The first Spiderman movie included a scene where Spiderman and Mary Jane finally kiss in romantic bliss. As they move hungrily into each other's mouths, a siren is heard in the background; Spiderman stops the kiss and looks to Mary Jane for her approval to go to work. She responds, "Go get 'em tiger."

Money also rocks wealthy lovers into reality. When George Clooney announced his engagement to Amal Alamuddin, questions were asked about a prenuptial agreement. While no source has produced the actual prenuptial agreement (if there is one), the entertainment media was abuzz with how Clooney would protect his $200 million wealth.

"Sources" say Clooney has a prenup, that it includes $1 million as a parting gift if the marriage lasts less than a year, and that more is added as long as the marriage continues. Whatever the case, it is clear that work and money are the backdrops of any relationship.

Rob Marmion/Shutterstock.com

9-1 Money and Relationships

"Marriage is about love; divorce is about money."

—UNKNOWN

Money is a concern for most individuals. When 487 men and 513 women were asked what area of their life they would most want to change (study conducted by Kelton Research), finances was at the top of the list (selected by 75% of men and 81% of women) over health, appearance, and self-esteem (Healy & Trap, 2012). Money affects the timing of one's marriage as high inflation creates economic uncertainty and reduces the marriage rate (Schellekens & Gliksberg, 2013). Whether a potential partner is in debt has an influence on the suitability of that partner as a potential mate. In a study of 5,500 never-married individuals,

almost two-thirds (65%) reported that they would not date someone with a credit card debt over $5,000 (Walsh, 2013). Debt is common among students—the median college debt is $13,000 (Fry & Caumont, 2014). Finally, arguing about money is a gateway to additional problems in the relationship, such as spending less time together and having disagreements about sex (Wheeler & Kerpelman, 2013).

Courtesy of Brittany Bolen

"How do I **love thee**? Let me count **your** money."

—UNKNOWN

Difficult economic times are also associated with positive consequences such as causing individuals to become less consumer oriented, more engaged in their relationships, and more involved in transcendental activities (religious, contemplative) (Etzioni, 2011). Indeed, the entrenched value of **consumerism**—to buy everything and to have everything now—has come under fire. The real stress that money inflicts on relationships is the result of internalizing the societal expectations of who one is, or should be, in regard to the pursuit of money. A wife whose husband had just bought a second McDonald's franchise said, "It's not fun anymore. We have money but I never see him. It isn't worth it." (The husband subsequently sold both stores, and while the couple had less money, their marriage recovered.)

9-1a **Money as Power in a Couple's Relationship**

Money is a central issue in relationships because of its association with power, control, and dominance. Generally, the more money a partner makes, the more power that person has in the relationship. Males in general make considerably more money than females and generally have more power in relationships. The average annual income of a male with some college education who is

Consumerism to buy everything and to have everything now.

poverty the lack of resources necessary for material well-being.

working fulltime is $52,580 compared with $36,553 for a female with the same education, also working full-time (*Statistical Abstract of the United States, 2012–2013*, Table 703).

When a wife earns an income, her power in the relationship increases. We (the authors of your text) know of a married couple in which the wife recently began to earn an income. Before doing so, her husband's fishing boat was stored in the couple's protected carport. Her new job/income resulted in her having more power in the relationship, as reflected in her parking her car in the carport and her husband putting his fishing boat underneath the pine trees to the side of the house.

A record 40% of all households with children under the age of 18 include mothers who are either the sole or primary source of income for the family. These "breadwinner moms" are made up of two very different groups: 5.1 million (37%) are married mothers who have a higher income than their husbands, and 8.6 million (63%) are single mothers (Wang et al., 2013).

To some individuals, money means love. While admiring the engagement ring of her friend, a woman said, "What a big diamond! He must really love you." The cultural assumption is that a big diamond equals expense and a lot of sacrifice and love. Similar assumptions are often made when gifts are given or received. People tend to spend more money on presents for the people they love, believing that the value of the gift symbolizes the depth of their love. People receiving gifts may make the same assumption. "She must love me more than I thought," mused one man. "I gave her a Blu-Ray movie for Christmas, but she gave me a Blu-Ray player. I felt embarrassed."

9-1b **Effects of Poverty on Marriages and Families**

In 2014, a two-person household with an income below $15,370 was defined as living in **poverty** (Poverty line, 2014). Such individuals have poorer physical and mental health, report lower personal and relationship satisfaction, and die at younger ages. The anxiety over lack of money may result in relationship conflict. Hardie and Lucas (2010) analyzed data from over 4,000 respondents in both married and cohabitation contexts and found that economic hardship was associated with more conflict in both sets of relationships. Predictably, money (the lack of it, disagreement over how it is spent) is a frequent problem reported by couples who become involved in marriage counseling.

Parenting is also negatively affected by economic pressure (e.g., cannot cover expenses, financial cutbacks, material needs). Parents experiencing economic pressure report more emotional distress and harsher parenting (Neppl, 2012). Relationship therapy may help to reduce the negative effects of economic pressures.

Work is often tiresome and relentless.

> "If I had a **dollar for every time** a random woman walked up to me and **tried to seduce me, I'd have 50 cents**. That's assuming drag queens are half price."
>
> —JAROD KINTZ, *THIS BOOK IS NOT FOR SALE*

Kahneman and Deaton (2010) defined emotional well-being as the quality of an individual's everyday experience—the frequency and intensity of experiences of joy, stress, sadness, anger, and affection that make one's life pleasant or unpleasant. They found that reporting emotional well-being rises with income and tops out at $75,000 a year (adjusting for inflation through 2015, the figure is around $80,000). Increases in income above this figure are not associated with increases in satisfaction.

9-2 Work and Marriage

> "I felt **poor** only when he started to **love his job more than** he loved his **family**."
>
> —DOREEN DUBREUIL

A couple's marriage is organized around the work of each spouse. Where the couple live is determined by where the spouses can get jobs. Jobs influence what time spouses eat, which family members eat with whom, when they go to bed, and when, where, and for how long they vacation. In this section, we examine some of the various influences of work on a couple's relationship. We begin by looking at the skills identified by dual earner spouses to manage their work/job/career so as to provide income for the family but minimal expense to their relationship.

9-2a Basic Rules for Managing One's Work Life to Have a Successful Marriage

Ma a tta and Uusiautti (2012) analyzed data from 342 married couples who explained their secrets for maintaining a successful relationship in the face of the demands of work. These secrets included turning a negative into a positive, being creative, tolerating dissimilarity (accepting the partner), and being committed to the relationship. The latter focused on accepting as a premise that the partners will work through whatever difficulties they encounter.

9-2b Employed Wives

Driven primarily by the need to provide income for the family, 77% of all U.S. wives with children are in the labor force. The time wives are most likely to be in the labor force is when their children are teenagers (between the ages of 14 and 17), the time when food and clothing expenses are the highest (*Statistical Abstract of the United States, 2012–2013*, Tables 599 and 600).

Some women prefer to be employed part-time rather than full-time (Parker & Wang, 2013). One option is the teaching profession, which allows employees to work about 10 months a year and to have 2 months free in the summer. Although many low-wage earners need two incomes to afford basic housing and a minimal standard of living, others have two incomes to afford expensive homes, cars, vacations, and educational opportunities for their children. Treas et al. (2011) found that homemakers (in 28 countries) are happier than full-time working wives. However, to be clear, many women enjoy their jobs and careers, and staying at home with children/being a full-time homemaker is not a role in which these women are interested.

Supertrooper/Shutterstock.com

Hoffnung and Williams (2013) assessed 200 female college seniors, most of whom wanted both a career and children, and followed up on them 16 years later. Though they could be divided into three role-status outcome groups—Have It All (mothers, employed full-time); Traditional (mothers, employed part-time or not at all), and Employed Only (childfree, employed full-time)—most of the women still wanted to "have it all." Many traditional women looked forward to returning to employment, and many of the employed only women wanted to have children. Some parents wonder if the money a wife earns by working outside the home is worth the sacrifices to earn it. Not only is the mother away from her children, but she must pay for others to care for the children.

According to the U.S. Bureau of Labor statistics, $62, 985.000 (in 2014) was the value of a stay-at-home mother per year in terms of what a dual-income family spends to pay for all the services that she provides (domestic cleaning, laundry, meal planning and preparation, shopping, providing transportation to activities, taking the children to the doctor, and running errands) (Carey & Trapp, 2014a). (The economic value of dad's work at home in 2014 was $24,103 (Insure.com, 2014).

Of interest is that working mothers give themselves slightly higher ratings than nonworking mothers for the job they are doing as parents. Among mothers with children under age 18 who work full-time or part-time, 78% say they are doing an excellent or very good job as parents. Among mothers who are not employed, 66% say the same (Parker & Wang, 2013).

The **mommy track** (stopping paid employment to spend time with young children) is

mommy track stopping paid employment to be at home with young children.

dual-career marriage one in which both spouses pursue careers.

another potential cost to those women who want to build a career. Taking time out to rear children in their formative years can derail a career. Unless a young mother has a supportive partner with a flexible schedule, family who live close and who can take care of her children, or money to pay for child care, she will discover that corporations need the work done and do not care about the kids. Kahn et al. (2014) reconfirmed that motherhood is costly to women's careers, but mostly to women who have three children. Those who have one or two children experience the greatest expense when the woman is young, but this disadvantage is eliminated by their forties and fifties.

9-2c Office Romance

The office or workplace is where people earn money to pay bills and pay down their debt. It is also a place where they meet and establish relationships, including love and sexual relationships. In a survey of over a thousand employees in the workplace, almost 60% (56%) said that they had been involved in an office romance (Vault Office Romance Survey, 2014). In an annual workplace romance survey (CareerBuilder.com, 2014), 38% of the respondents reported having dated someone who worked in the same company. Jane Merrill interviewed 70 adults who had been involved in an office romance. Of these interviews she reported that "…the office is still a scene of seduction and amorality. The hallmark of office romances is secrecy—either hiding a steamy short or long love relationship between two single people or where one or both is married or in a serious relationship" (Merrill & Knox, 2010). Undergraduates have also experienced romance on the job (see What's New? section).

9-2d Types of Dual-Career Marriages

A **dual-career marriage** is defined as one in which both spouses pursue careers and may or may not include dependents. A career is different from a job in that the former usually involves advanced education or training,

What's New?

UNDERGRADUATE LOVE AND SEX ON THE JOB

Seven-hundred seventy-four undergraduates in North Carolina, Florida, and California completed an Internet survey on office romance. The purpose was to identify the percent of the sample who had fantasized about love and sex on the job and the percent who actually experienced their fantasies. Other objectives included the percent reporting having kissed a coworker, whether the involvement was with someone of equal or higher status, how often they disclosed an office relationship to others, and the outcome of the office romance.

Over three-fourths (78%) of the respondents were female; 22% were male. Forty-seven percent were employed in a service job such as sales, fast food, or retail; 9% worked in an academic context, 12% in an office, and 4% in a medical context. Most (73%) regarded where they worked as a job, not a career, and 13% saw it as a place to meet a future spouse.

Over 40% of the student employees reported fantasies of both love and sex with a coworker in contrast to a quarter who *actually experienced* love and sex with a coworker.

Fantasies and Realities on the Job (N = 774)

	Fantasized About, %	Actually Experienced, %
Love relationship at the office	43	25
Sexual relationship at the office	41	24

Other findings included:

Kissing

Almost 30% (28.6%) reported that they had kissed a fellow employee at work.

Rank of Person Undergraduate Had Sex With

Almost 80% (79%) reported having had sex with a peer/coworker; 11% of those surveyed had sex with a boss, supervisor, or someone above them.

Telling Others about the Office Romance

About 30% (28.1%) reported that they told someone else about their office romance.

After the Office Romance Ends

Of the workplace romances that had ended, almost three-quarters (73%) ended positively with over half (54%) remaining friends. Thirteen percent still see each other, 5% are married to each other, and 1% live together. Less than 2% (1.7%) ended up losing their job at the place where they had the office romance.

The Society for Human Resources Management conducted a survey in which they interviewed 380 human resource professionals about office romance policies. Over half (54%) had no policy; 36% had a written policy; and 6% had a verbal policy (Carey & Trapp, 2013).

Sources: Carey, A. R., & P. Trapp. (2013, October 20). Companies' workplace policies. *USA Today*, p. A1.

Merrill, J., & D. Knox. (2010). *Finding love from 9 to 5: Trade secrets of an office love*. Santa Barbara, CA: Praeger.

full-time commitment, long hours/night/weekend work "off the clock," and a willingness to relocate. Dual-career couples typically operate without a person (e.g., a "wife") who stays home to manage the home and children. However, some couples hire a nanny.

Types of dual-career marriages include those in which both careers are equally important (**HIS/HER career marriage**). About 6 in 10 wives work today, nearly double the percentage in 1960. Sixty-two percent endorse the modern marriage (HIS/HER career marriage) in which the husband and wife both work and both take care of the household and children (Parker & Wang, 2013).

HIS/HER career marriage a husband's and wife's careers are given equal precedence.

Some HIS/HER career marriages are **commuter marriages**, a type of long-distance marriage where spouses live in different locations during the workweek (and sometimes for longer periods of time) to accommodate the careers of the respective spouses. The estimated number of commuter marriages is about 3.5 million—50 of the wives from which had been in a commuter marriage at least a year and were interviewed (another 25 wives were in focus groups) by McBridge and Bergen (2014).

One issue spouses coped with was the physical separation from the partner. One spouse said, "Say you have something great happen to you at work, and the first person you want to walk in and see is your husband or your wife and tell them what happened. Well, we don't have them at home." Another issue is the lack of understanding others have of such a "weird" arrangement ("If they really loved each other, they would not get jobs apart from each other" is the thinking). A third issue is managing fidelity. One wife warned, "Be sure that your marriage can take this...because if you have any doubt in your mind that he's not going to be faithful to you...or that the marriage is strong enough that it's going to make it through this commute...don't do it."

For couples who do not have traditional gender role attitudes, the wife's career may take precedence (**HER/HIS career marriage**). In such marriages, the husband is willing to relocate and to subordinate his career for his wife's. Such a pattern is also likely to occur when a wife earns considerably more money than her husband. In some cases, the husband who is downsized by his employer or who prefers the role of full-time parent becomes "Mr. Mom." According to Pew Research Center analysis of 2012 census data, stay-at-home days make up 16% of at-home parents (Carey & Trapp, 2014b). There are advantages to both the wife, husband, and children. Not only may the wife benefit in terms of her career advancement but the relationship of the father with his children will be closer than that of fathers in traditional marriages who spend less time with their children. More common, the husband's career takes precedence (HIS/HER career marriage), the traditional pattern. In some cases, both spouses share a career or work together (**THEIR career marriage**). Some news organizations hire both spouses to travel abroad to cover the same story. These careers are rare.

In the following sections, we look at the effects on women, men, their marriage, and their children when a wife is employed outside the home.

9-3 Effects of the Wife's Employment on the Spouses and Marriage

A major challenge is for women to combine a career and motherhood. While some new mothers enjoy their work role and return to work soon after their children are born, others anguish over leaving their baby to return to the workforce.

The new mother discovers that there are now two spheres to manage—work and family—which may result in **role overload**—not having the time or energy to meet the demands of their responsibilities in the roles of wife, parent, and worker. Because women have traditionally been responsible for most of the housework and child care, employed women come home from work to what Hochschild (1989) calls the second shift, housework and child care that have to be done after work. According to Hochschild, the **second shift** has the following result:

> Women tend to talk more intently about being overtired, sick, and "emotionally drained." Many women could not tear away from the topic of sleep. They talked about how much they could "get by on" . . . six and a half, seven, seven and a half, less, more . . . . Some apologized for how much sleep they needed . . . . They talked about how to avoid fully waking up when a child called them at night, and how to get back to sleep. These women talked about sleep the way a hungry person talks about food. (*p. 9*)

---

commuter marriages a type of long-distance marriage where spouses live in different locations during the workweek (and sometimes for longer periods of time) to accommodate the careers of the respective spouses.

HER/HIS career marriage a wife's career is given precedence over her husband's career.

THEIR career marriage a career shared by a couple who travel and work together (e.g., journalists).

role overload not having the time or energy to meet the demands of their responsibilities in the roles of wife, parent, and worker.

second shift housework and child care that are done when the parents return home after work.

Personal View: Going Back to Work—A New Mother's View

Some mothers anguish over leaving their newborn and returning to full-time employment.

You probably get the most advice of your life during pregnancy, especially if it's your first. Some advice is very welcome and some, not so much. Actually, I wanted as much advice as possible about anything and everything baby. I figured it would better prepare me . . . HA! "Be prepared" is something I heard repeatedly after I answered yes to "Are you going back to work?" I knew I would be very sad, but I also knew there was no other choice. I was too busy worrying about making sure all the cute pink things were washed and folded (and refolded) in her drawer as I anticipated her arrival. Thinking about having her and then having to leave her were distant thoughts. I mean, 8 to 10 weeks of time off is a really long time isn't it? More time off than a summer between grade school. It seemed like an eternity. But there is no way to prepare yourself for the reality. She is born, your life is forever different. Your time in the hospital is a whirlwind. You blow in, and blow out.

Then you are home, with your precious baby and you have 10 weeks . . . 10 long weeks. But no matter how much time you take with your newborn, it flies by. It's never enough time. You are trying to heal from delivery and learn about being a new mother at the same time. You're losing sleep and using more energy. A spinning tornado of being tired and falling in love . . . and POOF . . . it's gone. You go through periods of guilt for leaving her, sadness, missing her, worrying to death and even a slight bit of anger at your spouse. Hey, "if you made more money, we could afford for me to stay home."

Your first day back at work will be very hard. If you work somewhere with many employees like I did, I had people coming to my office every few minutes, "It's nice to have you back!" What can you possibly say to that? It's good to be back . . . hell no. But you are not going to tell them how much you would rather be home with your baby. If you are lucky, you have an amazing mother who always says the right thing at the right time. When I was pouring my heart out to her over the phone my last day at the house with my daughter she said, "Yes, I couldn't imagine leaving my babies with anyone else. It's going to be really hard when she does everything first with them and not with you. I sure hope she says mama before her caregiver's name." Talk about a freak-out.

The best thing you can do if going back to work is hard for you is to talk to other mothers that went through it. One of my friends stayed home for a year before she put her son in day care. She says now that she would feel like a bad mother if she took him out of his day care because he learns so much and really enjoys it. When I'm feeling really sad about being at work I try to think of all the things we will be able to afford for her because I work. Like sports teams, music lessons, dance, vacations, etc., that are also important for her growth and development. Also remember that no matter what, you will always be mommy, and no one can take that away.

—Amanda Kinsch
(Subsequent to writing the above, Amanda quit her job and she and her husband have another baby.)

Part of this role overload may be that women have more favorable attitudes toward housework and child care than men. Two researchers assessed the attitudes of 732 spouses and found that women had more favorable attitudes toward cleaning, cooking, and child care than did men—women enjoyed these activities more, set higher standards for them, and felt more responsible for them (Poortman & Van der Lippe, 2009).

Another stressful aspect of employment for employed mothers in dual-earner marriages is **role conflict**, being confronted with incompatible role obligations. For example, the role of a career woman is to stay late and prepare a report for the following day. However, the role of a mother is to pick up her child from day care at 5 P.M. When these roles collide, there is role conflict. Although most women resolve their role conflicts by giving preference to the mother role, some give priority to the career role and feel guilty about it.

Role strain, the anxiety that results from being able to fulfill only a limited number of role obligations, occurs for both women and men in dual-earner marriages. No one is at home to take care of housework and children while they are working, and they feel strained at not being able to do everything.

Konstantin Gushcha/Shutterstock.com

9-3a Effects of the Wife's Employment on Her Husband

Husbands also report benefits from their wives' employment. These include being relieved of the sole responsibility for the financial support of the family and having more freedom to quit his job, change jobs, or go to school. Since men traditionally had no options but to work full-time, they now benefit by having a spouse with whom to share the daily rewards and stresses of employment.

9-3b Effects of the Wife's Employment on the Marriage

Helms et al. (2010) analyzed the relationship between employment patterns and marital satisfaction of 272 dual-earner couples and found that coprovider couples (in contrast to those where there were distinct primary and secondary providers) reported the highest marital satisfaction. In addition, these couples reported the most equitable division of housework. Lam et al. (2012) confirmed that employment of the wife was associated with greater household participation on the part of the husband.

role conflict being confronted with incompatible role obligations.

role strain the anxiety that results from being able to fulfill only a limited number of role obligations.

However, Minnotte et al. (2013) found reduced marital satisfaction for more egalitarian husbands and wives, suggesting that husbands may have resented the intrusion of family life into their work roles. More traditional relationships were also associated with higher sex frequency (Kornrich et al., 2013).

Are marriages in which the wife has her own income more vulnerable to divorce? Not if the wife is happy. But if she is unhappy, her income will provide her a way to take care of herself when she leaves. Schoen et al. (2002) wrote, "Our results provide clear evidence that, at the individual level, women's employment does not destabilize happy marriages but increases the risk of disruption in unhappy marriages" (p. 643). Hence, employment would not affect a happy marriage but it can affect an unhappy one. Further research by Schoen et al. (2006) revealed that full-time employment of the wife is actually associated with marital stability.

In one-fourth of marriages, wives earn higher incomes than their husbands. In a survey of 4,606 adult workers, 59% of the men agreed that "I am comfortable if my spouse earns more than I do" (Yang & Ward, 2010). Meisenbach (2010) interviewed 15 U.S. female breadwinners (FBWs) and identified their issues, concerns, and worries. They enjoyed having control in their relationships but were ambivalent about whether the control was good for the relationship.

What is the effect of control over one's time devoted to work? Olsen and Dahl (2010) confirmed that having control over the hours one works is related to positive family functioning. Sandberg et al. (2013) analyzed data on 281 couples and found that negative couple interaction was associated with less work satisfaction and elevated depression scores; hence, a marriage-to-work spillover can be costly for families, organizations, and governments.

9-4 Work and Family: Effects on Children

Dual-earner parents want to know how children are affected by maternal employment. Johnson et al. (2013) found that the number of hours worked by the mother was unrelated to the behavior of middle school children. However, where the single mother or both the mother and father work nonstandard hours (nights/weekends) there is an association with lower levels of parent–child closeness with adolescents (Hendrix & Parcel, 2014).

However, Bauer et al. (2012) confirmed that the more hours a mother worked outside the home (regardless of when the hours were) the less often the family shared family meals and the less often adolescents were encouraged to eat fruits and vegetables. And there is a positive effect of wife's employment on the relationship of the father with his children. Meteyer and Perry-Jenkins (2012) confirmed that the more hours the wife worked, the greater the father was involved with his children. However, a disadvantage for children of two-earner parents is that they receive less supervision. Letting children come home to an empty house is particularly problematic. The issue of self-care children is discussed shortly.

9-4a Quality Time with Children

The term *quality time* has become synonymous with good parenting. Dual-income parents struggle with not having enough quality time with their children. Mormons typically set aside "Monday home evenings" as a time to bond, pray, and sing together. Other parents (child-centered parents) noted that quality time occurred when they were having heart-to-heart talks with their children. Still other parents believed that all the time they were with their children was quality time. Whether they were having dinner together or riding to the post office, quality time was occurring if they were together. As might be expected, mothers assumed greater responsibility for quality time.

9-4b Day Care Considerations

While most parents prefer relatives (spouse or partner or another relative) for the day care arrangement for their children, more than half of U.S. children are in center-based child care programs. Forty-three percent of 3-year-olds and 69%

Parents are always apprehensive about the care their babies and children are getting when they are in day care.

Courtesy of Brittany Bolen

of 5-year-old children are in center-based day care programs (*Statistical Abstract of the United States 2012–2013*, Table 578).

Employed parents are concerned that their children get good-quality care. Their concern is warranted. Cortisol is a steroid hormone that plays an important role in adaptation to stress. Higher levels of cortisol reflect higher levels of stress experienced by the individual. Gunnar et al. (2010) assessed the cortisol stress levels of 151 children in full-time, home-based day care centers and compared these levels to those of children the same age who were cared for in their own home. Increases were noted in the majority of children (63%) at day care, with 40% classified as a stress response. These increased cortisol levels began in the morning and continued throughout the afternoon.

Vandell et al. (2010) examined the effects of early child care 15 years later and found that higher quality care predicted higher cognitive–academic achievement at age 15, with escalating positive effects at higher levels of quality. These findings do not mean that day care is bad for children, but suggest the need for caution in selecting a day care center.

Wagner et al. (2013) confirmed the exhaustion/stress experienced by full-time day care workers. However, De Schipper et al. (2008) noted that day care workers who engage in high-frequency positive behavior engender secure attachments with the children they work with. Hence,

MidoSemsem/Shutterstock.com

children of such workers do not feel they are on an assembly line but bond with their caretakers. Parents concerned about the quality of day care their children receive might inquire about the availability of webcams. Some day care centers offer full-time webcam access so that parents or grandparents can log onto their computers and see the interaction of the day care worker with their children.

Day care costs are a factor in whether a low-income mother seeks employment, because the cost can absorb her paycheck. Even for dual-earner families, cost is a factor in choosing a day care center. Day care costs vary widely from nothing, where friends trade off taking care of the children, to very expensive institutionalized day care in large cities. The cost of high-quality infant care can be as high as $2,000 per month. Day care for children ages 2 and over ranged from $300 to $1,600 a month in 2014. These costs are for one child; most spouses have two children. Do the math.

> **"I balance work with raising 2 kids with great precision.** *I plan everything in advance: who's picking up. We have charts, maps and lists on the fridge, all over the house. I sometimes feel like I'm with the CIA.* **"**
>
> —KATE WINSLET, ACTRESS

9-5 Balancing Work and Family Life

Work is definitely stressful on individuals, spouses, and relationships. Stulhofer et al. (2013) analyzed data on 2,112 men and found that the chance of experiencing one or more sexual health difficulties in the past 12 months were about 1.8 times higher among men who reported the highest levels of workplace difficulties than among men who experienced no such difficulties.

Hoser (2012) examined work and family life and found that the **spillover thesis** worked in only one direction—work spread into family life in the form of doing overtime, taking work home, attending seminars organized by the company, and being "on call" on the weekend/during vacation. Rarely did one's family life dictate the work role.

Virtual work is one answer. Based on a survey of 2,500 adults age 18 and older conducted by Ricoh Americas and Harris International, 64% of the respondents reported a preference for working "virtually" compared to 34% who preferred to work in an office (Yang & Gonzalez, 2013).

One of the major concerns of employed parents and spouses is how to juggle the demands of work and family. When there is conflict between work and family, various strategies are employed to cope with the stress of role overload and role conflict, including (1) the superperson strategy, (2) cognitive restructuring, (3) delegation of responsibility, (4) planning and time management, and (5) role compartmentalization (Stanfield, 1998).

9-5a Superperson Strategy

The **superperson strategy** involves working as hard and as efficiently as possible to meet the demands of work and family. A superperson often skips lunch and cuts back on sleep and leisure to have more time available for

spillover thesis work spreads into family life in the form of the worker/parent doing overtime, taking work home, attending seminars organized by the company, being "on call" on the weekend/during vacation, and always being on the computer in reference to work.

superperson strategy involves working as hard and as efficiently as possible to meet the demands of work and family.

David Crockett/Shutterstock.com

Technology has made it increasingly difficult to separate work and play. This woman is on vacation but checking and responding to work emails.

work and family. Women are particularly vulnerable because they feel that if they give too much attention to child care concerns, they will be sidelined into lower-paying jobs with no opportunities.

Hochschild (1989) noted that the terms **superwoman** or **supermom** are cultural labels that allow a woman to regard herself as very efficient, bright, and confident. However, Hochschild noted that this is a "cultural cover-up" for an overworked and/or frustrated woman. As noted earlier, not only does the woman have a job in the workplace (first shift), she comes home to another set of work demands in the form of house care and child care (second shift). Finally, she has a third shift (Hochschild, 1997).

The **third shift** is the expense of emotional energy by a spouse or parent in dealing with various issues in family living. Although young children need time and attention, responding to conflicts and problems with teenagers also involves a great deal of emotional energy. Minnottea et al. (2010) studied 96 couples and found that women perform more "emotion work." Opree and Kalmijn (2012) noted that employed women who also take care of an adult child or aging parents are more likely to report negative changes in their own mental health.

Mothers who try to escape the supermom trap through eliciting the help of their husbands may experience a downside. Sasaki et al. (2010) studied 78 dual-career couples with an 8-month-old infant and found that a greater husband's contribution to care giving was associated with the wife's lower self-competence. The authors concluded that "despite increasingly egalitarian sex roles, employed mothers seem to be trapped between their desire for help with childrearing and the threat to their personal competence posed by failure to meet socially constructed ideals of motherhood."

9-5b Cognitive Restructuring

Another strategy used by some women and men experiencing role overload and role conflict is **cognitive restructuring**, which involves viewing a situation in positive terms. Exhausted dual-career earners often justify their time away from their children by focusing on the benefits of their labor: their children live in a nice house in a safe neighborhood and attend the best schools. Whether these outcomes offset the lack of quality time may be irrelevant; the beliefs serve simply to justify the two-earner lifestyle.

9-5c Delegation of Responsibility and Limiting Commitments

A third way couples manage the demands of work and family is to delegate responsibility to others for performing certain tasks. Because women tend to bear most of the responsibility for child care and housework, they may choose to ask their partner to contribute more or to take responsibility for these tasks.

Another form of delegating responsibility involves the decision to reduce one's current responsibilities and not take

superwoman (supermom) a cultural label that allows a mother who is experiencing role overload to regard herself as particularly efficient, energetic, and confident.

third shift the emotional energy expended by a spouse or parent in dealing with various family issues.

cognitive restructuring viewing a situation in positive terms.

Christopher Boswell/Shutterstock.com

on additional ones. For example, women and men may give up or limit committing to volunteer responsibilities. One woman noted that her life was being consumed by the responsibilities of her church; she had to change churches because the demands were relentless. In the realm of paid work, women and men can choose not to become involved in professional activities beyond those that are required.

9-5d **Time Management**

While two-thirds of women prefer to work part-time as opposed to full-time, Vanderkam (2010) argues that women who work part-time end up spending just 41 more minutes daily on child care and 10 minutes more per day playing with the child than if they worked full-time. By working full-time, she says the woman affords high-quality day care and can focus on the child when she is not at work. The full-time worker is not exhausted from being with the children all day but may be more "emotionally ready" to spending dinner, bath, and reading time with their children at night.

Other women use time management by prioritizing and making lists of what needs to be done each day. This method involves trying to anticipate stressful periods, planning ahead for them, and dividing responsibilities with the spouse. Such division of labor allows each spouse to focus on an activity that needs to be done (grocery shopping, picking up children at day care) and results in a smoothly functioning unit.

Having flexible jobs and/or careers is particularly beneficial for two-earner couples. Being self-employed, telecommuting, or working in academia permits flexibility of schedule so that individuals can cooperate on what needs to be done. Alternatively, some dual-earner couples attempt to solve the problem of child care by having one parent work during the day and the other parent work at night so that one of them can always be with the children. Shift workers often experience sleep deprivation and fatigue, which may make fulfilling domestic roles as a parent or spouse difficult for them. Similarly, **shift work** may have a negative effect on a couple's relationship because of their limited time together.

shift work having one parent work during the day and the other parent work at night so that one parent can always be with the children.

role compartmentalization separating the roles of work and home so that an individual does not dwell on the problems of one role while physically being at the place of the other role.

photosync/Shutterstock.com

"It **doesn't interest** me **what you do for a living**. I **want to know** what you ache for, and if you **dare to dream** of meeting your heart's desire."

—ORIAH MOUNTAIN DREAMER, *THE INVITATION*

9-5e **Role Compartmentalization**

Some spouses use **role compartmentalization** separating the roles of work and home so that they do not think about or dwell on the problems of one when they are at the other. Spouses unable to compartmentalize their work and home feel role strain, role conflict, and role overload, with the result that their efficiency drops in both spheres. Some families look to the government and their employers for help in balancing the demands of family and work.

9-6 Trends in Money, Work, and Family Life

Families will continue to be stressed by work. Employers will, increasingly, ask employees to work longer and do more without the commensurate increases in salary or benefits. Businesses are struggling to stay solvent and workers will take the brunt of the instability.

The number of wives who work outside the home will increase—the economic needs of the family will demand that they do so. Husbands will adapt, most willingly, some reluctantly. Children will become aware that budgets are tight, tempers are strained, and leisure with the family in the summer may not be as expansive as previously. As children go to college they can benefit from exposure to financial management information—being careful about credit card debt and how they spend money (Fiona et al., 2012).

While the percent of wives in the workforce may increase, the percent of mothers who do not work outside the home is increasing. The percent in 2012 was 29%, up from 23% in 1999 (Cohn et al., 2014). Difficulty finding employment and concerns about the employment effects on children are explanations.

STUDY TOOLS 9

Ready to study? In the book, you can:

- ⮑ Rip out the chapter review card at the back of the book for a handy summary of the chapter and key terms.

- ⮑ Assess your attitudes toward working mothers with the Self-Assessment card at the back of the book.

Online at CENGAGEBRAIN.COM you can:

- ⮑ Prepare for tests with quizzes.

- ⮑ Review the key terms with Flash Cards.

- ⮑ Play games to master concepts.

Abuse in Relationships

> "Each time he came he **would twist my** defenseless **body** into a different pose, **as if I were** his very own **doll**."
>
> —RACHEL ABBOTT, *ONLY THE INNOCENT*

Television media regularly feature horror stories of women who are beaten up or killed by their partners. Whether the motive is revenge, to be free for another partner, financial gain, or jealousy, the result is the same. What began as an intimate love relationship ends in the death of one's former partner or spouse. In this chapter, we examine the other side of intimacy in one's relationships. We begin by looking at the nature of abuse and defining some terms.

10-1 Types of Relationship Abuse

There are several types of abuse in relationships.

10-1a Violence as Abuse

Also referred to as physical abuse, **violence** is defined as physical aggression with the purpose of controlling or intimidating the partner. Examples of physical violence include pushing, throwing something at the partner, slapping, hitting, and forcing the partner

violence physical aggression with the purpose to control, intimidate, and subjugate another human being.

lofilolo/Getty Images

to have sex. Eleven percent of 3,573 undergraduate females and 3% of 1,149 undergraduate males reported that they had been in a physically abusive relationship (Hall & Knox, 2015). One in four women in the United States experiences interpersonal partner violence in her lifetime (Liu et al., 2013). **Intimate partner violence** (IPV) is an all-inclusive term that refers to crimes committed against current or former spouses, boyfriends, or girlfriends.

There are two types of violence. One type is **situational couple violence** (SCV) where conflict escalates over an issue (e.g., money, sex) and one or both partners lose control. The person feels threatened and seeks to defend himself or herself. Control is lost and the partner strikes out. Both partners may lose control at the same time, so it is symmetrical. A second type of violence,

intimate partner violence an all-inclusive term that refers to crimes committed against current or former spouses, boyfriends, or girlfriends.

situational couple violence conflict escalates over an issue and one or both partners lose control.

referred to as **intimate terrorism** (IT), is designed to control the partner (Brownridge, 2010).

Individuals who have been victims of interpersonal violence report that they have not found it easy to discern when a social encounter was becoming dangerous and was moving toward violence (Witte & Kendra, 2010). The authors emphasized that learning to recognize cues (e.g., the partner is overly aggressive or encourages alcohol consumption) is essential to avoiding such experiences.

Battered woman syndrome (also referred to as **battered woman defense**) is a legal term used in court that the person accused of murder was suffering from abuse so as to justify his or her behavior ("I shot him because he raped me"). While there is no medical or psychological term, the "syndrome" refers to frequent, severe maltreatment which often requires medical treatment. Therapists define battering as physical aggression that results in injury and is accompanied by fear and terror (Jacobsen & Gottman, 2007).

Klipfel et al. (2014) assessed the occurrence of emotional, physical, and sexual interpersonal aggression reported by 161 individuals within various levels of relationships. The relationship and percent of reported intimate partner violence in the various relationships follow: committed romantic relationships (69%), casual dating relationships (33%), friends with benefits relationships (31%), booty calls (36%), and one-night stands (35%). The take home message is that the greater the commitment in the relationship, the greater the reported IPV (intimate partner violence).

Battering may lead to murder. **Uxoricide** is the murder of a woman by a romantic partner. Paralympics superstar Oscar Pistorius was sentenced to five years in prison for the murder of his super model girlfriend Reeva Steenkamp. Thirty percent of homicides of women 18–24 are by an intimate partner (Teten et al., 2009). Most perpetrators are a male partner. Blacks and Hispanic women have higher rates of being killed by their intimate partners than non-Hispanic women (Azziz-Baumgartner et al., 2011). **Intimate partner homicide** is the murder of a spouse.

Other forms of murder in the family are **filicide** (murder of an offspring by a parent), **parricide** (murder of a parent by an offspring), and **siblicide** (murder of a sibling). Family homicide is rare—less than 1 in 100,000 (Diem & Pizarro, 2010). Regarding filicide, parental mental illness, domestic violence, and alcohol/substance abuse are common contexts (Sidebotham, 2013).

A great deal more abuse occurs than is reported. As few as 5% of rapes are reported to the authorities (Heath et al., 2013). The primary reasons rape victims do not report their experience to the police are not wanting others to know, nonacknowledgment of rape (e.g., "It wasn't really rape...I just gave in") and criminal justice concerns ("It won't do any good," "I will be blamed") (Cohn et al., 2013). Yet the psychological damage which results from rape is enormous. Johnson et al. (2008) emphasized that IPV is associated with post-traumatic stress disorder (PTSD) and results in social maladjustment and personal or social resource loss.

Regardless of the type, violence is not unique to heterosexual couples (Porter & Williams, 2011). In a study of violence in a sample of 284 gay and bisexual men, the researchers found that almost all reported psychological abuse, more than a third reported

Physical violence includes choking the partner.

Courtesy of Chelsea Curry

intimate terrorism behavior designed to control the partner.

battered woman syndrome (battered woman defense) legal term used in court that the person accused of murder was suffering from to justify their behavior. Therapists define battering as physical aggression that results in injury and accompanied by fear and terror.

uxoricide the murder of a woman by a romantic partner.

intimate partner homicide murder of a spouse.

filicide murder of an offspring by a parent.

parricide murder of a parent by an offspring.

siblicide murder of a sibling.

physical abuse, and 10% reported being forced to have sex. Abuse in gay relationships was less likely to be reported to the police because some gays did not want to be "outed" (Bartholomew et al., 2008). Differences between violence in other-sex couples versus same-sex couples include that the latter is more mild, the threat of outing is present, and the violence is more isolated because it often occurs in a context of being in the closet. Finally, victims of same-sex relationship abuse lack legal protections and services (most battered women's shelters are not available to lesbians).

10-1b Emotional Abuse

> **"Emotional abuse** is underneath all other types of abuse—the most damaging aspect of physical, sexual, mental, etc., abuse is the **trauma to** our **hearts** and souls from being **betrayed by the people** that we **love and trust."**
>
> —ROBERT BURNEY, THERAPIST

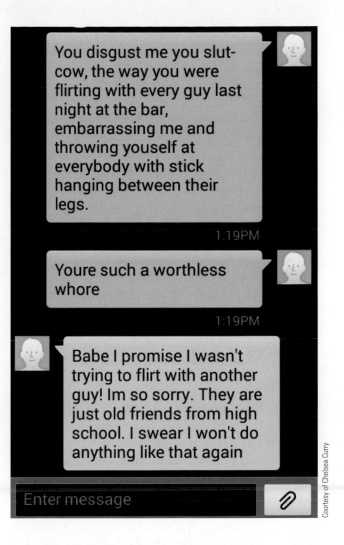

You disgust me you slut-cow, the way you were flirting with every guy last night at the bar, embarrassing me and throwing youself at everybody with stick hanging between their legs.

1:19PM

Youre such a worthless whore

1:19PM

Babe I promise I wasn't trying to flirt with another guy! Im so sorry. They are just old friends from high school. I swear I won't do anything like that again

Enter message

Courtesy of Chelsea Curry

Emotional abuse (also known as psychological abuse, verbal abuse, or symbolic aggression) is non-physical behavior designed to denigrate and control the partner. Follingstad and Edmundson (2010) identified the top emotionally abusive behaviors of one's partner in a national sample of adults in the United States: refusal to talk to the partner as a way of punishing the partner, making personal decisions for the partner (e.g., what to wear, what to eat, whether to smoke), throwing a temper tantrum and breaking things to frighten the partner, criticizing/belittling the partner to make him or her feel bad, and acting jealous when the partner was observed talking or texting a potential romantic partner. Other emotionally abusive behaviors include:

Yelling and screaming—as a way of intimidating
Staying angry/pouting—until the partner gives in
Requiring an accounting—of the partner's time
Criticism—treating the partner with contempt
Ridicule—making the partner feel stupid

Withholding—emotional and physical contact
Isolation—prohibiting the partner from spending time with friends, siblings, and parents
Threats—threatening the partner with abandonment or threats of harm to oneself, family, or partner's pets
Public demeaning behavior—insulting the partner in front of others
Restricting behavior—restricting the partner's mobility (e.g., being told not to spend time with friends, siblings, parents) or telling the partner what to do
Demanding behavior—requiring the partner to do as the abuser wishes (e.g., have sex)

emotional abuse nonphysical behavior designed to denigrate the partner, reduce the partner's status, and make the partner feel vulnerable to being controlled by the partner.

A final example of emotional abuse is **revenge porn**—posting on the Internet nude photos of one's ex-partner. This is both emotional and sexual abuse. Some states are considering legislation against this behavior.

Thirty-six percent of 3,574 undergraduate females (and 22% of 1,148 undergraduate males) reported that they "had been involved in an *emotionally* abusive relationship" (Hall & Knox, 2015). According to Follingstad and Edmundson (2010), while both partners in a relationship may report that they engage in emotionally abusive behavior, each partner tends to report that he or she engages in less frequent and less severe emotionally abusive behavior. In effect, the respondent creates "a picture of their own use of psychological abuse as limited in scope and not harmful in nature" (p. 506).

10-1c **Mutual or Unilateral Abuse?**

Marcus (2012) assessed the frequency of unilateral or mutual partner violence in a sample of 1,294 young adults. A quarter of the couples reported a mutually violent pattern compared to three-quarters who evidenced a unilateral pattern. Mathes (2013) noted that among those couples where violence is a part of their relationship, individuals select each other to play their game—to be violent. In all cases, the more violence, the lower the relationship quality.

10-1d **Female Abuse of Partner**

Swan et al. (2008) revealed that females who abuse their partners are more likely to be motivated by self-defense and fear, whereas male abusers are more likely to be driven by the desire to control. Hence, women who are violent toward their partners tend to strike back rather than throw initial blows. But abusive females may have more variable motives. Whitaker (2014) noted that women were more likely to express violence for reasons of losing their temper, to try to get their partner to listen, to make their partner do as they wanted, or to punish their partner. When

revenge porn posting nude photos of ex; a form of emotional and sexual abuse; some states considering legislation against this behavior.

stalking unwanted following or harassment of a person that induces fear in the victim.

Most female abuse toward males is reactive—they are hit and so they hit back.

Courtesy of Chelsea Curry

women are abusive, they are rarely arrested by the police. Men do not want others to know that they "aren't man enough" to stand up to an abusive woman.

10-1e **Stalking**

Abuse may take the form of stalking. **Stalking** is defined as unwanted following or harassment that induces fear in the victim. This pathological state is characterized by obsessional thinking (Duntley & Buss, 2012).

Five types of stalkers have been identified by Mullen et al. (1999):

1. **Rejected stalkers.** Ex-partners of a terminated romantic relationship seeking revenge. They often want the partner back. These individuals may be dangerous/kill the partner.

2. **Resentful stalkers.** I Individuals who have a vendetta against the victim (e.g., jealous of the person for stealing a boyfriend). Their goal is to frighten or cause the person distress.

3. **Intimacy stalker.** Seeks to have an intimate relationship with the victim. Celebrities David Letterman and Jodi Foster have been stalked by persons who wanted a relationship with them.

4. **Incompetent suitor.** Stalker feels bested by a competitor but still seeks the attention of the victim. The stalker may feel entitled to be loved by the victim.

5. **Predatory stalker.** Sexual predator who is planning to attack the victim.

Twenty-six percent of 3,562 undergraduate females and 18% of 1,150 undergraduate males reported that they had been stalked (followed and harassed) (Hall & Knox, 2015). Most (77%) stalkers are male (Lyndon et al., 2012).

Less threatening than stalking is **obsessive relational intrusion** (ORI), the relentless pursuit of intimacy with someone who does not want it. The person becomes a nuisance but does not have the goal of harm as does the stalker. People who cross the line in terms of pursuing an ORI relationship or responding to being rejected (stalking) engage in a continuum of behavior. Spitzberg and Cupach (2007) have identified ORI behaviors:

1. **Hyperintimacy.** Telling a person that they are beautiful or desirable to the point of making them uncomfortable.

2. **Relentless electronic contacts.** Flooding the person with text or email messages and phone calls. These contacts may reach the extent of **cyber victimization**, which includes being sent threatening text/email, unsolicited obscene email, computer viruses, or junk mail (spamming). Short of cyber victimization is **cyber contol**, whereby individuals use communication technology, such as cell phones, email, and social networking sites to monitor or control romantic partners. In one study of 745 undergraduates, close to 25% of female college students monitored their partner's behavior by checking emails, even password-protected accounts, versus only 6% of males. In addition, more females than males thought it was appropriate to check email and cell phone call histories of their partners. A sociobiological explanation can be used to explain that more females are involved in monitoring their partner's behavior than vice versa—males are more likely to cheat and women are protecting their relationship (Burke et al., 2011).

3. **Interactional contacts.** Showing up at the person's workplace or gym. The intrusion may also include joining the same volunteer groups as the pursued.

4. **Surveillance.** Monitoring the movements of the pursued (e.g., by following the person or driving past the person's house).

5. **Invasion.** Breaking into the person's house and stealing objects that belong to the person; stealing the person's identity; or putting Trojan horses (viruses) on the person's computer. One woman downloaded child pornography on her boyfriend's computer and called the authorities to arrest him. After a jury trial he was sent to prison.

6. **Harassment or intimidation.** Leaving unwanted notes on the person's desk or a dead animal on the doorstep.

7. **Threat or coercion.** Threatening physical violence or harm to the person's family or friends.

8. **Aggression or violence.** Carrying out a threat by becoming violent (e.g., kidnapping or rape).

SkyLine/Fotolia

10-1f **Reacting to the Stalker**

Being very controlling in an existing relationship is predictive that the partner will become a stalker when the relationship ends. Ending a relationship with a potential stalker in a way that does not spark stalking is important. Some guidelines include:

1. Make a direct statement to the person ("I am not interested in a relationship with you, my feelings about you will not change, and I know that you will respect my decision and direct your attention elsewhere").

2. Seek protection through formal channels (police or court restraining order).

obsessive relational intrusion the relentless pursuit of intimacy with someone who does not want it.

cyber victimization harassing behavior which includes being sent threatening email, unsolicited obscene email, computer viruses, or junk mail (spamming); can also include flaming (online verbal abuse) and leaving improper messages on message boards.

cyber control use of communication technology, such as cell phones, email, and social networking sites, to monitor or control partners in intimate relationships.

3. Avoid the perpetrator (ignore text messages/emails, do not walk with or talk to, hang up if the person calls).

4. Use informal coping methods (block text messages/phone calls).

5. Stay away. Do not offer to be friends since the person may misinterpret this as a romantic overture.

> ## "If he mistreated and **abused his last** girlfriend, **why would you want to be** his **new** girlfriend?"
>
> —KAREN E. QUINONES MILLER, AUTHOR

Radius Images/Getty Images

10-2 Reasons for Violence and Abuse in Relationships

Research suggests that numerous factors contribute to violence and abuse in intimate relationships. These factors operate at the cultural, community, individual, and family levels.

10-2a Cultural Factors

In many ways, American culture tolerates and even promotes violence (Walby, 2013). Violence in the family stems from the acceptance of violence in our society as a legitimate means of enforcing compliance and solving conflicts at interpersonal, familial, national, and international levels. Violence and abuse in the family may be linked to such cultural factors as violence in the media, acceptance of corporal punishment, gender inequality, and the view of women and children as property.

corporal punishment the use of physical force with the intention of causing a child to experience pain, but not injury, for the purpose of correction or control of the child's behavior.

vvita/Shutterstock.com

Violence in the Media One need only watch boxing matches, football, or the evening news to see the violence in war, school shootings, and domestic murders. New films and TV movies and TV programs (e.g., *Dateline*) regularly feature themes of violence.

Corporal Punishment of Children The use of physical force with the intention of causing a child pain, but not injury, for the purposes of correcting or controlling the child's behavior and/or making the child obedient is **corporal punishment**. In the United States, corporal punishment in the form of spanking, hitting, whipping, and paddling or otherwise inflicting pain on the child is legal in all states (as long as the corporal punishment does not meet the individual state's definition of child abuse). Seventy percent of U.S. respondents agreed that spanking is necessary to discipline children (Ma et al., 2012). Unlike Sweden, Italy, Germany, and 12 other countries which have banned corporal punishment in the home, violence against children has become a part of U.S. cultural heritage.

Corporal punishment is associated with negative outcomes. Ma et al. (2012) found that the use of corporal punishment on adolescents by mothers and fathers was associated with a greater proportion of youth externalizing behaviors (e.g., breaking rules at home/school, hanging around kids who get into trouble, being mean to others, and threatening to hurt people).

Other researchers have found that children who are victims of corporal punishment display more violence, have an increased incidence of depression as adults, report lower relationship quality as adult college students (Larsen et al., 2011), and are more likely to experience abuse in adult relationships (particularly women) (Maneta et al., 2012). Male exposure to harsh physical punishment (when combined with exposure to sexually explicit pornography) is associated with sexual aggression against women (Simons et al., 2012). Child development specialists recommend an end to corporal punishment to reduce the risk of physical abuse, harm to other children, and to break the cycle of abuse.

Gender Inequality Domestic violence and abuse may also stem from traditional gender roles. Traditionally, men have also been taught that they are superior to women and that they may use their aggression toward women, believing that women need to be "put in their place." The greater the inequality and the more the woman is dependent on the man, the more likely the abuse. Conversely, women with a higher income and education than their partner report more frequent abuse since these achievements may be a threat to the male's masculinity (Anderson, 2010).

Some occupations, such as police officers and military personnel, lend themselves to contexts of gender inequality. In military contexts, men notoriously devalue, denigrate, and sexually harass women. In spite of the rhetoric about gender equality in the military, women in the Army, Navy, and Air Force academies continue to be sexually harassed. In 2012, according to the Department of Defense, there were 26,000 sexual assaults in the military, yet only about 14% filed a complaint. Fear of reporting sexual harassment is a primary reason for not reporting such sexual harassment (Lawrence & Penaloza, 2013). Male perpetrators may not separate their work roles from their domestic roles. One student in our classes noted that she was the ex-wife of a Navy Seal and that "he knew how to torment someone, and I was his victim."

10-2b **Community Factors**

Community factors that contribute to violence and abuse in the family include social isolation, poverty, and inaccessible or unaffordable health care, day care, elder care, and respite care services and facilities.

Diversity in Other Countries

In cultures where a man's honor is threatened if his wife is unfaithful, the husband's violence toward her is tolerated by the legal system. Unmarried women in Jordan who have intercourse (even if through rape) are viewed as bringing shame on their parents and siblings and may be killed; this action is referred to as an **honor crime** or **honor killing**. The legal consequence is minimal to nonexistent. For example, a brother may kill his sister if she has intercourse with a man she is not married to and spend no more than a month or two in jail as the penalty.

Social Isolation Living in social isolation from extended family and community members increases the risk of being abused. Spouses whose parents live nearby are least vulnerable.

Poverty Abuse in adult relationships occurs among all socioeconomic groups. However, poverty and low socioeconomic status are a context of high stress which lends itself to expression of this stress by violence and abuse in interpersonal relationships.

Inaccessible or Unaffordable Community Services Failure to provide medical care to children and elderly family members sometimes results from the lack of accessible or affordable health care services in the community. Without elder care and respite care facilities, families living in social isolation may not have any help with the stresses of caring for elderly family members and children.

10-2c **Individual Factors**

Individual factors associated with domestic violence and abuse include psychopathology, personality characteristics, and alcohol or substance

honor crime (honor killing) refers to unmarried women who are killed because they bring shame on their parents and siblings; occurs in Middle Eastern countries such as Jordan.

abuse. A number of personality characteristics have also been associated with people who are abusive in their intimate relationships. Some of these characteristics follow:

1. **Dependency.** Therapists who work with batterers have observed that they are overly dependent on their partners. Because the thought of being left by their partners induces panic and abandonment anxiety, batterers use physical aggression and threats of suicide to keep their partners with them.

2. **Jealousy.** Along with dependence, batterers exhibit jealousy, possessiveness, and suspicion. An abusive husband may express his possessiveness by isolating his wife from others; he may insist she stay at home, not work, and not socialize with others. His extreme, irrational jealousy may lead him to accuse his wife of infidelity and to beat her for her presumed affair.

3. **Need to control.** Abusive partners have an excessive need to exercise power over their partners and to control them. The abusers do not let their partners make independent decisions (including what to wear), and they want to know where their partners are, whom they are with, and what they are doing. Abusers like to be in charge of all aspects of family life, including finances and recreation.

4. **Unhappiness and dissatisfaction.** Abusive partners often report being unhappy and dissatisfied with their lives, both at home and at work. Many abusers have low self-esteem and high levels of anxiety, depression, and hostility. They may take out their frustration with life on their partner.

5. **History of aggressiveness.** Abusers often have a history of interpersonal aggressive behavior. They have poor impulse control and can become instantly enraged and lash out at the partner. Battered women report that episodes of violence are often triggered by minor events, such as a late meal or a shirt that has not been ironed.

6. **Quick involvement.** Because of feelings of insecurity, the potential batterer will rush his partner quickly into a committed relationship. If the woman tries to break off the relationship, the man will often try to make her feel guilty for not giving him and the relationship a chance.

Andris Torms/Shutterstock.com

7. **Blaming others for problems.** Abusers take little responsibility for their problems and blame everyone else. For example, when they make mistakes, they will blame their partner for upsetting them and keeping them from concentrating on their work. A man may become upset because of what his partner said, hit her because she smirked at him, and kick her in the stomach because she poured him too much (or not enough) alcohol.

8. **Jekyll-and-Hyde personality.** Abusers have sudden mood changes so that a partner is continually confused. One minute an abuser is nice, and the next minute angry and accusatory. Explosiveness and moodiness are the norm.

9. **Isolation.** An abusive person will try to cut off a partner from all family, friends, and activities. Ties with anyone are prohibited. Isolation may reach the point at which an abuser tries to stop the victim from going to school, church, or work.

10. **Alcohol and other drug use.** Whether alcohol reduces one's inhibitions to display violence, allows one to avoid responsibility for being violent, or increases one's aggression, it is associated with violence and abuse (even if the partner is pregnant; Eaton et al., 2012). Younger individuals with more severe drug addictions (e.g., a history of overdose) are more likely to be violent (Fernandez-Montalvo et al., 2012).

11. **Criminal/psychiatric background.** Eke et al. (2011) examined the characteristics of 146 men who murdered or attempted to murder their intimate partner. Of these, 42% had prior criminal charges, 15% had a psychiatric history, and 18% had both Shorey et al. (2012) identified the mental health problems in men arrested for domestic violence and found high rates of PTSD, depression, generalized anxiety disorder (GAD), panic disorder, and social phobia.

12. **Impulsive.** Miller et al. (2012) identified one of the most prominent personality characteristics associated with aggression/abuse. "Impulsive behavior in the context of negative affect" was consistently related to aggression across multiple indices.

10-2d Relationship Factors

Halpern-Meekin et al. (2013) studied a sample of 792 relationships and found that "relationship churning" was associated with both physical and emotional violence. The researchers compared relationships which had ended, those which were still together, and those which had an on-and-off pattern (churners). Individuals in the latter type relationship were twice as likely to report physical violence and half again as likely to report emotional abuse compared to the other two patterns.

10-2e Family Factors

Family factors associated with domestic violence and abuse include being abused as a child, having parents who abused each other, and not having a father in the home.

Child Abuse in Family of Origin Individuals who were abused as children are more likely to be abusive toward their intimate partners as adults.

Family Conflict Children learn abuse from their family context. Children whose fathers were not affectionate were more likely to be abusive to their own children.

Parents Who Abuse Each Other Parents who are aggressive toward each other create a norm of aggression in the family and are more likely to be aggressive toward their children (Graham et al., 2012). While most parents are not aggressive toward each other in front of their children (Pendry et al.,

2011), Dominguez et al. (2013) reconfirmed the link between witnessing parental abuse (either father or mother abusive toward the other) and being in an abusive relationship as an adult. However, a majority of children who witness abuse do not continue the pattern—a family history of violence is only one factor out of many associated with a greater probability of adult violence.

10-3 Sexual Abuse in Undergraduate Relationships

Some women experience sexual abuse in addition to a larger pattern of physical abuse. In a sample of 3,572 undergraduate females at two large universities, 39% reported being pressured by a partner they were dating to have sex (Hall & Knox, 2015). Men are also pressured to have sex (20% of 1,150 undergraduate males) (Hall & Knox, 2015). In other research, in a sample of 1,400 men ages 18–24, 6% reported that they were coerced to have vaginal sex with a female; 1% by a male to have oral or anal sex (Smith et al., 2010).

Sexual abuse is a worldwide phenomenon. In 2012, a 23-year-old college student was gang raped on a bus in New Delhi, India. She and her fiancé were beaten and thrown off the bus and left for dead. "A woman [here] is a possession, like a piece of land," said Amon Deol, general secretary of a Punjab women's rights group (Mahr, 2013, p. 12).

10-3a Acquaintance and Date Rape

About 85% of rapes are perpetrated by someone the woman knows (12% of victims have been raped by both an acquaintance and a stranger—double victims) (Hall & Knox, 2012). The type of rape by someone the victim knows is referred to as **acquaintance rape**, which is defined as nonconsensual sex between adults (of the same or other sex) who know each other. The behaviors of sexual coercion occur on a continuum from verbal pressure and threats to use of physical force to obtain sexual acts, such as oral sex, sexual intercourse, and anal sex.

acquaintance rape nonconsensual sex between adults (of same or other sex) who know each other.

What's New?

DOUBLE VICTIMS: SEXUAL COERCION BY A DATING PARTNER AND STRANGER

This study identified the characteristics of victims of sexual coercion who were double victims—on one occasion by a dating partner and on another occasion by a stranger. In a sample of 2,747 undergraduates the percent reporting experiencing sexual coercion in the respective contexts was 23.9% date, 6.4% stranger, 11.2% both, and 58.5% neither. Those who reported pressure to have sex from both a stranger and a date—**double victims**—were more likely to be a white female who had been abused as a child, to have been emotionally abused by a partner, to have cohabited, used alcohol/drugs, used the Internet to find a partner, dated interracially, hooked up, been in a friends with benefits relationship, and lied to a partner. The profile that emerged was that of an early victimized female (child abuse, possible prior emotional partner abuse) who lowers inhibitions via alcohol/drug use and engages in sex in noncommitted relationships (e.g., hooking up, friends with benefits). Lying to one's partner suggests a less than ideal relationship.

Source: Adapted and abridged from Hall, S., & D. Knox. (2011, November). Double victims: Sexual coercion by a dating partner and a stranger. Poster presented at the annual meeting of the National Council on Family Relations, Orlando, FL.

double victims individuals who report being a victim of forced sex by both a stranger and by a date or acquaintance.

female rape myths beliefs that deny victim injury or cast blame on the woman for her own rape.

male rape myths beliefs that deny victim injury or make assumptions about his sexual orientation.

date rape one type of acquaintance rape which refers to nonconsensual sex between people who are dating or are on a date.

The perpetrator of a rape is likely to believe in various rape myths. **Female rape myths** are beliefs about the female that deny that she was raped or cast blame on the woman for her own rape. These beliefs are false, widely held, and justify male aggression. Examples include: "women deserve to be raped" (particularly when they drink too much and are provocatively dressed); "women fantasize about and secretly want to be raped"; and "women who really don't want to be raped resist more—they could stop a guy if they really wanted to." McMahon (2010) found that female rape myths were more likely to be accepted by males, those pledging a fraternity/sorority, athletes, those without previous rape education, and those who did not know someone who had been sexually assaulted. Mouilso and Calhoun (2013) noted that being mentally disturbed is associated with belief in these myths.

There are also **male rape myths** such as "men can't be raped" and "men who are raped are gay." Not only is there a double standard of perceptions that only women can be raped but a double standard is operative in the stereotypical perception of the gender of the person engaging in sexual coercion (e.g., men who rape are aggressive; women who rape are promiscuous [Oswald & Russell, 2006]).

One type of acquaintance rape is **date rape**, which refers to nonconsensual sex between people who are dating or on a date. The woman (while both women and men may be raped on a date, we will refer to the female since male–female rape is more prevalent) has no idea that rape is on the agenda. She may have gone out to dinner, had a pleasant evening, and gone back to her apartment with her date only to be raped.

10-3b Sexual Abuse in Same-Sex Relationships

Rape also occurs in same-sex relationships. Gay, lesbian, and bisexual individuals are not immune to experiencing sexual abuse in their relationships. Rothman et al. (2011) reviewed studies which involved a total sample of 139,635 gay, lesbian, and bisexual women and men and found that the highest estimates reported for lifetime sexual assault were for lesbian and bisexual women (85%). In regard to childhood sexual assault of lesbian and bisexual women the percentage was 76%. For childhood sexual assault of gay and bisexual men, it was 59%.

10-3c Alcohol and Rape

Alcohol is the most common rape drug. A person under the influence cannot give consent. Hence, a person who has sex with someone who is impaired is engaging in rape. In a study of 340 college rape victims, 41% reported being impaired and 21% reported being incapacitated—hence over 60% were in an altered state (Littleton et al., 2009). Farris et al. (2010) also found that alcohol dose was related to men interpreting a woman's friendliness as sexual interest.

Tasteless and odorless, Rophypnol, the "date rape" drug, results in victims losing their memory for 8 to 10 hours.

Rommel Canlas/Shutterstock.com

controlled substance to anyone without his or her knowledge and with the intent of committing a violent crime (such as rape). Violation of this law is punishable by up to 20 years in prison and a fine of $250,000.

The effects of rape include loss of self-esteem, loss of trust, and the inability to be sexual. Zinzow et al. (2010) studied a national sample of women and found that forcible rape was more likely to be associated with PTSD and with major depression than was drug- or alcohol-facilitated or incapacitated rape.

Acknowledging that one is a rape victim (rather than keeping it a secret) is associated with fewer negative psychological symptoms (e.g., depression) and increased coping (e.g., moving beyond the event to a new relationship) (Clements & Ogle, 2009). Denial, not discussing it with others, and avoiding the reality that one has been raped prolongs one's recovery.

10-3d Rophypnol: The Date Rape Drug

Of those occasions in which a woman is incapacitated due to alcohol or drugs, 85% are voluntary (the woman is willingly drinking or doing drugs). In 15% of the cases, however, drugs are used against her will (Lawyer et al., 2010).

Rophypnol—also known as the date rape drug, rope, roofies, Mexican Valium, or the "forget (me) pill"—causes profound, prolonged sedation and short-term memory loss. Similar to Valium but 10 times as strong, Rophypnol is a prescription drug used as a potent sedative in Europe. It is sold in the United States for about $5, is dropped in a drink (where it is tasteless and odorless), and causes victims to lose their memory for 8 to 10 hours. During this time, victims may be raped yet be unaware until they notice signs of it (e.g., blood in panties) the next morning.

The Drug-Induced Rape Prevention and Punishment Act of 1996 makes it a crime to give a

10-4 Abuse in Marriage Relationships

The chance of abuse in a relationship increases with marriage. Indeed, the longer individuals know each other and the more intimate their relationship, the greater the abuse.

10-4a General Abuse in Marriage

Abuse in marriage may differ from unmarried abuse in that the husband may feel ownership of the wife and feel the need to control her. But the behaviors of abuse are the

Rophypnol causes profound, prolonged sedation and short-term memory loss; also known as the date rape drug, roofies, Mexican Valium, or the "forget (me) pill."

Personal View:
Not 12 Years a Slave but 13 Years an Abused Wife

I was verbally, physically, and sexually abused for 13 years in my marriage from ages 22–34. I never had any hint during the two years we were dating that the man I loved (he was my soul mate) would end up making my life a living hell. He was a little jealous but I thought this was cute and that he cared about me. In fact he made me feel special…he made me feel LOVED…we could talk for hours and hours.

The very next day after we were married, when I came home from work he had put all my dresses and skirts on the bed with the hem taken out of them…I wondered what in the world I had gotten myself into.

The abuse occurred mostly when he had been drinking:

- **Accusations** …accusing me of sleeping with anyone breathing.

- **Fear** …he jumped at me…got into my face…he threatened me with bodily harm…anything to intimidate me.

- **Shame** …he belittled me, called me names, cussed at me.

- **Control** …he got mad at me for visiting my family

- **Physical abuse** …he pushed, pinched, squeezed, spit, and pulled a gun on me.

- **Rape** …he forced me to have sex with him (when he was drunk he would last forever).

He would stay out all night; come home and wake me up…and accuse me of going out and being with someone else…(looked for stamps on my hand from a club).

Black eyes were common…I kept sunglasses on for about 5–6 years…I will never forget one night when he was on top of me…he had me pinned down just hitting me in the face with his fist just as hard as he could.

He once backhanded me and knocked four of my front teeth loose (they later died and had to be pulled). I now have a plate in my mouth.

Another time he picked me up and slung me across the room with both of my legs hitting the coffee table and seriously bruising the front shins of my legs. Today, I still have problems…broken veins and poor circulation in my lower legs.

I left him many times…he would always talk me into coming back…buying me back…he would start back to church with me and we would go to church for months…then he would start his drinking again.

I once called the police and he was arrested…but I dropped the charges. The reason I stayed is because I loved him and when he wasn't drinking he was perfect and treated me right. But his abuse was changing my personality and one time I lost it and pulled a gun on him…so I knew it was time to go…I have since forgiven him and moved on. The message to my daughter and other young people is to make a deliberate choice to get out immediately after the abuse starts.

Source: Abridged and used with the permission of Teresa Carol Wimberly.

same—belittling the spouse, controlling the spouse, physically hurting the spouse, etc. The Personal View section provides the horror of abuse inside one marriage.

10-4b Men Who Abuse

Henning and Connor-Smith (2011) studied a large sample of men who were recently convicted of violence toward a female intimate partner (N = 1,130). More than half of the men (59%) reported that they were continuing or planning to continue the relationship with the partner they had abused. Reasons included being older (e.g., too late to start over), being married to the victim, having children together, attributing less blame to the victim for the recent offense, and having a childhood history of family violence.

10-4c Rape in Marriage

Marital rape, now recognized in all states as a crime, is forcible rape (includes vaginal, oral, anal) by one's spouse. Some states (Washington) recognize a marital defense exception for third-degree rape in which force is not used even though there is no consent. Over 30 countries (e.g., China, Afghanistan, Pakistan) have no laws against marital rape.

10-5 Effects of Abuse

Abuse has devastating effects on the physical and psychological well-being of victims. Abuse between parents also affects the children.

> "The **attacks** of which I have been the object **have broken the spring of life in me** .... People don't realize what it feels like to be constantly insulted."
>
> —EDOUARD MANET, FRENCH PAINTER

10-5a Effects of Partner Abuse on Victims

Effects of abuse are in reference to perception. Rhatigan and Nathanson (2010) found in a study of 293 college women that women took into account their own behavior in evaluating their boyfriends' abusive behavior. If they felt they had set him up by their own behavior, they were more understanding of his aggressiveness. Additionally, those with low self-esteem were more likely to take some responsibility for their boyfriend's abusive behavior. While some partners will rationalize or overlook their partner's abusive behavior, others are hurt by it and the effect is negative.

In general, abuse has devastating consequences. Becker et al. (2010) confirmed that being a victim of intimate partner violence is associated with symptoms of PTSD—loss of interest in activities/life in general, feeling detached from others, inability to sleep, irritability, etc. Sarkar (2008) confirmed that IPV (intimate partner violence) affected the woman's physical (increased the risk for unintended pregnancy/multiple abortions) and mental health (higher levels of anxiety and drug abuse).

10-5b Effects of Partner Abuse on Children

The most dramatic effects of abuse occur on pregnant women, which include increased risk of miscarriage, birth defects, low birth weight, preterm delivery, and neonatal death. Negative effects may also accrue to children who witness domestic abuse. Russella et al. (2010) found that children who observed parental domestic violence were more likely to be depressed as adults. Hence, a child need not be a direct target to abuse, but merely a witness to incur negative effects. However, Kulkarni et al. (2011) compared the effects of witnessing domestic violence and actually experiencing it on PTSD in later life and found that witnessing alone was not associated with PTSD.

10-6 The Cycle of Abuse

The cycle of abuse begins when a person is abused and the perpetrator feels regret, asks for forgiveness, and engages in positive behavior (gives flowers). The victim, who perceives few options and feels anxious terminating the relationship with the abusive partner, feels hope for the relationship at the contriteness of the abuser and does not call the police or file charges.

After the forgiveness, couples usually experience a period of making up or honeymooning, during which the victim feels good again about the partner and is hopeful for a nonabusive future. However, stress, anxiety, and tension mount again in the relationship, which the abuser relieves by violence toward the victim. Such violence is followed by the familiar sense of regret and pleadings for forgiveness, accompanied by a new round of positive behavior (flowers and candy).

marital rape
forcible rape by one's spouse—a crime in all states.

> "You don't have to wait for someone to treat you bad repeatedly. *All it takes is once, and if they get away with it that once, if they know they can treat you like that, then it sets the pattern for the future.*"
>
> —JANE GREEN, *BOOKENDS*

Figure 10.1
The Cycle of Abuse

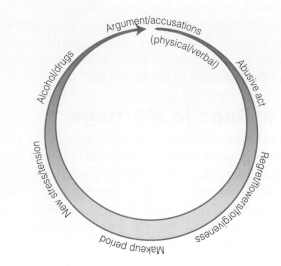

Source: *Choices in Relationships*, 11 ed.

As the cycle of abuse reveals, some victims do not prosecute their partners who abuse them. In response to this problem, Los Angeles has adopted a zero tolerance policy toward domestic violence. Under the law, an arrested person is required to stand trial and his victim required to testify against the perpetrator. The sentence in Los Angeles County for partner abuse is up to six months in jail and a fine of $1,000.

Figure 10.1 illustrates this cycle, which occurs in clockwise fashion. In the rest of this section, we discuss reasons why people stay in an abusive relationships and how to get out of them.

periodic reinforcement reinforcement that occurs every now and then (unpredictable). The abused victim never knows when the abuser will be polite and kind again (e.g., flowers and candy).

10-6a **Why Victims Stay in Abusive Relationships**

One of the most frequently asked questions to people who remain in abusive relationships is, "Why do you stay?" Alexander et al. (2009) noted that the primary reason a woman returns again and again to an abusive relationship is the emotional attachment to her partner—she is in love. While someone who criticizes ("you're ugly/stupid/pitiful"), is dishonest (sexually unfaithful), or physically harms a partner will create a context of interpersonal misery, the victim might still love that person.

Another explanation for why some people remain with abusive partners is that the abuse is only one part of the relationship. When such partners are not being abusive, they are kind, caring, and loving. It is these positive behaviors which keep the victim hooked. The psychological term is a **periodic reinforcement** which means that every now and then the abusive person "floods" the partner with strong love/positives which keep the partner in the relationship. In effect, the person who is abused stays in the relationship since it includes positives such as flowers, gifts, or declarations of love to entice a partner to stay in the relationship.

Whether the abused has children may have a positive or negative effect in terms of staying in an abusive relationship (Rhodes et al., 2010). Some mothers will leave their abusive partners so as not to subject their children to the abuse. But others will stay in an abusive marriage since the mothers want to keep the family together (and keep a father close for their children).

In earlier research, Few and Rosen (2005) interviewed 25 women who had been involved in abusive dating relationships from three months to nine years (average = 2.4 years) to find out why they stayed. The researchers conceptualized the women as **entrapped**—stuck in an abusive relationship and unable to extricate oneself from the abusive partner. Indeed, these women escalated their commitment to stay in hopes that doing so would eventually pay off. In effect, they had invested time with a partner they were in love with and wanted to do whatever was necessary to maintain the relationship. Some of the factors which drive the entrapment of these women include:

- Love ("I love him.")

- Fear of loneliness ("I'd rather be with someone who abuses me than alone.")

- Emotional dependency ("I need him.")

- Commitment to the relationship ("I took a vow 'for better or for worse.'")

- Hope ("He will stop being abusive—he's just not himself lately.")

- A view of violence as legitimate ("All relationships include some abuse.")

- Guilt ("I can't leave a sick man.")

- Fear for one's life ("He'll kill me if I leave him.")

- Economic dependence ("I have no money and no place to go.")

- Isolation ("I don't know anyone who can help me.")

Halligan et al. (2013) noted the impact of technology on the maintenance of abusive relationships. She and her colleagues found that undergraduates in abusive relationships could not stop themselves from checking their text messages—even if the message was from an abusive partner. Some of the messages from the abusive partners included an apology, a promise to never be abusive again, and a request to resume the relationship.

Nordling/Shutterstock.com

Battered women also stay in abusive relationships because they rarely have escape routes related to educational or employment opportunities, they do not want to disrupt the lives of their children, and they may be so emotionally devastated by the abuse (anxious, depressed, or suffering from low self-esteem) that they feel incapable of planning and executing their departure.

Disengaging from an abusive relationship is a process that takes time. One should not be discouraged when they see a friend return to an abusive context but recognize that progress in disengagement is occurring (she did it once) and that positive movement is predictive of eventually getting out.

10-6b Fighting Back? What Is the Best Strategy?

How might a partner respond when being physically abused? First, the goal is to avoid serious physical injury. Such injury is increased if the partner is drunk, on drugs, or has a weapon so forceful physical resistance (pushing, striking, struggling) to the mind-altered partner should be avoided. Verbal resistance (pleading, crying, or trying to assuage the offender) or nonforceful physical resistance (fleeing or hiding) might be more helpful in avoiding injury. Fortunately, the frequency of severe injury is low (7%) with most (69%) abusive incidents resulting in no injury (Powers & Simpson, 2012). Of course, any abuse in an intimate relationship may be too much and should be avoided or the relationship ended.

10-6c How to Leave an Abusive Relationship

Couple therapy to reduce abuse in an abusive relationship is hampered by the fact of previous abuse. Being a victim of previous psychological aggression (female or male) is related to fear of speaking in front of the partner and fear of being in therapy with

entrapped stuck in an abusive relationship and unable to extricate oneself from the abusive partner.

the partner (O'Leary et al., 2013). Hence, gains may be minimal.

> "**Leaving** an abusive partner is a **very difficult** thing to do. It frequently feels like you are **failing**, or destroying your family, or not trying to work things out, or not giving your partner 'a second chance.' It **hurts**, and it's **scary**."
>
> —BLAINE NELSON

Leaving an abusive partner begins with the person making a plan and acting on the plan—packing clothes/belongings, moving in with a sister, mother, or friend, or going to a homeless shelter. If the new context is better than being in the abusive context, the person will stay away. Otherwise, the person may go back and start the cycle all over. As noted previously, this leaving and returning typically happens seven times.

Sometimes the woman does not leave while the abuser is at work but calls the police and has the man arrested for violence and abuse. While the abuser is in jail, she may move out and leave town. In either case, disengagement from the abusive relationship takes a great deal of courage. Calling the National Domestic Violence Hotline (800-799-7233 [SAFE]), available 24 hours, is a point of beginning.

Kress et al. (2008) noted that involvement with an intimate partner who is violent may be life threatening. Particularly if the individual decides to leave the violent partner, the abuser may react with more violence and murder the person who has left. Indeed, a third of murders that occur in domestic violence cases occur shortly after a breakup. Specific actions that could be precursors to murder of an intimate partner are stalking, strangulation, forced sex, physical abuse, gun ownership, and drug or alcohol use on the part of the abuser. Moving quickly to a safe context (e.g., parents) is important.

Taking out a protective order whereby the accused is prohibited from being within close proximity of the victim partner is one option some abused partners take. While Kothari et al. (2012) confirmed that taking out such an order against a partner who has engaged in intimate partner violence was associated with a reduced frequency of repeated violence against the victim, other research is less clear. Ward et al. (2014) studied the effect of involving the police when interpersonal abuse occurs and found a different answer. When there was a history of abuse, there was an increased likelihood of future violent offending subsequent to police contact. When there was no history of abuse, there seemed to be no change, no increase in violence. So what is a person in an abusive relationship to do? The best answer seems to be to take the position that the abusive behavior will continue and to removing oneself from the abusive context and stay in a safe context.

10-7 Trends in Abuse in Relationships

Abuse in relationships will continue to occur behind closed doors, in private contexts where the abuse is undetected. Reducing such abuse will depend on prevention strategies focused at three levels: the general population, specific groups at high risk for abuse, and individuals/couples who have already experienced abuse. Public education and media campaigns aimed at the general population will continue to convey the criminal nature of domestic assault, suggest ways the abused might learn escape from abuse, and identify where abuse victims and perpetrators can get help. MacMillan et al. (2013) emphasized that children need to be socialized early that abuse, even that between their parents, is not acceptable.

Preventing or reducing abusive relationships through education necessarily involves altering aspects of American culture that contribute to such abuse. For example, violence in the media must be curbed (not easy, with nightly news clips of bombing assaults in other countries, gun shootings, violent films, etc.). Chavis et al. (2013) documented that parental exposure to a brief multimedia program designed to suggest alternatives to physical punishment resulted in parents making more frequent choices not to discipline their 6- to 24-month-old infants via spanking. Bennett et al. (2014) emphasized the need for bystander intervention programs which sensitized individuals to be aware of abuse and to intervene. Katz and Moore (2013) provided some data on the efficacy of such intervention programs. Finally, traditional gender roles and views of women and children as property must be replaced with egalitarian gender roles and respect for women and children.

Another important cultural change is to reduce violence-provoking stress by reducing poverty and unemployment and by providing adequate housing, nutrition, medical care, and educational opportunities for everyone. Integrating families into networks of community and kin would also enhance family well-being and provide support for families under stress.

STUDY TOOLS 10

Ready to study? In the book, you can:

- Rip out the chapter review card at the back of the book for a handy summary of the chapter and key terms.

- Learn more about your partner's behavior with the Self-Assessment card at the back of the book.

Online at CENGAGEBRAIN.COM you can:

- Prepare for tests with quizzes.

- Review the key terms with Flash Cards.

- Play games to master concepts.

Deciding About Children

"**Having children** has changed from an assumption to a **decision**."

—LAURA SCOTT, CHILDFREE WIFE

Of life's four critical decisions—whether to marry, whom to marry, whether to have a child, and which career path to pursue—only the decision to have children is the most life-changing irrevocable decision. The other choices, while difficult to make and having a significant impact on the individual, pale in comparison to the decision to have a child.

Most undergraduates have already decided to include having a child as part of their life plan. Ninety percent (an equal percent for women and men) of 4,692 undergraduates reported that they wanted to have a child (Hall & Knox, 2015). Yet the timing of having a baby/preventing an unwanted pregnancy is often not deliberate. Over half of all pregnancies are not intended (Finer & Zolna, 2014).

Your choices in regard to whether you want to have children and the use of contraception have important effects on your happiness, lifestyle, and resources. These choices, in large part, are influenced by social and cultural factors which often operate without your awareness.

11-1 Do You Want to Have Children?

Beyond a biological drive to reproduce (which not all adults experience), societies socialize their members to have children. Unless children are born, the society will cease to exist. In this section, we examine the social influences that motivate individuals to have children, the lifestyle changes that result from such a choice, and the costs of rearing children.

Courtesy of Chelsea Curry

11-1a Social Influences Motivating Individuals to Have Children

Our society tends to encourage child-bearing, an attitude known as **pronatalism**. Our family, friends, religion, and government encourage positive attitudes toward parenthood. Cultural observances also function to reinforce these attitudes.

karen roach/Shutterstock.com

Family Our experience of being reared in families encourages us to have families of our own. Our parents are our models. They married; we marry. They had children; we have children. We also expect to have a "happy family."

> **pronatalism**
> cultural attitude which encourages having children.

Friends Our friends who have children influence us to do likewise. Decisions are made in social context. Notice how many of your decisions are identical to those of your close friends—if you want to know, whether, when, and how many children a person will have, find out what decisions their three closest friends have made. Most undergraduates just out of high school have no interest in being married and having children while in school—a value shared by almost 100% of their peers.

Religion Religion is a strong influence on an individual's decision to have children. Catholics are taught that having children is the basic purpose (procreation) of marriage and gives meaning to the union. Mormons and ultra-Orthodox Jews also have a strong interest in having and rearing children.

Race Hispanics have the highest fertility rate of any racial/ethnic category.

Government The taxes (or lack of them) that our federal and state governments impose support parenthood. Although individuals have children for more emotional than financial reasons, married couples with children pay lower taxes than married couples without children. Assume there are two couples (one married couple with two children and one childfree married couple), each making $100,000. The couple with children would pay $1,950 less in federal tax than the childfree couple.

Cultural Observances Our society reaffirms its value for having children by identifying special days for Mom and Dad. Each year on Mother's Day and Father's Day (and now Grandparents' Day), parenthood is celebrated across the nation with cards, gifts, and embraces. People choosing not to have children have no cultural counterpart (e.g., Childfree Day). In addition to influencing individuals to have children, society and culture also influence feelings about the age parents should be when they have children. Recently, couples have been having children at later ages. Is this a good idea?

11-1b Individual Motivations for Having Children

While social influences are important, so are individual motivations in the decision to have children. Some of these inducements are obvious as in the desire to love and to be loved by one's own child, companionship, and the desire to be personally fulfilled as an adult by having a child. Some people also want to recapture their own childhood and youth by having a child. Motives that are less obvious may also be operative—wanting a child to avoid career tracking (e.g., "I'd rather have a baby than tenure") and to gain the acceptance and approval of one's parents and peers.

11-1c Evaluation of Lifestyle Changes

Becoming a parent often involves changes in lifestyle. Daily living routines become focused around the needs of the children. Living arrangements change to provide space for another person in the household. Some parents change their work schedule to allow them to be home more. Food shopping and menus change to accommodate the appetites of children. A major lifestyle change is the loss of freedom of activity and flexibility in one's personal schedule. Lifestyle changes are particularly dramatic for women. The time and effort required to be pregnant and rear children often compete with the time and energy needed to finish one's education or build one's career. Parents learn quickly that being both an involved parent and climbing the career ladder are difficult. The careers of women may suffer most.

Mother's Day reflects cultural support and approval for the role of mother.

DO PARENTS HAVE HIGHER LIFE SATISFACTION THAN NONPARENTS?

Since most individuals/couples have children, what is the outcome in terms of effect on life satisfaction? Data from Germany on 1,220 women and 1,107 men who became first-time parents and who reported their life satisfaction over a six-year period help answer the question.

Researcher Matthias Pollmann-Schult (2014) noted that during the first five years, children provide enormous positive emotional rewards and "psychological stimulation as well as pleasure, fun and excitement." They also "form the basis for relationships with other parents and thereby contribute to a sense of social connectedness."

Over time, however, children incur heavy burdens in terms of time costs, psychosocial stress, and financial costs which suppress life satisfaction for parents. Significant variables affecting life satisfaction include age of the child (higher life satisfaction with children under 5), marital status of the couple (higher for married couples than single parents), and financial arrangements (higher in male breadwinning households than dual-earner households). Additional children seem to further suppress the life satisfaction associated with parenting.

Courtesy of Brittany Bolen

Parents often experience immense joy in taking care of their infants and in watching them develop.

Source: Pollmann-Schult, M. (2014). Parenthood and life satisfaction: Why don't children make people happy? *Journal of Marriage and the Family*, 76: 310–336.

11-1d Awareness of Financial Costs

Meeting the financial obligations of parenthood is difficult for many parents. The costs begin with prenatal care and continue at childbirth. For an uncomplicated vaginal delivery, with a two-day hospital stay, the cost may total $10,000, whereas a Cesarean section birth may cost $14,000. The annual cost of a child less than 2 years old for middle-income parents ($57,600 to $99,730)—which includes housing ($3,870), food ($1,350), transportation ($1,540), clothing ($740), health care ($820), child care ($2,740), and miscellaneous ($890)—is $11,950. For a 15- to 17-year-old, the cost is $13,830 (*Statistical Abstract of the United States, 2012–2013,* Table 689). Lifetime costs of rearing a child born in 2012 to age 18 will be $241, 080 (U.S. Department of Agriculture, 2013).

11-2 How Many Children Do You Want?

Procreative liberty is the freedom to decide whether or not to have children. More women are deciding not to have children or to have fewer children (Department of Commerce et al., 2011).

> **procreative liberty** the freedom to decide to have children or not.

"As I grow older, I am more and more convinced that there are **many roads to** personal **fulfillment**— and that having **children** is **not for everyone.**"

—THEKLA RICHTER

11-2a **Childfree Marriage?**

About 20% of U.S. women over the age of 40 do not have children. The percent is expected to increase (Allen & Wiles, 2013). Of those who do not have children, 44% are voluntarily childfree, 40% are involuntarily childfree, and 16% plan to have a child in the future (Smock & Greenland, 2010).

What about the process of the decision to not have children—to remain childfree? Lee and Zvonkovic (2014) interviewed 20 childfree couples. The driving force behind most decisions to remain childfree was the importance the spouses gave to their relationship and the strength of their conviction. Three phases of the decision-making process were identified—agreement, acceptance, and closing the door (e.g., no longer capable of having a child). One of the respondents noted the difference between agreement and acceptance:

Agreeing was the first, really the first thing. Acceptance was the second thing . . . now that I think about it. It's like, true because she said, "I'm not gonna have kids" and I would go, "Okay." That's me agreeing. Accepting it took a little bit longer . . . I mean, I accept that's what it is, right? . . . the accepting means, we should not have kids. I mean, like by the time we are 30, it's like, "It's never going to happen." It's really total acceptance. "This is never gonna happen."

childlessness concerns the idea that holidays and family gatherings may be difficult because of not having children or feeling left out or sad that others have children.

The intentionally childfree may be viewed with suspicion ("they are selfish"), avoidance ("since they don't have children they won't like us or support our family values"), discomfort ("what would I have in common with these people?"), rejection ("they are wrong not to want children/I don't want to spend

Rearing children is hard work and often a thankless job—disciplining children and coping with unhappy/crying children.

Courtesy of Brittany Bolen

time with these people"), and pity ("they don't know what they are missing") (Scott, 2009). Indeed "the mere existence of a growing childless by choice population is a challenge to people who believe procreation is instinctive, intrinsic, biological, or obligatory" (p. 191).

Stereotypes about couples who deliberately elect not to have children include that they do not like kids, are immature, and are not fulfilled because they do not have a child to make their lives "complete." The reality is that such individuals may enjoy children and some deliberately choose careers to work with them (e.g., elementary school teacher). But they do not want the full-time emotional and economic responsibility of having their own children. Allen and Wiles (2013) interviewed 38 older (ages 63–93) childfree individuals and found an array of meanings (from grief to relief, from making a deliberate choice to do what felt "natural"—not have a child—to breaking a family of violence cycle).

Childlessness concerns is a concept which refers to the idea that holidays and family gatherings may be difficult because of not having children or feeling left out or sad that others have children. McQuillan et al. (2012) studied a representative sample of 1,180 women without children from the National Survey of Fertility Barriers and found that the degree to which women report these concerns is related to their reason for being childless; women who were voluntarily childless reported few concerns. Those most reactive were the childless due to biomedical causes or having delayed parenthood until they could no longer get pregnant.

Some people simply do not like children. Celebrity Bill Maher states, "I hate kids." Aspects of our society reflect **antinatalism** (a perspective against children). Indeed, there is a continuous fight for corporations to implement or enforce any family policies (from family leaves to flex time to on-site day care). Profit and money—

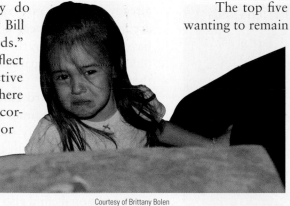
Courtesy of Brittany Bolen

not children—are priorities. In addition, although people are generally tolerant of their own children, they often exhibit antinatalistic behavior in reference to the children of others. Notice the unwillingness of some individuals to sit next to a child on an airplane. Asia Air has responded to this need by offering the first 12 rows in coach only for individuals ages 12 and above.

Laura Scott (2009), a childfree wife, set up the Childless by Choice Project and surveyed 171 childfree adults (ages 22 to 66, 71% female, 29% male) to identify their motivations for not having children. The categories of her respondents and the percentage of each follow:

Early articulators (66%)—these adults knew early that they did not want children
Postponers (22%)—adults who kept delaying when they would have children and remained childless
Acquiescers (8%)—those who made the decision to remain childless because their partner did not want children
Undecided (4%)—those who are childless but still in the decision-making process

The top five reasons her respondents gave for wanting to remain childfree were (Scott, 2009):

1. Current life/relationship satisfaction was great and they feared that parenthood would only detract from it.

2. Freedom and independence were strong values that they feared would be affected by children.

3. Avoidance of the responsibility for rearing a child.

4. No maternal/paternal instinct.

5. Accomplishing career and travel goals would be difficult as a parent.

Some of Scott's interviewees also had an aversion to children, having had a bad childhood or concerns about childbirth. But other individuals love children, enjoy them, and benefit from having them. Hoffnung and Williams (2013) analyzed data on 200 women and found that being a mother was associated with higher life satisfaction than being childfree.

11-2b **One Child?**

Only 3% of adults view one child as the ideal family size (Sandler, 2010). Those who have only one child may do so because they want the experience of parenthood without children markedly interfering with their lifestyle and careers. Still others have an only child

antinatalism
opposition to children.

Diversity in Other Countries

China has a one-child policy, which has led to forced abortions, sterilizations, and economic penalties for having more than one child. National visibility for this policy came in 2012 when Feng Jianmei could not pay the $6,300 fine for a second child and was forced to have an abortion to terminate her seven-month pregnancy. She was literally tied up, forced into a vehicle, and dragged to the hospital where she was given an injection to end the pregnancy (MacLeod, 2012). Feng et al. (2013) reviewed China's one-child policy and considered it an example of "extreme state intervention in reproduction and a serious and harmful policy mistake."

Most parents have two children. One advantage is that they can enjoy life together.

Courtesy of Joyce Chang

because of the difficulty in pregnancy or birthing the child. One mother said, "I threw up every day for nine months including on the delivery table. Once is enough for me." There are also those who have only one child because they can't get pregnant a second time.

11-2c **Two Children?**

The most preferred family size in the United States (for non-Hispanic White women) is the two-child family (1.9 to be exact!). Reasons for this preference include feeling that a family is "not complete" without two children, having a companion for the first child, having a child of each sex, and repeating the positive experience of parenthood enjoyed with their first child. Some couples may not want to "put all their eggs in one basket." They may fear that, if they have only one child and that child dies, they will not have another opportunity to enjoy parenting (unless they have a second child).

11-2d **Three Children?**

Couples are more likely to have a third child, and to do so quickly, if they

competitive birthing having the same number (or more) of children in reference to one's peers.

PAIRUT/Shutterstock.com

already have two girls rather than two boys. They are least likely to bear a third child if they already have a boy and a girl. One male said, "I have 12 older sisters...my parents kept having children till they had me."

Having a third child creates a middle child. This child is sometimes neglected because parents of three children may focus more on the baby and the firstborn than on the child in between. However, an advantage to being a middle child is the chance to experience both a younger and an older sibling. Each additional child also has a negative effect on the existing children by reducing the amount of parental time available to the other children. The economic resources available for each child are also affected by each subsequent child.

11-2e **Four Children**

More than three children are often born to parents who are immersed in a religion which encourages procreation. Catholics, Mormons, and ultra-Orthodox Jews typically have more children than Protestants. When religion is not a factor, **competitive birthing** may be operative whereby individuals have the same number (or more) of children as their peers.

The addition of each subsequent child dramatically increases the possible relationships in the family. For example, in a one-child family, four interpersonal relationships are possible: mother–father, mother–child, father–child, and father–mother–child. In a family of four, eleven relationships are possible.

11-2f **Contraception**

To achieve the desired family size and to prevent unwanted pregnancies, a good beginning is to be sober, plan whether or not sexual intercourse will be part of your relationship, and discuss contraception. Having a conversation about birth control is a good way to begin sharing responsibility for it; one can learn of the partner's interest in participating in the choice and use of a contraceptive method (see Table 11.1).

Table 11.1

Methods of Contraception and Sexually Transmitted Infections (STIs) Protection

Typical Use, Effectiveness[1]					
Method	**Rates, %**	**STI Protection**	**Benefits**	**Disadvantages**	**Cost[2]**
Oral contraceptive (the pill)	92	No	High effectiveness rate, 24-hour protection, and menstrual regulation	Daily administration, side effects possible, medication interactions	$10–$42 per month
Nexplanon/Implanon NXT® or Implanon® (3-year implant)	99.95	No	High effectiveness rate, long-term protection	Side effects possible, menstrual changes	$400–$600 insertion
Depo-Provera® (3-month injection) or Depo-subQ Provera 104®	97	No	High effectiveness rate, long-term protection	Decreases body calcium, not recommended for use longer than 2 years for most users, side effects likely	$45–$75 per injection
Ortho Evra® (transdermal patch)	92	No	Same as oral contraceptives except use is weekly, not daily	Patch changed weekly, side effects possible	$15–$32 per month
NuvaRing® (vaginal ring)	92	No	Same as oral contraceptives except use is monthly, not daily	Must be comfortable with body for insertion	$15–$48 per month
Male condom	85	Yes	Few or no side effects, easy to purchase and use	Can interrupt spontaneity	$2–$10 a box
Female condom	79	Yes	Few or no side effects, easy to purchase	Decreased sensation and insertion takes practice	$4–$10 a box
Spermicide	71	No	Many forms to choose, easy to purchase and use	Can cause irritation, can be messy	$8–$18 per box/tube/ can
Today® Sponge[3]	68–84	No	Few side effects, effective for 24 hours after insertion	Spermicide irritation possible	$3–$5 per sponge
Diaphragm and cervical cap[3]	68–84	No	Few side effects, can be inserted within 2 hours before intercourse	Can be messy, increased risk of vaginal/UTI infections	$50–$200 plus spermicide
Intrauterine device (IUD): Paraguard or Mirena	98.2–99	No/little maintenance, longer-term protection	Risk of PID increased, chance of expulsion		$150–$300
Withdrawal	73	No	Requires little planning, always available	Pre-ejaculatory fluid can contain sperm	$0
Periodic abstinence	75	No	No side effects, accepted in all religions/cultures	Requires a lot of planning, need ability to interpret fertility signs	$0
Emergency contraception	75	No	Provides an option after intercourse has occurred	Must be taken within 72 hours, side effects likely	$10–$32
Abstinence	100%	Yes	No risk of pregnancy or STIs	Partners both have to agree to abstain	$0

[1] Effectiveness rates are listed as percentages of women not experiencing an unintended pregnancy during the first year of typical use. Typical use refers to use under real-life conditions. Perfect use effectiveness rates are higher.

[2] Costs may vary. The Affordable Care Act, health care legislation passed by Congress and signed into law by President Obama on March 23, 2010, requires health insurance plans to cover preventive services and eliminate cost sharing for some services, including "All Food and Drug Administration approved contraceptive methods, sterilization procedures, and patient education and counseling for all women with reproductive capacity" (U.S. Department of Health and Human Services Health Resources and Services Administration, Women's Preventive Services Guidelines, http://www.hrsa.gov/womensguidelines, January 9, 2014).

[3] Lower percentages apply to women who have already given birth. Higher rates apply to nulliparous women (women who have never given birth).

Source: Developed and used by permission of Beth Credle Burt, MAEd, CHES, a health education specialist, Education Services Project Manager, Siemens Healthcare.

Men can also share responsibility by purchasing and using condoms, paying for medical visits and the pharmacy bill, reminding their partner to use the method, assisting with insertion of barrier methods, checking contraceptive supplies, and having a vasectomy if that is what the couple decide on. The partners should also remember that contraception provides no protection against STIs and to use condoms/dental dams.

Courtesy of Michelle North

11-2g Emergency Contraception

Emergency contraception (also called **postcoital contraception**) refers to various types of combined estrogen-progesterone morning-after pills or post-coital IUD insertion used primarily in three circumstances: when a woman has unprotected intercourse, when a contraceptive method fails (such as condom breakage, which occurs 7% of the time, or slippage), and when a woman is raped. "Better safe than sorry" requires immediate action because the sooner the EC pills are taken, the lower the risk of pregnancy—12 hours is best, and 72 hours at the latest (Tal et al., 2014).

While emergency contraception medication is available over the counter—no prescription is necessary (and no pregnancy test is required) for women age 17 and above—some parents feel their parental rights are being undermined. For females under age 17, a prescription is required. Although side effects (nausea, vomiting, and so on) may occur, they are typically over in a couple of days and the risk of being pregnant is minimal.

emergency contraception (postcoital contraception) refers to various types of morning-after pills.

infertility the inability to achieve a pregnancy after at least one year of regular sexual relations without birth control, or the inability to carry a pregnancy to a live birth.

11-2h Sex Selection

Some couples use sex selection technologies to help ensure a boy or a girl. MicroSort is the new preconception sperm-sorting technology, which allows parents to increase the chance of having a girl or a boy baby. The procedure is also called "family balancing" since couples that already have several children of one sex often use it. The eggs of a woman are fertilized and the sex of the embryos three to eight days old is identified. Only embryos of the desired sex are then implanted in the woman's uterus.

Puri and Nachtigall (2010) examined ethical considerations in regard to sex selection. One perspective held by sex selection technology providers argues that sex selection is an expression of reproductive rights and a sign of female empowerment to prevent unwanted pregnancies and abortions. A contrasting view is that sex selection contributes to gender stereotypes that could result in neglect of children of the lesser-desired sex.

11-3 Infertility

"It's **hard to wait** around for something you know **may never happen**; but it's even **harder to give up** when you know it's everything you want."

—UNKNOWN

Infertility is the inability to achieve a pregnancy after at least one year of regular sexual relations without birth control, or the inability to carry a pregnancy to a live birth.

11-3a Types of Infertility

Different types of infertility include the following:

1. **Primary infertility.** The woman has never conceived even though she wants to and has had regular sexual relations for the past 12 months.

2. **Secondary infertility.** The woman has previously conceived but is currently unable to do so

even though she wants to and has had regular sexual relations for the past 12 months.

3. **Pregnancy wastage.** The woman has been able to conceive but has been unable to produce a live birth.

Shapiro (2012) emphasized the importance of conceiving in one's twenties rather than delaying pregnancy until one's thirties or forties. The chance of conceiving per month in one's thirties is 20%; in one's forties, it is 10%. Dougall et al. (2013) interviewed women who delayed getting pregnant until age 40 or later—44% reported that they were "shocked" to learn that the difficulty in getting pregnant declines so steeply as a woman moves into her mid to late thirties. Schmidt et al. (2012) also noted that delaying pregnancy until 35 and beyond is associated with higher risk of preterm births and stillbirths.

11-3b Causes of Infertility

Although popular usage does not differentiate between the terms *fertilization* and the *beginning of pregnancy*, fertilization or **conception** refers to the fusion of the egg and sperm, whereas **pregnancy** is not considered to begin until five to seven days later, when the fertilized egg is implanted (typically in the uterine wall). Hence, not all fertilizations result in a pregnancy. An estimated 30% to 40% of conceptions are lost prior to or during implantation. Forty percent of infertility problems are attributed to the woman, 40% to the man, and 20% to both of them. Some of the more common causes of infertility in men include low sperm production, poor semen motility, effects of STIs (such as chlamydia, gonorrhea, and syphilis), and interference with passage of sperm through the genital ducts due to an enlarged prostate. Additionally, there is some association between high body mass index in men and sperm that is problematic in impregnating a female (Sandlow, 2013).

Infertility in the woman is related to her age, not having been pregnant before, blocked fallopian tubes, endocrine imbalance that prevents ovulation, dysfunctional ovaries, chemically hostile cervical mucus that may kill sperm, and effects of STIs (Van Geloven et al., 2013). Obesity in the woman is also related to her infertility. Frisco and Weden (2013) analyzed national data on 1,658 females at two different time periods and found that young women who were obese at baseline had higher odds of remaining childless and increased odds of underachieving fertility intentions than young women who were normal weight at baseline. The

researchers concluded that obesity has long-term ramifications for women's childbearing experiences with respect to whether and how many children women have in general relative to the number of children they want. Brandes et al. (2011) noted that unexplained infertility is one of the most common diagnoses in fertility care and is associated with a high probability of achieving a pregnancy—most spontaneously.

Difficulty conceiving and carrying a fetus to term is strenuous for both individuals and couples. Teskereci and Oncel (2013) reported data on 200 couples undergoing infertility treatment and found that the quality of life of the women in the study was lower than that of the men. Advanced age, low education level, unemployment status, lower income, and long duration of infertility were also associated with lower quality of life. Baldur-Felskov et al. (2013) studied a large sample of Danish women who were infertile and found a higher incidence of hospitalizations due to psychiatric disorders. Pritchard and Kort-Butler (2012) found involuntarily childless women less happy than biological or adoptive mothers.

An at-home fertility kit, Fertell, allows women to measure the level of their follicle-stimulating hormone on the third day of their menstrual cycles. An abnormally high level means that egg quality is low. The test takes 30 minutes and involves a urine stick. The same kit allows men to measure the concentration of motile sperm. Men provide a sample of sperm (e.g., masturbation) that swim through a solution similar to cervical mucus. This procedure takes about 80 minutes. Fertell has been approved by the Food and Drug Administration (FDA). No prescription is necessary and the cost for the kit is about $100.

11-3c Success Using Assisted Reproductive Technologies (ART)

Some infertile women seek various reproductive technologies to get pregnant. The cost of treating infertility is enormous. Katz et al. (2011) examined the costs for 398 women in eight infertility practices over an 18-month period. For the half who pursued IVF, the median per-person medication costs ranged from $1,182 for medications only to $24,373 and

conception refers to the fusion of the egg and sperm. Also known as fertilization.

pregnancy when the fertilized egg is implanted (typically in the uterine wall).

$38,015 for IVF and IVF-donor egg groups, respectively. In regard to the costs of successful outcomes (delivery or ongoing pregnancy by 18 months) for IVF, the cost was $61,377. Within the time frame of the study, costs were not significantly different for women whose outcomes were successful and women whose outcomes were not. Only 28% of couples who invest in a fertility clinic end up with a live birth (Lee, 2006). The sooner an infertility problem is suspected the more successful the intervention.

Researchers continue to investigate the reason for low birth weight in babies which result from ART. It is not known if such is due to ART procedures or to the fact of infertility itself (Kondapalli & Perales-Puchalt, 2013).

11-4 Adoption

"I have four children. **Two are adopted. I forget which** two."

—BOB CONSTANTINE

While adoption is one solution to infertility, it is rare for U.S. parents to adopt. Just over 1% of 18- to 44-year-old women reported having adopted a child (Smock & Greenland, 2010).

Angelina Jolie and Brad Pitt are celebrities who have given national visibility to adopting children. They are not alone in their desire to adopt children. The various routes to adoption are public (children from the child welfare system), private agency (children placed with nonrelatives through agencies), independent adoption (children placed directly by birth parents or through an intermediary such as a physician or attorney), kinship (children placed in a family member's home), and

stepparent (children adopted by a spouse). Motives for adopting a child include an inability to have a biological child, a desire to give an otherwise unwanted child a permanent loving home, or a desire to avoid contributing to overpopulation by having more biological children. Some couples may seek adoption for all of these motives. Fifteen percent of adoptions will be children from other countries.

11-4a Characteristics of Children Available for Adoption

Adoptees in the highest demand are healthy, white infants. Those who are older than 3, of a racial or ethnic group different from that of the adoptive parents, of a sibling group, or with physical or developmental disabilities, have been difficult to place. Sankar (2012) noted that only 1% and 2% of 200 individuals reported a preference for a child ages 6 to 10 and ages 11 to 14, respectively. She emphasized the importance of providing opportunities for potential parents to learn from adults who had been adopted as an older child and to provide competency parental training for parenting older children.

11-4b Children Who Are Adopted

Children who are adopted have an enormous advantage over those who are not adopted. Juffer et al. (2011) examined 270 research articles including more than 230,000 children to compare the physical growth, attachment, cognitive development, school achievement, self-esteem, and behavioral problems of adopted and nonadopted children. Results revealed that adopted children outperformed their nonadopted peers who remained in institutions and they showed a dramatic recovery in practically all areas of development.

"Who are your real parents?," "Why did your mother give you up?," and "Are those your real parents?" are questions children who are adopted must sometimes cope with. W.I.S.E. Up is a tool provided to adopted children to help them cope with these intrusive, sometimes uncomfortable questions (Singer, 2010). W.I.S.E. is an acronym for: **W**alk away, **I**gnore or change the subject, **S**hare what you are comfortable sharing, and **E**ducate about adoption in general. The tool emphasizes that adopted children are wiser about adoption than their peers and can educate them or remove themselves from the situation.

11-4c Costs of Adoption

Adopting from the U.S. foster care system is generally the least expensive type of adoption, usually involving little or no cost, and states often provide subsidies to adoptive parents. However, before the adoption process begins, a couple who are foster care parents to a child and who become emotionally bonded with the child, risk that the birth parents might reappear and request their child back.

Stepparent and kinship adoptions are also less costly and have less risk of the child being withdrawn. Agency and private adoptions can range from $5,000 to $40,000 or more, depending on travel expenses, birth mother expenses, and requirements in the state. International adoptions can range from $7,000 to $30,000 (see http://costs.adoption.com/).

11-4d Open versus Closed Adoptions

Another controversy is whether adopted children should be allowed to obtain information about their biological parents. In general, there are considerable benefits for having an open adoption, especially the opportunity for the biological parent to stay involved in the child's life. Adoptees learn early that they are adopted and who their biological parents are. Birth parents are more likely to avoid regret and to be able to stay in contact with their child. Adoptive parents have information about the genetic background of their adopted child. Goldberg et al. (2011) studied lesbian, gay, and heterosexual couples who were involved in an open adoption. While there were some tensions with the birth parents over time, most of the 45 adoptive couples reported satisfying relationships.

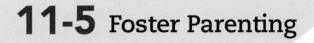

Denys Prykhodov/Shutterstock.com

11-5 Foster Parenting

"My **heart** was **bigger than** my **brain**."

—RESPONSE BY FOSTER PARENT WHEN ASKED WHY SHE FOSTERS

Some individuals seek the role of parent via foster parenting. A **foster parent**, also known as a family caregiver, is neither a biological nor an adoptive parent but is one who takes care of and fosters a child taken into custody. Foster care may be temporary or long term. A foster parent has made a contract with the state for the service, has judicial status, and is reimbursed by the state. Foster parents are screened for previous arrest records and child abuse and neglect. They are licensed by the state; some states require a "foster parent orientation" program.

Children placed in foster care have typically been removed from parents who are abusive, who are substance abusers, and/or who are mentally incompetent. Parents who are incarcerated or have mental health problems are more likely to have their parental rights terminated (Meyer et al., 2010). Although foster parents are paid for taking care of children in their home, they may also be motivated by their love of children. The goal of placing children in foster care is to improve their living conditions and then either return them to their family of origin or find a more permanent adoptive or foster home for them. Some couples become foster parents in the hope of being able to adopt a child who is placed in their custody.

Due to longer delays for foreign adoptions (e.g., it typically takes three years to complete a foreign adoption) and less availability of American infants, increasingly, couples are adopting a foster child. Waterman et al. (2011) studied behavior problems and parenting stress for children adopted from foster care homes and found that parents reported high adoption satisfaction despite ongoing behavioral problems with a third of the children in the sample.

11-5a Internet Adoption

Some couples use the Internet to adopt a baby. The Donaldson Adoption Institute surveyed 2,000 adoptees, adoptive parents, birth parents, and adoption professionals. The survey revealed the Internet's value of being able to make connections and provide information for parents seeking to adopt (Healy, 2013). But Roby and White (2010) noted the Internet's use to adopt a baby can pose serious problems

foster parent neither a biological nor an adoptive parent but a person who takes care of and fosters a child taken into custody.

of potential fraud, exploitation, and, most important, lack of professional consideration of the child's best interest. Couples should proceed with great caution. Policymakers should also be aware of the practice of "rehoming" where parents who have adopted a child use the Internet to place unwanted adopted children in new families. There is no monitoring or regulation of this practice (Healy, 2013).

11-6 Abortion

Abortion remains a controversial issue in the United States. Among American women, half will have an unintended pregnancy and 30% will have an abortion. About 60% of women having an abortion are in their twenties and unmarried (Guttmacher Institute, 2014). An abortion may be either an **induced abortion**, which is the deliberate termination of a pregnancy through chemical or surgical means, or a **spontaneous abortion (miscarriage)**, which is the unintended termination of a pregnancy. Miscarriages often represent a significant loss which is associated with depression/anxiety (Geller et al., 2010) and marital unhappiness (Sugiura-Ogasawara et al., 2013).

In this text the term *abortion* refers to induced abortion. In general, abortion is legal in the United States but it was challenged under the Bush administration. Specifically, federal funding was withheld if an aid group offered abortion or abortion advice. However, the Obama administration restored governmental approval for abortion. Obama said that denying such aid undermined "safe and effective voluntary family planning in developing countries."

11-6a Incidence of Abortion

When this country was founded abortion was legal and accepted by the public up until the time of "quickening"—the moment when a woman can feel the fetus inside her. By the time of

induced abortion the deliberate termination of a pregnancy through chemical or surgical means.

spontaneous abortion (miscarriage) the unintended termination of a pregnancy.

abortion rate the number of abortions per thousand women aged 15 to 44.

abortion ratio refers to the number of abortions per 1,000 live births. Abortion is affected by the need for parental consent and parental notification.

> **"I feel about Photoshop the way some people feel about abortion.** *It is appalling and a tragic reflection on the moral decay of our society... unless I need it, in which case, everybody be cool.* **"**
>
> —Tina Fey, comedian

the Civil War, one in five pregnancies was terminated by an abortion. Opposition to abortion grew in the 1870s, led by the American Medical Association launching a fierce campaign to make abortion illegal unless authorized and performed by a licensed physician. Abortion was made illegal in 1880, which did not stop the practice; a million abortions were performed illegally by the 1950s. In 1973, the Supreme Court upheld in *Roe v. Wade* the right of a woman to have a legal abortion. About 1.2 million abortions are performed annually in the United States. Although the number of abortions has been increasing among the poor (lower access to health care and health education), there has been a decrease among higher income women (due to increased acceptability of having a child without a partner and increased use of contraception). Ninety percent of abortions occur within the first three months of pregnancy (Guttmacher Institute, 2014).

The **abortion rate** (the number of abortions per 1,000 women ages 15 to 44) increased 1% between 2005 and 2008, from 19.4 to 19.6 abortions; the total number of abortion providers was virtually unchanged (Jones & Kooistra, 2011). About 40% of abortions are repeat abortions (Ames & Norman, 2012).

The **abortion ratio** refers to the number of abortions per 1,000 live births. Abortion is affected by the need for parental consent and parental

Diversity in Other Countries

There are wide variations in the range of cultural responses to the abortion issue. On one end of the continuum is the Kafir tribe in Central Asia, where an abortion is strictly the choice of the woman. In this preliterate society, there is no taboo or restriction with regard to abortion, and the woman is free to exercise her decision to terminate her pregnancy. One reason for the Kafirs' approval of abortion is that childbirth in the tribe is associated with high rates of maternal mortality. Because birthing children may threaten the life of significant numbers of adult women in the community, women may be encouraged to abort. Such encouragement is particularly strong in the case of women who are viewed as too young, too old, too sick, or too small to bear children.

A tribe or society may also encourage abortion for a number of other reasons, including practicality, economics, lineage, and honor. Abortion is practical for women in migratory societies. Such women must control their pregnancies, because they are limited in the number of children they can nurse and transport. Economic motivations become apparent when resources are scarce—the number of children born to a group must be controlled. Abortion for reasons of lineage or honor involves encouragement of an abortion in those cases in which a woman becomes impregnated in an adulterous relationship. To protect the lineage and honor of her family, the woman may have an abortion.

---

notification. **Parental consent** means that a woman needs permission from a parent to get an abortion if she is under a certain age, usually 18. **Parental notification** means that a woman has to tell a parent she is getting an abortion if she is under a certain age, usually 18, but she does not need parental permission. Laws vary by state. Call the National Abortion Federation Hotline at 1-800-772-9100 to find out the laws in your state.

11-6b Reasons for an Abortion

In a survey of 1,209 women who reported having had an abortion, the most frequently cited reasons were that having a child would interfere with a woman's education, work, or ability to care for dependents (74%); that she could not afford a baby now (73%); and that she did not want to be a single mother or was having relationship problems (48%). Nearly 4 in 10 women said they had completed their childbearing, and almost one-third of the women were not ready to have a child (Finer, 2005). Falcon et al. (2010) confirmed that the use of drugs was related to unintended pregnancy and the request for an abortion. Clearly, some women get pregnant when high on alcohol or other substances, regret the pregnancy, and want to reverse it.

Abortions performed to protect the life or health of the woman are called **therapeutic abortions**. However, there is disagreement over this definition. Garrett et al. (2001) noted, "Some physicians argue that an abortion is therapeutic if it prevents or alleviates a serious physical or mental illness, or even if it alleviates temporary emotional upsets. In short, the health of the pregnant woman is given such a broad definition

parental consent a woman needs permission from a parent to get an abortion if under a certain age, usually 18.

parental notification a woman has to tell a parent she is getting an abortion if she is under a certain age, usually 18, but she does not need parental permission.

therapeutic abortions abortions performed to protect the life or health of the woman.

that a very large number of abortions can be classified as therapeutic" (p. 218).

Some women with multifetal pregnancies (a common outcome of the use of fertility drugs) may have a procedure called *transabdominal first-trimester selective termination*. In this procedure, the lives of some fetuses are terminated to increase the chance of survival for the others or to minimize the health risks associated with multifetal pregnancy for the woman. For example, a woman carrying five fetuses may elect to abort three of them to minimize the health risks of the other two.

11-6c Pro-Life Abortion Position

A dichotomy of attitudes toward abortion is reflected in two opposing groups of abortion activists. Individuals and groups who oppose abortion are commonly referred to as "pro-life" or "antiabortion."

Of 4,700 undergraduates at two large universities, 25% reported that abortion was not acceptable under certain conditions (Hall & Knox, 2015). Pro-life groups favor abortion policies or a complete ban on abortion. They essentially believe the following:

1. The unborn fetus has a right to live and that right should be protected.

2. Abortion is a violent and immoral solution to unintended pregnancy.

3. The life of an unborn fetus is sacred and should be protected, even at the cost of individual difficulties for the pregnant woman.

Foster et al. (2013) studied the effect of pro-life protesters outside of abortion clinics on those individuals who came to the clinic to get an abortion. The researchers interviewed 956 women, 16% of whom said that they were very upset by the protesters. However, exposure to the protesters was not associated with differences in emotions one week after the abortion.

11-6d Pro-Choice Abortion Position

In the sample of 4,700 undergraduates referred to above, 62% reported that "abortion is acceptable

Martin Kubát/Shutterstock.com

under certain conditions" (Hall & Knox, 2015). Pro-choice advocates support the legal availability of abortion for all women. They essentially believe the following:

1. Freedom of choice is a central value—the woman has a right to determine what happens to her own body.

2. Those who must personally bear the burden of their moral choices ought to have the right to make these choices.

3. Procreation choices must be free of governmental control.

Although many self-proclaimed feminists and women's organizations, such as the National Organization for Women (NOW), have been active in promoting abortion rights, not all feminists are pro-choice.

11-6e Confidence in Making an Abortion Decision

To what degree do women considering abortion feel confident of their decision? Foster et al. (2012) studied 5,109 women who sought 5,387 abortions at one U.S. clinic and found that 87% of the women reported having high confidence in their decision before receiving counseling. Variables associated with uncertainty included being younger than 20, being Black, not having a high school diploma, having a history of depression, having a fetus with an anomaly, having general difficulty making decisions, having spiritual concerns, believing that abortion is murder, and fearing not being forgiven by God.

11-6f Physical Effects of Abortion

Part of the debate over abortion is related to its presumed effects. In regard to the physical effects, legal abortions, performed under safe medical conditions, are safer than continuing the pregnancy. The earlier in the pregnancy the abortion is performed, the safer it is. Vacuum aspiration, a frequently used method in early pregnancy, does not increase the risks to future childbearing. However, late-term abortions do increase the risks of subsequent miscarriages, premature deliveries, and babies of low birth weight.

Weitz et al. (2013) compared the outcome of 11,487 early aspiration abortions depending on who performed the abortion—nurse, certified nurse midwife, physician's assistant, and physician and found a complication rate of 1.8% with insignificant variation. These data support the adoption of policies to allow these providers to perform early aspirations to expand access to abortion care.

11-6g Psychological Effects of Abortion

Of equal concern are the psychological effects of abortion. The American Psychological Association reviewed all outcome studies on the mental health effects of abortion and concluded, "Based on our comprehensive review and evaluation of the empirical literature published in peer-reviewed journals since 1989, this Task Force on Mental Health and Abortion concludes that the most methodologically sound research indicates that among women who have a single, legal, first-trimester abortion of an unplanned pregnancy for nontherapeutic reasons, the relative risks of mental health problems are no greater than the risks among women who deliver an unplanned pregnancy" (Major et al., 2008, p. 71).

In contrast, Fergusson et al. (2013) surveyed eight publications looking at five outcome domains—anxiety, depression, alcohol misuse, illicit drug use/misuse, and suicidal behavior. Results revealed that abortion was associated with small to moderate increases in risks of anxiety, alcohol misuse, illicit drug use/misuse, and suicidal behavior. The researchers concluded that there is no available evidence to suggest that abortion has therapeutic effects in reducing the mental health risks of unwanted or unintended pregnancy.

However, in rebuttal, Steinberg (2013) disagreed with Fergusson's conclusions "because there are many poorly conducted studies in the literature and they cannot be combined with well-conducted studies" and noted that "there are some high-quality studies

Personal Views: Positive and Negative

Below are two abortion experiences reported by our students.

After going together for about three months, I got pregnant. I was a sophomore, could not tell my parents, and having a child was not an option. My "boyfriend" basically bailed . . . he just wasn't there for me. My roommate ended up driving me to another city where I had to walk through a bunch of protestors calling out that

"I didn't have to kill my baby." I was convinced what I had to do and went through with it. It was more painful than I thought it would be but I got through it and have never looked back. It was the right decision for me.

***

I was young and scared and didn't know what I was doing. I felt my parents would disown me; they knew I had gotten pregnant so I felt I didn't have any options—no money to care for me or the child, nobody to help me take care of the baby. I went through with the abortion but have always regretted it. I feel that I destroyed a human life and will never forgive myself. I think about how old my child would be, what it would be like if I had not had the abortion. I am now a staunch pro-life advocate and I am one of the people outside the abortion clinics trying to save a human life. Girls today think nothing of an abortion . . . they are not only killing their baby but also ruining their own life.

that provide evidence that abortion relative to birth of an unwanted pregnancy is not associated with higher rates of subsequent mental health problems, suggesting that abortion is not a cause of mental health problems."

Independent of the Fergusson and Steinberg standoff, Canario et al. (2013) analyzed data on 50 women (and 15 partners) one and six months after abortion and found a decrease in emotional disorder for all etiologies of abortion and an increase in perceived quality of a couple's relationship in therapeutic abortion over time. The researchers concluded that the psychological adjustment of an individual after abortion seems to be influenced by factors such as the couple's relationship.

11-6h Knowledge and Support of Male Partners of Women Who Have an Abortion

Jones et al. (2011) examined data from 9,493 women who had obtained an abortion to find out the degree to which their male partners knew of the abortion and their feelings about it. The overwhelming majority of women reported that the men by whom they got pregnant knew about the abortion, and most perceived these men to be supportive. Cohabiting men were particularly supportive. The researchers concluded that most women obtaining abortions are able to rely on male partners for social support.

11-7 Trends in Deciding about Children

While having children will continue to be the option selected by most couples, fewer will select this option, and being childfree will lose some of its stigma. Indeed, the pregnancy rate for women ages 15–29 has dropped steadily since 1990 (Curtin et al., 2013). Individualism and economics are the primary factors responsible for reducing the obsession to have children. To quote Laura Scott, "having children will change from an assumption to a decision." Once the personal, social, and economic consequences of having children come under close scrutiny the automatic response to have children will be tempered.

For the infertile, "google baby" will increasingly be considered as an option for having a child. *Google Baby* is the name of a documentary showing how infertile couples with a credit card can submit an order for a baby—the firm will find donor egg and donor sperm, fertilize the egg, and implant it in a surrogate mother in India. The couple need only fly to India, pick up their baby, and return to the states.

STUDY TOOLS 11

Ready to study? In the book, you can:

- Rip out the chapter review card at the back of the book for a handy summary of the chapter and key terms.

- Assess your attitudes toward having children with the Self-Assessment card at the back of the book.

Online at CENGAGEBRAIN.COM you can:

- Prepare for tests with quizzes.

- Review the key terms with Flash Cards.

- Play games to master concepts.

WHY CHOOSE?

Every 4LTR Press solution comes complete with a visually engaging textbook in addition to an interactive eBook. Go to CourseMate for **M&F** to begin using the eBook. Access at **www.cengagebrain.com**

Rearing Children

"**24/7** once you sign on to be a **mother**, that's the only shift they offer."

—JODI PICOULT, *MY SISTER'S KEEPER*

Angelina Jolie and Brad Pitt have three biological children and three adopted children. Of parenting, Angelina says, "… when you start your family, your life completely changes. And you completely live for someone else. From that moment on, they come first in every choice you make" (Green, 2014, p. 56). This chapter is about the transition from couplehood to parenthood and what is involved in rearing children. We begin by examining the choices parents make.

12-1 Parenting: A Matter of Choices

Parents endeavor to make the best decisions in rearing their children. Their choices have a profound impact on their children. In this section, we review the nature of parenting choices and identify some of the basic choices parents make.

12-1a Nature of Parenting Choices

Parents might keep the following points in mind when they make choices about how to rear their children:

1. **Not to make a parental decision is to make a decision.** Parents are constantly making choices even when they think they are not doing so. When a child makes a

Courtesy of Brittany Bolen

commitment ("I'll text you when I get to my friend's house") and does not do as promised, unless the parent addresses the issue (and provides a consequence such as grounding the child) the child learns nothing about taking commitments seriously.

2. **All parental choices involve trade-offs.** The decision to take on a second job or to work overtime to afford the larger house will come at the price of having less time to spend with one's children and being more exhausted when such time is

available. The choice to enroll a younger child in the highest-quality day care will mean less money for an older child's karate lessons.

3. **Reframe "regretful" parental decisions.** Parents sometimes chide themselves for some of their parenting decisions. Alternatively, they might emphasize the positive outcome of their choices. For example, not intervening quickly enough in a bad peer relationship with one child can be viewed as a "lesson" for such intervention with their other children.

12-1b Six Basic Parenting Choices

The six basic choices parents make include deciding (1) whether or not to have a child, (2) the number of children to have, (3) the interval between children, (4) the method of discipline, (5) the degree to which they will invest time with their children, and (6) whether or not to coparent (the parents cooperate in the development of the lives of their children). Though all of these decisions are important, the relative importance one places on parenting as opposed to one's career will have implications for the parents, their children, and their children's children.

12-2 Roles of Parents

Although finding one definition of **parenting** is difficult, there is general agreement about the various roles parents play in the lives of their children. New parents assume at least seven roles.

12-2a Caregiver

A major role of parents is the physical care of their children. From the moment of birth, when infants draw their first breath, parents stand ready to provide nourishment (milk), cleanliness (diapers), and temperature control (warm blanket). The need for

parenting defined in terms of roles including caregiver, emotional resource, teacher, and economic resource.

such sustained care in terms of a place to live/eat continues into adulthood as one-fourth of 18- to 34-year-olds have moved back in with their parents after living on their own (Parker, 2012).

12-2b Emotional Resource

Beyond providing physical care, parents are sensitive to the emotional needs of children in terms of their need to belong, to be loved, and to develop positive self-concepts. In hugging, holding, and kissing an infant, parents not only express their love for the infant but also reflect awareness that such displays of emotion are good for the child's sense of self-worth.

Parents also provide "emotion work" for children—listening to their issues, helping them figure out various relationships they are struggling with, etc. Minnottea et al. (2010) studied the emotion work of parents with their children in a sample of 96 couples and found that women did more of it. Indeed, the greater the number of labor hours by men, the fewer the number of emotion work hours for their children and the higher the number of hours for women.

12-2c Teacher

All parents think they have a philosophy of life or set of principles their children will benefit from. Parents soon discover that their children may not be interested in their religion or philosophy and, indeed, may rebel against it. This possibility does not deter parents from their role as teacher.

An array of self-help parenting books provide parents with ideas about the essentials children need that parents must teach. Galinsky (2010) identified seven skills all parents should be responsible for teaching their children: focus and self control, seeing someone else's point of view, communicating, making connections, critical thinking, taking on challenges, and self-directed engaged learning.

topseller/Shutterstock.com

HOW MUCH MONEY DO PARENTS GIVE THEIR ADULT CHILDREN?

One of the roles of parents is to provide financial assistance to their children. Indeed, 62% of young adults (ages 19–22) report that they receive money from their parents—mostly to pay bills. How much assistance are parents providing? A great deal, is the general answer—an average of $12,185 annually (Wightman et al., 2012). Padilla-Walker et al. (2012) studied a sample of 401 undergraduates and at least one of their parents and discovered that over half reported paying between $5,000 and $30,000 (30% paid $30,000 or more and 20% paid less than $5,000) yearly. Most felt it was their obligation as parents to help their children financially, particularly while their children were "emerging adults" (between 18 and late twenties).

When the parents were asked how long they should provide financial support, about half (49%) said until their offspring gets a job, about a quarter (23%) said when they graduate from college, and 6% said never. It seems that "lower levels of financial support may indeed facilitate a greater perception of oneself as an adult and promote more adult-like behaviors including fewer risky behaviors (e.g., drinking/binge drinking), greater identity development (at least in the domain of occupational identity), and higher numbers of work hours per week" (Padilla-Walker et al. 2012, p. 56). Hence, underwriting children completely seemed to interfere with the transition of those children to more responsible adult behavior.

Sources: Padilla-Walker, L. M., L. J. Nelson, & J. S. Carroll. (2012). Affording emerging adulthood: Parental financial assistance of their college-aged children. *Journal of Adult Development, 12:* 50–58.

Wightman, P., R. Schoeni, & K. Robinson. (2012, May 3). Familial financial assistance to young adults. Paper presented at the annual meeting of the Population Association of America, New Orleans, LA.

12-2d Economic Resource

New parents are also acutely aware of the costs for medical care, food, and clothes for infants, and seek ways to ensure that such resources are available to their children. Working longer hours, taking second jobs, and cutting back on leisure expenditures are attempts to ensure that money is available to meet the needs of the children.

Boomerang generation children are those young adults who leave home for college or marriage and come back, primarily for economic reasons. The What's New? section details how much money parents provide for their adult children.

12-2e Protector

Parents also feel the need to protect their children from harm. This role may begin in pregnancy in that the pregnant mother to be may stop smoking. Other expressions of the protective role include insisting that children wear seatbelts in cars and jackets in cold weather, protecting them from violence or nudity in the media, and protecting them from strangers. Increasingly, parents are joining the technological age and learning how to text message. In their role as protector, this ability allows parents to text message their children to tell them to come home, to phone home, or to work out a logistical problem—"meet me at the food court in the mall."

12-2f Health Promoter

The family is a major agent for health promotion—not only in promoting healthy food choices, responsible use of alcohol, nonuse of drugs, and safe driving skills, but also in ending smoking behavior. Knog et al. (2012) observed that parents who were most successful in getting their adolescents to stop smoking were positive models (they did not smoke themselves) and disapproved of their adolescent smoking.

boomerang generation adult children who return to live with their parents.

The parental protective function is universal. This sign is in Iceland.

12-2g Ritual Bearer

To build a sense of family cohesiveness, parents often foster rituals to bind members together in emotion and in memory. Prayer at meals and before bedtime, birthday, and holiday celebrations, and vacations at the same place (beach, mountains, and so on) provide predictable times of togetherness and sharing.

> "The **value of marriage** is not that adults produce children, but that **children produce adults**."
>
> —PETER DE VRIES, NOVELIST

transition to parenthood period from the beginning of pregnancy through the first few months after the birth of a baby during which the mother and father undergo changes.

oxytocin hormone from the pituitary gland during the expulsive stage of labor that has been associated with the onset of maternal behavior in lower animals.

12-3 Transition to Parenthood

The **transition to parenthood** refers to that period from the beginning of pregnancy through the first few months after the birth of a baby. The mother, father, and two of them as a couple undergo changes and adaptations during this period.

> "It (having a **baby**) **changes everything**."
>
> —ISABELLE TROADEC, NEW MOTHER OF 10-DAY-OLD BABY

12-3a Transition to Motherhood

Being a mother, particularly a biological or adoptive mother (in contrast to a stepmother), is a major social event for a woman (Pritchard & Kort-Butler, 2012). Indeed, a baby changes a woman's life forever. Her initial reaction is influenced by whether the baby was intended. Su (2012) studied 825 women and found that unintended births were associated with decreased happiness among mothers. Nevertheless, whatever her previous focus, her life will now include her baby.

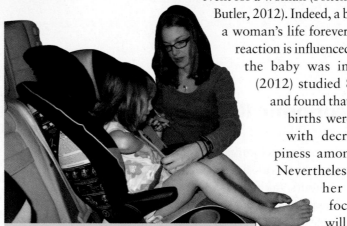

Parents do not start the car until their children are secure in their car seats.

One woman expressed: "Our focus shifts from our husbands to our children, who then take up all our energy. We become angry and lonely and burdened... and that's if we LIKE it." Of course, the new role of parent can become exhausting. Kotila et al. (2013) asked 182 dual-earner parents to keep diaries on their time involvement with their children. While both parents were highly involved with their infants, mothers were more involved than fathers in routine child care.

Sociobiologists suggest that the attachment between a mother and her offspring has a biological basis (one of survival). The mother alone carries the fetus in her body for nine months, lactates to provide milk, and, during the expulsive stage of labor, produces **oxytocin**, a hormone from the pituitary gland that has been associated with the onset of maternal behavior in lower animals. Most new mothers become emotionally bonded with their babies and resist separation.

Not all mothers feel joyous after childbirth. Naomi Wolf uses the term "the conspiracy of silence" to note motherhood is "a job that sucks 80 percent of the time" (quoted in Haag, 2011, p. 83). Some mothers do not bond immediately and feel overworked, exhausted, mildly depressed, and irritable; they cry, suffer loss of appetite, and have difficulty in sleeping. Many new mothers experience **baby blues**—transitory symptoms of depression 24 to 48 hours after the baby is born.

About 10% to 15% experience **postpartum depression**, a more severe reaction than baby blues (usually occurs within a month of the baby's birth). Postpartum depression is more likely in the context of a low-quality relationship when the partner is not supportive (Wickel, 2012). A complicated delivery is also associated with postpartum depression. Gelabert et al. (2012) looked at various personality traits that were associated with postpartum depression. They found that women who were perfectionist (characterized by concern over mistakes, personal standards, parental expectations, parental criticism, doubt about actions and organization) were more likely to report postpartum depression. To minimize baby blues and postpartum depression, antidepressants such as Zoloft and Prozac are used. Most women recover within a short time. With prolonged negative feelings, a clinical psychologist should be consulted.

Postpartum psychosis, a reaction in which a woman wants to harm her baby, is even more rare than postpartum depression. While having misgivings about a new infant is normal, the parent who wants to harm the infant should make these feelings known to the partner, a close friend, and a professional.

"I always thought I was born to be a musician, but I realize now that I was born to be a dad."

—DAVE JOHNSON (ON THE BIRTH OF HIS BABY EVIE)

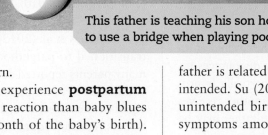

This father is teaching his son how to use a bridge when playing pool.

12-3b Transition to Fatherhood

Over two million fathers (21% of fathers who do not work outside the home) say they are home to take care of their children (Livingston, 2014). Just as some mothers experience postpartum depression, fathers, too, may become depressed following the birth of a baby. Quing et al. (2011) examined mothers' and fathers' postnatal (378 pairs) reactions and found that some parents of both genders reported postnatal depression (15% for mothers and 13% for fathers). The personal reaction of the male to the role of becoming a new father is related to whether the birth of the baby was intended. Su (2012) studied 889 men and found that unintended births were associated with depressive symptoms among fathers, particularly where financial strain was involved.

Regardless of whether the father gets depressed or the birth was intended, there is an economic benefit to the family for the husband becoming a father. Killewald (2013) noted that fathers who live with their wives and biological children become more focused and committed to work and their wage gains increase 4%. Regardless, mothers are typically disappointed in the amount of time the father helps with the new baby (Biehle & Michelson, 2012). Nomaguchi et al. (2012) confirmed that fathers' spending time with children, engagement, and cooperative coparenting were related to less parenting stress for mothers who were married to, cohabiting with, or dating the father.

Schoppe-Sullivan et al. (2008) emphasized that mothers are the gatekeepers of the father's involvement with his children. A

baby blues transitory symptoms of depression in a mother 24 to 48 hours after her baby is born.

postpartum depression a more severe reaction following the birth of a baby which occurs in reference to a complicated delivery as well as numerous physiological and psychological changes occurring during pregnancy, labor, and delivery; usually in the first month after birth but can be experienced after a couple of years have passed.

postpartum psychosis a reaction in which a woman wants to harm her baby.

Mothers can be the gatekeepers of the father's involvement with his children.

father may or may not be involved with his children to the degree that a mother encourages or discourages his involvement. The **gatekeeper role** is particularly pronounced after a divorce in which the mother receives custody of the children (the role of the father may be severely limited).

The importance of the father in the lives of his children is enormous and goes beyond his economic contribution (Dearden et al., 2013; McClain, 2011). While the role of father is not clearly defined and positive models are lacking (Ready et al., 2011), children who have a father who maintains active involvement in their lives tend to:

- Make good grades
- Be less involved in crime
- Have good health/self-concept
- Have a strong work ethic
- Have durable marriages
- Have a strong moral conscience
- Have higher life satisfaction

- Have higher incomes as adults
- Have higher education levels
- Have higher cognitive functioning
- Have stable jobs
- Have fewer premarital births
- Have lower incidences of child sex abuse
- Exhibit healthier/on time physical development

gatekeeper role
term used to refer to the influence of the mother on the father's involvement with his children.

Daughters may have an extra benefit of a close relationship with fathers. Byrd-Craven et al. (2012) noted that such a relationship was associated with the daughters' having lower stress levels which assisted them in coping with problems and in managing interpersonal relationships. To foster the relationship between fathers and daughters, the Indian Princess program has emerged. Google Indian Princess program for an organization in your community.

12-3c Transition from a Couple to a Family

The birth of a baby is usually a stressor event for each parent and for their relationship regardless of whether the child is biological or adopted and regardless of whether the parents are heterosexual or homosexual (Goldberg et al., 2010). However, not all couples report problems.

Holmes et al. (2013) studied 125 couples as they transitioned to parenthood and found that although many parents reported declines in love and increases in conflict, 23% of mothers and 37% of fathers reported equal or increased love; 20% of mothers and 28% of fathers reported equal or lower conflict. Durtschi and Soloski (2012) found that coparenting was associated with less parental stress and greater relationship quality. Similarly, salivary cortisol (stress hormone) levels of parents with babies was lower in those parents who worked together at night to get the children into bed (McDaniel et al., 2012). Hence, coparenting was related to lower stress levels.

Do women who are cohabiting experience the same negative change in relationship satisfaction as do married women? Yes. Mortensen et al. (2012) examined the data on 71,504 women who transitioned into motherhood over a two-year period and observed a similar negative change in relationship satisfaction during the transition to parenthood. However, cohabiting women started off and stayed less satisfied throughout the transition period.

Regardless of how children affect relationship satisfaction, spouses report more commitment to their relationship once they have children (Stanley & Markman, 1992). Figure 12.1 illustrates

A baby changes lovers into parents.

Figure 12.1
Percentage of Couples Getting Divorced by Number of Children

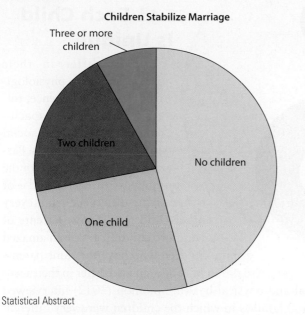

Children Stabilize Marriage

Statistical Abstract

that the more children a couple has, the more likely the couple will stay married. A primary reason for this increased commitment is the desire on the part of both parents to provide a stable family context for their children. In addition, parents of dependent children may keep their marriage together to maintain continued access to and a higher standard of living for their children. Finally, people (especially mothers) with small children feel more pressure to stay married (if the partner provides sufficient economic resources) regardless of how unhappy they may be. Hence, though children may decrease happiness, they increase stability, because pressure exists to stay together (the larger the family is, the more difficult the economic postdivorce survival is).

12-4 Parenthood: Some Facts

Parenting is only one stage in an individual's or couple's life (children typically live with an individual 30% of that person's life and with a couple 40% of their marriage). Parenting involves responding to the varying needs of children as they grow up, and parents require help from family and friends in rearing their children.

Some additional facts of parenthood follow.

12-4a Views of Children Differ Historically

Although children of today are thought of as dependent, playful, and adventurous, they were historically viewed quite differently (Beekman, 1977). Indeed, the concept of childhood, like gender, is a social construct rather than a fixed life stage. From the 13th through 16th centuries, children were viewed as innocent, sweet, and a source of amusement for adults. From the 16th through the 18th centuries, they were viewed as being in need of discipline and moral training. In the 19th century, whippings were routine as a means of breaking a child's spirit and bringing the child into submission. Although remnants of both the innocent and moralistic views of children exist today, the lives of children are greatly improved. Child labor laws in the United States protect children from early forced labor; education laws ensure a basic education; and modern medicine has been able to increase the life span of children.

Children of today are the focus of parental attention. In some families, everything children do is "fantastic" and deserves a gold star. The result is a generation of young adults who feel that they are special and should be catered to; they are entitled. Nelson (2010) noted that some parents have become "helicopter parents" (also referred to as hovercrafts and PFH—Parents from Hell) in that they are constantly hovering overprotectively at school and elsewhere with excessive interest to ensure their child's success.

12-4b Parents Create Diverse Learning Contexts in Which to Rear Their Children

The parents children have will markedly effect their life experience. Amish parents rear their children in homes without electricity (no television, cell phones, or video games). Charismatics and evangelicals (members of the conservative religious denominations) frequently send their children to Christian schools which emphasize a biblical and spiritual view of life. More secular parents tend to bring up their children with individualistic liberal views.

12-4c Parents Are Only One Influence in a Child's Development

Although parents often take the credit and the blame for the way their children turn out, they are only one among many influences on child development. Although parents are the first significant influence, peer influence becomes increasingly important during adolescence. For example, Ali and Dwyer (2010) studied a nationally representative sample of adolescents and found that having peers who drank alcohol was related to the adolescent drinking alcohol.

Siblings have also an important and sometimes lasting effect on each other's development. Siblings are social mirrors and models (depending on the age) for each other. They may also be sources of competition and can be jealous of each other. Teachers are also significant influences in the development of a child's values. As noted above, some parents send their children to religious schools to ensure that they will have teachers with conservative religious values. Selecting this structure for a child's education may continue into college.

Media in the form of television, replete with MTV and "parental discretion advised" movies are a major source of language, values, and lifestyles for children that may be different from those of the parents. Parents are also concerned about the sexuality and violence that their children are exposed to through television.

Another influence of concern to parents is the Internet. Though parents may encourage their children to conduct research and write term papers using the Internet, they may fear their children are also looking at pornography. Parental supervision of teenagers and their privacy rights on the Internet remain potential sources of conflict. Parents are also concerned about **sextortion** (online sexual extortion). Sextortion often takes the form of teenage girls with a webcam at a party who will visit an Internet chat room and flash their breasts. A week later they will get an email from a stranger who informs them that he has captured their video from the Internet and will post their

sextortion online sexual extortion.

responsiveness refers to the extent to which parents respond to and meet the needs of their children.

demandingness the manner in which parents place demands on children in regard to expectations and discipline.

photos on Facebook if they do not send him explicit photos (Wilson et al., 2010).

12-4d Each Child Is Unique

Children differ in their genetic makeup, physiological wiring, intelligence, tolerance for stress, capacity to learn, comfort in social situations, and interests. Parents soon become aware of the uniqueness of each child—of her or his difference from every other child they know. Parents of two or more children are often amazed at how children who have the same parents can be so different. Children also differ in their mental and physical health. Solomon (2012) interviewed 300 families in which the children were very different from their parents—autistic , gay, deaf, schizophrenic, etc. He entitled his book *Far from the Tree*.

12-4e Each Gender Is Unique

Wiseman (2013) reminds parents that the genders are also different and encourages them not to assume that boys are always emotionally resilient. Just because girls are socialized to be more open about their feelings (e.g., sad and depressed), boys may have these same feelings without a way to cope with them. She recommends that parents stay close to the emotional worlds of boys too—their falling in love, having their hearts broken, being bullied.

12-4f Parenting Styles Differ

Diana Baumrind (1966) developed a typology of parenting styles that has become classic in the study of parenting. She noted that parenting behavior has two dimensions: responsiveness and demandingness. **Responsiveness** refers to the extent to which parents respond to and meet the emotional needs of their children. In other words, how supportive are the parents? Warmth, reciprocity, person-centered communication, and attachment are all aspects of responsiveness. **Demandingness**, on the other hand, is the manner in which parents place demands on children in regard to expectations and discipline. How much control do they exert over their children? Monitoring and confrontation are also aspects of demandingness.

Albachiaraa/Shutterstock.com

Personal View: My Children Are the Yin and the Yang

My daughter as a baby was the type of child who would be categorized as "spirited" or "strong willed." The real definition without the fluff is that she was difficult to parent most of the time. As a baby, she had colic and reflux and was hard to soothe. She did not sleep well and demanded a great deal of attention. As a toddler, she was a perfectionist and intense in her play. She did not like to be told "no," was confrontive, and she liked to have control over her environment.

Today, my daughter (age 11) is the epitome of a "tween." She is moody, intense, entitled, and concerned that her peers like her. She is also amazingly creative and smart, has a good sense of humor, is loving, and craves my attention and affection after a rough day at school. She is becoming more responsible and contributing to the family. I am able to reframe in my own mind that some of the traits that almost put me over the edge when she was a baby (like her assertiveness) will serve her well as she matures into an adult.

My son was born when my daughter was 4 years old and I often wonder how I have another child that is so completely different from my daughter. My son was an "easy" baby. He had no eating or sleeping difficulties and his temperament was very laid back. When I tell people my son literally slept through the night—every night—from 6 months onward, they find it hard to believe. My son could play in his playpen without being entertained and was the type of baby that you could take out to a dinner and he was not fussy or disruptive. He was the preschool teacher's "favorite" student. Parenting him almost felt easy.

Now age 7, my son remains very easy to parent. He is goofy and smart and loves video games and doing creative things. Because he is so well behaved, I have to remind myself to give him as much attention as his sister. I can see him as a young man and I think that he will make a wonderful husband and father because he has so many amazing qualities. On days when I am feeling overwhelmed or the kids are being loud when I've already had a rough day at work, I try to remind myself that I have the best of both worlds. I have a girl and a boy—I have a yin and a yang.

—VICKI OLIVER

The yin and the yang.

Courtesy of Vicki Oliver

Categorizing parents in terms of their responsiveness and their demandingness creates four categories of parenting styles: permissive (also known as indulgent), authoritarian, authoritative, and uninvolved:

1. **Permissive parents are high on responsiveness and low on demandingness.** These parents are very lenient and allow their children to largely regulate their own behavior. Walcheski and Bredehoft (2010) found that the permissive parenting style is associated with overindulgence. For example, these parents state punishments but do not follow through; they give in to their children. These parents act out of fear that disciplining the child will cause the child to dislike his or her parents. The parents are controlled by the potential disapproval of the child.

2. **Authoritarian parents are high on demandingness and low in responsiveness.** They feel that children should obey their parents no matter what. Surjadi et al. (2013) confirmed

that harsh, inconsistent discipline from parents was related to poor relationship quality with romantic partners during cohabitation or the early years of marriage.

3. **Authoritative parents are both demanding and responsive.** They impose appropriate limits on their children's behavior but emphasize reasoning and communication. Authoritative parenting offers a balance of warmth and control and is associated with the most positive outcome for children. Examples of this style include parents telling the child their expectations of the child's behavior before the child engages in the activity, giving the child reasons why rules should be obeyed, talking with the child when he or she misbehaves, and explaining consequences (Walcheski & Bredehoft, 2010). Panetta et al. (2014) studied outcomes of various parenting styles of mothers and fathers on 195 seventh to eleventh grade adolescents. Temperament of adolescent (e.g., mood, flexibility–rigidity) explained half of the variance in adolescent outcomes. Parenting styles contributed a smaller but significant role. When both parents were authoritative, it was associated with more optimal outcomes in adolescents' personal adjustment than any other parenting style combination. In addition, when both parents were permissive and neglectful, these parenting styles were associated with poorer adolescent outcomes. Similar positive outcomes for authoritative parenting were observed in Japanese children by Uji et al. (2014).

4. **Uninvolved parents are low in responsiveness and demandingness.** These parents are not invested in their children's lives. Panetta et al. (2014) confirmed that permissive and neglectful parenting has negative outcomes for adolescents.

McKinney and Renk (2008) identified the differences between maternal and paternal parenting styles, with mothers tending to be authoritative and fathers tending to be authoritarian. Mothers and fathers also use different parenting styles for their sons and daughters, with fathers being more permissive with their sons than with their daughters. Overall, this study emphasized the importance of examining the different parenting styles on adolescent outcome and suggested that having at least one authoritative parent may be a protective factor for older adolescents. In summary, the authoritative parenting style is the combination of warmth, guidelines, and discipline—"I love you but you need to be in by midnight or you'll lose the privileges of having a cell phone and a car."

We have been discussing parenting styles of married parents. What about the divorced dad? Nonresidential divorced fathers tend to operate from the parenting style of being permissive and uninvolved which is associated with negative outcomes for children. Fathers in a coparenting arrangement are more likely to have an authoritative parenting style which is associated with high self-esteem and life satisfaction from their children. These findings are based on research by Bastaits et al. (2014) and their study of 684 divorced fathers.

> "The surest **way to a child's heart** is to **spend time** with them."
>
> —KEVIN HEATH

Diversity in Other Countries

Sweden provides a glimpse of parenting styles in other countries and how they have changed. Young to middle-age adults (sample of 1,015) living in a suburb of Stockholm answered questions about how they were reared. Their answers revealed a dramatic decrease in direct control by parents. Over time authoritarian patterns changed to more egalitarian family environments—from "stereotyped versions of fathers as decision makers and mothers as caregivers to both parents sharing decisions and garnering respect from children" (Trifan et al., 2014).

12-5 Principles of Effective Parenting

Dumka et al. (2010) noted that **parenting self-efficiency** (feeling competent as a parent) was associated with parental control. In effect, parents who viewed themselves as good teachers felt confident in their role which, in turn, translated into their ability to effectively manage the behavior of their adolescents. Numerous principles are involved in being effective parents (Keim & Jacobson, 2011). We begin with the most important of these principles, which involves giving time/love to your children as well as praising and encouraging them.

12-5a Give Time, Love, Praise, Encouragement, and Acceptance

Children need to feel that they are worth spending time with and that they are loved. Because children

Even taking one's child to the grocery store is to spend time with the child. The message is "I am important."

depend first on their parents for the development of their sense of emotional security, it is critical that parents provide a warm emotional context in which the children develop. Feeling loved as an infant also affects one's capacity to become involved in adult love relationships.

Abundant evidence from children reared in institutions where nurses attended only to their physical needs testify to the negative consequences of early emotional neglect. **Reactive attachment disorder** is common among children who were taught as infants that no one cared about them. Such children have no capacity to bond emotionally with others since they have no learning history of the experience and do not trust adults/caretakers/parents. A 5:1 ratio of positive to negative statements to the child is also associated with positive outcomes for the child (Armstrong & Clinton, 2012).

12-5b Avoid Overindulgence

Overindulgence may be defined as more than just giving children too much; it includes overnurturing and not providing enough structure. Using this definition, a study of 466 participants revealed that those who are overindulged tend to hold materialist values for success, are not able to delay gratification, and are less grateful for things and to others. Another result of overindulgence is that children grow up without consequences so they avoid real jobs where employers expect them to show up at 8:00 A.M. Indeed, not being overindulged promotes the ability to delay gratification, be grateful, and experience a view (nonmaterialism) that is associated with happiness (Slinger & Bredehoft, 2010).

Parents typically overindulge their children because they feel guilty for not spending time with them or because they (the parents) did not have certain material goods in their own youth. In a study designed to identify who overindulged, Clarke (2004) found that mothers were four times more likely to overindulge than fathers.

parenting self-efficiency feeling competent as a parent.

reactive attachment disorder common among children who were taught as infants that no one cared about them; these children have no capacity to bond emotionally with others since they have no learning history of the experience and do not trust adults, caretakers, or parents.

overindulgence defined as more than just giving children too much; includes overnurturing and providing too little structure.

Mothers spend more hours in child care so they are more vulnerable to being worn down—they overindulge their children as a coping mechanism for stress (give them expensive toys, etc., to entertain them).

12-5c Monitor Child's Activities/Drug Use

Abundant research suggests that parents who monitor their children and teens and know where their children are, who they are with, and what they are doing are less likely to report that their adolescents receive low grades, or are engaged in early sexual activity, delinquent behavior, and alcohol or drug use (Crosnoe & Cavanagh, 2010). Crutzen et al. (2012) assessed the effects of parental approval of children's drinking alcohol at home on their subsequent drinking behavior. Of 1,500 primary school children, those who were allowed to drink alcohol around their parents were less likely to consume alcohol when they were away from their parents.

Parents who drank alcohol under age and who used marijuana or other drugs wonder how to go about encouraging their own children to be responsible alcohol users and drug free. Drugfree.org has some recommendations for parents, including being honest with their children about previous alcohol and drug use, making clear that they do not want their children to use alcohol or drugs, and explaining that although not all alcohol or drug use leads to negative consequences, staying clear of alcohol/drug use is the best course of action.

12-5d Monitor Television and Pornography Exposure

Some parents monitor the amount and content of media exposure their children experience. A Kaiser Family Foundation (2010) study of media use of 8- to 18-year-olds found that they were watching television an average of 4.5 hours per day. Nonschool computer use was an hour and a half a day—much of which was unsupervised (two-thirds of the students said there were no rules about Internet use).

Children and youth watching pornography is not unusual. Flood (2009) noted the negative effects of pornography exposure among children including their developing sexist and unhealthy notions of sex and relationships. In addition, among boys and young men, frequent consumption of pornography is associated with supportive attitudes of sexual coercion and increases their likelihood of perpetrating assault.

Pornography teaches unrealistic expectations (e.g., females are expected to have bodies like porn stars, to enjoy "facials," and anal sex) and is devoid of integrating intimacy in sexual expression (Knox et al., 2014).

12-5e Use Technology to Monitor Cell Phone/Text Messaging Use

"Technological advances" was identified by 2,000 parents as the top issue that makes parenting children today tougher than in previous years. Sixty-one percent identified this issue; 57% noted increased violence (Payne & Trapp, 2013). While media technology can be used to enhance the bond between parents and children (Coyne, 2011), some parents are concerned about their child's use of a cell phone and texting.

Thirteen is the average age at which parents feel that it is appropriate for their child to own a cell phone (Gibbs, 2012). Not only may predators contact children/teenagers without the parents' awareness, but also 39% of teens report having sent sexually suggestive text messages and 20% have sent nude or seminude photos or videos of themselves. Parents are responsible for their children and should regularly go through their cell phone text messages, photos, and apps. New technology called SMS Tracker for Android permits parents to effectively take over their child's mobile phone. A parent can see all incoming and outgoing calls, text messages, and photos. Another protective device on the market is MyMobileWatchdog (known as MMWD), which monitors a child's cell phone use and instantly alerts the parents online if their son or daughter receives unapproved email, text messages, or phone calls. SecuraFone can reveal how fast the car in which the cell phone of the user is moving and alert parents. If a teen is speeding, the parents will know. One version shuts the texting capability down if the phone is going faster than five miles an hour. Mobiflock for Android allows parents to lock their child's phone for a predetermined time (e.g., 7 to 10 P.M. when studying is scheduled). In effect, these smart phones have web filters (which can block inappropriate websites), app filters (which make sure apps are kid-friendly), and contact filters (which can prevent harassing calls or texts by blocking certain numbers), which allow parents to monitor their child's location, texts, calls, browsing histories, app downloads, and photos they send and receive.

Dangerous apps that parents should not allow their children to have on their phones include Snapchat (photos disappear), Kik Messenger (hides content from parents), Yik Yak (has GPS so predator can locate), Poof (hides apps), O Mengle (can't track who is messaging your child), and Down (sex focused). Children/adolescents may say "you are invading my privacy" and "you don't trust me." The parents response is "I'm the parent, I love you, and I am responsible for you. When you are 18, you will have privacy. Until them, I'm on duty."

> ## "Successful teaching is not head-to-head; it is heart-to-heart."
>
> —TAMARA L. CHILVER

lucadp/Fotolia

12-5f Set Limits and Discipline Children for Inappropriate Behavior

The goal of setting limits and disciplining children is their self-control. Parents want their children to be able to control their own behavior and to make good decisions on their own. Parental guidance involves reinforcing desired behavior and punishing undesirable behavior. Unless parents levy negative consequences for lying, stealing, and hitting, children can grow up to be dishonest, to steal, and to be inappropriately aggressive. **Time-out** (a noncorporal form of punishment that involves removing the child from a context of reinforcement to a place of isolation for one minute for each year of the child's age) has been shown to be an effective consequence for inappropriate behavior (Morawska & Sanders, 2011). Withdrawal of privileges (use of cell phone, watching television, being with friends), pointing out the logical consequences of the misbehavior ("you were late; we won't go"), and positive language ("I know you meant well but . . .") are also effective methods of guiding children's behavior (more about time-out below).

Physical punishment is less effective in reducing negative behavior; it teaches the child to be aggressive and encourages negative emotional feelings toward the parents. When using time-out or the withdrawal of privileges, parents should make clear to the child that they disapprove of the child's behavior, not the child. A review of some of the alternatives to corporal punishment includes the following (Crisp & Knox, 2009):

1. **Be a positive role model.** Children learn behaviors by observing their parents' actions, so parents must model the ways in which they want their children to behave. If a parent yells or hits, the child is likely to do the same.

2. **Set rules and consequences.** Make rules that are fair, realistic, and appropriate to a child's level of development. Explain the rules and the consequences of not following them. If children are old enough, they can be included in establishing the rules and consequences for breaking them.

3. **Encourage and reward good behavior.** When children are behaving appropriately, give them verbal praise and reward them with tangible objects (occasionally), privileges, or increased responsibility.

4. **Use charts.** Charts to monitor and reward behavior can help children learn appropriate behavior. Charts should be simple and focus on one behavior at a time, for a certain length of time.

5. **Use time-out.** Time-out involves removing children from a situation following a negative behavior. This can help children calm down, end an inappropriate behavior, and reenter the situation in a positive way. Explain

time-out a noncorporal form of punishment that involves removing the child from a context of reinforcement to a place of isolation.

what the inappropriate behavior is, why the time-out is needed, when it will begin, and how long it will last. Set an appropriate length of time for the time-out based on age and level of development, usually one minute for each year of the child's age. Awareness of these alternatives is associated with a reduction in the use of spanking. Mothers exposed to child development information report a 30% reduction in the use of spanking (Mayer & Blome, 2013).

making decisions on how long to provide free room and board for an adult child.

"Sign on bedroom door of children: Checkout Time— Age 18."

—ERMA BOMBECK

12-5g Have Family Meals

Parents who stay emotionally connected with their children build strong relationships with them and report fewer problems. Bisakha (2010) found that families who have regular meals with their adolescents report fewer behavioral problems, such as less substance use and running away from home for females. For males, there was also less running away as well as less drinking, less physical violence, less property destruction, and less stealing. The researcher recommends that family meals be made a regular family ritual.

12-5h Encourage Responsibility

Giving children increased responsibility encourages the autonomy and independence they need to be assertive and self-governing. Giving children more responsibility as they grow older can take the form of encouraging them to choose healthy snacks and letting them decide what to wear and when to return from playing with a friend (of course, the parents should praise appropriate choices).

Children who are not given any control over, and responsibility for, their own lives remain dependent on others. A dependent child is a vulnerable child. Successful parents can be defined in terms of their ability to rear children who can function as independent adults.

Parents also recognize that there is a balance they must strike between helping their children and impeding their own growth and development. One example is

12-5i Adult Children Living with Parents

Forty-two percent of adult children ages 20–24 live with their parents (El Nasser, 2012). The recession (and resulting unemployment), college debt, and divorce all point to the primary reason—it is cheaper at mom's house. And, based on an Ameritrade Survey of 1,000 Gen Z members, adult children's level of embarrassment at living with their parents varies based on their own age—49% at age 25, 88% at age 30, and 91% at age 35 report hanging their head at living at mom's (Payne & Trap, 2013). Parents vary widely in how they view and adapt to adult children living with them. Some parents prefer that their children live with them, enjoy their company, and hope the arrangement continues indefinitely (this type of arrangement is often the norm in Italy). They have no rules about their children living with them and expect nothing from them. The adult children can come and go as they please, pay for nothing, and have no responsibilities or chores.

Other parents develop what is essentially a rental agreement, whereby their children are expected to pay rent, cook and clean the dishes, mow the lawn, and service the cars. No overnight guests are allowed, and a time limit is specified as to when the adult child is expected to move out.

A central issue from the point of view of young adults who live with their parents is whether their parents perceive them as adults or children. This perception has implications for whether they are free to come and go as well as to make their own decisions (Sassler et al., 2008). Males are generally left alone whereas females are often under more scrutiny.

Adult children also vary in terms of how they view the arrangement. Some enjoy living with their parents, volunteer to pay rent (though most do not), take care of their own laundry, and participate in cooking and housekeeping. Others are depressed that they are economically forced to live with their parents, embarrassed that they do so, pay nothing, and do nothing to contribute to the maintenance of the household. In a study of 30 young adults who had returned to the parental home, two-thirds paid nothing to live there and wanted to keep it that way. Those who did pay contributed considerably less than what rent would cost on the open market. Most paid for their cell phones, long-distance charges, and personal effects like clothing and toiletries (Sassler et al., 2008).

Whether parents and adult children discuss the issues of their living together will depend on the respective parents and adult children. Although there is no best way, clarifying expectations will help prevent misunderstandings. For example, what is the norm about bringing new pets into the home? Take, for example, a divorced son who moved back in with his parents along with his six-year-old son *and* a dog. His parents enjoyed being with them but were annoyed that the dog chewed on the furniture. Parental feelings eventually erupted that dismayed their adult child. He moved out, and the relationship with his parents became very strained. However, about three-fourths of the respondents in the Sassler et al. (2008) study reported generally satisfactory feelings about returning to their parental home. Most looked forward to moving out again but, in the short run, were content to stay to save money. In the sample of 1,144 undergraduate males, 14% (and 14% of 3,556 females) said that they would likely live with their parents after college (Hall & Knox, 2015).

12-5j Establish Norm of Forgiveness

Carr and Wang (2012) emphasized the importance of forgiveness in family relationships and the fact that forgiveness is a complex time-involved process rather than a one-time cognitive event (e.g., "I forgive you"). Respondents revealed in interviews with the researchers that forgiveness involved head over heart (finding ways to explain the transgression), time (sometimes months and years), and distance (giving the relationship a rest and coming back with renewed understanding).

12-5k Teach Emotional Competence

Wilson et al. (2012) emphasized the importance of teaching children **emotional competence**—experiencing , expressing, and regulating emotion. Being able to label when one is happy or sad (experiencing emotion), expressing emotion ("I love you"), and regulating emotion (e.g., reducing anger) assists children in getting in touch with their feelings and being empathetic with others. Wilson et al. (2012) reported on "Tuning in the Kids," a training program for parents to learn how to teach their children to be emotionally competent. Follow-up data on parents who took the six-session, two-hour-a-week program revealed positive outcomes/changes.

12-5l Provide Sex Education

Hyde et al. (2013) interviewed 32 mothers and 11 fathers about the sexuality education of their children. Most gave little to no explicit information about safe sex—they assumed that the school had done so and that their son or daughter was not in a romantic relationship. Some felt that talking to their children about sex might be viewed as encouraging sexual behavior.

Karen Rayne, PhD, is a specialist in talking with teens about sex. The mother of two teenagers, she recommends the following:

1. **Know yourself.** What are your expectations, your hopes, and your fears about your teenager's sexual and romantic development? You will have far more control over yourself and your interactions if you have a full understanding of these things.

2. **Remember that it's not about you.** Your teenager is, in fact, discovering sex for the first time. They want to talk about their current exciting, overwhelming path. So let them! That's how you'll find out what you can do to help your teenager walk this path—and remember, that's what matters most.

3. **Stop talking!** As the parent of a teenager, you are in the business of getting to know your teenager, not to give information. If you're talking, you can't hear anything your teenager is trying to tell you.

4. **Start listening!** Stop talking. Start listening. Remember what's most important about your role as a parent? And that can't happen if you don't really, really listen.

5. **You get only one question.** Since there's only one, you had better make

emotional competence teaching the child to experience emotion, express emotion, and regulate emotion.

it a good one that can't be answered with a yes or a no. Spend some time mulling over it. You can ask it when you're sure it's a good one. Open ended questions are best…"What was it like when…."

6. **Do something else.** Anything else. Many teenagers, especially boys, will have an easier time talking about sexuality and romance if you're doing something "side by side" like driving, walking, or playing a game rather than sitting and looking at each other.

7. **About pleasure and pain.** You have to talk about both. If you don't acknowledge the physical and emotional pleasure associated with sexuality, your teenager will think you're completely out of touch, and so you will be completely out of touch.

8. **Be cool like a cucumber.** It is only when you manage to have a calm, loving demeanor that your teenager will feel comfortable talking with you. And remember—you're in the business of getting to know your teenager—and the only way to do that is if your teenager keeps talking.

9. **Bring it on!** Your teenagers have tough questions, some of them quite specific and technical. If you're able to answer these questions with honesty, humor, and without judgment, your teenager will feel much more at home coming to you with the increasingly difficult emotional questions that touch their heart.

10. **Never surrender.** There may be times you feel like quitting. Like the millionth time when you've tried to have an actual conversation with your teenager—about anything, much less sex!—and your teenager has once again completely avoided eye contact and has not even acknowledged your existence. But you can't. You're still doing some good, so keep going. Trust me. (Source: Karen Rayne, PhD)

12-5m Express Confidence

"One of the greatest mistakes a parent can make," confided one mother, "is to be anxious all the time about your child, because the child interprets this as your lack of confidence in his or her ability to function independently." Rather, this mother noted that it is best to convey to one's child that you know that he or she will be fine because you have confidence in him/her. "The effect on the child," said this mother, "is a heightened sense of self-confidence." Another way to conceptualize this parental principle is to think of the self-fulfilling prophecy as a mechanism that facilitates self-confidence. If parents show their child that they have confidence in him or her, the child begins to accept these social definitions as real and becomes more self-confident.

12-5n Respond to the Teen Years Creatively

Parenting teenage children presents challenges that differ from those in parenting infants and young children. Teenagers literally have altered brains that have lower amounts of dopamine, which may disrupt their reward function and make them less responsive to social stimuli (Forbes et al., 2012). Teenagers are more likely to engage in a high rate of risky behavior—smoking, alcohol consumption, hazardous driving, drug use, delinquency, dares, sporting risks, rebellious behavior, and sexual intercourse (Becker, 2010). Seeking novelty, peer influences, genetic factors, and brain function are among the elements accounting for the vulnerability of adolescents.

> **"You can tell a child is growing up when he stops asking where he came from and starts refusing to tell where he is going."**
>
> —UNKNOWN

Conflicts between parents and teenagers often revolve around money and independence. The desired cell phone, designer clothes, and car can outstrip the budgets of many parents. Teens also want increasingly more freedom. However, neither of these issues needs result in conflict. When they do, the effect on the parent-child relationship may be inconsequential. One parent tells his children, "I'm just being the parent, and you're just being who you are; it is okay for us to disagree but you can't go." The following suggestions can help keep conflicts with teenagers at a lower level:

1. **Catch them doing what you like rather than criticizing them for what you do not like.** Adolescents are like everyone else—they do not like to be criticized, but they do like to be noticed for what they do that is exemplary.

2. **Be direct when necessary.** Though parents may want to ignore some behaviors of their children, addressing certain issues directly is necessary. Regarding the avoidance of sexually transmitted infections (STIs) or HIV infection and pregnancy, parents should inform their teenagers of the importance of using a condom or dental dam before sex.

3. **Provide information rather than answers.** When teens are confronted with a problem, try to avoid making a decision for them. Rather, provide information on which they may base a decision. What courses to take in high school and what college to apply to are decisions that might be made primarily by the adolescent. The role of the parent might best be to listen.

4. **Be tolerant of high activity levels.** Some teenagers are constantly listening to loud music, going to each other's homes, and talking on cell phones for long periods of time. Parents often want to sit in their easy chairs in peace. Recognizing that it is not realistic to always expect teenagers to be quiet and sedentary may be helpful in tolerating their disruptions.

5. **Engage in leisure activity with your teenagers.** Whether renting a DVD, eating a pizza, or taking a camping trip, structuring some activities with your teenagers is important. Such activities permit a context in which to communicate with them and their emotional well-being. Offer (2013) found that having family meals was associated with higher emotional well-being of adolescents.

   Sometimes teenagers present challenges with which their parents feel unable to cope. Aside from monitoring their behavior closely, family therapy may be helpful. Two major goals of such therapy are to increase the emotional bond between the parents and the teenagers and to encourage positive consequences for desirable behavior (e.g., concert tickets for good grades) and negative consequences for undesirable behavior (e.g., loss of cell phone/car privileges).

6. **Use technology to encourage safer driving.** GPS devices are now available which can tell a parent where their teenager is, record any sudden stops, and record speed. Some teenagers scream foul and accuse their parents of not trusting them. One answer is that of a father who required his daughter to pay for her car insurance but told her she could get a major discount if the GPS device was installed. She then thought the GPS was a good idea (Copeland, 2012).

12-6 Single-Parenting Issues

Forty percent of births in the United States are to unmarried mothers. Distinguishing between a single-parent "family" and a single-parent "household" is important. A **single-parent family** is one in which there is only one parent—the other parent is completely out of the child's life through death or complete abandonment or as a result of sperm donation, and no contact is ever made. In contrast, a **single-parent household** is one in which one parent typically has primary custody of the child or children, but the parent living out of the house is still a part of the child's life. This arrangement is also referred to as a binuclear family. In most divorce cases in which the mother has primary physical custody of the child, the

single-parent family family in which there is only one parent and the other parent is completely out of the child's life through death, sperm donation, or abandonment, and no contact is made with the other parent.

single-parent household one parent has primary custody of the child/children with the other parent living outside of the house but still being a part of the child's family; also called binuclear family.

child lives in a single-parent household because the child is still connected to the father, who remains part of the child's life. In cases in which one parent has died, the child, or children live with the surviving parent in a single-parent family because there is only one parent.

12-6a Single Mothers by Choice

Single parents enter their role through divorce or separation, widowhood, adoption, or deliberate choice to rear a child or children alone. Diane Keaton, never married, has two adopted daughters. An organization for women who want children and who may or may not marry is Single Mothers by Choice.

12-6b Challenges Faced by Single Parents

The single-parent lifestyle involves numerous challenges. See singleparent.lifetips.com for some interesting advice. Challenges associated with being a single parent include the following:

1. **Responding to the demands of parenting with limited help.** Perhaps the greatest challenge for single parents is taking care of the physical, emotional, and disciplinary needs of their children alone. Depression and stress are common among single parents (Hong & Welch, 2013).

2. **Resolving the issue of adult sexual needs.** Some single parents regard their parental role as interfering with their sexual relationships. They may be concerned that their children will find out if they have a sexual encounter at home or be frustrated if they have to go away from home to enjoy a sexual relationship. Some choices with which they are confronted include, "Do I wait until my children are asleep and then ask my lover to leave before morning?" or "Do I openly acknowledge my lover's presence in my life to my children and ask them not to tell anybody?" and "Suppose my kids get attached to my lover, who may not be a permanent part of our lives?"

3. **Coping with lack of money.** Single-parent families, particularly those headed by women, report that money is always lacking. The median income of a single-woman householder is $33,637, much lower than that of a single-man householder ($49,567) or a married couple ($74,130) (*ProQuest Statistical Abstract of the United States, 2014,* Table 722).

4. **Ensuring guardianship.** If the other parent is completely out of the child's life, the single parent needs to appoint a guardian to take care of the child in the event of the parent's death or disability.

5. **Obtaining prenatal care.** Single women who decide to have a child have poorer pregnancy outcomes than married women. Their children are likely to be born prematurely and to have low birth weight (Mashoa et al., 2010). The reason for such an association may be the lack of economic funds (no male partner with economic resources available) as well as the lack of social support for the pregnancy or the working conditions of the mothers.

6. **Coping with the absence of a father.** Another consequence for children of single-parent mothers is that they often do not have the opportunity to develop an emotionally supportive relationship with their father. Barack Obama noted, "I know what it is like to grow up without a father." We earlier detailed the value of an involved father for a child's life.

Courtesy of Brittany Bolen

Single parenting involves one person doing all the work of parenting, including feeding and bathing children every day.

7. **Avoiding negative life outcomes for the child in a single-parent family.** Researcher Sara McLanahan, herself a single mother, set out to prove that children reared by single parents were just as well off as those reared by two parents. McLanahan's data on 35,000 children of single parents led her to a different conclusion: children of only one parent were twice as likely as those reared by two married parents to drop out of high school, get pregnant before marriage, have drinking problems, and experience a host of other difficulties, including getting divorced themselves (McLanahan & Booth, 1989; McLanahan, 1991). In addition, Freeman and Temple (2010) found that adolescents from single-parent homes were more likely to be raped than those from two-parent homes. Lack of supervision, fewer economic resources, and less extended family support were among the culprits.

8. **Perpetuating a single-family structure.** Growing up in a single-family home increases the likelihood that the adult child will have a first child while unmarried and in a cohabitation relationship, thus perpetuating the single-family structure.

Though the risk of negative outcomes is higher for children in single-parent homes, most are happy and well adjusted. Maier and McGeorge (2013) emphasized that single parents (both mothers and fathers) are victims of negative stereotyping. The reality is that there are numerous positives associated with being a single parent. These include having a stronger bonding experience with one's children since they "are" the family, a sense of pride and self-esteem for being independent, and being a strong role model for offspring who observe their parent being able to "wear many hats."

12-7 Trends in Parenting

Trends in parenting will involve new contexts for children and new behaviors that children learn and parents tolerate. While parents will continue to be the primary context in which their children are reared, because the financial need for both parents to earn an income will increase, children will, increasingly, end up in day care, afterschool programs, and day camps during the summer. There will also be an increasing number of children reared in single-family contexts. These changed contexts will result in new parental norms where a wider range of behaviors on the part of their children will be accepted. Hence, since parents will be increasingly preoccupied with their jobs/careers, the norms their children learn in day care and other contexts will be more readily accepted since parents will have less time and energy trying to reverse them.

Hence, children may be less polite, less obedient, and less compliant to authority resulting more often in parental acceptance than addressing or correcting these behaviors. Teachers in the public school system often comment that "lack of parental involvement" is reflected in higher frequencies of student misbehavior and lack of respect for authority. We are not suggesting that children will become wild hellions, only that the contexts in which children will spend an increasing amount of time may be concerned about the well-being of the child than the family context (secondary groups, not primary groups).

STUDY TOOLS 12

Ready to study? In the book, you can:

⮑ Rip out the chapter review card at the back of the book for a handy summary of the chapter and key terms.

⮑ Assess how traditional your views of motherhood and methods of child discipline are with the Self-Assessment cards at the back of the book.

Online at CENGAGEBRAIN.COM you can:

⮑ Prepare for tests with quizzes.

⮑ Review the key terms with Flash Cards.

⮑ Play games to master concepts.

Stress and Crisis in Relationships

"I make the **most** of **all that comes** and the **least** of **all that goes**."

—SARA TEASDALE, POET

"Life is pain, highness. Anyone who says differently is selling something," said William Goldman, American novelist. His words crackle with reality as we trudge though the daily issues with which we are confronted. While life is also about joy, and relationships are a major source of joy, the downside is stress and crisis management—the focus of this chapter. We begin with defining stress and crisis.

13-1 Definitions and Sources of Stress and Crisis

stress is a reaction of the body to substantial or unusual demands (physical, environmental, or interpersonal). Psychological stress is higher among women than men; persons who were younger, less educated, and with lower incomes also report more stress (Cohen & Janicki-Deverts, 2012). Stress is also associated with irritability, high blood pressure, and depression (Barton & Kirtley, 2012), lower relationship satisfaction (Bodenmann et al., 2010), and sexual desire (Stephenson & Meston, 2012).

stress reaction of the body to substantial or unusual demands (physical, environmental, or interpersonal).

Nicki Pardo/Getty Images

> "You must learn to let go. *Release the stress. You were never in control anyway.*"
>
> —STEVE MARABOLI, *LIFE, THE TRUTH, AND BEING FREE*

Stress is a process rather than a state. For example, a person will experience different levels of stress throughout a divorce—acknowledging that one's marriage is over, telling the children, leaving the family residence, getting the final decree, and seeing one's ex with a new partner will result in varying levels of stress.

A **crisis** is a situation that requires changes in normal patterns of response

crisis a crucial situation that requires change in one's normal pattern of behavior.

Studying is an externally induced stress for most students.

Courtesy of Rachel Calisto

Stressors or crises may also be categorized as expected or unexpected. Expected family stressors include the need to care for aging parents and the death of one's parents. Unexpected stressors include contracting a sexually transmitted infection, becoming aware of an unplanned pregnancy, or experiencing the death/suicide of a friend.

Both stress and crisis events are normal aspects of family life and sometimes reflect a developmental sequence. Pregnancy, childbirth, job change or loss, children's leaving home, retirement, and widowhood are predictable for most couples and families. Crisis events may have a cumulative effect: the greater the number in rapid succession, the greater the stress.

13-1a **Resilient Families**

Just as the types of stress and crisis events vary, individuals and families vary in their ability to respond

behavior. A family crisis is a situation that upsets the normal functioning of the family and requires a new set of responses to the stressor. Sources of stress and crises can be external, such as the hurricanes that annually hit our coasts or devastating tornadoes in the spring. Other examples of an external crisis are economic recession, downsizing, or military deployment.

The source of stress and crisis events may also be internal (e.g., alcoholism, an extramarital affair, or Alzheimer's disease of one's spouse or parents). Another internal source of a crisis event is inheriting money from one's deceased parents. One spouse reported a dramatic change in the relationship when the partner suddenly "became rich" and did not "share the wealth."

"Depressed? Of course, we're all depressed. We've been so quickly, violently, and irreconcilably **plucked from** nature, from physical labor, from kinship and village mentality, from **every natural** and primordial **antidepressant**."

—M. ROBIN D'ANTAN

Bouncing back and staying back is the modus operandi of this man who has continued an involved and meaningful life in spite of a diving accident that resulted in his being in a wheelchair for more than 20 years.

successfully to crisis events. **Resiliency** refers to a family's strengths and ability to respond to a crisis in a positive way. Lane et al. (2012) identified the various aspects of **family resiliency** as beliefs (e.g., finding meaning in diversity, a positive outlook, spirituality), organizational patterns (e.g., flexibility, connectedness), and communication process (e.g., clarity, open emotional sharing). A family's ability to bounce back from a crisis (from negative health diagnosis, loss of one's job to the death of a family member) reflects its level of resiliency.

> ## "When it is **dark enough**, you **can see** the stars."
>
> —CHARLES BEARD, HISTORIAN

13-1b A Family Stress Model

Various theorists have explained how individuals and families experience and respond to stressors. The ABCX model of family stress was developed by Reuben Hill in the 1950s. The model can be explained as follows:

A = stressor event
B = family's management strategies, coping skills
C = family's perception, definition of the situation
X = family's adaptation to the event

A is the stressor event, which interacts with B, the family's coping ability or crisis-meeting resources. Both A and B interact with C, the family's appraisal or perception of the stressor event. X is the family's adaptation to the crisis. Thus, a family that experiences a major stressor (e.g., a spouse with a spinal cord injury) but has great coping skills (e.g., religion or spirituality, love, communication, and commitment) and perceives the event to be manageable will experience a moderate crisis. However, a family that experiences a less critical stressor event (e.g., their child makes Cs and Ds in school) but has minimal coping skills (e.g., everyone blames everyone else) and perceives the event to be catastrophic will experience an extreme crisis.

13-2 Positive Stress-Management Strategies

Researchers Burr and Klein (1994) administered an 80-item questionnaire to 78 adults to assess how families experiencing various stressors such as bankruptcy, infertility, a disabled child, and a troubled teen used various coping strategies and how they felt these strategies worked. In the following sections, we detail some helpful stress-management strategies.

13-2a Scaling Back and Restructuring Family Roles

Higgins et al. (2010) studied 1,404 men and 1,623 women in dual-earner families. These spouses evidenced two stress-reducing strategies: scaling back and restructuring family roles. Men were more likely to scale back than women. Scaling back means being selective in volunteering for various committees or activities so as not to overextend oneself. Restructuring family roles means abandoning rigid roles and recruiting children to help.

> ## "If there is a **sin against life**, it consists perhaps not so much in despairing of life as in **hoping for another** life and in **eluding** the implacable grandeur of **this life**."
>
> —ALBERT CAMUS

13-2b Choosing a Positive Perspective

The strategy that most respondents reported as being helpful in coping with a crisis was choosing a positive view of the crisis situation. Survivors of hurricanes, tornados, and earthquakes routinely focus on the fact that they and their loved ones are alive rather than the loss of their home or material possessions. Buddhists have the saying, "Pain is inevitable; suffering is not." This perspective emphasizes that how one views a situation, not the situation itself, determines its impact on you. One Chicago woman said after a pipe burst and caused $30,000 worth of damage to her home: "If it's not about your health, it's irrelevant." A team of researchers (Baer et al., 2012) confirmed the importance of selecting positive

resiliency a family's strength and ability to respond to a crisis in a positive way.

family resiliency the successful coping of family members under adversity that enables them to flourish with warmth, support, and cohesion.

cognitions for 87 adults who enrolled/participated in an eight-week program on mindfulness-based stress reduction (MBSR). Mindfulness can be thought of as choosing how you view a phenomenon. These adults were coping with chronic illness and chronic pain and reduced their stress by being mindful to frame their situation positively: "we are alive," "we are coping," and "we have each other."

> "Since we **cannot change reality**, let us **change the way we see** reality."
>
> —NIKOS KAZANTZAKIS, GREEK PHILOSOPHER

Regardless of the crisis event, one can view the crisis positively. A betrayal can be seen as an opportunity for forgiveness, unemployment as a stage to spend time with one's family, and ill health as a chance to appreciate one's inner life. Positive cognitive functioning is associated with keeping the crisis in perspective and moving on (Phillips et al., 2012).

Exercise has an enormous positive effect on cutting stress.

Courtesy of Rachel Calisto

13-2c **Exercise**

The Centers for Disease Control and Prevention (CDC) and the American College of Sports Medicine (ACSM) recommend that people ages 6 years and older engage regularly, preferably daily, in light to moderate physical activity for at least 30 minutes at a time. These recommendations are based on research that has shown the physical, emotional, and cognitive benefits of exercise. Researchers (Mata et al., 2013; Silveira et al., 2013; & Vina et al., 2012) emphasized the psychological benefits of exercise (including its role in reducing depression and viewing exercise as a proactive drug). Exercise is the most important of all behaviors for good physical and mental health.

> "A **faithful friend** is the **medicine** of life."
>
> —UNKNOWN

13-2d **Family Cohesion, Friends, and Relatives**

Mendenhall et al. (2013) found that individuals who are imbedded in a network of close family relationships tend to cope well with stress. A network of friendships and relatives is also associated with coping with various life transitions. Such relationships provide both structural and emotional support.

13-2e **Love**

A love relationship also helps individuals cope with stress. Being emotionally involved with another and sharing the crisis experience with that person helps insulate individuals from being devastated by a traumatic event. Love is also viewed as helping resolve relationship problems. Over 86% of 1,137 undergraduate males and 89% of 3,548 undergraduate females agreed with the statement, "A deep love can get a couple through any difficulty or difference" (Hall & Knox, 2015).

13-2f **Religion and Spirituality**

A strong religious belief is associated with coping with stress (Mendenhall et al., 2013). Goodman et al. (2013) conducted interviews with spouses representing different religions throughout the United States and found that religion was functional in assisting the spouses to cope with stress and crisis events. Ellison et al. (2011) previously examined the role of religion and marital satisfaction/coping with stress. They found that **sanctification** (viewing the marriage as having divine character or significance) was associated both with predicting positive marital quality and providing a buffer for financial and general stress on the marriage. Green and Elliott (2010) also noted that those who identify as religious report better health and marital satisfaction. Not only does religion provide a rationale for one's plight ("It is God's will"), but it also offers a mechanism to ask for help. Religion is a social institution which connects one to others who may offer both empathy and physical assistance.

"Sometimes life is full of laughs. Sometimes it ain't funny."

—HOYT AXTON, COUNTRY WESTERN SINGER

Courtesy of Rachel Calisto

Pets are companions and associated with stress reduction.

13-2g Laughter and Play

A sense of humor is related to lower anxiety (Grases et al., 2012). Forty-one undergraduates took part in a study in which they watched a 25-minute drama video and a 25-minute humor video. The latter was related to reducing their level of anxiety.

13-2h Sleep

Getting an adequate amount of sleep is associated with lower stress levels. Midday naps are also associated with positive functioning, particularly memory and cognitive function (Pietrzak et al., 2010).

13-2i Pets

Hughes (2011) emphasized that having an animal as a pet is associated with reducing blood pressure and stress, preventing heart disease, and fighting depression. Veterinary practices are encouraged to increase the visibility of this connection so that more individuals might benefit.

13-3 Harmful Stress-Management Strategies

Some coping strategies not only are ineffective for resolving family problems but also add to the family's stress by making the problems worse. Respondents in the Burr and Klein (1994) research identified several strategies they regarded as harmful to overall family functioning. These included keeping feelings inside, taking out frustrations on or blaming others, and denying or avoiding the problem.

Burr and Klein's research also suggested that women and men differ in their perceptions of the usefulness of various coping strategies. Women were more likely than men to view as helpful such strategies as sharing concerns with relatives and friends, becoming more involved in religion, and expressing emotions. Men were more likely than

sanctification viewing the marriage as having divine character or significance.

women to use potentially harmful strategies such as using alcohol, keeping feelings inside, or keeping others from knowing how bad the situation was.

13-4 Five Individual, Couple, and Family Crisis Events

Some of the more common crisis events that spouses and families face include physical illness, mental illness, an extramarital affair, unemployment, substance abuse, and death.

13-4a Physical Illness and Disability

When one partner has a debilitating illness, there are profound changes in the roles of the respective partners and their relationship. Mutch (2010) interviewed eight partners who took care of a spouse with multiple sclerosis (MS) after 20 years of marriage. The partners reported experiencing a range of feelings, including a sense of duty and a sense of loss, as they prioritized the health and needs of their spouse above their own. Partners reported losing their sense of identity as a spouse as their caretaking role became "the career." Some partners also felt out of control due to the unpredictable and progressive nature of MS and because it consumed their life 24 hours every day. Some felt guilty at not being satisfied with their life and wanting some freedom and independence.

While 24 million adults in the United States report that they have a disability, only 6.8 million of these use a visible assistive device. Hence, individuals may have a hidden disability such as chronic back pain, multiple sclerosis, rheumatoid arthritis, or chronic fatigue syndrome (Lipscomb, 2009). These illnesses are particularly invasive in that conventional medicine has little to offer besides pain medication. For example, spouses with chronic fatigue syndrome may experience financial

iofoto/Shutterstock.com

consequences ("I could no longer meet the demands of my job so I quit"), gender role loss ("I couldn't cook for my family" or "I was no longer a provider"), and changed perceptions by their children ("They have seen me sick for so long they no longer ask me to do anything").

In those cases in which the illness is fatal, **palliative care** is helpful. This term describes the health care for the individual who has a life-threatening illness (focusing on relief of pain and suffering) and support for the individual and his or her loved ones. Such care may involve the person's physician or a palliative care specialist who works with the physician, nurse, social worker, and chaplain. Pharmacists or rehabilitation specialists may also be involved. The goals of such care are to approach the end of life with planning (how long should life be sustained on machines?) and forethought to relieve pain (medication) and provide closure.

We have been discussing reacting to the crisis of a spouse with a disability. Smith and Grzywacz (2013) focused on the crisis of having a child with special health care needs and found negative physical and mental health effects on parents. Their study covered a 10-year period.

13-4b Mental Illness

Mental illness is defined as "alterations in thinking, mood or behavior that are associated with distress and impaired function" (Marshall et al., 2010). Eight percent of adults 18–29 have a serious mental illness. This percentage drops as individuals age to 1.4% of those over the age of 65 (Hudson, 2012). Depression is common among college students. Field et al. (2012) found an unusually high percentage (52%) of depression among the 238 respondents who also reported anxiety, intrusive thoughts, and sleep disturbances.

The toll of mental illness on a relationship can be immense. A major initial attraction of partners to each other includes intellectual and emotional qualities. Butterworth and Rodgers (2008) surveyed 3,230 couples to assess the degree to which mental illness of a spouse or spouses affects divorce and found that couples in which either men or women reported mental health problems had higher rates of marital disruption than couples in which neither spouse experienced

palliative care
health care for the individual who has a life-threatening illness which focuses on relief of pain/ suffering and support for the individual.

Black Hole: A Spouse Talks about Being Depressed

If you have experienced life in the Black Hole, then you don't need an explanation of it. If you have never experienced life in the Black Hole, it is impossible for anyone to explain it to you, and even if someone could explain it to you, you still wouldn't understand it.

The Black Hole is, by definition, an irrational state. The only thing you can comprehend about it is that you cannot comprehend it. Offering ANY advice, judgmental comments, suggestions like—"If you would only . . . ," "You've got to want to help yourself . . . ," "I know you are depressed, BUT . . . ," "You are not being rational . . ."—to someone who is in the Black Hole is not helpful.

On the contrary, offering advice is very destructive. It may temporarily relieve YOUR frustration with the person, but all it does for them is to give them a serving of guilt to deal with as they wait for the Black Hole to pass. If you really want to help somebody who is in the Black Hole, there IS one thing you can do—and that one thing is NOTHING.

After living in the Black Hole for a lifetime, a person has pretty much heard everything that you plan to tell them about it. Eventually, people who visit the Black Hole learn that it is a Monster, which comes without warning or invitation, it stays for a while, and it leaves when it is ready. The Monster doesn't time its visits to avoid holidays, vacations, or rainy days. It just barges in.

A person also knows what works for them while they are in the Black Hole. For some it may be exercise, or fishing, or sex (if they are still physically able), or music, or going to the beach, or talking about it, or employing logic to deal with it, or prayer, or reciting positive affirmations. For others, like me, there are only two things that help—complete solitude and sleep. Those two things do not provide a cure, but they do help you cope with the Monster until he leaves.

Nobody asks why you wear glasses. They just assume that you wear glasses because you need to, they don't ask questions about it, they don't offer suggestions and they don't try to fix it. And they don't assume that you wear glasses because of something they did. What a blessing it would be, if people treated those who struggle with clinical depression with the same respect.

Source: Former student of the author. Name withheld by request.

---

mental health problems. For couples in which both spouses reported mental health problems, rates of marital disruption reflected the additive combination of each spouse's separate risk. See the Black Hole for insight into depression.

13-4c Middle-Age Crazy (Midlife Crisis)

The stereotypical explanation for a middle-aged man who buys a convertible, has an affair, or marries a 20-year old, is that he is "having a midlife crisis." The label conveys that such individuals feel old, think that life is passing them by, and seize one last great chance to do something they have always wanted to do. Indeed, one father (William Feather) noted, "Setting a good example for your children takes all the fun out of middle age."

nycshooter/iStockphoto.com

"Middle age is the awkward period when Father Time catches up with Mother Nature."

—HAROLD COFFIN

However, a 10-year study of close to 8,000 U.S. adults ages 25 to 74 by the MacArthur Foundation Research Network on Successful Midlife Development revealed that, for most respondents, the middle years brought no crisis at all but a time of good health, productive activity, and community involvement. Less than a quarter (23%) reported a crisis in their lives. Those who did experience a crisis were going through a divorce. Two-thirds were accepting of getting older; one-third did feel some personal turmoil related to the fact that they were aging (Goode, 1999).

Of those who initiated a divorce in midlife, 70% had no regrets and were confident that they did the right thing. This fact is the result of a study of 1,147 respondents ages 40 to 79 who experienced a divorce in their forties, fifties, or sixties. Indeed, midlife divorcers' levels of happiness or contentment were similar to those of single individuals their own age and those who remarried (Enright, 2004).

Some people embrace middle age. The Red Hat Society (http://www.redhatsociety.com/) is a group of women who have decided to "greet middle age with verve, humor, and élan. We believe silliness is the comedy relief of life [and] share a bond of affection, forged by common life experiences and a genuine enthusiasm for wherever life takes us next." The society traces its beginning to when Sue Ellen Cooper bought a bright red hat because of a poem Jenny Joseph wrote in 1961, titled the "Warning Poem." The poem reads:

> When I am an old woman I shall wear purple
> With a red hat which doesn't go and doesn't suit me.

Cooper then gave red hats to friends as they turned 50. The group then wore their red hats and purple dresses out to tea, and that is how it got started. Now there are over 1 million members worldwide.

In the rest of this chapter, we examine how spouses cope with the crisis events of an extramarital affair, unemployment, drug abuse, and death. Each of these events can be viewed either as devastating and the end of meaning in one's life or as an opportunity and challenge to rise above.

extramarital affair refers to a spouse's sexual involvement with someone outside the marriage.

13-4d Extramarital Affair

Affairs are not unusual. Of spouses in the United States, about one-fourth of husbands

Husband's text: *Darling, I've been hit by a car outside the office. Paula brought me to the hospital. They have been making tests and taking X-rays. The blow to my head has been very strong but fortunately it seems that did not cause any serious injury. However, I have three broken ribs, a compound fracture in the left leg, and they may have to amputate the right foot.*

Wife's response: *Who's Paula?*

—INTERNET HUMOR

and one-fifth of wives report ever having had intercourse with someone to whom they were not married (Russell et al., 2013).

"When a **man cheats**, it is said it is because **he is a dog**. When a **woman cheats**, it is said it is **because her man is a dog**."

—MOKOKOMA MOKHONOANA, *DIVIDED AND CONQUERED*

Types of Extramarital Affairs The term **extramarital affair** refers to a spouse's sexual involvement with someone outside the marriage. Affairs are of different types, which may include the following:

1. **Brief encounter.** A spouse meets and hooks up with a stranger. In this case, the spouse is usually out of town, and alcohol is often involved.

2. **Paid sex.** A spouse seeks sexual variety with a prostitute who will do whatever he wants (as happened in the case of former New York governor Eliot Spitzer). These encounters usually

go undetected unless there is a sexually transmitted infection (STI), the person confesses, or the prostitute exposes the client.

3. **Instrumental or utilitarian affair.** This is sex in exchange for a job or promotion, to get back at a spouse, to evoke jealousy, or to transition out of a marriage.

4. **Coping mechanism.** Sex can be used to enhance one's self-concept or feeling of sexual inadequacy, compensate for failure in business, cope with the death of a family member, or test one's sexual orientation.

5. **Paraphiliac affairs.** In these encounters, the on-the-side sex partner acts out sexual fantasies or participates in sexual practices that the spouse considers bizarre or abnormal, such as sexual masochism, sexual sadism, or transvestite fetishism.

6. **Office romance.** Two individuals who work together may drift into an affair. David Petraeus (former CIA director) and John Edwards (former presidential candidate) became involved in affairs with women they met on the job.

7. **Internet use.** Internet usage now tops 1.6 billion people (Hertlein & Piercy, 2012). Although, legally, an extramarital affair does not exist unless two people (one being married) have intercourse, Internet use can be disruptive to a marriage or a couple's relationship.

8. **Facebook infidelity.** While Internet affairs via Match.com is common (persons in a "monogamous" relationship simply look for someone new), Facebook (boasting 1.2 billion active users) allows one to connect with old partners, friends of friends, persons at work, etc. and start a relationship. Cravens and Whiting (2014) identified a range of Facebook behaviors that are variously defined as cheating: friending an ex-partner or ex-spouse, sending private messages to a partner, listing one's relationship status as single, not friending one's partner on Facebook, having multiple accounts, etc. Infidelty is so rampant that Facebookcheating.com features stories of how relationships have ended because of affairs begun on Facebook.

While men and women agree that actual kissing, touching breasts/genitals, and sexual intercourse constitute infidelity, they disagree about the degree to which online behaviors constitute cheating. Based on data collected by Hines (2012), men are less likely than women to define as cheating emailing a person online for relationship advice (27% versus 51%), having a friendly conversation with someone in a chat room called "Married and Lonely" (64% versus 84%), and creating a pet name for a person he/she met in an Internet chat room (40% versus 70%). Males also are less likely to view online pornography as cheating (27% versus 64%).

Computer friendships may move to feelings of intimacy, involve secrecy (one's partner does not know the level of involvement), include sexual tension (even though there is no overt sex), and take time, attention, energy, and affection away from one's partner. Cavaglion and Rashty (2010) noted the anguish embedded in 1,130 messages on self-help chat boards from female partners of males involved in cybersex relationships and pornographic websites. The females reported distress and feelings of ambivalent loss that had an individual, couple, and sexual relationship impact. Cramer et al. (2008) also noted that women become more upset when their man was *emotionally* unfaithful with another woman (although men become more upset when their partner was *sexually* unfaithful with another man).

Cybersex can also be problematic. Examples of cybersex include "…participating in sexual acts through the use of webcams, playing sexual computer games, and participating in sexual dialogue in chat rooms. These activities may or may not end in sexual climax" (Jones & Tuttle, 2012, p. 275). Spouses who discover their partner spending increasing amounts of time engaging in cybersex feel angry, betrayed, and depressed.

Unhappy spouses often turn to the Internet to find a new partner.

Courtesy of Brittany Bolen

Extradyadic involvement, or extrarelational involvement, refers to the sexual involvement of a pair-bonded individual with someone other than the partner. Extradyadic involvements are not uncommon. Of 1,147 undergraduate males, 21% agreed with the statement, "I have cheated on a partner I was involved with" (23% of 3,563 undergraduate females). When the statement was "A partner I was involved with cheated on me," 40% of the males and 50% of the females agreed (Hall & Knox, 2015).

Men are more upset if their wife has a heterosexual than a homosexual affair while women are equally upset if their spouse has a homosexual or a heterosexual affair (Confer & Cloud, 2011). Characteristics associated with spouses who are more likely to have extramarital sex include male gender, a strong interest in sex, permissive sexual values, low subjective satisfaction in the existing relationship, employment outside the home, low church attendance, greater sexual opportunities, higher social status (power and money), and alcohol abuse (Hall et al., 2008).

Traditional marriage scripts fidelity. Traditional wedding vows state, "Hold myself only unto you as long as we both should live." Table 13.1 identifies the alternatives for resolving the transition from multiple to one sexual partner till death.

Unless a couple have a polyamorous or swinging relationship, if they are to have sex outside their marriage, it is via cheating.

Reasons for Extramarital Affairs Reasons spouses give for becoming involved with someone other than their mate include (Jeanfreau et al., 2014; MaddoxShaw et al., 2013; Omarzu et al., 2012):

extradyadic involvement refers to sexual involvement of a pair-bonded individual with someone other than the partner; also called extrarelational involvement.

Coolidge effect term used to describe waning of sexual excitement and the effect of novelty and variety on increasing sexual arousal.

1. **Variety, novelty, and excitement.** Most spouses enter marriage having had numerous sexual partners. Extradyadic sexual involvement may be motivated by the desire for continued variety, novelty, and excitement. One of the characteristics of sex in long-term committed relationships is the tendency for it to become routine. Early in a relationship, the partners cannot seem to have sex

Table 13.1
Till Death Do Us Part?

Monogamy	Cheating	Swinging	Polyamory
Spouse is only sex partner	Husband and/or wife cheats	Spouses agree to multiple sex partners	Spouses agree to multiple love and sex partners

often enough. However, with constant availability, partners may achieve a level of satiation, and the attractiveness and excitement of sex with the primary partner seem to wane.

The **Coolidge effect** is a term used to describe this waning of sexual excitement and the effect of novelty and variety on sexual arousal:

One day President and Mrs. Coolidge were visiting a government farm. Soon after their arrival, they were taken off on separate tours. When Mrs. Coolidge passed the chicken pens, she paused to ask the man in charge if the rooster copulated more than once each day. "Dozens of times," was the reply. "Please tell that to the President," Mrs. Coolidge requested. When the President passed the pens and was told about the

rooster, he asked, "Same hen every time?" "Oh no, Mr. President, a different one each time." The President nodded slowly and then said, "Tell that to Mrs. Coolidge." (*Bermant, 1976, pp. 76–77*)

Whether or not individuals are biologically wired for monogamy continues to be debated. Monogamy among mammals is rare (from 3% to 10%), and monogamy tends to be the exception more often than the rule (Morell, 1998). Pornography use, which involves viewing a variety of individuals in sexual contexts, is associated with extramarital sex (Wright, 2013). Even if such biological wiring for plurality of partners does exist, it is equally debated whether such wiring justifies nonmonogamous behavior—that individuals are responsible for their decisions.

2. **Workplace friendships.** Drifting from being friends to lovers is not uncommon in the workplace (Merrill & Knox, 2010). We noted in the last chapter that about 60% of persons in the workforce reported having become involved with someone at work. Coworkers share the same world 8 to 10 hours a day and, over a period of time, may develop good feelings for each other that eventually lead to a sexual relationship. Tabloid reports regularly reflect that romances develop between married actors making a movie together (e.g., Brad Pitt and Angelina Jolie met on a movie set).

These swans are one of the few lower animals who form pair bonds for long periods of time.

3. **Relationship dissatisfaction.** It is commonly believed that people who have affairs are not happy in their marriage. Spouses who feel misunderstood, unloved, and ignored sometimes turn to another who offers understanding, love, and attention. An affair is a context where a person who feels unloved and neglected can feel loved and important.

4. **Sexual dissatisfaction.** Some spouses engage in extramarital sex because their partner is not interested in sex. Others may go outside the relationship because their partners will not engage in the sexual behaviors they want and enjoy. The unwillingness of the spouse to engage in oral sex, anal intercourse, or a variety of sexual positions sometimes results in the other spouse's looking elsewhere for a more cooperative and willing sexual partner.

5. **Revenge.** Some extramarital sexual involvements are acts of revenge against one's spouse for having an affair. When partners find out that their mate has had or is having an affair, they are often hurt and angry. One response to this hurt and anger is to have an affair to get even with the unfaithful partner.

6. **Homosexual relationship.** Some individuals marry as a front for their homosexuality. Cole Porter, known for such songs as "I've Got You Under My Skin," "Night and Day," and "Every Time We Say Goodbye," was a homosexual who feared no one would buy his music if his sexual orientation were known. He married Linda Lee Porter (alleged to be a lesbian), and their marriage lasted until Porter's death 30 years later.

Other gay individuals marry as a way of denying their homosexuality. These individuals are likely to feel unfulfilled in their marriage and may seek involvement in an extramarital homosexual relationship. Other individuals may marry and then discover later in life that they desire a homosexual relationship. Such individuals may feel that (1) they have been homosexual or bisexual all along, (2) their sexual orientation has changed from heterosexual to homosexual or bisexual, (3) they are unsure of their sexual orientation and want to explore a homosexual relationship, or (4) they are predominantly heterosexual but wish to experience a homosexual relationship for variety. The term **down low** refers to African American married men who have sex with men and hide this behavior from their spouse.

7. **Aging.** A frequent motive for intercourse outside marriage is the desire to return to the feeling of youth. Ageism, which is discrimination against the elderly, promotes the idea that being young is good and being old is bad. Sexual attractiveness is equated with youth, and having an affair may confirm to older partners that they are still sexually desirable. Also, people may try to recapture the love, excitement, adventure, and romance associated with youth by having an affair.

Hemera Technologies/Photos.com/Thinkstock

8. **Absence from partner.** One factor that may predispose a spouse to an affair is prolonged separation from the partner. Some wives whose husbands are away for military service report that the loneliness can become unbearable. Some husbands who are away say that remaining faithful is difficult. Partners in commuter relationships may also be vulnerable to extradyadic sexual relationships.

down low term refers to African American married men who have sex with men and hide this from their spouse.

snooping investigating (without the partner's knowledge or permission) a romantic partner's private communication (e.g., text messages, email, and cell phone use) motivated by concern that the partner may be hiding something.

alienation of affection law which gives a spouse the right to sue a third party for taking the affections of a spouse away.

Revealing One's Affair by Confession/Partner Snooping Walters and Burger (2013) identified how individuals revealed their infidelity—in person (38%), over the phone (38%), by a third partner (12%), via email (6%), and through text messaging (6%). A primary motivation for disclosure was respect either for the primary partner or for the history of the primary relationship. While some felt the need to confess because they were guilty or felt the need to be honest, others felt the relationship would benefit from openness/honesty.

In other cases, the cheating was discovered by snooping. **Snooping**, also known as covert intrusive behavior, is defined as investigating (without the partner's knowledge or permission) a romantic partner's private communication (e.g., text messages, email, and cell phone use) motivated by concern that the partner may be hiding something. Derby et al. (2012) analyzed snooping behavior in 268 undergraduates and found that almost two-thirds (66%) reported that they had engaged in snooping behavior, most often when the partner was taking a shower. Primary motives were curiosity and suspicion that the partner was cheating. Being female, being jealous, and having cheated were associated with higher frequencies of snooping behavior.

> **"I should have known** then **it wasn't *nothing*,** as he called it. But I was eight months pregnant.... I **swallowed the *nothing*,** straightaway after the usual tears and denial."
>
> —SUZANNE FINNAMORE, *SPLIT: A MEMOIR OF DIVORCE*

Other Effects of an Affair Reactions to the knowledge that one's unfaithfulness vary. Some relationships end. Negash et al. (2014) found that 36% of 539 young adult females reported emotional or sexual extradyadic involvement (EDI) in the last two months. Such behavior was particularly predictive of ending one's primary relationship if the partner thought the relationship was of high quality and felt particularly disillusioned and betrayed by the EDI.

Seven states (Hawaii, Illinois, North Carolina, Mississippi, New Mexico, South Dakota, and Utah) recognize **alienation of affection** lawsuits which give a spouse the right to sue a third party for taking the affections of a spouse away. Alienation of affection claims evolved from common law, which considered women property of their husbands. The reasoning was if another man was accused of stealing his "property," a husband could sue him for damages. The law applies to both women and men, so a woman who steals another woman's man can be sued for taking her property away. Such was the case of Cynthia Shackelford who sued

REACTIONS TO DISCOVERING ONE'S PARTNER IS CHEATING

"How could you?" asked an angry wife who had discovered her husband in a secluded parked car with a woman from his office. The scene was on *Cheaters*, a television program which features spouses caught in the act of cheating on their mates. Cheating also occurs in undergraduate relationships. The purpose of this research was to identify how undergraduates react to the knowledge that one's romantic partner has cheated—had sexual intercourse with someone else.

The Survey

A 47-item questionnaire was posted on the Internet and completed by 244 undergraduates (83% of the survey respondents were female, 69% were White, and 52% were in their first year). Over 60% were emotionally involved with one person.

Results: Gender Differences in Reactions

Over half (51%) reported they had been cheated on by a partner with whom they were in a romantic relationship (17% of the sample reported that they had cheated on a romantic partner). Significantly more women than men (55% versus 31%) reported that a romantic partner had cheated on them.

In addition to a significant gender difference with men cheating more than women, there were significant gender differences in the reactions to a partner's cheating. Women were more likely to cry, put their partner under surveillance, confront their partner, and get tested for an STI.

Results: Differences in "Unhealthy" and "Healthy" Reactions by Gender

A subset of 19 "unhealthy" and 9 "healthy" reactions to a partner's cheating were identified to ascertain if there were gender differences. For example, increasing one's alcohol consumption, becoming suicidal, and having an affair out of revenge were identified as "unhealthy" reactions, while confronting the partner, forgiving the partner, and seeing a therapist were regarded as more "healthy" reactions. For each item, participants were assigned a score of 1 point if they agreed (somewhat agree, agree, strongly agree) with the reaction (e.g., "I drank more alcohol" or "I forgave my partner"). Scores on the unhealthy and healthy reactions were summed to create an overall count of demonstrated unhealthy and healthy reactions.

Females reported a significantly higher percentage of healthy reactions/behaviors than males; females averaged 3.71 healthy behaviors (SD = 1.46) compared to males, who averaged 2.44 healthy behaviors (SD = 1.58).

Theoretical Framework and Discussion

Symbolic interaction theory and social exchange theory provide frameworks for understanding reactions to the knowledge that one's partner has cheated. Symbolic interaction posits that a couple's relationship is created and maintained on agreed-upon meanings of various behaviors. A major concept inherent in this theoretical framework is definition of the situation. In reference to cheating, partners in a relationship have definitions about the meaning of one partner having sexual intercourse with someone outside the dyad. Among the undergraduates in the current study, most of whom expected fidelity, the definition of the situation was a monogamous relationship with the expectation of fidelity. As such, a feeling of betrayal occurred when fidelity was breeched.

The social exchange framework views the interaction between partners in a romantic relationship in terms of profit and cost. Both partners enter the relationship with promised love and fidelity and expect the same in return. When one partner does not exchange fidelity for fidelity, there is a significant cost to the faithful partner for remaining in a relationship where the partner has been unfaithful. Indeed a common reaction among the sample of undergraduates in this study to the knowledge that their partner had cheated was to terminate the relationship; almost half (47%) ended the relationship with the partner who cheated.

Implications of the Study

There are three implications of this study. One, cheating in romantic relationships among undergraduates is not uncommon. Over half (51%) reported having been cheated on. Two, the knowledge that one's partner has cheated is traumatic. Feeling betrayed (6.24 out of 7) was the most common reaction and was often accompanied by crying (5.81 out of 7), depression (4.65 out of 7), and increased drinking (2.92 out of 7). Three, the range of alternative reactions to the knowledge of infidelity was extensive, including many healthy alternatives including forgiveness, exercise, and seeing a therapist (with women more likely than men to select healthy alternatives).

Source: Abridged and adapted from Barnes, H., D. Knox, & J. Brinkley. (2012, March). CHEATING: Gender differences in reactions to discovery of a partner's cheating. Paper presented at the annual meeting of the Southern Sociological Society, New Orleans, LA.

Anne Lundquist in 2010 for "alienating" her husband from her and breaking up her 33-year marriage. A jury awarded Cynthia Shackelford $4 million in punitive damages and $5 million in compensatory damages. The decision has been appealed. In North Carolina, about 200 alienation of affection lawsuits are filed annually. The infraction must have occurred while the couple was still married (not during the separation period) and there is a three-year statute of limitations.

valdis torms/Shutterstock.com

Successful Recovery from Infidelity

When an affair is discovered, a sense of betrayal pervades the nonoffending spouse or partner (Barnes et al., 2012). While an affair is a high-frequency cause of a couple deciding to divorce, keeping the relationship together (with forgiveness and time) is the most frequent outcome. Abrahamson et al. (2012) interviewed seven individuals who had experienced an affair in their relationship and who were still together two years later. The factors involved in rebuilding their relationship included:

1. **Motivation to stay together.** Having been together several years, having children, having property jointly, not wanting to "fail," and fearing life alone were factors which motivated the partners to stay together. The basic feeling is that we have a lot to gain by working this out.

2. **Taking joint responsibility.** The betrayed partner found a way to acknowledge she or he had contributed to the affair so that there was joint responsibility for the affair.

3. **Forgiveness, counseling, and not referring to the event again.** Forgiveness involved letting go of one's resentment, anger, and hurt; accepting that we all need forgiveness; and moving forward (Hill, 2010). Bagarozzi (2008) noted that the offending spouses must take responsibility for the affair, agree not to repeat the behavior, and grant their partners the right to check up on them to regain trust. The Personal View section focuses on the decision to end a relationship rather than work through it when a partner has an affair.

4. **Vicarious learning.** Noting that others who ended a relationship over an affair were not necessarily happier/better off.

5. **Feeling pride in coming through a difficult experience.** An affair is a major crisis. Coming through it together can actually strengthen the couple's relationship.

13-4e Unemployment

The unemployment rate is defined as the number of people actively looking for a job divided by the labor force. Changes in unemployment depend mostly on inflows made up of nonemployed people starting to look for jobs, of employed people who lose their jobs and look for new ones, and of people who stop looking for employment. In early 2014, the U.S. unemployment rate was 6.6. African Americans, Latinos, young workers, and less-educated workers are the most vulnerable to being chronically unemployed, which will affect their economic, social, and personal health (Schmitt & Jones, 2012).

Unemployment/economic stress is associated with a decline in relationship satisfaction (Williamson et al., 2013), child maltreatment (Euser et al., 2010), and divorce (Eliason, 2012). The personal effects of unemployment are more severe for men than for women (Backhans & Hemmingsson, 2012). Our society expects men to

My husband had an extended affair with my best friend for seven years. I found out by reading a text message she had sent him. I was devastated. We had three children.

We went to a marriage counselor who told me, "You're going to need to resolve this in a way that it does not wreck your life whether or not you stay married…one option is to resolve it AND stay married…to keep your family together." This was probably the best advice I could have received. I was bitterly angry, hurt that he would deceive me, and wanted revenge. But the cost would be breaking up our family and messing up the lives of my children with their father.

While he is the one who had the affair, I was not without blame. When the children came, I put them first and our lives as lovers came to a halt. My husband asked me for time together alone but I told him we should always do things as a family. Since I was not there for him, he sought someone else…could I blame him completely…I think not.

He ended the relationship with the other woman (as did I) and we resolved to devote time to ourselves as well as to our children. The result is that we avoided a divorce, our family remains together, and we showed our children "you can work through anything if you want to." No question but what this was the right decision for us.

—NAME WITHHELD BY REQUEST

be the primary breadwinners in their families and equates masculine self-worth and identity with job and income.

When spouses or parents lose their jobs as a result of physical illness or disability, the family experiences a double blow—loss of income combined with higher medical bills. Unless an unemployed spouse is covered by the partner's medical insurance, unemployment can result in loss of health insurance for the family. Insurance for both health care and disability is very important to help protect a family from an economic disaster.

> "I felt **empty and sad** for years, and for a long, **long time, alcohol worked**. I'd drink, and all the sadness would go away. Not only did the sadness go away, but I was also fantastic. I was beautiful, funny, I had a great figure, and I could do math. But **at some point**, the **booze stopped working**."
>
> —DINA KUCERA, *EVERYTHING I NEVER WANTED TO BE*

13-4f Alcohol/Substance Abuse

A person has a problem with alcohol or a substance if it interferes with his or her health, job, or relationships. Spouses, parents, and children who abuse alcohol and/or drugs contribute to the stress and conflict experienced in their respective marriages and families. Although some individuals abuse drugs to escape from unhappy relationships or the stress of family problems, substance abuse inevitably adds to the individual's marital and family problems when it results in health and medical problems, legal problems, loss of employment, financial ruin, school failure, emotionally distant relationships (Cox et al., 2013; Lotspeich-Younkin & Bartle-Haring, 2012), and divorce. Table 13.2 reflects substance abuse at various age categories.

13-4g Death of Family Member

Even more devastating than drug abuse are family crises involving death—of one's child, parent, or loved one (we discuss the death of one's spouse in Chapter 15 on relationships in the later years). The crisis is particularly acute when the death is a suicide.

Alcohol becomes a serious problem for some individuals.

Table 13.2

Current Drug Use by Type of Drug and Age Group

Type of Drug Used	Age 12 to 17, %	Age 18 to 25, %	Age 26 to 34, %
Marijuana and hashish	7.4	18.5	10.5
Cocaine	.2	1.5	1.1
Alcohol	13.6	61.5	No data
Cigarettes	8.3	34.2	No data

Source: Adapted from *ProQuest Statistical Abstract of the United States, 2014*, 2nd ed., Tables 112 and 215. Bethesda, MD.

Death of One's Child A parent's worst fear is the death of a child. Most people expect the death of their parents but not the death of their children. Amy Winehouse died at the age of 27 in 2011. Her distraught parents, Mitch and Janis, said that they were "left bereft" at her death. Grief feelings may be particularly

chronic sorrow
grief-related feelings that occur periodically throughout the lives of those left behind.

acute on the anniversary of the death of an individual with whom one was particularly close.

"The **room fills** up with my **absent child**."

—SHAKESPEARE

Maple et al. (2013) interviewed 22 parents following the death of a young adult child and found that the parents needed to maintain a relationship with their child including public and private memorials to internal dialogues. Mothers and fathers sometimes respond to the death of their child in different ways. When they do, the respective partners may interpret these differences in negative ways, leading to relationship conflict and unhappiness. To deal with these differences, spouses need to be patient and practice tolerance in allowing each to grieve in her or his own way. Men typically become work focused and women become focused on the grieving sibling (Alam et al., 2012).

Death of One's Parent Terminally ill parents may be taken care of by their children. Such care over a period of years can be emotionally stressful, financially draining, and exhausting. Hence, by the time the parent dies, a crisis has already occurred.

Reactions to the death of a loved one (whether parent or partner) is not something one "gets over." Burke et al. (1999) noted that grief is not a one-time experience that people adjust to and move on. Rather, for some, there is **chronic sorrow**, where grief-related feelings occur periodically throughout the lives of those left behind. Grief feelings may be particularly acute on the anniversary of the death or when the bereaved individual thinks of what might have been had the person lived. Burke et al. (1999) noted that 97% of the individuals in one study who had experienced the death of a loved one 2 to 20 years earlier met the criteria for chronic sorrow.

Boss (2013) also noted that closure may not be a realistic goal but to learn to cope with the death of a loved one by finding meaning and dropping the expectation that one should "get over it." Actor Liam Neeson noted that three years after the death of his wife, actress Natasha Richardson, he still experienced waves of grief. Particularly difficult are those cases of ambiguous loss where the definitions of death become muddled, as when a person disappears or has a disease such as Alzheimer's.

13-5 Marriage (Relationship) Therapy

University students have a positive view of marriage therapy. In a study of 288 undergraduate and graduate students, 93% of the females and 82% of the males agreed that, "I would be willing to see a marriage counselor before I got a divorce" (Dotson-Blake et al., 2010).

Undergraduate couples might consider consulting a marriage therapist (who specializes in relationships… being married is irrelevant) about their relationship rather than remaining in an unhappy relationship or ending a relationship that can be improved. Signs to look for in your own relationship that suggest you might consider seeing a therapist include chronic arguing, feeling distant/avoiding each other, being unable or unwilling to address issues which create tension/dissatisfaction, feeling depressed, drifting into a relationship with someone else, increased drinking, and privately contemplating separation or breaking up. Relationship therapy may help identify behaviors which create unhappiness for the respective partners, identify new behaviors to replace negative ones, make commitments to change, and begin new behaviors so that the partners can start feeling better about each other.

13-5a Availability of Marriage/Relationship Therapists

There are around 50,000 marriage and family therapists in the United States. Whatever marriage therapy costs, it can be worth it in terms of improved relationships. If divorce can be averted, both spouses and children can avoid the trauma and thousands of dollars will be saved. Effective marriage therapy usually involves seeing the spouses together (conjoint therapy).

> "If you **don't have time** to work on your **marriage, when** will you **have time** to work on your **divorce**?"
>
> —LOUISE SAMMONS, MARRIAGE THERAPIST

13-5b Effectiveness of Behavioral Couple Therapy

Marriage therapists use more than 20 different treatment approaches, most of which attempt to be evidence based (Lindblad-Goldberg & Northey, 2013). If you and your partner are seeing a marriage/relationship therapist, the two of you should clearly identify your goal to the therapist: "Our goal is to feel better about each other, improve our communication with each other, and spend more time together doing things of mutual enjoyment," and ask the therapist if and how he or she can help you achieve your goals. If the answer you receive is not satisfactory, find another therapist.

Most therapists (31%) report that they use either a behavioral or cognitive-behavioral approach. A behavioral approach—also referred to as **behavioral couple therapy (BCT)**—means that the therapist focuses on behaviors the respective spouses want increased or decreased, initiated or terminated, and then negotiates behavioral exchanges between the partners.

Some therapists use behavior contracts which are agreements partners make of new behaviors to engage in between sessions. The following is an example and assumes that the partners argue frequently, never compliment each other, no longer touch each other, and do not spend time together. The contract calls for each partner to make no negative statements to the other, give two compliments per day to the other, hug or hold each other at least once a day, and allocate Saturday night to go out to dinner alone with each other. On the contract, under each day of the week, the partners check that they did what they agreed to; the contract is given to the therapist at the next appointment. Partners who change their behavior toward each other often discover that the partner changes also and there is a new basis for each to feel better about each other and their relationship.

Sometimes clients do not like behavior contracts and say to the behavior therapist, "I want my partner to compliment me and hug me because my partner wants to, not because you wrote it down on this silly contract." The behavior therapist acknowledges the desire for the behavior

behavioral couple therapy therapeutic focus on behaviors the respective spouses want increased or decreased, initiated or terminated.

Behavior Contract for Partners

Name of Partners _____ *Date:_Week of June 8–14*

Behaviors each partner agrees to engage in and days of week:

	Mon	*Tues*	*Wed*	*Thurs*	*Fri*	*Sat*	*Sun*
1. No negative statements to partner.	☐	☐	☐	☐	☐	☐	☐
2. Compliment partner twice each day.	☐	☐	☐	☐	☐	☐	☐
3. Hug or hold partner once a day.	☐	☐	☐	☐	☐	☐	☐
4. Out to dinner Saturday night.	☐	☐	☐	☐	☐	☐	☐

Source: From Crisp, B., & D. Knox. (2009). *Behavioral family therapy*. Durham, NC: Carolina Academic Press.

to come from the heart of the partner and points out that the partner is making a choice to engage in new behavior to please the partner.

Cognitive-behavioral therapy, also referred to as **integrative behavioral couple therapy (IBCT)**, emphasizes a focus on cognitions. South et al. (2010) found that spouses were most happy when they perceived the positive and negative behaviors occurred at the desirable frequency—high frequency of positives and low frequency of negatives.

13-5c **Telerelationship (Skype) Therapy**

An alternative to face-to-face therapy is **telerelationship therapy**, which uses the Internet (Skype). Both therapist and couple log on to Skype, where each can see and hear the other while the session is conducted online. Terms related to telerelationship therapy are telepsychology, telepsychiatry, virtual therapy, and video interaction guidance (VIG) (Magaziner, 2010; Doria et al., 2014). Telerelationship therapy allows couples to become involved in relationship therapy independent of where they live (e.g., isolated rural areas), the availability of transportation, and time (i.e., sessions can be scheduled outside the 9 to 5 block). While the efficacy of telerelationship therapy compared with face-to-face therapy continues to be researched, evidence regarding its value in

Increasingly, couples are discovering that relationship therapy is available online.

individual (Nelson & Bui, 2010) and family therapy (Doria et al., 2014) has been documented. Regarding the latter, Doria et al. (2014) provided content analysis of 15 therapeutic sessions (by three therapists) which improved family happiness, parental self-esteem and self-efficacy, and attitude–behavior change. These data emphasize that therapy can be effectively conducted over the Internet.

integrative behavioral couple therapy therapy which focuses on the cognitions or assumptions of the spouses, which impact the way spouses feel and interpret each other's behavior.

telerelationship therapy therapy sessions conducted online, often through Skype, where both therapist and couple can see and hear each other.

13-6 Trends Regarding Stress and Crisis in Relationships

Stress and crisis will continue to be a part of relationships. No spouse, partner, marriage, family, or relationship is immune. A major source of stress will be economic—the difficulty in securing and maintaining employment and sufficient income to take care of the needs of the family.

Most relationship partners will also show resilience to rise above whatever crisis happens. The motivation to do so is strong, and having a partner to share one's difficulties reduces the sting. As noted earlier, it is always one's perception of an event, not the event itself, which will determine the severity of a crisis and the capacity to cope with and overcome it.

STUDY TOOLS 13

Ready to study? In the book, you can:

- ⮑ Rip out the chapter review card at the back of the book for a handy summary of the chapter and key terms.

- ⮑ Assess the "hardiness" of your family with the Self-Assessment card at the back of the book.

Online at CENGAGEBRAIN.COM you can:

- ⮑ Prepare for tests with quizzes.

- ⮑ Review the key terms with Flash Cards.

- ⮑ Play games to master concepts.

Divorce and Remarriage

"Some people think that it's **holding on** that makes one **strong; sometimes it's letting go.**"

—UNKNOWN

Just as weddings are a time of joy replete with thoughts of a wonderful life together, divorces are occasions for thoughts of doom and fear for the future. **Divorce** (there are about 1 million divorces each year in the United States) is the legal ending of a valid marriage contract. The lifetime probability of couples getting divorced today is between 40% and 50% (Cherlin, 2010). But the individual's age at marriage, education, race, and religion influence probability of divorce for that person. For college graduates who wed in the 1980s at age 26 or older, 82% were still married 20 years later. For college grads who married when the partners were less than age 26, 65% were still married 20 years later. If the couple were high school graduates and married when the partners were below age 26, 49% were still married. Hence, education and age at marriage influence one's chance of divorce. In addition, race (Whites are less likely to divorce), religion (less devout are more likely to divorce), and previous marriage (previously married are more likely to divorce) also influence one's chance of divorce (Parker-Pope, 2010).

Divorce rates have been relatively stable in recent years. The principal factor for the lack of increase in the divorce rate is that people are delaying marriage so that they are older at the time of marriage. Indeed,

divorce the legal ending of a valid marriage contract.

Bacho/Shutterstock.com

the older a person at the time of marriage, the less likely the person is to divorce. In this chapter we examine the issues to be considered before getting divorced or ending a relationship, the process of adjustment, and movement into a new relationship or marriage.

14-1 Deciding Whether to Continue or End a Relationship/Get a Divorce

Several factors are predictive of maintaining a relationship or letting it go. In addition, there is a process in terms of what should be considered before deciding to end a relationship/divorce.

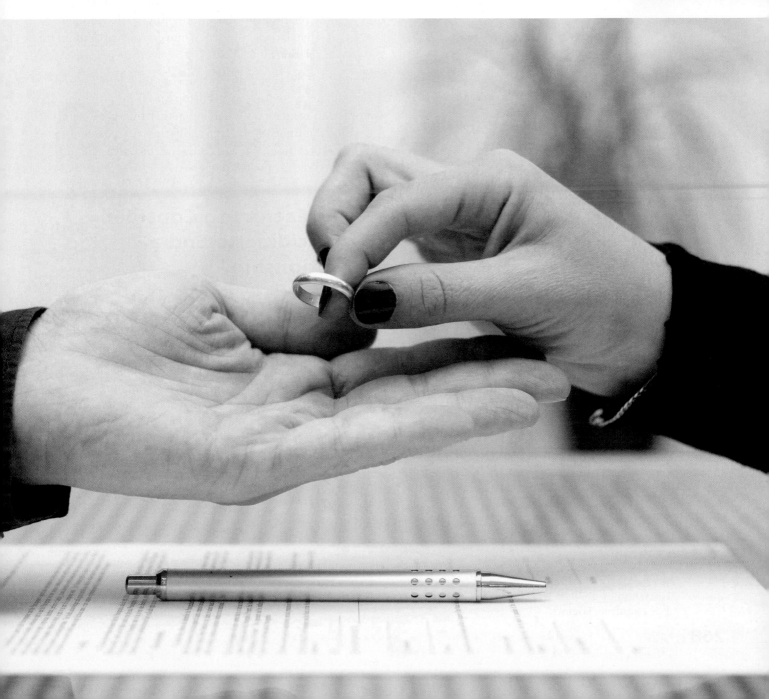

14-1a Deal-Breakers and Factors Predictive of Continuing or Ending a Relationship

Walsh (2013) reported on Match.com's survey of 5,500 singles who identified "deal-breakers." Deal-breakers for women were "dating someone who was secretive with their texts" (77%), dating someone who was lazy (72%), dating someone who was shorter (71%), or dating someone who was a virgin (51%). Deal-breakers for men were dating someone who was disheveled/unclean (63%), someone who was lazy (60%), or someone who did not care about her career (46%).

Rhoades et al. (2010) identified the following four factors involved in whether a person continues or ends a relationship:

1. **Dedication.** Motivation to build/maintain a high-quality relationship, have a long-term relationship.

2. **Perceived constraints.** Factors operating to keep the couple together. For example, social pressure to stay together from parents/friends/church group, intangible investments (e.g., years together) lost if relationship ends, belief that one's quality of life would deteriorate should the relationship end, concern for the welfare of one's partner, fear that taking the steps (e.g., talking to partner) to end the relationship would be difficult, or fear of finding a suitable replacement if the relationship ends.

3. **Material constraints.** Observable changes which would occur if the relationship ends, for example, dealing with shared debt, apartment, furniture, and pets.

4. **Feeling trapped.** Individuals rate the degree to which they feel trapped by the investments in the relationship. The more individuals who felt trapped, the slower they moved toward ending a relationship.

The researchers analyzed data on 1,184 unmarried adults in a romantic relationship and found that each of the four factors predicted whether the relationship would continue or end. Basically the partners choose to end a relationship when the rewards of leaving outweighed the costs of staying.

14-1b Factors to Consider in Deciding to End a Relationship

All relationships have difficulties, and all necessitate careful consideration of various issues before making a decision to end the relationship or to get a divorce.

1. **Consider improving the relationship rather than ending it.** Some people end a relationship and later regret having done so. Setting unrealistically high standards may end a relationship/marriage prematurely. Particularly in our individualistic society with the fun/love/sex focus of relationships, anything that deviates is considered for dumping. Becoming involved in marriage therapy before consulting an attorney may be a wise choice. Of course, we do not recommend giving an abusive relationship more time, as abuse, once started, tends to increase in frequency and intensity.

Sashkin/Shutterstock.com

The new buzzword for divorce is "conscious uncoupling," which was used by Gwyneth Paltrow and Chris Martin of their divorce in 2014 (the term was developed by psychotherapist Katherine Woodward Thomas). Of course, couples may also choose to continue their coupling (and improve their relationship).

"Commitment means we haven't left ourselves an escape hatch."

—JOHN FOWLES

2. **Acknowledge and accept that terminating a relationship will be difficult and painful.** Yazedjian and Toews (2010) studied college breakups and found that they were associated with depression and a drop in grades. Both partners are usually hurt, although the person with the least interest in maintaining the relationship will suffer less.

3. **Select your medium of breaking up.** While some break up face to face, others do so in a text message, in an email, in a voice mail, in a phone conversation, or on Facebook. People have different ideologies and values about the use of various technologies in breaking up (Gershon, 2010). Most prefer that their partner break up with them face to face rather than in a text message (Faircloth et al., 2012). Be sensitive to how your partner will view getting the bad news.

4. **In talking with your partner, blame yourself for the breakup.** One way to end a relationship is to blame yourself by giving a reason that is specific to you (e.g., "I need more freedom," "I want to go to graduate school," or "I'm not ready to settle down"). If you blame your partner, the relationship may continue because your partner may promise to make the changes and you may

dencg/Shutterstock.com

feel obligated to give your partner a second chance. For example, if you say "You drink too much," the partner may say, "I will stop drinking" and you may feel obligated to "give the person a chance."

5. **Cut off the relationship completely.** If you are the person ending the relationship, it will probably be easier for you to continue to see the other person without feeling too hurt. However, the other person will probably have a more difficult experience and will heal faster if you stay away completely. Alternatively, some people are skilled at ending love relationships and turning them into friendships. Though this is difficult and infrequent, it can be rewarding across time for the respective partners.

6. **Learn from the terminated relationship.** Included in the reasons a relationship may end are behaviors of being too controlling; being oversensitive, jealous, or too picky; cheating; fearing commitment; and being unable to compromise and negotiate conflict. Some of the benefits of terminating a relationship are recognizing one's own contribution to the breakup and working on any behaviors that might be a source of problems. Otherwise, one might repeat the process in the next relationship.

7. **Allow time to grieve over the end of the relationship.** Ending a love relationship is painful. Allowing yourself time to experience such grief will help you heal for the next relationship. Recovering from a serious relationship can take 12 to 18 months.

8. **Clean your Facebook page.** Angry spouses sometimes post nasty notes about their ex on their Facebook page that can be viewed by the ex's lawyer. Also, if there are any incriminating photos of indiscreet encounters, drug use, wild parties or the like, these can be used in court and should be purged. Twitter and blog postings should also be scrutinized.

Break It to Me Gently

Researcher Sprecher and her colleagues (2014) surveyed 335 undergraduates and identified compassionate strategies for breaking up. Two of the most commonly used strategies were "I verbally explained to my partner in person my reasons for desiring to break up," and "I told my partner that I didn't regret the time we had spent together in the relationship." Both of these were compassionate strategies which more often resulted in the partners remaining friends. More brutal ways of breaking up included manipulation (e.g., starting an argument/blaming it on the partner) or distant communication (e.g., text message).

14-2 Macro Factors Contributing to Divorce

Sociologists emphasize that social context creates outcome. This concept is best illustrated by the fact that from 1639 to 1760, the Puritans in Massachusetts averaged only one divorce per year (Morgan, 1944). The social context of that era involved strong profamily values and strict divorce laws, with the result that divorce was almost nonexistent for over 100 years. In contrast, divorce occurs more frequently today as a result of various structural and cultural factors, also known as macro factors.

14-2a Increased Economic Independence of Women

In the past, an unemployed wife was dependent on her husband for food and shelter. No matter how unhappy her marriage was, she stayed married because she was economically dependent on her husband. Her husband was her lifeline. But finding gainful employment outside the home made it possible for a wife to afford to leave her husband if she wanted to. Now that three-fourths of wives are employed, fewer women are economically trapped in unhappy marriages. A wife's employment does not increase the risk of divorce in a happy marriage. However, it does provide an avenue of escape for women in unhappy or abusive marriages (Kesselring & Bremmer, 2006).

Employed wives are also more likely to require an egalitarian relationship; although some husbands prefer this role relationship, others are unsettled by it. Another effect of a wife's employment is that she may meet someone new in the workplace and become aware of an alternative to her current partner. Finally, unhappy husbands may be more likely to divorce if their wives are employed and able to be financially independent (requiring less alimony and child support).

14-2b Changing Family Functions and Structure

Many of the protective, religious, educational, and recreational functions of the family have been largely taken over by outside agencies. Family members may now look to the police for protection, the church or synagogue for meaning, the school for education, and commercial recreational facilities for fun rather than to each other within the family for fulfilling these needs. The result is that, although meeting emotional needs remains an important and primary function of the family, fewer reasons exist to keep a family together.

In addition to the change in functions of the family brought on by the Industrial Revolution, family structure has changed from that of the larger extended family in a rural community to a smaller nuclear family in an urban community. In the former, individuals could turn to a lot of people in times of stress; in the latter, more stress necessarily falls on fewer shoulders. Also, with marriages more isolated and scattered, kin may not live close enough to express their disapproval for the breakup of a marriage. With fewer social consequences for divorce, Americans are more willing to escape unhappy unions.

14-2c Liberal Divorce Laws/ Social Acceptance

All states recognize some form of **no-fault divorce** in which neither party is identified as the guilty party or the cause of the divorce (e.g., committing adultery). In effect, divorce is granted after a period of separation (typically 12 months). Nevada requires the shortest waiting period of six weeks. Most other states require from 6 to 12 months. The goal of no-fault divorce is to make

no-fault divorce neither party is identified as the guilty party or the cause of the divorce.

Mihail Jershov/Shutterstock.com

divorce less acrimonious. However, this objective has not been achieved as spouses who divorce may still fight over custody of the children, child support, spouse support, and division of property. Nevertheless, social acceptability as well as legal ease may affect the frequency of divorce. Frimmel et al. (2013) noted that in Austria the increase in divorce is greatly influenced by its acceptance.

14-2d Prenuptial Agreements and the Internet

New York family law attorney Nancy Chemtob notes that those who have prenuptial agreements are more likely to divorce, since one can cash out without economic devastation. In addition, she suggested that the Internet contributes to divorce since a bored spouse can go online to the various dating sites and see what alternatives are out there. Spinning up a new relationship online before dumping the spouse of many years is not uncommon. Brown and Lin (2012) documented that divorce is becoming more common (1 in 4) in adults age 50 and older.

14-2e Fewer Moral and Religious Sanctions

While previously some churches denied membership to the divorced, today many priests and clergy recognize that divorce may be the best alternative in particular marital relationships. Churches increasingly embrace single and divorced or separated individuals, as evidenced by divorce adjustment groups.

14-2f More Divorce Models

The prevalence of divorce today means that most individuals know someone who is divorced. The more divorced people a person knows, the more normal divorce will seem to that person. The less deviant the person perceives divorce to be, the greater the probability the person will divorce if that person's own marriage becomes strained. Divorce has become so common that numerous websites for the divorced are available.

14-2g Mobility and Anonymity

When individuals are highly mobile, they have fewer roots in a community and greater anonymity. Spouses who move away from their respective families and friends often discover that they are surrounded by strangers who do not care if they stay married or not. Divorce thrives when promarriage social expectations are not operative. In addition, the factors of mobility and anonymity also result in the removal of a consistent support system to help spouses deal with the difficulties they may encounter in marriage.

14-2h Social Class, Ethnicity, and Culture

Charles Murray (2012) argues that social class influences who stays married and points out that educated White Americans are the least likely to divorce. Indeed, the less educated with less income are more likely to divorce.

Asian Americans and Mexican Americans also have lower divorce rates than European Americans or African Americans because they consider the family unit to be of greater value (familism) than their individual interests (individualism). Unlike familistic values in Asian cultures, individualistic values in U.S. culture emphasize the goal of personal happiness in marriage. When spouses stop having fun (when individualistic goals are no longer met), they sometimes feel no reason to stay married. Of 4,691 undergraduates, only 10% agreed that "I would not divorce my spouse for any reason" (Hall & Knox, 2015).

14-3 Micro Factors Contributing to Divorce

Macro factors are not sufficient to cause a divorce. One spouse must choose to divorce and initiate proceedings. Such a view is micro in that it focuses on individual decisions and interactions within a specific relationship. A partner whose

Growing apart, falling out of love, sometimes starts a process of filling out papers and entering the legal system.

behavior does not meet one's expectations (infidelity, abuse, value parents over spouse) spells marital dissatisfaction (Dixon et al., 2012). A discussion of micro factors follows.

14-3a Growing Apart/ Differences

The top reason for seeking divorce given by a sample of 886 divorcing individuals was "growing apart" (55%) (Hawkins et al., 2012). The individuals found that they no longer had anything in common.

14-3b Falling Out of Love

Benjamin et al. (2010) noted that the absence of love in a relationship was associated with an increased chance of divorce. Forty-one percent of 4,730 undergraduates reported that they would divorce a spouse they no longer loved (Hall & Knox, 2015). No couple is immune to falling out of love and getting divorced. Lavner and Bradbury (2010) studied 464 newlyweds over a four-year period and found that, even in those cases of reported satisfaction across the four years, some couples abruptly divorced. Whether the divorce was triggered by a personal indiscretion (e.g., infidelity) or crisis (e.g., death of child), the point is that years of satisfaction do not make a couple immune to divorce.

14-3c Limited Time Together: Video Game Addict Widow

Some spouses do not spend any time together. While children and careers typically impact couple time, some partners are also online video game addicts. Northrup and Shumway (2014) interviewed 10 wives who revealed that their game-addicted husbands spent 40 plus hours a week online playing video games—to the exclusion of themselves, their children, and their friends. The impact on their marriage was devastating since their husbands flew into a rage whenever their wives confronted them about their video gaming.

14-3d Low Frequency of Positive Behavior

People marry because they anticipate greater rewards from being married than from being single. During courtship, each partner engages in a high frequency of positive verbal (compliments) and nonverbal (eye contact, physical affection) behavior toward the other. The good feelings the partners experience as a result of these positive behaviors encourage them to marry to "lock in" these feelings across time. Mitchell (2010) interviewed 390 married couples and found that intimacy was associated with marital happiness. Just as love feelings are based on partners making the choice to engage in a high frequency of positive behavior toward each other, defeatist feelings are created when these positive behaviors stop and negative behaviors begin. Thoughts of divorce then begin (to escape the negative behavior).

14-3e Having an Affair

In a survey of U.S. individuals, half of those who reported having participated in extramarital sex also reported that they were either divorced or separated (Allen & Atkins, 2012). In an Oprah.com survey of 6,069 adults, an affair was the top reason respondents said they would seek a divorce (Healy & Salazar, 2010). Thirty-three percent said an affair would be the deal-breaker. Other top responses were chronic fighting (28%), no longer being in love (24%), boredom (8%), and sexual incompatibility (4%).

14-3f Poor Communication/ Conflict Resolution Skills

The second most frequent reason for seeking divorce given by a sample of 886 divorcing individuals was "not able to talk together" (Hawkins et al., 2012). Not only do individuals distance themselves from each other by not talking, they further complicate their relationship since they have no way to reduce conflict.

Managing differences and conflict in a relationship helps to reduce the negative feelings that develop in a relationship. Some partners respond to conflict by withdrawing emotionally from their relationship; others respond by attacking, blaming, and failing to listen to their partner's point of view.

14-3g Changing Values

Both spouses change throughout the marriage. "He's not the same person I married" is a frequent observation of people contemplating divorce. One minister married and decided seven years later that he did not like the confines of his religious or marital role. He left the ministry, earned a PhD, and began to drink and have affairs. His wife now found herself married to a clinical psychologist who spent his evenings at bars with other women. The couple divorced.

Because people change throughout their lives, the person selected at one point in life may not be the same partner one would select at another point. Margaret Mead, the famous anthropologist, noted that her first marriage was a student marriage; her second, a professional partnership; and her third, an intellectual marriage to her soul mate, with whom she had her only child. At each of several stages in her life, she experienced a different set of needs and selected a mate who fulfilled those needs.

14-3h Onset of Satiation

Satiation, also referred to as habituation, refers to the state in which a stimulus loses its value with repeated exposure. Spouses may tire of each other. Their stories are no longer new, their sex is repetitive, and their presence for each other is no longer exciting as it was at the beginning of the relationship. Some people who feel trapped by the boredom of constancy decide to divorce and seek what they believe to be more excitement by returning to singlehood and new partners. One man said, "I traded something good for something new." A developmental task of marriage is for couples to enjoy being together and not demand a constant state of excitement (which is not possible over a 50-year period). The late comedian George Carlin said, "If all of your needs are not being met, drop some of your needs." If spouses did not expect so much of marriage, maybe they would not be disappointed.

> "**Marriage** is like a **high-fiber diet**. You know it is probably **good for you**, but it **doesn't excite** the taste buds."
>
> —ANONYMOUS

14-3i Having the Perception That One Would Be Happier If Divorced

Women file most divorce applications. Their doing so may be encouraged by their view that they will achieve greater power over their own life. They feel that by getting a divorce they will have their own money (in the form of child support and/or alimony) without having a man they do not want in the house. In addition, they will have greater control over their children, since women are more often awarded custody (see Section 14-9 of this chapter which reveals that this is changing).

14-3j Top 20 Factors Associated with Divorce

Researchers have identified the characteristics of those most likely to divorce (Park & Raymo, 2013; Djamba et al., 2012; Amato, 2010; Nunley & Seals, 2010; Chiu & Busby, 2010).

Some of the more significant associations include the following:

1. Less than two years of hanging out together (partners know little about each other)

2. Having little in common (similar interests serve as a bond between people)

3. Marrying at age 17 and younger (associated with low education and income and lack of maturity)

4. Being different in race, education, religion, social class, age, values, and libido (widens the gap between spouses)

5. Not being religiously devout (less bound by traditional values)

6. Having a cohabitation history with different partners (pattern of establishing and breaking relationships)

satiation a stimulus loses its value with repeated exposure; also called habituation.

What's New?

TERMINATED UNDERGRADUATE RELATIONSHIPS: REASONS AND REACTIONS

Most undergraduates have experienced the end of a romantic relationship. The purpose of this study was to investigate the strategies they used or were subjected to in regard to the end of their last romantic relationship and to identify the outcomes for the individuals and their relationship. Basically we wanted to know why relationships ended (e.g., infidelity, boredom), how (e.g., face to face or text message), and the post-breakup relationship outcomes (e.g., enemies, friends).

Sample

A convenience sample from a large southeastern university completed a voluntary, anonymous 25-item online survey on relationship breakups. The sample (N = 478) was predominately female (70%), White (71%), and heterosexual (80%).

Research Questions and Findings

Two major questions guided this research.

Research Question 1

What reasons did undergraduates give for ending their last romantic relationship? See Table 14.1 for the

Table 14.1
Main Reasons for Breaking Up (N = 478)

	Male, %	Female, %	Total, %
Bored in the relationship/not happy	32	22	23
Betrayal of partner	9	17	14
Different interests	11	11	11
I met someone new	12	10	10
Different values	13	7	8
Moved away	6	9	8

reasons given by 478 respondents. There was no significant difference between sex of the respondent and the reason given for ending the relationship.

Research Question 2

Following the ending of the romantic relationship, what were the effects and feelings associated with the breakup (e.g., depression, guilt, relief)? And to what degree did these vary by sex of respondent (see Table 14.2)?

"Feeling initially upset but recognizing the breakup was for the best" was the primary reaction to breaking up. "I was glad it was over" was the second most frequent response. Chi-square analysis showed that for females there was a significant difference between who initiated the breakup and the reactions to those breakups (χ^2 = 55.51, df = 8, p < .001). For example, if the woman ended the relationship she was more likely to report "I'm glad it was over" and "I was initially upset but felt it was for the best." In contrast, males showed

Table 14.2
Reactions to the Breakup

	I Ended the Relationship		My Partner Ended the Relationship		It Was Mutual		Total
	Male	Female	Male	Female	Male	Female	
I was glad it was over	16	44	2	4	9	9	84
I was initially upset but feel it was for the best	27	78	11	43	16	48	223
I was depressed	5	11	3	30	5	7	61
I saw a counselor to help with the breakup	0	0	1	2	0	0	3
I felt suicidal	0	2	0	1	0	1	4
Total	48	135	17	80	30	65	375

no significant difference in effect regardless if they were the initiators of the breakup or not.

Theoretical Framework

Symbolic interactionism provided the theoretical framework for viewing the findings of this study. Symbolic interactionism is a micro-level theory that focuses on the meanings individuals attribute to phenomena. Symbolic interactionists focus on the importance of symbols, subjective versus objective reality, and the definition of social situations. The respondents identified "feeling bored/unhappy" and "betrayal of partner" as reasons to terminate a relationship. The fact that Americans are socialized to think in individualist rather than Asian familistic terms emphasizes the cultural backdrop on which romantic breakup decisions are made. In addition, U.S. youth are taught to get upset and end a romantic relationship in response to a partner's cheating. French lovers are less quick to end a relationship over an indiscretion or an affair.

Similarly, these undergraduates viewed the ending of their romantic relationships as an undesirable event which resulted in the culturally scripted response— being sad, being upset, being depressed. Older individuals might take another view—that the end of an unfulfilling romantic relationship is an opportunity to meet a new partner and create a more fulfilling relationship.

Implications

There are three implications of the data. First, the ending of romantic relationships is filled with angst. Respondents spoke of feeling unhappy, betrayed, and replaced by another lover. They also revealed feelings of anger and jealousy.

Second, the aftermath of a romantic relationship may have negative consequences. Thirty percent of the females reported feeling depressed when their partner ended the relationship. Over 40% reported that they were initially upset.

Third, romantic breakups are not serious enough to induce thoughts of suicide or to seek counseling. Only 4 of the 478 respondents revealed that they felt suicidal; only three sought counseling. Indeed, most undergraduates were resilient and moved on. Almost half (46%) said that while they were initially upset, they believed the ending of the romantic relationship was for the best.

Source: Updated, abridged, and adapted from "Saying goodbye in romantic relationships: Strategies and outcomes," by Brackett, A., J. Fish, & D. Knox. (2013, February 21–23). Poster presented at the Southeastern Council on Family Relations, Birmingham, AL.

7. Having been previously married (less fearful of divorce)

8. Having no children or fewer children (less reason to stay married)

9. Having limited education (associated with lower income, more stress, less happiness)

10. Falling out of love (spouses have less reason to stay married)

11. Being unfaithful (broken trust, emotional reason to leave relationship)

12. Growing up with divorced parents (models for ending rather than repairing relationship; may have inherited traits such as alcoholism that are detrimental to staying married)

13. Having poor communication skills (issues go unresolved and accumulate)

14. Having mental problems (bipolar, depression, anxiety) or physical disability (chronic fatigue syndrome)

15. Having seriously ill child (impacts stress, finances, couple time)

16. Having premarital pregnancy or unwanted child (spouses may feel pressure to get married; stress of parenting unwanted child)

17. Emotional/physical abuse (relationship is aversive)

18. Lacking commitment (for nontraditional spouses, divorce is seen as an option if the marriage does not work out)

19. Unemployment (finances decrease, stress increases)

20. Alcoholism/substance abuse (partner no longer dependable)

The more of these factors that exist in a marriage, the more vulnerable a couple is to divorce. Regardless of the various factors associated with divorce, there is debate about the character of people who divorce. Are they selfish, amoral people who are incapable of making good on a commitment to each other and who wreck the lives of their children? Or are they individuals who care a great deal about relationships and would not settle for a bad marriage? Indeed, they may divorce precisely because they value marriage and want to rescue their children from being reared in an unhappy home.

14-4 Consequences of Divorce for Spouses/Parents

"It is much **better to come from** a **'broken home' than** to **live in** one."

—PAM LEWIS, DIVORCED MOTHER OF TWO AND MENTAL HEALTH PROFESSIONAL

For both women and men, divorce is often an emotional and financial disaster. After death of a spouse, separation and divorce are among the most difficult of life's crisis events (notice that all three top life crisis events are in reference to the end of a love relationship).

14-4a Financial Consequences of Divorce

"**Marriage** is **grand**. **Divorce** is **twenty grand**."

—JAY LENO, FORMER *TONIGHT SHOW* HOST

Both women and men experience a drop in income following divorce, but women may suffer more (Warrener et al., 2013). Because men usually have greater financial resources, they may take all they can with them when they leave. The only money they may continue to give to an ex-wife is court-ordered child support or spousal support (alimony). Republican Senator John McCain pays his former wife $17,000 in alimony annually. While most states do not provide alimony, they provide for an equitable distribution of property.

Remarriage generally restores a woman's economic stability. When remarried mothers and fathers are compared there is a "matrilineal tilt" in terms of money transfers to children (Clark & Kenny, 2010). In

The real cost of divorce is emotional (shattered image), not money.

effect, while single divorced fathers give more money to their children than single divorced mothers, remarried mothers give more money to their children than remarried men. According to their data, 21% of divorced mothers gave financial transfers to their biological children over the past two years as compared to 16% of divorced fathers doing so.

Remarried mothers prefer to give their resources to their biological children with whom they are more likely to have maintained a relationship since the divorce. In contrast, remarried fathers are less likely to make money transfers since they have a less close relationship to their children since the divorce and they have a new wife monitoring their spending behavior. The new wife of a remarried father said to him when his biological son asked him for money, "We don't want to encourage his economic dependence . . . do we?" The son did not get the money.

How money is divided at divorce depends on whether the couple had a prenuptial agreement or a

Divorce often means dads sometime only get to see their children four days a month . . . every other weekend.

Brittany Bolen

postnuptial agreement. Such agreements are most likely to be upheld if an attorney insists on four conditions—full disclosure of assets by both parties, independent representation by separate counsel, absence of coercion or duress, and terms that are fair and equitable.

14-4b Fathers' Separated from Their Children

Mercadante et al. (2014) studied fathers going through divorce in Western Australia and found that they were "at an emotional disadvantage during separation, not only grieving the loss of their former marital relationship, but also their simultaneous loss of contact with their children, their fathering role, and their former family routine." While fathers in the United States are also disadvantaged, they have a choice of how involved they want to be with their children postdivorce. Increasingly, judges today recognize the value of the father for children and are more open to awarding joint custody. But fathers must be aggressive and get a legal contract which gives them equal joint and **physical custody**. Otherwise, they may be cut out of the lives of their children. Trusting the former spouse to be amicable about access to the children is a mistake.

14-4c One Parent May Alienate Their Children from the Other Parent

Parental alienation syndrome (PAS) is an alleged disturbance in which children are obsessively preoccupied with deprecation and/or criticism of a parent, denigration that is unjustified and/or exaggerated (Gardner, 1998). Meier (2009) reviewed the history of parental alienation syndrome and parental alienation (PA) and found the former to be questioned by most researchers (e.g., PAS is not a medical psychosis with specific criteria). However, **parental alienation** can be defined as an alliance between a parent and a child that isolates the other parent (Godbout & Parent, 2012).

The following are examples of behaviors that either parent may engage in to alienate a child from the other parent (Godbout & Parent, 2012; Schacht, 2000; Teich, 2007):

1. Minimizing the importance of contact and the relationship with the other parent, including moving far away with the child so as to make regular contact difficult

2. Exhibiting excessively rigid boundaries; rudeness or refusal to speak to or inability to tolerate the presence of the other parent, even at events important to the child; refusal to allow the other parent near the home for visitation drop-off or pickup

3. Having no concern about missed visits with the

postnuptial agreement an agreement about how money is to be divided should a couple later divorce, which is made after the couple marry.

physical custody the distribution of parenting time between divorced spouses.

parental alienation syndrome (PAS) an alleged disturbance in which children are obsessively preoccupied with deprecation and/or criticism of a parent; denigration that is unjustified and/or exaggerated.

parental alienation estrangement of a child from a parent due to one parent turning the child against the other.

other parent (e.g., taking the child out of town when the other parent is supposed to have legal access)

4. Showing no positive interest in the child's activities or experiences during visits with the other parent and withholding affection if the child expresses positive feelings about the absent parent

5. Granting autonomy to the point of apparent indifference ("It's up to you if you want to see your dad. I don't care.")

6. Overtly expressing dislike of a visitation ("If you loved me you would not want to see your mom.")

7. Refusing to discuss anything about the other parent ("I don't want to hear about . . . ") or showing selective willingness to discuss only negative matters

8. Using innuendo and accusations against the other parent, including statements that are false, and blaming the other parent for the divorce

9. Portraying the child as an actual or potential victim of the other parent's behavior

10. Demanding that the child keep secrets from the other parent

11. Destroying gifts from or memorabilia of the other parent

12. Promoting loyalty conflicts (e.g., offering an opportunity for a desired activity that conflicts with scheduled visitation)

The most telling sign that children have been alienated from a parent is the irrational behavior of the children, who for no properly explained reason say that they want nothing further to do with one of the parents. Indeed, such children have a lack of ambivalence toward the alienation, lack of guilt or remorse about the alienation, and always take the alienating parent's side in the conflict (Baker & Darnall, 2007).

Children who are alienated from one parent are sometimes unable to see through the alienation process and regard their negative feelings as natural. Such children are similar to those who have been brainwashed by cult leaders to view outsiders negatively. Godbout and Parent (2012) interviewed children who had been alienated from a parent and observed "difficulties at school, internal and external behavior problems, and a search for identity after reaching adulthood." Toren et al. (2013) reported on a 16-session parallel group therapy program for 22 children with parental alienation and their parents. The children's level of anxiety and depression decreased significantly following the therapeutic intervention.

14-5 Negative and Positive Consequences of Divorce for Children

Over a million children annually experience the divorce of their parents. Like all life experiences, there are both negative and positive outcomes. Negative outcomes include lower psychological well-being, more likely to drop out of first year of college or make lower grades (Soria & Linder, 2014), have earlier sexual behavior (Carr & Springer, 2010), have higher rates of alcohol/marijuana use (Arkes, 2013), exhibit depressive symptoms or antisocial behavior (Strochschein, 2012), and have lower commitment in romantic relationships (Cui, 2011).

But there are also positive aspects of divorce. Halligan et al. (2014) analyzed data from 336 undergraduates who were asked to identify positive outcomes they experienced from the divorce of their parents. Table 14.3 provides the percentages "agreeing" or "strongly agreeing" with various outcomes.

Table 14.3 reflects that children of divorce may choose to notice the positive aspects of divorce rather than buy into the cultural script that "divorce is terrible and my life will be ruined because of my parent's divorce." Guinart and Grau (2014) reemphasized that while divorce is an unsettling trauma for children, they adapt and long-term negative consequences are not inevitable. A stable, loving, involved parent is the key to positive child development whether the parents are together or divorced. South (2013) also found that some children of divorced parents are very deliberate in trying not to repeat the mistakes of their parents. Some of their respondents reported that "they were hard on their romantic partners," meaning that they were unwilling to let issues slide but addressed them quickly even though doing so might cause conflicts.

Although a major factor that determines the effect of divorce on children is the degree to which the

Table 14.3
Twenty Positive Effects of Divorce: Percentage of Respondents Agreement

	Percent
Since my parents' divorce, I am more compassionate for people who are going through a difficult time.	65.63
I have greater tolerance for people with different viewpoints since my parents' divorce.	63.16
Since my parents' divorce, I have been exposed to different family values, tradition, and lifestyles.	60.01
I have liked spending time alone with my mother since my parents' divorce.	57.71
My mother is happier since the divorce.	57.20
I rely less on my parents for making decisions since my parents' divorce.	53.51
I have liked spending time alone with my father since my parents' divorce.	45.61
My mother has made a greater effort to spend quality time with me since the divorce.	45.61
I can spend more time with the parent I prefer since my parents' divorce.	45.37
Since my parents' divorce, I have felt closer to my mother.	44.98
My relationship with my mother has improved since my parents' divorce.	44.74
My father is happier since the divorce.	43.85
My parents' divorce has made me closer to my friends.	42.54
Since my parents' divorce, I have greater appreciation for my siblings.	40.78
My father has made a greater effort to spend quality time with me since the divorce.	38.60
Since my parents' divorce, I feel closer to my siblings.	35.96
After my biological parents' divorce, I noticed that I was exposed to less conflict between my parents on a daily basis.	34.93
My relationship with my father has improved since my parents' divorce.	34.65
I think my parents have a "good" divorce.	34.35
My parents are more civil to each other since the divorce.	34.21

divorcing parents are civil, legal and physical custody are important issues. The following section discusses how judges go about making this decision.

14-5a How Parents Can Minimize Negative Effects of Divorce for Children

Researchers have identified the following conditions under which a divorce has the fewest negative consequences for children:

1. **Healthy parental psychological functioning.** Children of divorced parents benefit to the degree that their parents remain psychologically fit and positive, and socialize their children to view the divorce as a "challenge to learn from." Parents who nurture self-pity, abuse alcohol or drugs, and socialize their children to view the divorce as a tragedy from which they will never recover create negative outcomes for their children. Some divorcing parents can benefit from therapy as a method for coping with their anger or depression and to ensure that they make choices in the best interests of their children (e.g., encouraging the child's relationship with the other parent). Some parents become involved in divorce education programs. Crawford (2012) observed a reduction in conflict between parents who completed a divorce education program.

2. **A cooperative relationship between the parents.** The most important variable in a child's positive adjustment to divorce is when the child's parents continue to maintain a cooperative relationship throughout the separation, divorce, and postdivorce period. It is not divorce, but the postdivorce relationship with the parents which has a negative effect on children. Bitter parental conflict places the children in the middle. One daughter of divorced parents said, "My father told me, 'If you love me, you would come visit me,' but my mom told me, 'If you love me, you won't visit him.'" Numerous states mandate parenting classes as part of the divorce process.

3. **Parental attention to the children and allowing them to grieve.** Children benefit when both the custodial and the noncustodial parent continue to spend time with them and to communicate to them that they love them and

Personal View: What a Mother Told Her Sons about the Impending Divorce

The following are the words of a mother of two children (8 and 12) as she tells her children of the pending divorce. It assumes that both parents take some responsibility for the divorce and are willing to provide a united front to the children. The script should be adapted for one's own unique situation.

Daddy and I want to talk to you about a big decision that we have made. A while back we told you that we were having a really hard time getting along, and that we were having meetings with someone called a therapist who has been helping us talk about our feelings, and deciding what to do about them.

We also told you that the trouble we are having is not about either of you. Our trouble getting along is about our grown-up relationship with each other. That is still true. We both love you very much, and love being your parents. We want to be the best parents we can be.

Daddy and I have realized that we don't get along so much, and disagree about so many things all the time, that we want to live separately, and not be married to each other anymore. This is called getting divorced. Daddy and I care about each other but we don't love each other in the way that happily married people do. We are sad about that. We want to be happy, and

want each other to be happy. So to be happy we have to be true to our feelings.

It is not your fault that we are going to get divorced. And it's not our fault. We tried for a very long time to get along living together but it just got too hard for both of us.

We are a family and will always be your family. Many things in your life will stay the same. Mommy will stay living at our house here, and Daddy will move to an apartment close by. You both will continue to live with mommy and daddy but in two different places. You will keep your same rooms here, and will have a room at Daddy's apartment. You will be with one of us every day, and sometimes we will all be together, like to celebrate somebody's birthday, special events at school, or scouts. You will still go to your same school, have the same friends, go to soccer, baseball, and so on. You will still be part of the same family and will see your aunts, uncles, and cousins.

The most important things we want you both to know are that we love you, and we will always be your mom and dad . . . nothing will change that. It's hard to understand sometimes why some people stop getting along and decide not to be friends anymore, or if they are married decide to get divorced. You will probably have lots of different feelings about this. While you can't do anything to change the decision that daddy and I have made, we both care very much about your feelings. Your feelings may change a lot. Sometimes you might feel happy and relieved that you don't have to see and feel daddy and me not getting along. Then sometimes you might feel sad, scared, or angry. Whatever you are feeling at any time is OK. Daddy and I hope you will tell us about your feelings, and it's OK to ask us about ours. This is going to take some time to get used to. You will have lots of questions in the days to come. You may have some right now. Please ask any question at any time.

Daddy and I are here for you. Today, tomorrow, and always. We love you with our heart and soul.

are interested in them. Parents also need to be aware that their children do not want the divorce and to allow them to say so. A parental response might be, "I know you wanted us to stay together and we tried, but decided that we just couldn't live together any more. You will find that life is never easy."

4. **Encouragement to see noncustodial parent.** Children benefit when custodial parents (usually mothers) encourage their children to maintain regular and stable visitation schedules with the noncustodial parent.

5. **Attention from the noncustodial parent.** Children benefit when they receive frequent and consistent attention from noncustodial parents, usually the fathers. Noncustodial parents who do not show up at regular intervals exacerbate their children's emotional insecurity by teaching them, once again, that parents cannot be depended on. Parents who show up consistently teach their children to feel loved and secure.

6. **Assertion of parental authority.** Children benefit when both parents continue to assert their parental authority and continue to support the discipline practices of each other.

7. **Regular and consistent child support payments.** Support payments (usually from the father to the mother) are associated with economic stability for the child.

8. **Stability.** Not moving and keeping children in the same school system is beneficial to children. Some parents—called **latchkey parents**—spend every other week with the children in the family home so the children do not have to alternate between residences. Moving to a new location causes children to be cut off from their friends, neighbors, and teachers. It is important to keep their life as stable as possible during a divorce.

9. **Children in a new marriage.** Manning and Smock (2000) found that divorced noncustodial fathers who remarried and who had children in the new marriage were more likely to shift their emotional and economic resources to the new family unit than were fathers who did not have a new marriage and new biological children. Fathers might be alert to this potential and consider each child, regardless of when or with whom the child was born, as worthy of continued love, time, and support.

10. **Age and reflection on the part of children of divorce.** Sometimes children whose parents are divorced benefit from growing older and reflecting on their parents' divorce as adults rather than as children. Nielsen (2004) emphasized that daughters who feel distant from their fathers can benefit from examining the divorce from the viewpoint of the father (was he alienated by the mother?), the cultural bias against fathers (they are maligned as "deadbeat dads" who "abandon their families for a younger woman"), and the facts about divorced dads (they are more likely than mothers to be depressed and suicidal following divorce). In general, age and temperament impact the adjustment of a child to the divorce of his or her parents. Young children have fewer cultural negative labels than teenagers, and highly sensitive children react more negatively than those who seem unaffected by whatever.

14-6 Prerequisites for Having a "Successful" Divorce

Although the concept of a "successful" divorce has been questioned (Amato et al., 2011), some divorces may cause limited damage to the partners and their children. Indeed, most people are resilient and "are able to adapt constructively to their new life situation within two to three

latchkey parents divorcing parents who spend every other week with the children in the family home so the children do not have to alternate between residences.

Divorce Mediation Not Litigation

Divorce mediation is a process in which spouses who have decided to separate or divorce meet with a neutral third party (mediator) to negotiate four issues: (1) how they will parent their children, which is referred to as child custody and visitation; (2) how they are going to financially support their children, referred to as child support; (3) how they are going to divide their property, known as property settlement; and (4) how each one is going to meet their financial obligations, referred to as spousal support.

Pixland/Thinkstock

Benefits of Mediation

There are enormous benefits from avoiding litigation and mediating one's divorce:

1. **Better relationship.** Spouses who choose to mediate their divorce have a better chance for a more civil relationship because they cooperate in specifying the conditions of their separation or divorce. Mediation emphasizes negotiation and cooperation between the divorcing partners. Such cooperation is particularly important if the couple has children, in that it provides a positive basis for discussing issues in reference to the children and how they will be parented across time.

2. **Economic benefits.** Mediation is less expensive than litigation. The combined cost (total cost to both spouses) of hiring attorneys and going to court over issues of child custody and division of property is around $35,000. A mediated divorce typically costs less than $5,000.

3. **Less time-consuming process.** Although a litigated divorce can take two to three years, a mediated divorce takes two to three months; for highly motivated individuals, a "mediated settlement conference" "... can take place in one session from 8:00 A.M. until both parties are satisfied with the terms" (Amato).

4. **Avoidance of public exposure.** Some spouses do not want to discuss their private lives and finances in open court. Mediation occurs in a private and confidential setting.

5. **Greater overall satisfaction.** Mediation results in an agreement developed by the spouses, not one imposed by a judge or the court system. A comparison of couples who chose mediation with couples who chose litigation found that those who mediated their own settlement were more satisfied with the conditions of their agreement. In addition, children of mediated divorces were exposed to less marital conflict, which may facilitate their long-term adjustment to divorce.

Basic Mediation Guidelines

Divorce mediators conduct mediation sessions with certain principles in mind:

1. **Children.** What is best for a couple's children should be the major concern of the parents because they know their children far better than a judge or a mediator. Children of divorced parents adjust best under three conditions: (1) that both parents have regular and frequent access to the children; (2) that the children see the parents relating in a polite and positive way; and (3) that

divorce mediation meeting with a neutral professional who negotiates child custody, division of property, child support, and alimony directly with the divorcing spouses.

each parent talks positively about the other parent and neither parent talks negatively about the other to the children.

Sometimes children are included in the mediation. They may be interviewed without the parents present to provide information to the mediator about their perceptions and preferences. Such involvement of the children has superior outcomes for both the parents and the children (McIntosh et al., 2008).

2. **Fairness.** It is important that the agreement between the soon to be ex-spouses be fair, with neither party being exploited or punished. It is fair for both parents to contribute financially to the children and to have regular access to their children.

3. **Open disclosure.** The spouses will be asked to disclose all facts, records, and documents to ensure an informed and fair agreement regarding property, assets, and debts.

4. **Other professionals.** During mediation, spouses may be asked to consult an accountant regarding tax laws. In addition, each spouse is encouraged to consult an attorney throughout the mediation and to have the attorney review the written agreements that result from the mediation. However, during the mediation sessions, all forms of legal action by the spouses against each other should be stopped.

5. **Confidentiality.** The mediator will not divulge anything spouses say during the mediation sessions without their permission. The spouses are asked to sign a document stating that, should they not complete mediation, they agree not to empower any attorney to subpoena the mediator or any records resulting from the mediation for use in any legal action. Such an agreement is necessary for spouses to feel free to talk about all aspects of their relationship without fear of legal action against them for such disclosures.

Divorce mediation is not for every couple. It does not work where there is a history of spouse abuse, where the parties do not disclose their financial information, where one party is controlled by someone else (e.g., a parent of one of the divorcing spouses) or where there is the desire for revenge. Mediation should be differentiated from **negotiation** (where spouses discuss and resolve the issues themselves), **arbitration** (where a third party, an arbitrator, listens to both spouses and makes a decision about custody, division of property, and so on), and **litigation** (where a judge hears arguments from lawyers representing the respective spouses and decides issues of custody, child support, division of property, and spousal support).

The following chart identifies a continuum of consequences from negotiation to litigation.

negotiation identifying both sides of an issue and finding a resolution that is acceptable to both parties.

arbitration third party listens to both spouses and makes a decision about custody, division of property, child support, and alimony.

litigation a judge hears arguments from lawyers representing the respective spouses and decides issues of custody, child support, division of property, etc.

Negotiation	Mediation	Arbitration	Litigation
Cooperative			Competitive
Low cost			High cost
Private			Public
Protects relationships			Damages relationships
Focus on the future			Focus on the past
Parties in control			Parties lose control

years following divorce, a minority being defeated by the marital breakup, and a substantial group of women being enhanced" (Hetherington, 2003, p. 318). The following are some of the behaviors spouses can engage in to achieve a successful divorce:

1. **Mediate rather than litigate the divorce.** Divorce mediators encourage a civil, cooperative, compromising relationship while moving the couple toward an agreement on the division of property, custody, and child support. In contrast, attorneys make their money by encouraging hostility so that spouses will prolong the conflict, thus running up higher legal bills. In addition, the couple cannot divide money spent on divorce attorneys (average is $17,500 for *each* side so a litigated divorce cost will start at $35,000). Benton (2008) noted that the worst thing divorcing spouses can do is for each to hire the "meanest, nastiest, most expensive yard dog lawyer in town" because doing so will only result in a protracted expensive divorce in which neither spouse will win. Because the greatest damage to children from a divorce is caused by a continuing hostile and bitter relationship between their parents, some states require divorce mediation as a mechanism to encourage civility in working out differences and to clear the court calendar of protracted legal battles. Research confirms there are enormous benefits of mediation versus litigation (Amato, 2010).

2. **Coparent with your ex-spouse.** Setting aside negative feelings about your ex-spouse so as to cooperatively coparent not only facilitates parental adjustment but also takes children out of the line of fire. Such coparenting translates into being cooperative when one parent needs to change a child care schedule, sitting together during a performance by the children, and showing appreciation for the other parent's skill in responding to a crisis with the children.

3. **Take some responsibility for the divorce.** Because marriage is an interaction between spouses, one person is seldom totally to blame for a divorce. Rather,

content valence
positive or negative emotions associated with the content of a thought.

both spouses share reasons for the demise of the relationship. Take some responsibility for what went wrong. What did *you* do wrong that you could correct in a subsequent relationship?

> ## "Your relationship may be 'Breaking Up,' but you won't be 'Breaking Down.' If anything you're **correcting a mistake** that **was hurting four people.**"
>
> —D. IVAN YOUNG, *BREAK UP, DON'T BREAK DOWN*

4. **Be deliberate in negative and positive thoughts about the ex.** Divorced people are susceptible to feeling as though they are failures—they see themselves as Divorced with a capital D, a situation sometimes referred to as "hardening of the categories" disease. They also have difficulty moving on. Brenner and Vogel (2014) studied the **content valence** (positive or negative emotions associated with the content of a thought) of ex-relationship thoughts, and emphasized that moving on was related to thinking about negative aspects of the relationship (e.g., "lied to me," "cheated on me," "emotionally unsupportive") and minimizing positive thoughts (e.g., "wonderful lover," "loved me," "special times at the beach") about the ex.

> ## "Nothing **erases unpleasant thoughts** more effectively than **concentration on pleasant ones.**"
>
> —HANS SELYE, STRESS RESEARCHER

One technique (called the stop-think technique) for being deliberate about one's thoughts is to write down positive statements about why breaking up is a good thing ("I am free of living with a lying cheater and have an opportunity to be involved with someone who will be honest with and respect me/our relationship" and focusing on these rather than the positive aspects of the former partner/relationship or the negative aspects of breaking up ("I will be lonely).

5. **Avoid alcohol and other drugs.** The stress and despair that some people feel during and following the divorce process sometimes make them vulnerable to the use of alcohol or other drugs. These should be avoided because they produce an endless negative cycle. For example, stress is relieved by alcohol; alcohol produces a hangover and negative feelings; the negative feelings are relieved by more alcohol, producing more negative feelings, etc.

6. **Engage in aerobic exercise.** Exercise helps one to not only counteract stress but also avoid it. Jogging, swimming, riding an exercise bike, or engaging in other similar exercise for 30 minutes every day increases oxygen to the brain and helps facilitate clear thinking. In addition, aerobic exercise produces endorphins in the brain, which create a sense of euphoria (runner's high).

7. **Continue interpersonal connections.** Adjustment to divorce is facilitated by continuing relationships with friends and family. These individuals provide emotional support and help buffer the feeling of isolation and aloneness. First Wives World (www.firstwivesworld.com) is a new interactive website that provides an Internet social network for women transitioning through divorce.

8. **Let go of the anger for your ex-partner.** Former spouses who stay negatively attached to an ex by harboring resentment and trying to get back at the ex prolong their adjustment to divorce. The old adage that "you can't get ahead by getting even" is relevant to divorce adjustment.

9. **Allow time to heal.** Because self-esteem usually drops after divorce, a person is often vulnerable to making commitments before working through feelings about the divorce. Most individuals need between 12 and 18 months to adjust to the end of a marriage. Although being available to others may help to repair one's self-esteem, getting remarried during this time should be considered cautiously. Two years between marriages is recommended.

14-7 Remarriage

"Sometimes the **wrong choices** bring us to the **right places.**"

—UNKNOWN

Divorced spouses are not sour on marriage. Although they may want to escape from the current spouse, they are open to having a new spouse. In the past, two-thirds of divorced females and three-fourths of divorced males remarried (Sweeney, 2010), with the new unions occurring between two and four years of the previous divorce (Brown & Porter, 2013). When comparing divorced individuals who remarried and divorced individuals who have not remarried, the remarried individuals reported greater personal and relationship happiness.

The majority of divorced people remarry for many of the same reasons as those for a first marriage—love, companionship, emotional security, and a regular sex partner. Other reasons are unique to remarriage and include financial security (particularly for a wife with children), help in rearing one's children, the desire to provide a "social" father or mother for one's children, escape from the stigma associated with the label "divorced person," and legal threats regarding the custody of one's children. With regard to the latter, the courts view a parent seeking custody of a child more favorably if the parent is married. The religiously devout remarry, in part because religion is profamily and they seek the context which reflects those values (Brown & Porter, 2013).

If a single mother's goal is to be remarried, how does the involvement of the nonresident father with her children impact the likelihood of remarriage? The answer from 882 divorced mothers is that such involvement increases her chances of remarriage. A team of researchers (McNamee et al., 2014) noted that such nonresident father contact creates leisure time for the mother and opportunities for her to develop new relationships "by providing private time (without children) on a regular basis. Moreover, men may be more willing to adopt the stepfather role when biological fathers remain physically present in their children's lives because this signals fewer parenting responsibilities for the stepfather."

14-7a Issues for Those Who Remarry

Several issues challenge people who remarry (Ganong & Coleman, 1999; Goetting, 1982; Kim, 2011; Martin-Uzzi & Duval-Tsioles, 2013; Scarf, 2013):

1. **Boundary maintenance.** Ghosts of the first marriage, in terms of the ex-spouse, must be dealt with. A parent must decide how to relate to an ex-spouse to maintain a good parenting relationship for the biological children while keeping an emotional distance from the ex to prevent problems from developing with the new partner.

   Some spouses continue to be emotionally attached to and have difficulty breaking away from an ex-spouse. These former spouses have what Masheter (1999) terms a **negative commitment** whereby such individuals "have decided to remain [emotionally] in this relationship and to invest considerable amounts of time, money, and effort in it . . . [T]hese individuals do not take responsibility for their own feelings and actions, and often remain 'stuck,' unable to move forward in their lives" (p. 297).

   In some cases an uncooperative ex-spouse can be a source of bonding for the remarried spouses—but they may also be frustrated in dealing with the ex (Martin-Uzzi & Duval-Tsioles, 2013). For example, the newly remarried couple may have planned an adult couple vacation and the ex might call the night before saying she or he cannot take the children for their regular weekend.

2. **Emotional remarriage.** Remarriage involves beginning to trust and love another person in a new relationship. Such feelings may come slowly as a result of negative experiences in a previous marriage.

The spouses of this remarried couple both brought children into their marriage.

Colin Oates

3. **Psychic remarriage.** Divorced individuals considering remarriage may find it difficult to give up the freedom and autonomy of being single and to develop a mental set conducive to pairing. This transition may be particularly difficult for people who sought a divorce as a means to personal growth and autonomy. These individuals may fear that getting remarried will put unwanted constraints on them.

4. **Community remarriage.** This aspect involves a change in focus from single friends to a new mate and other couples with whom the new pair will interact. The bonds of friendship established during the divorce period may be particularly valuable because they have given support at a time of personal crisis. Care should be taken not to drop these friendships.

5. **Parental remarriage.** Because most remarriages involve children, people must work out the nuances of living with someone else's children. Mothers are usually awarded primary physical custody, and this circumstance translates into a new stepfather adjusting to the mother's children and vice versa. The late film critic Roger Ebert married a woman with two children and said of his new stepchildren.

   > I took joy in the role and loved the children. They represented children I believed I might never have. I never saw them as competing with me for their mother's attention, but as sharing their family with me. (Ebert, 2011, p. 568)

negative commitment spouses who continue to be emotionally attached to and have difficulty breaking away from ex-spouses.

6. **Economic and legal remarriage.** A second marriage may begin with economic responsibilities to a first marriage. Alimony and child support often threaten the harmony and sometimes even the economic survival of second marriages. Although the income of a new wife is not used legally to decide the amount her new husband is required to pay in child support for his children of a former marriage, his ex-wife may petition the court for more child support. The ex-wife may do so, however, on the premise that his living expenses are reduced with a new wife and that, therefore, he should be able to afford to pay more child support. Although an ex-wife is not likely to win, she can force the new wife to go to court and to disclose her income (all with considerable investment of time and legal fees for a newly remarried couple).

csm_weh/iStockphoto.com

Economic issues in a remarriage may become evident in another way. A remarried woman who receives inadequate child support from an ex-spouse and needs money for her child's braces, for instance, might wrestle with how much money to ask her new husband for.

There may also be a need for a marriage contract to be drawn up before the wedding. Suppose a wife moves into the home of her new husband. If he has a will stating that his house goes to his children from a former marriage at his death and no marriage contract that either gives his wife the house or allows her to stay in the house rent free until her death, his children can legally throw her out of the house. The same is true for their beach house which he brought into the marriage. If his will gives the beach house to his children, his wife may have no place to live.

"Marriage is the triumph of imagination over intelligence; second marriage is the triumph of hope over experience."

—SAMUEL JOHNSON, ENGLISH WRITER

14-7b Stability of Remarriages

Fox and Shriner (2014) studied remarried couples involved in premarital education and found that they feared another marital failure, which in conjunction with stepfamily formation, promoted attachment insecurities. In a comparison of first and remarried individuals (Whitton, 2013), the latter had more positive attitudes toward divorce and a weaker commitment to marriage. Indeed, when marital quality was low in both first and second marriages, the latter were much more likely to feel divorce was an option. This finding was true independent of whether or not the remarried adults brought children into the new marriage.

Sweeney (2010) confirmed that remarriages are more likely than first marriages to end in divorce in the early years of remarriage. Reck and Higginbotham (2012) found that remarried men and women reported significant difficulties within the first three years of remarriage. For example, men reported their role as husband, father, and stepfather was particularly challenging due to the difficulties in gaining new familial expectations and parenting their own biological and new stepchildren (e.g., disciplining, establishing trust). Women reported similar difficulties in being a wife, mother, and stepmother; however, women generally reported higher levels of difficulty in these roles.

Mirecki et al. (2013) confirmed lower marital satisfaction of spouses in second marriages. Since 65% of second marriages include the presence of stepchildren, integrating the various individuals into a functioning family is challenging. Higher education on the part of the spouses seemed to help (higher education is associated with higher income). That second marriages, in general, are more susceptible to divorce than first

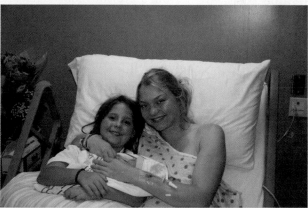
Courtesy of Rodney Sanders

This woman is with her stepdaughter and they are holding the new baby with her now husband (previously married).

marriages may also be related to the fact that divorced individuals are less fearful of divorce (e.g., they know they can survive divorce) than individuals who have never divorced.

Though remarried people are more vulnerable to divorce in the early years of their subsequent marriage, they are less likely to divorce after 15 years of staying in the second marriage than those in first marriages (Clarke & Wilson, 1994). Hence, these spouses are likely to remain married because they want to, not because they fear divorce.

14-8 Stepfamilies

More than half of all children will spend some time in a family arrangement other than the traditional family (Crosnoe & Cavanagh, 2010). At any given time 5.3 million children under age 18 are living with a biological parent and a married or cohabiting stepparent (Sweeney, 2010). Stepfamilies, also known as blended, binuclear, remarried, or reconstituted families, represent the fastest-growing type of family in the United States. A **blended family** is one in which spouses in a new marriage relationship blend their respective children from at least one other spouse from a previous marriage. The term **binuclear family** refers to a family that spans two households; when a married couple with children divorce, their family unit typically spreads into two households. There is a movement away from the use of the term *blended* because stepfamilies really do not blend. The term **stepfamily** (sometimes referred to as step relationships) is the term currently in vogue. Although there are various types of stepfamilies, the most common is a family in which the partners bring children from

One advantage to children of their parents getting divorced and remarried is that they may get a new sibling. This younger daughter got a big sister.

Colin Oates

previous relationships into the new home, where they may also have a child of their own. The couple may be married or living together, heterosexual or homosexual, and of any race. Nuru and Wang (2014) emphasized the need to include step relationships for those in cohabiting relationships.

Various myths abound regarding stepfamilies, including that new family members will instantly bond emotionally, that children in stepfamilies are damaged and do not recover, that stepmothers are "wicked home-wreckers," that stepfathers are uninvolved with their stepchildren, and that stepfamilies are not "real" families. Stepfamilies are also stigmatized. **Stepism** is the assumption that stepfamilies are inferior to biological families. Like racism, heterosexism, sexism, and ageism, stepism involves prejudice and discrimination.

Stepfamilies differ from nuclear families in a number of ways. These are identified in Table 14.4. These changes impact the parents, their children, and their stepchildren and require adjustment on the part of each member.

14-8a Developmental Tasks for Stepfamilies

A **developmental task** is a skill that, if mastered, allows a family to grow as a cohesive unit. Developmental tasks that are not mastered will move the family closer to the point of disintegration. Some of the more

---

blended family family wherein spouses in a remarriage bring their children to live with the new partner and at least one other child.

binuclear family family that lives in two households as when parents live in separate households following a divorce.

stepfamily family in which spouses in a new marriage bring children from previous relationships into the new home.

stepism the assumption that stepfamilies are inferior to biological families.

developmental task a skill that, if mastered, allows a family to grow as a cohesive unit.

Table 14.4
Differences between Nuclear Families and Stepfamilies

Nuclear Families	Stepfamilies
1. Children are (usually) biologically related to both parents.	1. Children are biologically related to only one parent.
2. Both biological parents live together with children.	2. As a result of divorce or death, one biological parent does not live with the children. In the case of joint physical custody, children may live with both parents, alternating between them.
3. Beliefs and values of members tend to be similar.	3. Beliefs and values of members are more likely to be different because of different backgrounds.
4. The relationship between adults has existed longer than relationship between children and parents.	4. The relationship between children and parents has existed longer than the relationship between adults.
5. Children have one home they regard as theirs.	5. Children may have two homes they regard as theirs.
6. The family's economic resources come from within the family unit.	6. Some economic resources may come from an ex-spouse.
7. All money generated stays in the family.	7. Some money generated may leave the family in the form of alimony or child support.
8. Relationships are relatively stable.	8. Relationships are in flux: new adults adjusting to each other; children adjusting to a stepparent; a stepparent adjusting to stepchildren; stepchildren adjusting to each other.
9. No stigma is attached to nuclear family.	9. Stepfamilies are stigmatized.
10. Spouses had a childfree period.	10. Spouses had no childfree period.
11. Inheritance rights are automatic.	11. Stepchildren do not automatically inherit from stepparents.
12. Rights to custody of children are assumed if divorce occurs.	12. Rights to custody of stepchildren are usually not considered.
13. Extended family networks are smooth and comfortable.	13. Extended family networks become complex and strained.
14. Nuclear family may not have experienced loss.	14. Stepfamily has experienced loss.
15. Families experience a range of problems.	15. Stepchildren tend to be a major problem.

important developmental tasks for stepfamilies are discussed in this section.

1. Nurture the new marriage relationship.

2. Be patient for stepparent/stepchild relationships to develop.

3. Have realistic expectations.

4. Accept your stepchildren.

5. Give parental authority to your spouse/coparent.

6. Establish your own family rituals.

7. Support the children's relationship with their absent parent.

8. Cooperate with the children's biological parent and coparent.

9. Use living apart together (LAT) to reduce the strain and conflict of stepfamily living.

LAT is a structural solution to many of the problems of stepfamily living (see Chapter 2 for a more thorough discussion). By getting two condos (side by side or one on top of the other), a duplex, or two small houses and having the respective biological parents live in each of the respective units with their respective children, everyone wins. The children and biological parent will experience minimal disruption as the spouses transition to the new marriage. This arrangement is particularly useful where both spouses have children ranging in age from 10 to 18. Situations where only one spouse has children from a former relationship or those in which the children are very young will have limited benefit. The new spouses can still spend plenty of time together to nurture their relationship without spending all of their time trying to manage the various issues that come up with the stepchildren.

14-9 Trends in Divorce and Remarriage

Divorce remains stigmatized in our society, as evidenced by the term **divorcism**—the belief that divorce is a disaster. In view of this cultural attitude, a number of attempts will continue to be made to reduce divorce rates. Couple/relationship education (CRE) provides guidelines for instruction in communication, conflict resolution, and parenting skills. Lucier-Greer et al. (2012) found positive outcomes for remarried couples with children who were involved in these programs that were targeted toward stepfamily issues.

Divorced fathers will also demand to be treated equally in custody decisions. The shift is a result of judges and lawyers going through divorce themselves as children or as adults and being aware of the unfair treatment of divorced dads (Jayson, 2014). Some law firms specialize in working with dads to ensure their equal treatment in the courts.

The remarriage rate is dropping fast. In 1990, 50 of 1,000 divorced or widowed remarried. By 2011, the number had dropped to 29. Cohabitation is the alternative many once-married individuals seek (Jayson, 2013).

divorcism the belief that divorce is a disaster.

STUDY TOOLS 14

Ready to study? In the book, you can:

- Rip out the chapter review card at the back of the book for a handy summary of the chapter and key terms.

- If you have experienced divorce in your family, explore your attitudes about it with the Self-Assessment card at the back of the book.

Online at CENGAGEBRAIN.COM you can:

- Prepare for tests with quizzes.

- Review the key terms with Flash Cards.

- Play games to master concepts.

ONE APPROACH.
70 UNIQUE SOLUTIONS.

The Later Years

"Want to know **who** is **living** a long **healthy life**? Ask one question, '**Is** the person **married**?'"

—AMY RAUER, RAND CORPORATION

Traditional college students live in a context of youth. Aside from relationships with grandparents, undergraduates rarely think of issues related to aging and often make fun of the elderly when they mimic, "Help, I've fallen and can't get up." Yet the elderly represent the one minority group everyone (unless they are unfortunate and die young) eventually joins. Understanding the process of and issues related to aging is the focus of this chapter.

15-1 Age and Ageism

In 2015, about 14.8% of the 325 million individuals in the United States are age 65 and older. By 2020, this percentage will grow to 16.8. By 2025, almost one in five Americans (18.8%) will be over the age of 65 (*ProQuest Statistical Abstract of the United States, 2014*, Table 9). Researcher Amy Rauer (2013) noted that we are moving toward a society of more "walkers than strollers."

All societies have a way to categorize their members by age. And all societies provide social definitions for particular ages.

> **"Years** may **wrinkle** the skin, but to **give up interests** wrinkles the **soul."**
>
> —DOUGLAS MACARTHUR, AMERICAN AUTHOR

15-1a **Defining Age**

A person's **age** may be defined chronologically, physiologically, psychologically, sociologically, and culturally. Chronologically, an "old" person is defined as one who has lived a certain number of years. The concept has obvious practical significance in everyday life. Bureaucratic organizations and social programs identify chronological age as a criterion of certain social rights and responsibilities. One's age determines the right to drive, vote,

age term which may be defined chronologically (number of years), physiologically (physical decline), psychologically (self-concept), sociologically (roles for the elderly/retired), and culturally (meaning age in one's society).

Table 15.1

Life Expectancy

Year	White Males	Black Males	White Females	Black Females
2015	77.7	72.7	82.3	78.7
2020	78.5	73.82.6	83	79.5

Source: *Proquest Statistical Abstract of the United States, 2014*, 2nd ed., Table 112. Bethesda, MD.

buy alcohol or cigarettes, and receive Social Security and Medicare benefits.

Age has meaning in reference to the society and culture of the individual. In ancient Greece and Rome, where the average life expectancy was 20 years, an individual was old at 18; similarly, one was old at 30 in medieval Europe and at 40 in the United States in 1850. In the United States today, however, people are usually not considered old until they reach age 65. However, our society is moving toward new chronological definitions of "old." Three groups of the elderly have emerged—the "young-old," the "middle-old," and the "old-old." The young-old are typically between the ages of 65 and 74; the middle-old, 75 to 84; and the old-old, 85 and beyond. Current life expectancy is shown in Table 15.1.

Gilbert Meilaender's (2013) book *Should We Live Forever? The Ethical Ambiguities of Aging* emphasizes the preoccupation with longevity. Research is being conducted in the Department of Genetics, University of Cambridge that will hopefully add hundreds of years to one's life. And, if case science does not work fast enough, the Cryonics Institute (http://www.cryonics.org/) offers to freeze your body and wake you up (hopefully) in the future when technology is available to replace human body parts indefinitely and to stop the aging process. The promise of never growing old and living forever is here. This thought certainly gives a new slant on "till death do us part."

There are different ways of defining age. In the following sections we look at several ways.

ageism the systematic persecution and degradation of people because they are old.

1. **Physiologically.** People are old when their auditory, visual, respiratory, and cognitive capabilities decline significantly. Becoming disabled is associated with being old. Sleep changes also occur for the elderly, including going to bed earlier, waking up during the night, and waking up earlier in the morning, as well as such disorders such as snoring and obstructive sleep apnea.

2. **Physical dependence.** People who need full-time nursing care for eating, bathing, and taking medication properly and who are placed in nursing homes are thought of as being old. Indeed, successful aging is culturally defined as maintaining one's health, independence, and cognitive ability. It is not death but the slow deterioration from aging that brings the most fear.

3. **Psychologically.** A person's self-concept is important in defining how old that person is. Once they see themselves as old, they are.

4. **Sociologically.** Once individuals occupy roles such as retiree, grandparent, and Social Security recipient, others begin to see them as old.

5. **Culturally.** The society in which an individual lives defines when and if a person becomes old and what being old means. In U.S. society, the period from age 18 through 64 is generally subdivided into young adulthood, adulthood, and middle age. Cultures also differ in terms of how they view and take care of their elderly. Spain is particularly noteworthy in terms of care for the elderly, with most elderly people receiving care from family members and other relatives.

"Youth has no age."

—PABLO PICASSO

15-1b Ageism

Every society has some form of **ageism**—the systematic persecution and degradation of people because they are old. Ageism is reflected in negative stereotypes of the elderly, such as they are slow, they do not change their ways, they are grumpy, they are poor drivers, they cannot/do not want to learn new things, they are incompetent, and they are physically and/or cognitively impaired (Nelson, 2011). Ageism also occurs when older individuals are treated differently because of their age, such as when they are spoken to loudly in simple language, when it is assumed they cannot understand normal speech, or when they are denied employment due to their age. Another form of ageism—**ageism by invisibility**—occurs when older adults are not included in advertising and educational materials. Ageism is similar to sexism, racism, and heterosexism. The elderly are shunned, discriminated against in employment, and sometimes victims of abuse.

Media not only emphasize youth and beauty (Haboush et al., 2012) but also contribute to the negative image of the elderly who are portrayed as difficult, complaining, and burdensome. Some individuals with ageist attitudes attempt to distance themselves from being regarded as old by engaging in risky behaviors (e.g., alcohol, drugs, cigarettes) (Popham et al., 2011).

Negative stereotypes and media images of the elderly engender **gerontophobia**—a shared fear or dread of the elderly. Such a negative view may create a self-fulfilling prophecy. For example, an elderly person forgets something and attributes forgetting to age. A younger person, however, is unlikely to attribute forgetfulness to age, given cultural definitions surrounding the age of the onset of senility.

The negative meanings associated with aging underlie the obsession of many Americans to conceal their age by altering their appearance. With the hope of holding on to youth a little bit longer, aging Americans spend billions of dollars each year on plastic surgery, exercise equipment, hair products, facial creams, and Botox injections.

> "For the unlearned, **old age** is winter; **for the learned**, it is the season of the **harvest**."
>
> —HASIDIC SAYING

Fibobjects/Dreamstime.com

15-1c Theories of Aging

Gerontology, the study of aging, has various theories to help in its understanding (North & Fiske, 2012). Table 15.2 identifies several theories, the level (macro or micro) of the theory, and the theorists typically associated with the theory, assumptions, and criticisms. As noted earlier, there are diverse ways of conceptualizing the elderly.

15-2 Caregiving for the Frail Elderly: The "Sandwich Generation"

About 20 million Americans provide care to their frail parents (Leopold et al., 2014). An elderly parent is defined as **frail** if he or she has difficulty with at least one personal care activity or other activity related to independent living; the severely disabled are unable to complete three or more personal care activities. These personal care activities include bathing, dressing, getting in and out of bed, shopping for groceries, and taking medications. Most (over 90%) frail elderly do not have long-term health care insurance, and many children choose to take care of their elderly parents. The term *children* typically means female adult children taking care of their mothers (fathers often have a spouse or are deceased) (Leopold et al., 2014). These women provide **family caregiving** and are known as the **sandwich generation** because they take care of their parents and their children simultaneously. Of course, male adult children also take care of their mothers or fathers or both.

ageism by invisibility when older adults are not included in advertising and educational materials.

gerontophobia fear or dread of the elderly, which may create a self-fulfilling prophecy.

gerontology the study of aging.

frail term used to define elderly people if they have difficulty with at least one personal care activity (feeding, bathing, toileting).

family caregiving adult children providing care for their elderly parents.

sandwich generation generation of adults who are "sandwiched" between caring for their elderly parents and their own children.

Table 15.2
Theories of Aging

Name of Theory	Level of Theory	Theorists	Basic Assumptions	Criticisms
Disengagement	Macro	Elaine Cumming, William Henry	The gradual and mutual withdrawal of the elderly and society from each other is a natural process. It is also necessary and functional for society that the elderly disengage so that new people can be phased in to replace them in an orderly transition.	Not all people want to disengage; some want to stay active and involved. Disengagement does not specify what happens when the elderly stay involved.
Activity	Macro	Robert Havighurst	People continue the level of activity they had in middle age into their later years. Though high levels of activity are unrelated to living longer, they are related to reporting high levels of life satisfaction.	Ill health may force people to curtail their level of activity. The older a person, the more likely the person is to curtail activity.
Conflict	Macro	Karl Marx, Max Weber	The elderly compete with youth for jobs and social resources such as government programs (Medicare).	The elderly are presented as disadvantaged. Their power to organize and mobilize political resources such as the American Association of Retired Persons is underestimated.
Age stratification	Macro	M. W. Riley	The elderly represent a powerful cohort of individuals passing through the social system that both affect and are affected by social change.	Too much emphasis is put on age, and little recognition is given to other variables within a cohort such as gender, race, and socioeconomic differences.
Modernization	Macro	Donald Cowgill	The status of the elderly is in reference to the evolution of the society toward modernization. The elderly in premodern societies have more status because what they have to offer in the form of cultural wisdom is more valued. The elderly in modern technologically advanced societies have low status because they have little to offer.	Cultural values for the elderly, not level of modernization, dictate the status of the elderly. Japan has high respect for the elderly and yet is highly technological and modernized.
Symbolic	Micro	Arlie Hochschild	The elderly socially construct meaning in their interactions with others and society. Developing social bonds with other elderly can ward off being isolated and abandoned. Meaning is in the interpretation, not in the event.	The power of the larger social system and larger social structures to affect the lives of the elderly is minimized.
Continuity	Micro	Bernice Neugarten	The earlier habit patterns, values, and attitudes of the individual are carried forward as a person ages. The only personality change that occurs with aging is the tendency to turn one's attention and interest on the self.	Other factors than one's personality affect aging outcomes. The social structure influences the life of the elderly rather than vice versa.
Interpersonal	Micro	Julian Palmore III Jean-Pierre Langlois	Negative assumptions based on physical apperance (droopy eyes means sad person).	Some elderly are in good physical condition.

ANGER OF FAMILY CARETAKERS WHEN TAKING CARE OF ELDERLY RELATIVES

Researchers Crespo and Fernandez-Lansac (2014) interviewed 129 caregivers who had primary responsibility for taking care of an elderly (age 60 or older) dependent family member for a period of at least six months. Most of the caregivers were women and were ether an older adult child or the spouse of the care recipient. The diagnosis of the person they took care of was most often dementia, Alzheimer's type.

The caregiving experience was extensive—about 16 hours a day for about 4.5 years. Most of these caregivers had some help from other family members or some kind of formal service.

The purpose of the research was to find out how often the caregivers felt mad/furious and how often they expressed their anger. Results revealed that around 40% of the caregivers could be categorized as having moderate to severe anger levels and a similar percent showed moderate to severe levels of anger expressions. Higher levels of anger were evident if there was a nonloving or negative relationship with the elderly person they were taking care of. Over time, feelings of resentment about taking care of the elderly family member could lead to interpersonal conflict and further deterioration of the bond between the caregiver and the care recipient. Particularly problematic was the cognitive confusions of the care recipient and repetitive questions to the caregiver which made anger control challenging.

On most occasions, the caregivers controlled their expressions of anger since they felt their own anger was an unacceptable emotion. Indeed, the caregivers made an effort to deliberately control and suppress their anger. Both children and spouses who cared for the elderly patient had similar levels of anger or anger expression. The research emphasized the difficulty of caring for an elderly family member, the struggle family members experience in doing so and the effort they expend to try and do the right thing. Also, evident is the degree to which help from other family members or formal services are valuable in reducing anger.

Source: Crespo, M., & V. Fernandez-Lansac. (2014). Factors associated with anger and anger expression in caregivers of elderly relatives. *Aging & Mental Health*, 18: 454–462.

Caregiving for an elderly parent has two meanings. One form of caregiving refers to providing personal help with the basics of daily living, such as getting in and out of bed, bathing, toileting, and eating. A second form of caregiving refers to performing instrumental activities, such as shopping for groceries, managing money (including paying bills), and driving the parent to the doctor.

The typical caregiver is a middle-aged married woman/mother who works outside the home. High levels of stress and fatigue may accompany caring for one's elders. Lee et al. (2010) studied family members (average age 46) providing care for elderly parents as well as children and found that the respective responsibilities resulted in not having enough time for both and in an increase in problems associated with stress. When the two roles were compared, taking care of the parents often translated into missing more workdays. The number of individuals in the sandwich generation will increase for the following reasons:

1. **Longevity.** The over-85 age group, the segment of the population most in need of care, is the fastest-growing segment of our population.

2. **Chronic disease.** In the past, diseases took the elderly quickly. Today, diseases such as arthritis and Alzheimer's are associated not

Members of the sandwich generation not only care for aging parents but also for children and, sometimes, grandchildren.

Sandor Kacso/Fotolia

with an immediate death sentence but with a lifetime of managing the illness and being cared for by others.

3. **Fewer siblings to help.** The current generation of elderly have fewer children than the elderly in previous generations. Hence, the number of siblings available to help look after parents is more limited. Children without siblings are more likely to feel the weight of caring for elderly parents alone.

4. **Commitment to parental care.** Contrary to the myth that adult children in the United States abrogate responsibility for taking care of their elderly parents, most children institutionalize their parents only as a last resort. Indeed, most adult children want to take care of their aging parents either in the parents' own home or in the adult child's home. When parents can no longer be left alone and can no longer cook for themselves, full-time nursing care is sought. Asian children, specifically Chinese children, are socialized to expect to take care of their elderly in the home.

5. **Lack of support for the caregiver.** Caring for a dependent, aging parent requires a great deal of effort, sacrifice, and decision making on the part of more than 20 million adults in the United States who are challenged with this situation. The emotional toll on the caregiver may be heavy. Guilt (over not doing enough), resentment (over feeling burdened), and exhaustion (over the relentless care demands) are common feelings.

Some people reduce the strain of caring for an elderly parent by arranging for home health care. This involves having a nurse go to the home of a parent and provide such services as bathing and giving medication. Other services include taking meals to the elderly (e.g., through Meals on Wheels). The National Family Caregiver Support Program provides support services for individuals (including grandparents) who provide family caregiving services. Such services include elder-care resource and referral services, caregiver support groups, and classes on how to care for an aging parent. In addition, states increasingly provide family caregivers with a tax credit or deduction.

age discrimination
a situation where older people are often not hired and younger workers are hired to take their place.

15-3 Issues Confronting the Elderly

Numerous issues become concerns as people age. In middle age, the issues are early retirement (sometimes forced), job layoffs (recession-related cutbacks), **age discrimination** (older people are often not hired and younger workers are hired to take their place), separation or divorce from a spouse, and adjustment to children leaving home. For some in middle age, grandparenting is an issue if they become the primary caregiver for their grandchildren. As couples move from the middle to the later years, the issues become more focused on income, health, retirement, and sexuality.

"It isn't at what age you want to retire but at what income."

—GEORGE FOREMAN, FORMER HEAVYWEIGHT CHAMPION

15-3a **Income**

A regret of many retirees is that they did not save enough money for retirement—that they spent more than they should have during their peak earning years. Indeed, for most elderly, the end of life is characterized by reduced income. Social Security and pension benefits, when they exist, are rarely equal to the income a retired person formerly earned. Many elderly continue working since they cannot afford to quit. The median annual income of men aged 65 and older is $27,707; women, $15,362 (*ProQuest Statistical Abstract of the United States, 2014*, Table 732).

Elderly women are particularly disadvantaged because their out-of-home employment has often been discontinuous, part-time, and low-paying. Social Security and private pension plans favor those with continuous, full-time work histories. Hence, their retirement incomes are considerably lower than the retirement income of males.

> "I think the **main problem** people have **getting older**, whether they know it or not, is that you're **closer to dying**."
>
> —ELIZABETH LESSER, EDUCATOR

15-3b Physical Health

There is a gradual deterioration in physical well-being as one ages (Proulx & Ermer, 2013). Park-Lee et al. (2013) reported on national data of 8,875 old-old individuals (85 years or older) in home health care, nursing homes, or hospice. Over two-thirds needed assistance in performing three or more activities of daily living (ADLs) and were bladder incontinent. Hypertension and heart disease were the two most common chronic health conditions. Reported physical health also varies by race/ethnicity. Park et al. (2013) compared elderly (65 and older) White, African Americans, Cuban, and non-Cuban Hispanics and found that racial/ethnic minority older adults rated their health more poorly.

Good physical health is the single most important determinant of an elderly person's reported happiness. Weight has an effect on one's health as one ages. Felding et al. (2011) studied 1,600 sedentary elderly individuals and found that inability to walk 400 meters was associated with negative outcomes which could include higher morbidity, mortality, hospitalizations, and a poorer quality of life (depression). Although most (over 90%) of the elderly do not exercise, walking at least 20 minutes a day results in physical and cognitive benefits for them (Aoyagi & Shephard, 2011). Johnson et al. (2014) reconfirmed the association between greater activity and higher quality of life.

> "One minute I've got bright Auburn hair down to the small of my back, *I'm riding on a motorcycle, and doing speed with a really dangerous guy; the next minute I'm a pleasant-enough lady in her forties biting into a gooey dessert and settling my broadening butt onto a couch with a couple of cats. Well, go ahead, is what I tell myself when buying a sack of Hershey's Kisses.*"
>
> —ASHLEY WALKER

Good physical health has an effect on marital quality. Iveniuk et al. (2014) noted that wives with husbands in fair or poor physical health were more likely to report higher levels of marital conflict, but the reverse was not true. The authors suggested that men with compromised health have limited leverage (in terms of reduced resources and status) to resist changes expected by their wives, so the wives are asked to give in with fewer rewards.

Body image/physical appearance are also concerns of the elderly. Roy and Payette (2012) reviewed the literature on body image among Western seniors and found that while most were not as concerned about physical appearance as youth, they did feel captive to their aging bodies (there was a discrepancy in their inner self and outward appearance). Women experienced more dissatisfaction about their physical appearance/bodies than did men.

15-3c Mental Health

Mental health may worsen for some elderly. Mood disorders, with depression being the most frequent, are more common among the elderly. Scheetz et al. (2012) found higher rates of depression among centenarians. Bereavement over the death of a spouse, loneliness, physical illness, and institutionalization may be the culprits. Foreclosures, associated with depression, during the recession were particularly difficult for the elderly (Cagney et al., 2014). Insomnia in the elderly may also be a precursor to depression (Nadorff et al., 2013).

Dementia, which includes Alzheimer's disease, is the mental disorder most associated with aging. Worldwide, 36 million individuals live with dementia-like disease (Hogsnes et al., 2014).

There has been considerable cultural visibility in regard to the medical use of marijuana. Ahmend et al. (2014) reported data on older adults in regard to its use with Alzheimer's patients; behavioral disturbances and nighttime agitation were reduced.

Hogsnes et al. (2014) interviewed 11 spouses of persons with dementia before and after relocating them to a nursing home. Feelings of shame and guilt and feelings of isolation preceded relocating the spouse to a nursing home. The event which triggered the move was threats and physical violence (some with a knife) directed toward the caring partner. After relocating the spouse to a nursing home, partners described feelings of both guilt and freedom, living with grief, feelings of loneliness in the spousal relationship, and striving for acceptance despite a lack of completion.

15-3d Divorce

Middle-aged and older adults do not always live "happily ever after." According to Brown and Lin (2012), there is a gray divorce revolution occurring in that the divorce rate for those age 50 and older doubled from 1990 to 2009 (now one in four divorces or 600,000 annually). Blacks, formerly married, noncollege, and those married less than 10 years were most likely to divorce. Reasons include that older people are more likely to be remarried and this group is more likely to divorce, greater acceptance of divorce (e.g., friends likely to be divorced), and wife more likely to be economically independent. Other reasons for later life divorce may include that children are grown (less concern about the effect of divorce on them), health (time left for a second life), prenuptial agreements (limit one's economic liability to spouse being divorced), and dating sites (find a new partner quickly).

Some spouses are literally dumped as they age. According to son and biographer Peter Ford, Cynthia Hayword, the third wife of Glenn Ford (famous movie star of the fifties) enjoyed being on his arm for 11 years of traveling the world and parties with the rich and famous in Hollywood. When he became ill in his mid-eighties, she checked him into a dependency care facility (Pasadena Las Encinas Hospital) and divorced him (Ford, 2011).

15-3e Retirement

Retirement represents a rite of passage through which most elderly pass. Pond et al. (2010) interviewed 60 individuals from 55 to 70 who revealed their reasons for retirement. These included poor health and the "maximization of life." The latter referred to retiring while they were healthy and could enjoy/fulfill other life goals. Another reason was "health protection"—decisions motivated by health protection and promotion. Being able to retire since they could afford to do so was also a reason.

The retirement age in the United States for those born after 1960 is 67. Individuals can take early retirement at age 62, with reduced benefits. Retirement

This retired couple travel the globe. They are at Victoria Falls in Africa.

Courtesy of Bob Bradley

dementia loss of brain function that occurs with certain diseases. It affects memory, thinking, language, judgment, and behavior.

affects an individual's status, income, privileges, power, and prestige. One retiree noted that he was being waited on by a clerk who looked at his name on his check, thought she recognized him and said, "Didn't you use to be somebody?"

People least likely to retire are unmarried, widowed, single-parent women who need to continue working because they have no pension or even Social Security benefits—if they do not work or continue to work, they will have no income, so retirement is not an option. Some workers experience what is called **blurred retirement** rather than a clear-cut one. A blurred retirement means the individual works part-time before completely retiring or takes a "bridge job" that provides a transition between a lifelong career and full retirement. Others may plan a **phased retirement** whereby an employee agrees to a reduced work load in exchange for reduced income. This arrangement is beneficial to both the employee and employer.

Wang et al. (2011) identified five variables associated with enjoying retirement—individual attributes (e.g., physical and mental health/financial stability), pre-retirement job-related variables (e.g., escape from work stress/job demands), family-related variables (e.g., being happily married), retirement-transition-related issues (e.g., planned voluntary retirement), and postretirement activities (e.g., bridge employment, volunteer work). Indeed, individuals who have a positive attitude toward retirement are those who have a pension waiting for them, are married (and thus have social support for the transition), have planned for retirement, are in good health, and have high self-esteem. Price and Nesteruk (2013) emphasized that identifying interests to share in retirement is important for couple satisfaction.

Paul Yelsma is a retired university professor. For a successful retirement, he recommends the following:

- Just like going to high school, the military, college, or your first job, know that the learning curve is very sharp. Don't think retirement is a continuation of what you did a few years earlier. Learn fast or you will be bored.

- For every 10 phone calls you make expect about 1 back; your status has dwindled and you have less to offer those who want something from you.

- Have at least three to five hobbies that you can do almost any time and any place. Feed your hobbies or they will die.

- Having several things to look forward to is so much more exciting than looking backward. If you looked backward to high school while in college, your life would be boring.

- Develop a new physical exercise program that you want to do. Don't expect others to support your activities. The "pay off" in retirement is vastly different from what colleagues expected of you.

- Find new friends who have time to share (colleagues are often too busy).

- Retirement is a *new* time to live and *new* things have to happen.

- Travel—we have traveled to seven new countries in the past few years.

Some retired individuals volunteer—giving back their time and money to attack poverty, illiteracy, oppression, crime, and so on. Examples of volunteer organizations are the Service Corps of Retired Executives, known as SCORE (www.score.org), Experience Works (www.experienceworks.org), and Generations United (www.gu.org).

15-3f Retirement Communities: Niche Aging

The stereotype of the elderly residing in an assisted living facility has been replaced by the concept of "niche aging," whereby retirement communities are set up for those with a particular interest (Barovick, 2012). For those into country music, there is a community in Franklin, Tennessee, which offers not only the array of housing alternatives but also recording studios and performance venues; for Asian Americans, there is Aegis Gardens in Freemont, California; for gay and lesbian retirees, there is Fountaingrove Lodge in Santa Rosa, California. There are now 100 such communities including those set up close to universities so individuals can continue to take classes (Oberlin and Dartmouth). Some adults seek these contexts and check in as early as age 40.

> **"Life is short—smile while you still have your teeth."**
>
> —UNKNOWN

blurred retirement an individual working part-time before completely retiring or taking a "bridge job" that provides a transition between a lifelong career and full retirement.

phased retirement an employee agreeing to a reduced work load in exchange for reduced income.

15-3g Sexuality

There are numerous changes in the sexuality of women and men as they age (Vickers, 2010). Frequency of intercourse drops from about once a week for those 40 to 59 to once every six weeks for those 70 and older. Changes in men include a decrease in the size of the penis from an average of 5 to about 4½ inches. Elderly men also become more easily aroused by touch rather than visual stimulation, which was arousing when they were younger. Erections take longer to achieve, are less rigid, and it takes longer for the man to recover so that he can have another erection. "It now takes me all night to do what I used to do all night" is the adage aging men become familiar with.

Levitra, Cialis, and Viagra (prescription drugs that help a man obtain and maintain an erection) are helpful for about 50% of men. Others with erectile dysfunction may benefit from a pump that inflates two small banana-shaped tubes that have been surgically implanted into the penis. Still others benefit from devices placed over the penis to trap blood so as to create an erection.

Women also experience changes including menopause, which is associated with a surge of sexual libido, an interest in initiating sex with her partner, and greater orgasmic capacity. Not only are they free from worry about getting pregnant, but also estrogen levels drop and testosterone levels increase. A woman's vaginal walls become thinner and less lubricating; the latter issue can be resolved by lubricants like KY Jelly.

Table 15.3 describes these and other physiological sexual changes that occur as individuals age.

"It is **not sex** that gives the **pleasure** but **the lover**."

—MARGE PIERCY, AMERICAN POET

Some spouses are sexually inactive. Karraker and DeLamater (2013) analyzed data on 1,502 men and women ages 57 to 85 and found that 29% reported no sexual activity for the past 12 months or more. The longer the couple had been married, the older the spouse, and the more compromised the health of the spouse, the more likely the individual was to report no sexual activity. Syme et al. (2013) reported on the sexuality of adults aged 63 to 67. Factors associated with lack of sexual satisfaction included spouse in poor health, history of diabetes, and fatigue. Elderly women were less sexually satisfied than men.

As noted earlier, the most sexually active are in good health. Diabetes and hypertension are major

Table 15.3
Physiological Sexual Changes as Individuals Age

Physical Changes in Sexuality as Men Age	Physical Changes in Sexuality as Women Age
1. Delayed and less firm erection.	1. Reduced or increased sexual interest.
2. More direct stimulation needed for erection.	2. Possible painful intercourse due to menopausal changes.
3. Extended refractory period (12 to 24 hours before rearousal can occur).	3. Decreased volume of vaginal lubrication.
4. Fewer expulsive contractions during orgasm.	4. Decreased expansive ability of the vagina.
5. Less forceful expulsion of seminal fluid and a reduced volume of ejaculate.	5. Possible pain during orgasm due to less flexibility.
6. Rapid loss of erection after ejaculation.	6. Thinning of the vaginal walls.
7. Loss of ability to maintain an erection for a long period.	7. Shortening of vaginal width and length.
8. Decrease in size and firmness of the testes.	8. Decreased sex flush, reduced increase in breast volume, and longer postorgasmic nipple erection.

Source: Adapted from Janell L. Carroll, *Sexuality Now: Embracing Diversity*, 4th ed., p. 273. © 2012. Wadsworth, a part of Cengage Learning, Inc. Reproduced by permission. www.cengage.com/permissions.

causes of sexual dysfunction. Incontinence (leaking of urine) is particularly an issue for older women and can be a source of embarrassment. The most frequent sexual problem for men is erectile dysfunction; for women, the most frequent sexual problem is the lack of a partner.

> "You can **live to** be a **hundred** if you **give up** all the **things** that **make you want to** live to be a hundred."
>
> —WOODY ALLEN, FILM DIRECTOR

15-4 Successful Aging

There is considerable debate about the meaning of successful aging. Torres and Hammarström (2009) noted that at least three definitions have been used: absence of physical problems/disability (no diseases or cognitive impairment), presence of strategies to cope with aging that results in positive emotions, sense of well-being, not feeling lonely, and having a positive view of aging.

Claudia Kawas of the National Institutes of Health studied 1,600 plus individuals who were over the age of 90. The goal of the research was to find out those factors associated with successful aging (Stahl, 2014). These included not smoking, exercising 45 minutes a day, being socially active (book clubs, being with friends), drinking alcohol moderately, drinking coffee (1–3 cups a day), and maintaining or gaining weight. Taking vitamins was not associated with longevity.

Boyes (2013) also identified the value of outdoor adventure for the elderly. In interviews and surveys of 80 elderly individuals, he emphasized the physical (healthier), social (connections with other), and psychological (stronger sense of well-being) benefits of involvement in an outdoor adventure program. Physical exercise also helps promote cognitive function (Wendell et al., 2014). Having a happy marriage was also important for successful aging. Indeed, the elderly who were identified as being "happy and well" were six times more likely to be in a good marriage than those who were identified as "sad and sick."

A positive attitude is also critical to aging successfully. One of Clarke and Bennett's (2013) respondents

(75-year-old woman with back problems, bursitis, lupus, and arthritis) said:

> If you don't have a good attitude, that's it. You know? If you're feeling sorry for yourself and you're sitting there thinking, "Oh this is awful today. My knee hurts. I can't walk. I can't do this." Then sure you're going to sit there and you're going to feel worse.

There are also gender differences in how men and woman age. Men feel a keen sense of loss of control and their increasing dependence on women. Women feel dismayed at their altered physical appearance and how their illness will impact others (Clarke & Bennett, 2013).

15-5 Relationships and the Elderly

Relationships in the later years vary. Some elderly men and women are single and date.

15-5a Dating

Brown and Shinohara (2013) studied the dating behavior of 3,005 individuals ages 57–85 and found that 14% of singles were in a dating relationship. Dating was more common among men than women and declined with age. Compared to nondaters, daters were more socially advantaged, college educated, and had more assets, better health, and more social connectedness. Some elderly

One aspect of successful aging is enduring friendships.

seek dating partners through Internet sites which cater to older individuals. Ourtime.com is an example.

Alterovitz and Mendelsohn (2013) noted the importance of health in seeking a partner for the elderly. In their study of 450 personal ads, they compared the middle aged (40–45), young-old (60–74), and old-old (75 plus) and found that the two younger cohorts focused on a partner for adventure, romance, sex, and a soul mate; the older group was more likely to mention the importance of a healthy partner. We (authors of your text) know of an example where a woman in her sixties met a man in his sixties via Match.com. On their second date she asked him to walk with her, which ended up being 1.5 miles. She later said, "I was testing to see how healthy he was…if he could not walk a mile and a half it would have been our last date."

15-5b Use of Technology to Stay Connected

Youth regularly text friends and romantic partners throughout the day. Madden (2010) noted that older adults and senior citizens are increasing their use of technology to stay connected as well. Over 40% of adults over the age of 50 use email. And almost half (47%) of Internet users 50–64 and 25% of users 65 and older use social networking sites such as Facebook. Indeed these older adults and seniors view themselves among the Facebooking and LinkedIn masses. These figures have doubled since 2009.

Use of Twitter has not caught on among older adults and senior citizens, with only 11% of online adults ages 50–64 reporting such usage. Email and online news are the most frequent Internet behaviors of this age group.

"The years teach much which the days never knew."

—RALPH WALDO EMERSON, AMERICAN POET

15-5c Relationships between Elderly Spouses

Marriages that survive into late life are characterized by little conflict, considerable companionship, and mutual supportiveness. All but one of the 31 spouses over age 85 in the Johnson and Barer (1997) study reported "high expressive rewards" from their mate. Walker and Luszcz (2009) reviewed the literature on elderly

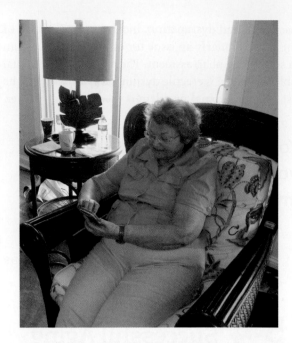

couples and found marital satisfaction related to equality of roles and marital communication. Health may be both improved by positive relationships and decreased by negative relationships. Rauer (2013) reported on 64 older couples (married an average of 42 years) and emphasized that being married in late life was the best of all contexts in terms of social/emotional, economic, and behavioral resources (e.g., taking care of each other). She noted that taking care of a spouse actually benefits the caregiver with feelings of self-esteem (but the outcome for the spouse being cared for may be negative since he or she may feel dependent).

The wedding is the first page in a novel—you don't know where it is going and you don't know how it will end.

—JACK TURNER, PSYCHOLOGIST

Only a small percentage (8%) of individuals older than 100 are married. Most married centenarians are men in their second or third marriage. Many have outlived some of their children. Marital satisfaction in these elderly marriages is related to a high frequency of expressing love feelings to one's partner. Though it is assumed that spouses who have been married for a long time should know how their partners feel, this is often not the case. Telling each other "I love you" is very important to these elderly spouses.

Courtesy of Joyce Chang

Grandfathers communicate joy in the relationship they have with their grandchildren.

15-5d **Grandparenthood**

Grandparent is a significant role for the elderly. Among adults aged 40 and older who had children, close to 95% are grandparents and most have, on average, five or six grandchildren. College students report that their grandparents express affection to them via concern/interest, gifts, interactions, verbal, nonverbal, and support/encouragement (Mansson, 2012). Factors influencing the quality of the grandparent/grandchild relationship include distance, age of the child/parent, and relationship the grandparent has with the parents (Dunifon & Bajracharya, 2012).

Some grandparents actively take care of their grandchildren full-time, provide supplemental help in a multigenerational family, assist on an occasional part-time basis, or occasionally visit their grandchildren. Grandparents see themselves as caretakers, emotional and/or economic resources, teachers, and historical connections on the family tree.

About 4% of children live with their grandparents or other relatives (Carr & Springer, 2010). Almost 8% (7.8%) of children grow up in three-generation households. Almost half (45%) of single parents live in three-generation households at some time (Pilkauskas, 2012).

15-5e **Styles of Grandparenting**

Grandparents have different styles of relating to grandchildren. Although some grandparents are formal and rigid, others are informal and playful, and authority lines are irrelevant. Still others are surrogate parents providing considerable care for working mothers and/or single parents.

Drew and King (2012) surveyed 289 individuals 18–25 in regard to their relationship with their grandparents and confirmed most were closer to their grandmother but satisfied with both sets of relationships. While some grandparents are close with their grandchildren, others are distant and show up only for special events like birthdays and holidays. Email and webcams help grandparents stay connected to their grandchildren.

Finally, the quality of the grandparent-grandchild relationship can be affected by the parents' relationship to their own parents. If a child's parents are estranged from their parents, it is unlikely that the child will have an opportunity to develop a relationship with the

Grandmothers are often the workhorse of the family—taking care of infants and helping with artwork.

grandparents. Divorce often shatters the relationship grandparents have with their grandchildren (Doyle et al., 2010).

15-5f Effect of Divorce on the Grandparent-Child Relationship

Divorce often has a negative effect on grandparents seeing their grandchildren. The situation most likely to produce this outcome is when the children are young, the wife is granted primary custody, and the relationship between the spouses is adversarial.

Some grandparents are not allowed to see their grandchildren. In all 50 states, the role of the grandparent has limited legal and political support. By a vote of six to three, the Supreme Court *(Troxel v. Granville)* sided with the parents and virtually denied 60 million grandparents the right to see their grandchildren. The court viewed parents as having a fundamental right to make decisions about with whom their children could spend time—including keeping them from grandparents. However, some courts have ruled in favor of grandparents. Stepgrandparents have no legal rights to their stepgrandchildren.

15-5g Benefits of Grandparents to Grandchildren

Of a sample of 4,690 university students, 83% agreed with the statement, "I have a loving relationship with my grandmother" (72% reported having had a loving relationship with their grandfather) (Hall & Knox, 2015). Grandchildren report enormous benefits from having a close relationship with grandparents, including development of a sense of family ideals, moral beliefs, and a work ethic. Feeling loved is also a major benefit. This poem reflects the love of a grandmother.

15-6 The End of One's Life

"I thought about these **bodies** we call home: how amazingly **reliable** and how appallingly **unreliable** they are, such **miracles** of nature and such **natural disasters**....**At the end** of a long hallway a door opens. Someone calls your name out. No, there's been **no mistake**. The mistake was in imagining this would never happen."

—SY SAFRANSKY, EDITOR, *SUN MAGAZINE*

Thanatology is the examination of the social dimensions of death, dying, and bereavement (Bryant, 2007). The end of one's life sometimes involves the death of one's spouse.

thanatology the examination of the social dimensions of death, dying, and bereavement.

> "**Love** is so **short**, **forgetting** is so long."
>
> —PABLO NERUDA, POET AND DIPLOMAT

15-6a Death of One's Spouse

The death of a spouse is one of the most stressful life events a person experiences. Compared to both the married and the divorced, the widowed are the most lonely and have the lowest life satisfaction (Ben-Zur, 2012; Hensley, 2012). Because women tend to live longer than men, they are more likely to experience the role of being without a spouse.

Although individual reactions and coping mechanisms for dealing with the death of a loved one vary, several reactions to death are common. These include shock, disbelief and denial, confusion and disorientation, grief and sadness, anger, numbness, physiological symptoms such as insomnia or lack of appetite, withdrawal from activities, immersion in activities, depression, and guilt. Eventually, surviving the death of a loved one involves the recognition that life must go on, the need to make sense out of the loss, and the establishment of a new identity.

Women and men tend to have different ways of reacting to and coping with the death of a loved one. Women are more likely than men to express and share feelings with family and friends and are also more likely to seek and accept help, such as attending grief support groups. Initial responses of men are often cognitive rather than emotional. From early childhood, males are taught to be in control, to be strong and courageous under adversity, and to be able to take charge and fix things. Showing emotions is labeled weak.

Men sometimes respond to the death of their spouse in behavioral rather than emotional ways. Sometimes they immerse themselves in work or become involved in physical action in response to the loss. For example, a widower immersed himself in repairing a beach cottage he and his wife had recently bought. Later, he described this activity as crucial to getting him through those first two months. Another coping mechanism for men is the increased use of alcohol and other drugs.

Women's response to the death of their husbands may necessarily involve practical considerations. Johnson and Barer (1997) identified two major problems of widows—the economic effects of losing a spouse and the practical problems of maintaining a home alone. The latter involves such practical issues as cleaning the gutters, painting the house, and changing the air filters throughout the house.

Whether a spouse dies suddenly or after a prolonged illness has an impact on the reaction of the remaining spouse. The sudden death is associated with being less at peace with death and more angry at life. Death of one's spouse is also associated with one's own death—the widowed are more likely to die sooner than a spouse with a living partner (Simeonova, 2013).

The age at which one experiences the death of a spouse is also a factor in one's adjustment. People in their eighties may be so consumed with their own health and disability that they have little emotional energy to grieve. But even after the spouse's death, the emotional relationship with the deceased may continue. Some widows and widowers report a feeling that their spouses are with them for years after the death of their beloved. Some may also dream of their deceased spouses, talk to their photographs, and remain interested in carrying out their wishes. Such continuation of the relationship may be adaptive by providing meaning and purpose for the living or maladaptive in that it may prevent the surviving spouse from establishing new relationships.

15-6b Use of Technology and Death of One's Partner

Just as technology has changed the way relationships begin (e.g., text messaging), its use in memorializing the deceased is increasing. Some spouses are so distraught over the death of their partner that they cannot weather a traditional funeral. But they can upload a eulogy and videos to a website where others may pay their respects, post their own photos of the deceased, and blog comments of various experiences with the deceased. One such website for the deceased is http://www.respectance.com.

The widowed are lonely and grieve over the loss of their spouse.

Courtesy of Rachel Calisto

Grandchildren are often of enormous help and comfort to aging grandparents.

15-6c Preparing for Death

What is it like for those near the end of life to think about death? To what degree do they go about actually preparing for death? Johnson and Barer (1997) interviewed 48 individuals with an average age of 93 to find out their perspectives on death. Most interviewees were women (77%); of those, 56% lived alone, but 73% had some sort of support from their children or from one or more social support services. The following findings are specific to those who died within a year after the interview:

1. **The last year of life.** Most of the respondents had thought about death and saw their life as one that would soon end. Most did so without remorse or anxiety. With their spouses and friends dead and their health failing, they accepted death as the next stage in life. They felt like the last leaf on the tree (see Personal View section).

   The major fear these respondents expressed was not the fear of death but of the dying process. Dying in a nursing home after a long illness is a dreaded fear. Sadly, almost 60% of the respondents died after a long, progressive illness. They had become frail, fatigued, and burdened by living. They identified dying in their sleep as the ideal way to die. Some hastened their death by no longer taking their medications; others wished they could terminate their own life. Competent adults have the legal right to refuse or discontinue medical interventions. For incompetent individuals, decisions are made by a surrogate— typically a spouse or child (McGowan, 2011).

Personal View: Thoughts at the End of Life

Individuals with an average age of 93 were interviewed and shared their thoughts as they reached the end of life (Johnson & Barer, 1997).

If I die tomorrow, it would be all right. I've had a beautiful life, but I'm ready to go. My husband is gone, my children are gone, and my friends are gone.

***

That's what is so wonderful about living to be so old. You know death is near and you don't even care.

***

I've just been diagnosed with cancer, but it's no big deal. At my age, I have to die of something. (p. 205)

For some, the end is frightening. Movie star Glenn Ford spoke into a tape recorder at the end of his life:

Is this the way my life is going to end up, in deep sadness? I don't know if I'm even capable of working. I'm very insecure. I think I've lost it. I don't know if I can handle memorizing lines. My mind is so shook up, I don't know.... I've thought about taking my own life. It's a drastic thing to do—a horrible thing to do. But I don't think I can go on living like I am now ... I'm lost. I'm just completely washed up. I'm finished. (Ford, 2011, p. 296)

Source: Ford, P. 2011. Glenn Ford, A Life. Madison, Wisconsin: University of Wisconsin Press.

2. **Behaviors in the last year of life.** Aware that they are going to die, most simplify their life, disengage from social relationships, and leave final instructions. In simplifying their life, they sell their home and belongings and move to smaller quarters. One 81-year-old woman sold her home, gave her car away to a friend, and moved into a nursing home. The extent of her belongings became a chair, a lamp, and a TV.

Some divorce just before they die. Upon learning he had terminal cancer, actor Dennis Hopper filed for divorce from his fifth wife. One explanation for this behavior is that divorce removes a spouse from automatically getting part of a deceased spouse's estate and allows control of dispensing one's assets (often to one's own children) while one is alive.

Disengaging from social relationships is selective. Some maintain close relationships with children and friends but others "let go." They may no longer send out Christmas cards and stop sending letters. Phone calls become the source of social connections. Some leave final instructions in the form of a will or handwritten note expressing wishes of where to be buried, handling costs associated with disposal of the body (e.g., cremation), and directives about pets. One of Johnson and Barer's (1997) respondents left $30,000 each to specific caregivers to take care of several pets (p. 204).

Elderly who may have counted on children to take care of them may find that their children have scattered because of job changes, divorce, or both, and may be unavailable for support. Some children simply walk away from their parents or leave their care to their siblings. The result is that the elderly may have to fend for themselves.

One of the last legal acts of the elderly is to have a will drawn up. Stone (2008) emphasized how wills may stir up sibling rivalry (one sibling may be left more than another), be used as a weapon against a second spouse (by leaving all of one's possessions to one's children), or reveal a toxic secret (a mistress and several children named as heirs). Some elderly dictate how they wish to be buried. A new option is the "green burial" where the individual is returned to the earth without being embalmed, no casket, etc. It is a way for the soon to be deceased to "return to nature" (Kelly, 2012).

Some live full lives up until the very end. Ted Kennedy died of brain cancer at age 77. Up until the last week of his life, he was active in politics (he had sponsored over 300 bills in his Senate career). His words resonate, "For all those whose cares have been our concern, the work goes on, the cause endures, the hope still lives, and the dream must never die."

15-7 Trends and the Elderly in the United States

The elderly will increase in number and political clout. By 2030, 30% of the U.S. population will be over the age of 55 (the percentage is now 21). The challenges of old age will be the same—coping with dwindling income, declining health, and the death of loved ones. On the positive side, greater attention will be paid to the health needs of the elderly. Medicare will help pay for some of the medical needs. However, it alone will be inadequate and other private sources will be needed.

STUDY TOOLS 15

Ready to study? In the book, you can:

- Rip out the chapter review card at the back of the book for a handy summary of the chapter and key terms.
- Assess the well-being of your family with the Self-Assessment card at the back of the book.

Online at CENGAGEBRAIN.COM you can:

- Prepare for tests with quizzes.
- Review the key terms with Flash Cards.
- Play games to master concepts.

4LTR Press solutions are designed for today's learners through the continuous feedback of students like you. Tell us what you think about **M&F** and help us improve the learning experience for future students.

YOUR FEEDBACK MATTERS.

REFERENCES

Chapter 1

Amato, P. R., A. Booth, D. R. Johnson, & S. F. Rogers. (2007). *Alone together: How marriage in America is changing*. Cambridge, MA: Harvard University Press.

Blumer, H. G. (1969). The methodological position of symbolic interaction. In *Symbolic interactionism: Perspective and method*. Englewood Cliffs, NJ: Prentice-Hall.

Bunger, R. (2014). Department of Anthropology, East Carolina University, personal communication.

Cherlin, A. J. (2010). Demographic trends in the United States: A review of research in the 2000s. *Journal of Marriage and Family, 72*: 403–419.

Cohn, D. (2013). Love and marriage. Pew Research Center. Retrieved January 12, 2014, from http://www.pewsocialtrends.org/2013/02/13/love-and-marriage

Cooley, C. H. (1964). *Human nature and the social order*. New York: Schocken.

Curran, M. A., E. A. Utley, & J. A. Muraco. (2010). An exploratory study of the meaning of marriage for African Americans. *Marriage & Family Review, 46*: 346–365.

Darnton, K. (2012, February 12). Deception at Duke. *Sixty Minutes/CBS Television*.

Eagan, K., J. B. Lozano, S. Hurtado, & M. H. Case. (2013). *The American freshman: National norms fall 2013*. Los Angeles: Higher Education Research.

Elliott, L., B. Easterling, & D. Knox. (2012, March). Taking Chances in Romantic Relationships. Poster session presented at the annual meeting of the Southern Sociological Society, New Orleans, LA.

Girgis, S., R. P. George, & R. Anderson. (2011). What is marriage? *Harvard Journal of Law & Public Policy, 34*: 245–287.

Goldberg, J. S., & M. J. Carlson. (2014). Parents' relationship quality and children's behavior in stable married and cohabiting families. *Journal of Marriage and Family, 76*: 762–777.

Gregory, J. D. (2010). Pet custody: Distorting language and the law. *Family Law Quarterly, 44*: 35–64.

Hall, S., & D. Knox. (2015). Relationship and sexual behaviors of a sample of 4730 university students. Unpublished data collected for this text. Department of Family and Consumer Sciences, Ball State University and Department of Sociology, East Carolina University.

Holmes, E. K., T. Sasaki, & N. L. Hazen. (2013). Smooth versus rocky transitions to parenthood: Family systems in developmental context. *Family Relations, 62*: 824–837.

Isaacson, W. (2011). *Steve Jobs*. New York: Simon & Schuster.

Isaacson, W. (2007). *Einstein*. New York: Simon & Schuster.

James, S. D. (2008, May 7). Wild child speechless after tortured life. *ABC News*.

Kefalas, M., F. F. Furstenberg, P. J. Carr, & L. Napolitano. (2011). Marriage is more than being together: The meaning of marriage for young adults. *Journal of Family Issues, 32*: 845–875.

Lax, E. (1991). *Woody Allen: A biography*. New York: Knopf.

Lorber, J. (1998). *Gender inequality: Feminist theories and politics*. Los Angeles: Roxbury.

Marshall, J. (2013, November 5–9). Can we understand that which we cannot define? How marriage and family therapists define the family. Poster presented at the annual meeting of the National Council on Family Relations, San Antonio, TX.

Martin, B. A., & C. Dula. (2010). More than skin deep: Perceptions of, and stigma against, tattoos. *College Student Journal, 44*: 200–206.

McHale, S. M., K. A. Updegraff, & S. D. Whiteman. (2012). Sibling relationships and influences in childhood and adolescence. *Journal of Marriage and Family, 74*: 913–930.

Mead, G. H. (1934). *Mind, self, and society*. Chicago: University of Chicago Press.

Meinhold, J. L., A. Acock, & A. Walker. (2006, November). The influence of life transition statuses on sibling intimacy and contact in early adulthood. Paper presented at the annual meeting of the National Council on Family Relations, Orlando, FL.

Muraco, J. A., & M. A. Curran. (2012). Marriage for young adults in romantic relationships. *Marriage and Family Review, 48*: 227–247.

Murdock, G. P. (1949). *Social structure*. New York: Free Press.

National Marriage Project. (2012). *The state of our unions: Marriage in 2012*. The University of Virginia Center for Marriage and Families at the Institute for American Values.

Payne, C., & V. Bravo. (2013, November 11). Do you consider your pet to be part of the family? *USA Today*, p. A1.

Randall, B. (2008). *Songman: The story of an Aboriginal elder of Uluru*. Sydney, Australia: ABC Books.

Rauer, A. (2013, February 22). From golden bands to the golden years: The critical role of marriage in older adulthood. Presentation at the annual meeting of the Southeastern Council on Family Relations, Birmingham , AL.

Ryan, C., & C. Jetha. (2010). *Sex at dawn*. New York: Harper Perennial.

Severson, K. (2011, July 5). Systematic cheating is found in Atlanta's school system. *The New York Times*.

Silverstein, L. B., & C. F. Auerbach. (2005). (Post) modern families. In Jaipaul L. Roopnarine & U. P. Gielen (Eds.), *Families in global perspective*. Boston: Pearson Education.

Starnes, T. (2011, June 15). School surveys 7th graders on oral sex. Fox News.

Statistical Abstract of the United States, 2012–2013, 131th ed. Washington, DC: U.S. Bureau of the Census.

Vespa, J. (2013). Relationship transitions among older cohabiters: The role of health, wealth, and family ties. *Journal of Marriage and Family, 75*: 933–949.

Whitton, S. W., & B. A. Buzzella. (2012). Using relationship education programs with same-sex couples: A preliminary evaluation of program utility and needed modifications. *Marriage and Family Review, 48*: 667–688.

Zeitzen, M. K. (2008). *Polygamy: A cross-cultural analysis*. Oxford: Berg.

Chapter 2

Ashwin, S., & O. Isupova. (2014). "Behind every great man. . . .": The male marriage wage premium examined qualitatively. *Journal of Marriage and Family, 76*: 37–55.

Aubrey, J. S., & S. E. Smith. (2013). Development and validation of the endorsement of the Hookup Culture Index. *Journal of Sex Research, 50*: 435–448.

Barriger, M., & C. J. Velez-Blasini. (2013). Descriptive and injunctive social norm overestimation in hooking up and their role as predictors of hook-up activity in a college student sample. *Journal of Sex Research, 50*: 84–94.

Blackstrom, L., E. A. Armstrong, & J. Puentes. (2012). Women's negotiation of cunnilingus in college hookups and relationships. *Journal of Sex Research, 49*: 1–12.

Booth, A., E. Rustenbach, & S. McHale. (2008). Early family transitions and depressive symptom changes from adolescence to early adulthood. *Journal of Marriage and Family, 70*: 3–14.

Bradshaw, C., A. S. Kahn, & B. K. Saville. (2010). To hook up or date: Which gender benefits? *Journal Sex Roles, 62*: 661–669.

Chang, J., R. Ward, D. Padgett, & M. F. Smith. (2012, November 1). Do feminists hook up more? Examining pro-feminism attitude in the context of hooking-up. Paper presented at the National Council on Family Relations, Phoenix, AZ.

Cohn, D. (2013). Love and marriage. Pew Research Center. Retrieved from http://www.pewsocialtrends.org/2013/02/13/love-and-marriage/

Denney, J. T. (2010). Family and household formations and suicide in the United States. *Journal of Marriage and the Family, 72*: 202–213.

Dew, J. (2011). Financial issues and relationship outcomes among cohabiting individuals. *Family Relations, 60*: 178–190.

Eck, B. A. (2013). Identity twists and turns: How never-married men make sense of an unanticipated identity. *Journal of Contemporary Ethnography, 42*: 31–33.

Ellison, N. B., & J. T. Hancock. (2013). Profile as promise: Honest and deceptive signals in online dating. *Security & Privacy, IEEE, 11*: 85–88.

Gottman, Julie. (2013, June 27). The conversation—a discussion of living apart together relationships. NPR radio.

Guzzo, K. B. (2014). Trends in cohabitation outcomes: Compositional changes and engagement among never-married young adults. *Journal of Marriage and Family, 76*: 826–842.

Hall, J. A. (2014). First comes social networking, then comes marriage? Characteristics of Americans married 2005–2012 who met through social networking sites. *Cyberpsychology, Behavior, and Social Networking, 17*: 322–326.

Hall, J. A., N. Park, M. J. Cody, & H. Song. (2010). Strategic misrepresentation in online dating: The effects of gender, self-monitoring, and personality traits. *Journal of Social and Personal Relationships, 27*: 117–135.

Hall, S., & D. Knox. (2015). Relationship and sexual behaviors of a sample of 4730 university students. Unpublished data collected for this text. Department of Family and Consumer Sciences, Ball State University and Department of Sociology, East Carolina University.

Heino, R. D., N. B. Ellison, & J. L. Gibbs. (2010). Relations shopping: Investigating the market metaphor in online dating. *Journal of Social and Personal Relationships, 27*: 427–447.

Hess, J. (2012). Personal communication. Appreciation is expressed to Judye Hess for the development of this section. For more information about Judye Hess, see http://www.psychotherapist.com/judyehess/

Jacinto, E., & J. Ahrend. (2012). Living apart together. Unpublished data provided by Jacinto and Ahrend.

Jayson, S. (2011, March 30). Is dating dead? *USA Today*, p. A1.

_____. (2012, October 18). Couples of all kinds are cohabiting. *USA Today*, p. A1.

_____. (2013, February 14). The end of online dating. *USA Today*, p. 1A et passim.

Jose, A., K. D. O'Leary, & A. Moyer. (2010). Does premarital cohabitation predict subsequent marital stability and marital quality? A meta-analysis. *Journal of Marriage and Family, 72*: 105–116.

Kalish, R., & M. Kimmel. (2011). Hooking up. *Australian Feminist Studies, 26*: 137–151.

Kamiya, Y., M. Doyle, J. C. Henretta, et al. (2013). Depressive symptoms among older adults: The impact and later life circumstances and marital status. *Aging and Mental Health, 17*: 349–357.

Kasearu, K. (2010). Intending to marry . . . students' behavioral intention towards family forming. *TRAMES: A Journal of the Humanities & Social Sciences, 14*: 3–20.

Klinenberg, E. (2012). *Going solo: The extraordinary rise and surprising appeal of living alone.* New York: Penguin.

Kotlyar, I., & D. Ariely. (2013). The effect of nonverbal cues on relationship formation. *Computers in Human Behavior, 29*: 544–551.

Kuperberg, A. (2014). Age at coresidence, premarital cohabitation, and marital dissolution 1985–2009. *Journal of Marriage and the Family, 76*: 352–369.

LaBrie, J. W., J. F. Hummer, T. M. Gaidarov, A. Lac, & S. R. Kenney. (2014). Hooking up in the college context: The event-level effects of alcohol use and partner familiarity on hookup behaviors and contentment. *Journal of Sex Research, 51*: 62–73.

Lewis, M. A., D. C. Atkins, J. A. Blayney, D. V. Dent, & D. L. Kaysen. (2013). What is hooking up? Examining definitions of hooking up in relation to behavior and normative perceptions. *Journal of Sex Research, 50*: 757–766.

Liat, K., & H. Havusha-Morgenstern. (2011). Does cohabitation matter? Differences in initial marital adjustment among women who cohabited and those who did not. *Families in Society, 92*: 120–127.

Lo, S. K., A. Y. Hsieh, & Y. P. Chiu. (2013). Contradictory deceptive behavior in online dating. *Computers in Human Behavior, 29*: 1755–1762.

Lydon, J., T. Pierce, & S. O'Regan. (1997). Coping with moral commitment to long-distance dating relationships. *Journal of Personality and Social Psychology, 73*: 104–13.

Lyssens-Danneboom, V., S. Eggermont, & D. Mortelmans. (2013). Living apart together (LAT) and law: Exploring legal expectations among LAT individuals in Belgium. *Social & Legal Studies, 22*: 357–376.

Manning, W. D., & J. A. Cohen. (2012). Premarital cohabitation and marital dissolution: An examination of recent marriages. *Journal of Marriage and the Family, 74*: 377–387.

Manning, W. D., D. Trella, H. Lyons, & N. C. DuToit. (2010). Marriageable women: A focus on participants in a community Healthy Marriage Program. *Family Relations, 59*: 87–102.

Miller, A. J., S. Sassler, & D. Kusi-Appouh. (2011). The specter of divorce: Views from working- and middle-class cohabiters. *Family Relations, 60*: 602–616.

Muraco, J. A., & M. A. Curran. (2012). Associations between marital meaning and reasons to delay marriage for young adults in romantic relationships. *Marriage and Family Review, 48*: 227–247.

Murray, C. (2012). *Coming apart: The state of white America, 1960–2010.* New York: Crown Forum.

National Survey of Family Growth. (2014). Percent currently married by sex. Retrieved March 31, 2014, from http://www.statisticbrain.com/marriage-statistics/

Olmstead, S. B., P. N. E. Roberson, F. D. Fincham, & K. Pasley. (2012, November). Hooking up and risky sex behaviors among first semester college men: What is the role of pre-college experience? Presentation at the National Council on Family Relations, Phoenix, AZ.

Ortyl, T. A. (2013). Long-term heterosexual cohabiters and attitudes toward marriage. *Sociological Quarterly, 54*: 584–609.

Pew Research Center. (2010). Social & demographic trends: The decline of marriage and rise of new families. Published and retrieved on November 18 from http://pewresearch.org/pubs/1802/decline-marriage-rise-new-families

Pistole, M. C. (2010). Long-distance romantic couples: An attachment theoretical perspective. *Journal of Marital and Family Therapy, 36*: 115–125.

Proquest Statistical Abstract of the United States, 2014. 2nd ed. Bethesda, MD.

Pope, A. L., & C. S. Cashwell. (2013). Moral commitment in intimate committed relationships: A conceptualization from cohabiting same-sex and opposite sex partners. *Family Journal, 21*: 5–14.

Rauer, A. (2013, February 22). From golden bands to the golden years: The critical role of marriage in older adulthood. Annual meeting of the Southeastern Council on Family Relations, Birmingham, AL.

Reinhold, S. (2010). Reassessing the link between premarital cohabitation and marital instability. *Demography, 47*: 719–733.

Rhoades, G. K., S. M. Stanley, & H. J. Markman. (2012). A longitudinal investigation of commitment dynamics in cohabiting relationships. *Journal of Family Issues, 33*: 369–390.

Rosenfeld, M. J., & R. J. Thomas. (2012). Searching for a mate: The rise of the Internet as a social intermediary. *American Sociological Review, 77*: 523–547.

Rosin, H. (2010, August). The end of men. *The Atlantic*.

Ross, C., D. Knox, & M. Zusman. (2008, April). "Hey Big Boy": Characteristics of university women who initiate relationships with men. Poster presented at the annual meeting of the Southern Sociological Society, Richmond, VA.

Sassler, S., & A. J. Miller. (2011). Class differences in cohabitation processes. *Family Relations, 60:* 163–177.

Smith, A., & M. Duggan. (2013, October 21). Dating, social networking, mobile online dating & relationships. Pew Research Center. Retrieved from http://pewinternet.org/Reports/2013/Online-Dating.aspx

Sobal, J., & K. L. Hanson. (2011). Marital status, marital history, body weight, and obesity. *Marriage & Family Review, 47:* 474–504.

Sommers, P., S. Whiteside, and K. Abbott. (2013, November 5–9). Cohabitation: Perspectives of undergraduate students. Poster presented at the annual meeting of the National Council on Family Relations, San Antonio, TX.

Stafford, L. (2010). Geographic distance and communication during courtship. *Communication Research, 37:* 275–297.

Uecker, J., & M. Regnerus. (2011). *Premarital sex in America: How young Americans meet, mate, and think about marrying.* United Kingdom: Oxford University Press.

United Nations. (2011, December). World marriage patterns. *Population Facts.* Retrieved October 2, 2013, from http://www.un.org/en/development/desa/population/publications/pdf/popfacts/PopFacts_2011-1.pdf

Valle, G., & K. H. Tillman. (2014). Childhood family structure and romantic relationships during the transition to adulthood. *Journal of Family Issues, 35:* 97–124.

Vespa, J. (2014). Historical trends in the marital intentions of one-time and serial cohabitors. *Journal of Marriage & Family, 76:* 207–217.

Walsh, S. (2013). Match's 2012 singles in America survey. Retrieved from http://www.hookingupsmart.com/2013/02/07/hookinguprealities/matchs-2012-singles-in-america-survey/

Wang, W., & K. Parker. (2014, September 23). Record share of Americans have never married. Pew Research Center. Retrieved from http://www.pewsocialtrends.org/2014/09/24/record-share-of-americans-have-never-married/

Wienke, C., & G. J. Hill. (2009). Does the "Marriage Benefit" extend to partners in gay and lesbian relationships?: Evidence from a random sample of sexually active adults. *Journal of Family Issues, 30:* 259–273.

Willoughby, B. J., J. S. Carroll, & D. M. Busby. (2012). The different effects of "living together": Determining and comparing types of cohabiting couples. *Journal of Social and Personal Relationships, 29:* 397–419.

Wu, P., & W. Chiou. (2009). More options lead to more searching and worse choices in finding partners for romantic relationships online: An experimental study. *CyberPsychology & Behavior, 12:* 315–318.

Chapter 3

Abboud, P. (2013). The third gender: Samoa's Fa'afafine people. Retrieved September 29, 2014, from http://www.pedestrian.tv/features/arts-and-culture/meet-the-third-gender-samoas-faafafine-people/70b3c7c8-66fc-4453-9f06-1de911d249ee.htm

Barclay, L. (2013). Liberal daddy quotas: Why men should take care of the children, and how liberals can get them to do it. *Hypatia, 28:* 163–178.

Barnes, H., D. Knox, & J. Brinkley. (2012, March 23). CHEATING: Gender differences in reactions to discovery of a partner's cheating. Poster presented at the Southern Sociological Society, New Orleans. LA.

Bleske-Rechek, A., E. Somers, C. Micke, L. Erickson, L. Matteson, C. Stocco, B. Schumacher, & L. Ritchie. (2012). Benefit or burden? Attraction in cross-sex friendship. *Journal of Social and Personal Relationships, 29:* 569–596.

Brennan, C. (2014, April 14). Why the game of golf is in serious trouble. *USA Today,* p. 2A.

Bulanda, J. R. (2011). Gender, marital power, and marital quality in later life. *Journal of Women & Aging, 23:* 3–22.

Craig, L., & K. Mullan. (2010). Parenthood, gender and work-family time in the United States, Australia, Italy, France, and Denmark. *Journal of Marriage and Family, 72:* 1344–1361.

Darlow, S., & M. Lobel. (2010). Who is beholding my beauty? Thinness ideals, weight, and women's responses to appearance evaluation. *Sex Roles, 63:* 833–843.

DeMaris, A., L. A. Sanchez, & K. Knivickas. (2012). Developmental patterns in marital satisfaction: Another look at covenant marriage. *Journal of Marriage and the Family, 74:* 989–1004.

East, L., D. Jackson, L. O'Brien, & K. Peters. (2011). Condom negotiation: Experiences of sexually active young women. *Journal of Advanced Nursing, 67:* 77–85.

England, P. (2010). The gender revolution: Uneven and stalled. *Gender & Society, 24:* 149–166.

Eubanks Fleming, C. J. & J. V. Córdova. (2012). Predicting relationship help seeking prior to a marriage checkup. *Family Relations, 61:* 90–100.

Fisher, H. (2010). The new monogamy: Forward to the past: An author and anthropologist looks at the future of love. *The Futurist, 44:* 26–29.

Garfield, R. (2010). Male emotional intimacy: How therapeutic men's groups can enhance couples therapy. *Family Process, 49:* 109–122.

Gerding, A., & N. Signorielli. (2014). Gender roles in tween television programming: A content analysis of two genres. *Sex Roles, 70:* 43–56.

Gervais, S. J., A. M. Holland, & M. D. Dodd. (2013). My eyes are up here: The nature of the objectifying gaze toward women. *Sex Roles, 69:* 557–570.

Grogan, S. (2010). Promoting positive body image in males and females: Contemporary issues and future directions. *Sex Roles, 63:* 757–765.

Haag, P. (2011). *Marriage confidential.* New York: Harper Collins.

Hall, S., & D. Knox. (2015). Relationship and sexual behaviors of a sample of 1,155 undergraduate males and 3,581 undergraduate females. Unpublished data collected for this text. Department of Family and Consumer Sciences, Ball State University and Department of Sociology, East Carolina University.

Hollander, D. (2010). Body image predicts some risky sexual behaviors among teenage women. *Perspectives on Sexual & Reproductive Health, 42:* 67–87.

Jayson, S. (2014, January 21). Could a date get any more confusing? *USA Today,* p. 4B.

Kohlberg, L. (1966). A cognitive-developmental analysis of children's sex-role concepts and attitudes. In E. E. Macoby (Ed.), *The development of sex differences.* Stanford, CA: Stanford University Press.

———. (1969). State and sequence: The cognitive developmental approach to socialization. In D. A. Goslin (Ed.), *Handbook of socialization theory and research* (pp. 347–480). Chicago: Rand McNally.

Lease, S. H., A. B. Hampton, K. M. Fleming, L. R. Baggett, S. H. Montes, & R. J. Sawyer. (2010). Masculinity and interpersonal competencies: Contrasting White and African American men. *Psychology of Men & Masculinity, 11:* 195–207.

Lincoln, A. E. (2010). The shifting supply of men and women to occupations: Feminization in veterinary education. *Social Forces, 88:* 1969–1998.

Lucier-Greer, M., & F. Adler-Baeder. (2010, November 3–5). Gender role attitudes during divorce & remarriage: Plastic or plaster? Poster presented at the annual meeting of the National Council on Family Relations, Minneapolis, MN.

Maltby, L E., M. E. L. Hall, T. L. Anderson, & K. Edwards. (2010). Religion and sexism: The moderating role of participant gender. *Sex Roles, 62:* 615–622.

Mandara, J., F. Varner, & S. Richman. (2010). Do African American mothers really "love" their sons and "raise" their daughters? *Journal of Family Psychology, 24:* 41–50.

McGeorge, C. R., T. S. Carlson, & R. B. Toomey. (2012, November 3–5). Establishing the validity of the Feminist Couple Therapy Scale. Paper presented at the annual meeting of the National Council on Family Relations, Phoenix, AZ.

Mead, M. (1935). *Sex and temperament in three primitive societies.* New York: William Morrow.

Meliksah, D., O. Simsek, & A. Procsa. (2013). I am so happy 'cause my best friend makes me feel unique: Friendship, uniqueness and happiness. *Journal of Human Happiness, Studies,* 14: 1201–1224.

Michniewicz, K. S., J. A. Vandello, & J. K. Bosson. (2014). Men's (Mis)perceptions of the gender threatening consequences of unemployment. *Sex Roles,* 70: 88–97.

Minnottea, K. L., D. E. Pedersena, S. E. Mannonb, & G. Kiger. (2010). Tending to the emotions of children: Predicting parental performance of emotion work with children. *Marriage & Family Review,* 46: 224–241.

Monro, S. (2000). Theorizing transgender diversity: Towards a social model of health. *Sexual and Relationship Therapy,* 15: 33–42.

Murphy, M. J., L. Deets, & M. Peterson. (2010, November 3–5). The effect of emotional labor and emotional abuse on relationship satisfaction. Poster presented at the annual meeting of the National Council on Family Relations, Minneapolis, MN.

National Science Foundation, National Center for Science and Engineering Statistics. (2012). Doctorate recipients from U.S. universities. 2010. Special Report NSF 12-305. Arlington, VA.

Nelms, B. J., D. Knox, & B. Easterling. (2012). The relationship talk: Assessing partner commitment. *College Student Journal,* 46: 178–182.

Pew Research Center. (2012, April 19). A gender reversal on career trends. Young women now top young men in valuing a high-paying career. Published and retrieved from http://pewresearch.org/pubs/2248/gender-jobs-women-men-career-family-educational-attainment-labor-force-participation?src=prc-newsletter

Pfeffer, C. A. (2010). "Women's work"? Women partners of transgender men doing housework and emotion work. *Journal of Marriage & Family,* 72: 165–183.

Proquest Statistical Abstract of the United States, 2014. Bethesda, MD.

Read, S., & E. Grundy. (2011). Mental health among older married couples: the role of gender and family life. *Social Psychiatry & Psychiatric Epidemiology,* 46: 331–341.

Reece, M., D. Herbenick, J. D. Fortenberry et al. (2012). The National Survey of Sexual Health and Behavior. Retrieved January 20, 2014, from http://www.nationalsexstudy.indiana.edu/graph.html

Rees, C., & G. Pogarsky. (2011). One bad apple may not spoil the whole bunch: Best friends and adolescent delinquency. *Journal of Quantitative Criminology,* 27: 197–223.

Robnett, R. D., & C. Leaper. (2013). "Girls don't propose!: A mixed-methods examination of marriage tradition preferences and benevolent sexism in emerging adults. *Journal of Adolescent Research,* 28: 96–121.

Russell, V. M, L. R. Baker, & J. K. Mcnulty. (2013). Attachment insecurity and infidelity in marriage: Do studies of dating relationships really inform us about marriage? *Journal of Family Psychology,* 27: 241–251.

Sandberg, S. (2013). *Lean in: Women, work, and the will to lead.* New York: Alfred A. Knopf.

Sherman, A. M., & E. L. Zurbriggen. (2014). "Boys Can Be Anything": Effect of Barbie play on girls' career cognitions. *Sex Roles,* 70: 195–208.

Shin, K., H. Shin, J. A. Yang, & C. Edwards. (2010). Gender role identity among Korean and American college students: Links to gender and academic achievement. *Social Behavior & Personality: An International Journal,* 38: 267–272.

Simon, R. W., & K. Lively. (2010). Sex, anger, and depression. *Social Forces,* 88: 1543–1568.

Ter Gogt, T. F. M., R. C. M. E. Engles, S. Bogers, & M. Kloosterman. (2010). "Shake It Baby, Shake It": Media preferences, sexual attitudes and gender stereotypes among adolescents. *Sex Roles,* 63: 844–859.

Trucco, E. M., C. R. Colder, & W. F. Wieczorek. (2011). Vulnerability to peer influence: A moderated mediation study of early adolescent alcohol use initiation. *Addictive Behaviors,* 36: 729–736.

Wallis, C. (2011). Performing gender: A content analysis of gender display in music videos. *Sex Roles,* 64: 160–172.

Way, N. (2013). Boys' friendships during adolescence: Intimacy, desire, and loss. *Journal of Research on Adolescence,* 23: 201–213.

Weisgram, E. S., R. S. Bigler, & L. S. Liben. (2010). Gender, values, and occupational interests among children, adolescents, and adults. *Child Development,* 81: 778–796.

Wells, R. S., T. A. Seifert, R. D. Padgett, S. Park, & P. Umbach. (2011). Why do more women than men want to earn a four-year degree? Exploring the effects of gender, social origin, and social capital on educational expectations. *Journal of Higher Education,* 82: 1–32.

Witt, M. G., & W. Wood. (2010). Self-regulation of gendered behavior in everyday life. *Journal of Sex Roles,* 62: 9–10.

Yu, L., & D. Zie. (2010). Multidimensional gender identity and psychological adjustment in middle childhood: A study in China. *Sex Roles,* 62: 100–113.

Chapter 4

Abowitz, D., D. Knox, & K. Berner. (2011, March). Traditional and non-traditional husband preference among college women. Paper presented at annual meeting of the Eastern Sociological Society, Philadelphia.

Alford, J. J., P. K. Hatemi, J. R. Hibbing, N. G. Martin, & L. J. Eaves. (2011). The politics of mate choice. *The Journal of Politics,* 73: 362–379.

Allendorf, K. (2013). Schemas of marital change: From arranged marriages to eloping for love. *Journal of Marriage and the Family,* 75: 453–469.

Allgood, S. M., & J. Gordon. (2010, November 3–5). Premarital advice: Do engaged couples listen? Poster presented at the annual meeting of the National Council on Family Relations, Minneapolis, MN.

Amato, P. R., A. Booth, D. R. Johnson, & S. F. Rogers. (2007). *Alone together: How marriage in America is changing.* Cambridge, MA: Harvard University Press.

Barber, L. L., & M. L. Cooper. (2014). Rebound sex: Sexual motives and behaviors following a relationship breakup. *Archives of Sexual Behavior,* 43: 251–265.

Berscheid, E. (2010). Love in the fourth dimension. *Annual Review of Psychology,* 61:1–25.

Blair, S. L. (2010). The influence of risk-taking behaviors on the transition into marriage: An examination of the long-term consequences of adolescent behavior. *Marriage & Family Review,* 46: 126–146.

Blomquist, B. A., & T. A. Giuliano. (2012). Do you love me, too? Perceptions of responses to "I love you." *North American Journal of Psychology,* 14: 407–418.

Bradford, K., W. Stewart, B. Higginbotham, L. Skogrand, & C. Broadbent. (2012, November 1). Validating the Relationship Knowledge Questionnaire. Poster presented at the annual meeting of the National Council on Family Relations, Phoenix, AZ.

Bredow, C. A., T. L. Huston, & G. Noval. (2011). Market value, quality of the pool of potential mates, and singles' confidence about marrying. *Personal Relationships,* 18: 39–57.

Brown, A. (2010, April). How to accurately interpret a peer's social class: Symbols of class status and presentation of self in college students. Paper presented at the Southern Sociological Society, Atlanta, GA.

Brown, P. J., & J. Sweeney. (2009). The anthropology of overweight, obesity and the body. *AnthroNotes,* 30(1).

Burr, B. K, J. Viera, B. Dial, H. Fields, K. Davis, & D. Hubler. (2011, November 18). Influences of personality on relationship satisfaction through stress. Poster presented at the annual meeting of the National Council on Family Relations, Orlando, FL.

Carter, C. S., & S. W. Porges. (2013). The biochemistry of love: An oxyticin hypothesis. *Science & Society, 14*: 12–16.

Carter, G. L., A. C. Campbell, & S. Muncer. (2014). The dark triad personality: Attractiveness to women. *Personality and Individual Differences, 56*: 57–61.

Chaney, C., & K. Marsh. (2009). Factors that facilitate relationship entry among married and cohabiting African Americans. *Marriage & Family Review, 45*: 26–51.

Chapman, G. (2010). *The five love languages: The secret to love that lasts.* Chicago: Northfield Publishing.

Cohn, D. (2013). Love and marriage. Pew Research Center. Retrieved from http://www.pewsocialtrends.org/2013/02/13/love-and-marriage/

Dailey, R. M., A. A. McCracken, B. Jin, K. R. Rossetto, & E. W Green. (2013). Negotiating breakups and renewals: Types of on-again/off-again dating relationships. *Western Journal of Communication, 77*: 382–410.

Diamond, L. M. (2003). What does sexual orientation orient? A biobehavioral model distinguishing romantic love and sexual desire. *Psychological Review, 110*: 173–192.

Dijkstra, P., D. P. H. Barelds, H. A. K. Groothof, S. Ronner, & A. P. Nautal. (2012). Partner preferences of the intellectually gifted. *Marriage & Family Review, 48*: 96–108.

Dubbs, S. L., & A. P. Buunk. (2010). Sex differences in parental preferences over a child's mate choice: A daughter's perspective. *Journal of Social and Personal Relationships, 27*: 1051–1059.

Edwards, T. M. (2000, August 28). Flying solo. *Time*, 47–53.

Elliott, L., B. Easterling, & D. Knox. (2012, March). Taking chances in romantic relationships. Poster presented at the annual meeting of the Southern Sociological Society, New Orleans, LA.

Ellison, C. G., A. M. Burdette, & W. B. Wilcox. (2010). The couple that prays together: Race and ethnicity, religion, and relationship quality among working-age adults. *Journal of Marriage and Family, 72*: 963–975.

Fehr, B., C. Harasymchuk, & S. Sprecher. (2014). Compassionate love in romantic relationships: A review of some new findings. *Journal of Social and Personal Relationships, 31*: 575–600.

Finkenauer, C. (2010). Although it helps, love is not all you need: How Caryl Rusbult made me discover what relationships are all about. *Personal Relationships, 17*: 161–163.

Fisher, H. E., L. L. Brown, A. Aron, G. Strong, & D. Mashek. (2010). Reward, addiction, and emotion regulation systems associated with rejection in love. *Journal of Neurophysiology, 104*: 51–60.

Fisher, M. L., & C. Salmon. (2013). Mom, dad, meet my mate: An evolutionary perspective on the introduction of parents and mates. *Journal of Family Studies, 19*: 99–107.

Foster, J. (2010). How love and sex can influence recognition of faces and words: A processing model account. *European Journal of Social Psychology , 40*: 524–535.

Foster, J. D. (2008). Incorporating personality into the investment model: Probing commitment processes across individual differences in narcissism. *Journal of Social and Personal Relationships, 25*: 211–223.

Freud, S. (1905/1938). Three contributions to the theory of sex. In A. A. Brill (Ed.), *The basic writings of Sigmund Freud.* New York: Random House.

Frye-Cox, N. (2012, November). Alexithymia and marital quality: The mediating role of loneliness. Paper presented at the National Council on Family Relations, Phoenix, AZ.

Gonzaga, G. C., S. Carter, & J. Galen Buckwalter. (2010). Assortative mating, convergence, and satisfaction in married couples. *Personal Relationships, 17*: 634–644.

Greitemeyer, T. (2010). Effects of reciprocity on attraction: The role of a partner's physical attractiveness. *Personal Relationships, 17*: 317–330.

Haandrikman, K. (2011). Spatial homogamy: The geographical dimensions of partner choice. *Journal of Economic and Social Geography, 102*: 100–110.

Hall, S., & D. Knox. (2015). Relationship and sexual behaviors of a sample of 4,730 university students. Unpublished data collected for this text. Department of Family and Consumer Sciences, Ball State University and Department of Sociology, East Carolina University.

Hou, F., & J. Myles. (2013). Interracial marriage and status-caste exchange in Canada and the United States. *Ethnic & Racial Studies, 36*: 75–96.

Huston, T. L., J. P. Caughlin, R. M. Houts, S. E. Smith, & L. J. George. (2001). The connubial crucible: Newlywed years as predictors of marital delight, distress, and divorce. *Journal of Personality and Social Psychology, 80*: 237–252.

Jackson, J. (2011, November). Premarital counseling: An evidence-informed treatment protocol. Paper presented at the annual meeting of the National Council on Family Relations, Orlando, FL.

Jenks, R. J. (2014, July 7). An on-line survey comparing swingers and polyamorists. *Electronic Journal of Human Sexuality, 17*.

Johnson, M. D., & J. R. Anderson. (2011, November). The longitudinal association of marital confidence, time spent together and marital satisfaction. Paper presented at the annual meeting of the National Council on Family Relations, Orlando, FL.

Kem, J. (2010). Fatal lovesickness in Marguerite de Navarre's Heptaméron. *Sixteenth Century Journal, 41*: 355–370.

Kennedy, D. P., J. S. Tucker, M. S. Pollard, M. Go, & H. D. Green. (2011). Adolescent romantic relationships and change in smoking status. *Addictive Behaviors, 36*: 320–326.

Knox, D., & M. Zusman. (2007). Traditional wife? Characteristics of college men who want one. *Journal of Indiana Academy of Social Sciences, 11*: 27–32.

Lambert, N. M., S. Negash, T. F. Stillman, S. B. Olmstead, & F. D. Fincham. (2012). A love that doesn't last: Pornography consumption and weakened commitment to one's romantic partner. *Journal of Social & Clinical Psychology, 31*: 410–438.

Langeslag, S. J. E., P. Muris, & I. H. A. Fraken. (2013). Measuring romantic love: Psychometric properties of the infatuation and attachment scales. *Journal of Sex Research, 50*: 739–774.

Lee, J. A. (1973). *The colors of love: An exploration of the ways of loving.* Don Mills, Ontario: New Press.

———. (1988). Love-styles. In R. Sternberg & M. Barnes (Eds.), *The psychology of love*(pp. 38–67). New Haven, CT: Yale University Press.

Leno, J. (1996). *Leading with my chin.* New York: Harper Collins.

Maier, T. (2009). *Masters of sex.* New York: Perseus Books.

Markman, H. J., G. K. Rhoades, S. M. Stanley, E. P. Ragan, & S. W. Whitton. (2010). The premarital communication roots of marital distress and divorce: The first five years of marriage. *Journal of Family Psychology, 24*: 289–298.

McClure, M. J., & J. E. Lydon. (2014). Anxiety doesn't become you: How attachment anxiety compromises relational opportunities. *Journal of Personality and Social Psychology, 106*: 89–111.

Mcintosh, W.D., L. Locker, K. Briley, R. Ryan, & A. Scott. (2011). What do older adults seek in their potential romantic partners? Evidence from online personal ads. *International Journal of Aging & Human Development, 72*: 67–82.

Meltzer, A. L., & J. K. McNulty. (2014). "Tell me I'm sexy . . . and otherwise valuable": Body valuation and relationship satisfaction. *Personal Relationships, 21*: 68–87.

Moore, D., S. Wigby, S. English, S. Wong, T. Sekely, & F. Harrison. (2013). Selflessness is sexy: Reported helping behavior increases desirability of men and women as long-term sexual partners. *BMC Evolutionary Biology, 13*: 130–182.

Neto, F. (2012). Perceptions of love and sex across the adult life span. *Journal of Social and Personal Relationships, 29*: 760–775.

Oberbeek, G., S. A. Nelemans, J. Karremans, & Rutger C. M. E. Engels. (2012). The malleability of mate selection in speed-dating events. *Archives of Sexual Behavior, 42*: 1163–1171.

Ozay, B., D. Knox, & B. Easterling. (2012, March). You're Dating Who? Parental attitudes toward interracial dating. Poster presented at the annual meeting of the Southern Sociological Society, New Orleans, LA.

Perilloux, C., D. S. Fleischman, & D. M. Buss. (2011). Meet the parents: Parent-offspring convergence and divergence in mate preferences. *Personality & Individual Differences, 50:* 253–258.

Pew Research Center. (2011). The burden of student debt. Retrieved from http://pewresearch.org/databank/dailynumber/?NumberID =1257

Plant, E. A., J. Kunstman, & J. K. Maner. (2010). You do not only hurt the one you love: Self- protective responses to attractive relationship alternatives. *Journal of Experimental Social Psychology, 46:* 474–477.

Randler, C., & S. Kretz. (2011). Assortative mating in morningness–eveningness. *International Journal of Psychology, 46:* 91–96.

Reis, H. T., M. R. Maniaci, & R. D. Rogge. (2014). The expression of compassionate love in everyday compassionate acts. *Journal of Social and Personal Relationships, 31:* 651– 676.

Reiss, I. L. (1960). Toward a sociology of the heterosexual love relationship. *Journal of Marriage and Family Living,* 22:139–145.

Reynaud, M., L. Blecha, & A. Benyamina. (2011). Is love passion an addictive disorder? *American Journal of Drug & Alcohol Abuse, 36:* 261–267.

Richeimer, S. (2011). Love hurts. KABC, Los Angeles. Retrieved from http://abclocal.go.com/kabc/story?section=news/health/your _health&id=8039618

Riela, S., G. Rodriguez, A. Aron, X. Xu, & B. P. Acevedo. (2010). Experiences of falling in love: Investigating culture, ethnicity, gender, and speed. *Journal of Social and Personal Relationships,* 27: 473–493.

Ross, C. B. (2006, March 24). An exploration of eight dimensions of self-disclosure on relationship. Paper presented at the Southern Sociological Society, New Orleans, LA.

Sack, K. (2008, August 12). Health benefits inspire rush to marry, or divorce. *The New York Times.*

Sassler, S. (2010). Partnering across the life course: Sex, relationships, and mate selection. *Journal of Marriage and Family, 72:* 557–575.

Scheff, E. (2014). *The polyamorists next door.* Lanham, MD: Rowman & Littlefield.

Shelon, J. N., T. E. Trail, T. V. West, & H. B. Bergsieker. (2010). From strangers to friends: The interpersonal process model of intimacy in developing interracial friendships. *Journal of Social and Personal Relationships,* 27: 71–90.

Sprecher, S. (2002). Sexual satisfaction in premarital relationships: Associations with satisfaction, love, commitment, and stability. *Journal of Sex Research, 39:* 190–196.

Stanik, C. E., S. M. McHale, & A. C. Couter. (2013). Gender dynamics predict changes in marital love among African-American couples. *Journal of Marriage & Family, 75:* 795– 798.

Stanley, S. M., E. P. Ragan, G. K. Rhoades, & H. J. Markman. (2012). Examining changes in relationship adjustment and life satisfaction in marriage. *Journal of Family Psychology, 26:* 165–170.

Starr, L.R., J. Davila, C. B. Stroud, P. C. Clara Li, A. Yoneda, R. Hershenberg, & M. Ramsay Miller. (2012). Love hurts (in more ways than one): Specificity of psychological symptoms as predictors and consequences of romantic activity among early adolescent girls. *Journal of Clinical Psychology, 68:* 373–381.

Statistical Abstract of the United States, 2012–2013, 131th ed. Washington, DC: U.S. Bureau of the Census.

Sternberg, R. J. (1986). A triangular theory of love. *Psychological Review,* 93:119–135.

Stinehart, M. A., D. A. Scott, & H. G. Barfield. (2012). Reactive attachment disorder in adopted and foster care children: Implications for mental health professionals. *The Family Journal: Counseling and Therapy for Couples and Families, 20:* 355–360.

Toufexis, A. (1993, February15). The right chemistry. *Time,* 49–51.

Tsunokai, G. T., & A. R. McGrath. (2011). Baby boomers and beyond: Crossing racial boundaries in search for love. *Journal of Aging Studies,* 25: 285–294.

Tzeng, O. C. S., K. Wooldridge, & K. Campbell. (2003). Faith love: A psychological construct in intimate relations. *Journal of the Indiana Academy of the Social Sciences,* 7:11–20.

Vennum, A. (2011). Understanding young adult cyclical relationships. Dissertation, Florida State University, College of Home Economics.

Walsh, S. (2013). Match's 2012 singles in America survey. Retrieved from http://www.hookingupsmart.com/2013/02/07/hookinguprealities /matchs-2012-singles-in-america-survey/

Warren, J. T., S. M. Harvey, & C. R. Agnew. (2012). One love: Explicit monogamy agreements among heterosexual young couples at increased risk of sexually transmitted infections. *Journal of Sex Research ,* 49: 282–289.

Wiersma, J. D., J. L. Fischer, B. C. Bray, & J. P. Clifton. (2011, November 18). Young adult drinking partnerships: Where are couples 6 years later? Poster presented at the annual meeting of the National Council on Family Relations, Orlando, FL.

Yodanis, C., S. Lauer, & R. Ota. (2012). Interethnic romantic relationships: Enacting affiliative ethnic identities. *Journal of Marriage and Family,* 74: 1021–1037.

Yoo, H. C., M. F. Steger, & R. M. Lee. (2010). Validation of the subtle and blatant racism scale for Asian American college students (SABR-A^2). *Cultural Diversity and Ethnic Minority Psychology,* 16: 323–334.

Zamora, R., C. Winterowd, J. Koch, & S. Roring. (2013). The relationship between love styles and romantic attachment styles in gay men. *Journal of LGBT Issues in Counseling,* 7: 200–217.

Chapter 5

Aponte, R., & R. Pessagno. (2010). The communications revolution and its impact on the family: Significant, growing, but skewed and limited in scope. *Marriage & Family Review,* 45: 576–586.

Balderrama-Durbin, C. M., E. S. Allen, & G. K. Rhoades. (2012). Demand and withdraw behaviors in couples with a history of infidelity. *Journal of Family Psychology,* 26: 11– 17.

Bauerlein, M. (2010). Literary learning in the hyperdigital age. *Futurist,* 44: 24–25.

Bergdall, A. R., J. M. Kraft, K. Andes, M. Carter, K. Hatfield-Timajchy, & L. Hock-long. (2012). Love and hooking up in the new millennium: Communication technology and relationships among urban African American and Puerto Rican young adults. *Journal of Sex Research,* 49: 570–582.

Burke-Winkelman, S., K. Vail-Smith, J. Brinkley, & D. Knox. (2014, February 3). Sexting on the college campus. *Electronic Journal of Human Sexuality,* 17.

Carey, A. R., & V. Bravo. (2014, July 10). Obsessed with your cell phone? *USA Today,* p. 1A.

Carey, A. R., & V. Salazar. (2011, February 1). Women talk and text more. *USA Today,* p. 1.

Coyne, S. M., L. Stockdale, D. Busby, B. Iverson, & D. M. Grant. (2011). "I luv u :)!": A descriptive study of the media use of individuals in romantic relationships. *Family Relations,* 60: 150–162.

Dir, A. L., A. Coskunpinar, J. L. Steiner, & M. A. Cyders. (2013). Understanding differences in sexting behaviors across gender, relationship status, and sexual identity, and the role of expectancies in sexting. *Cyberpsychology, Behavior, and Social Networking,* 16: 568–574.

Easterling, B., D. Knox, & A. Brackett. (2012). Secrets in romantic relationships: Does sexual orientation matter? *Journal of GLBT Family Studies,* 8: 198–210.

Frye, N. E. (2011). Responding to problems: The roles of severity and barriers. *Personal Relationships,* 18: 471–486.

Furukawa, R., & M. Driessnack. (2013). Video-mediated communication to support distant family connectedness. *Clinical Nursing Research,* 22: 82–94.

Gallmeier, C. P., M. E. Zusman, D. Knox, & L. Gibson. (1997). Can we talk? Gender differences in disclosure patterns and expectations. *Free Inquiry in Creative Sociology, 25*: 129–225.

Ganong, L. H., M. Coleman, R. Feistman, T. Jamison, & M. S. Markham. (2011, November 18). Communication technology and post-divorce co-parenting. Poster presented at the annual meeting of the National Council on Family Relations, Orlando, FL.

Garfield, R. (2010). Male emotional intimacy: How therapeutic men's groups can enhance couples therapy. *Family Process, 49*: 109–122.

Gibbs, N. (2012, August 27). Your life is fully mobile: Time Mobility Survey. *Time Magazine,* 32 and following.

Gershon, I. (2010). *The breakup 2.0.* New York: Cornell University Press.

Gottman, J. (1994). *Why marriages succeed or fail.* New York: Simon & Schuster.

Hall, S., & D. Knox. (2015). Relationship and sexual behaviors of a sample of 4,730 university students. Unpublished data collected for this text. Department of Family and Consumer Sciences, Ball State University and Department of Sociology, East Carolina University.

Hertenstein, M. J., M. J. Hertenstein, J. M. Verkamp, A. M. Kerestes, & R. M. Holmes. (2007). The communicative functions of touch in humans, nonhuman primates, and rats: A review and synthesis of the empirical research. *Genetic Social and General Psychology Monographs, 132*: 5–94.

Hill, E. W. (2010). Discovering forgiveness through empathy: Implications for couple and family therapy. *Journal of Family Therapy, 32*: 169–185.

Impett, E. A., J. B. Breines, & A. Strachman. (2010). Keeping it real: Young adult women's authenticity in relationships and daily condom use. *Personal Relationships, 17*: 573–584.

Huang, H., & L. Leung. (2010). Instant messaging addiction among teenagers in China: Shyness, alienation and academic performance decrement. *CyberPsychology & Behavior, 12*: 675–679.

Kelley, K. (2010). *Oprah: A biography.* New York: Crown Publishers.

Kimmes, J. G., A. B. Edwards, J. L. Wetchler, & J. Bercik. (2014, June 18). Self and other ratings of dyadic empathy as predictors of relationship satisfaction. *The American Journal of Family Therapy.* Online.

King, A. L. S., A. M. Valenca, A. C. O. Silva, T. Baczynski, M. R. Carvalho, & A. E. Nardi. (2013). Nomophobia: dependency on virtual environments or social phobia? *Computers in Human Behavior, 29*: 140–144.

Kurdek, L. A. (1994). Areas of conflict for gay, lesbian, and heterosexual couples: What couples argue about influences relationship satisfaction. *Journal of Marriage and the Family, 56*: 923–934.

———. (1995). Predicting change in marital satisfaction from husbands' and wives' conflict resolution styles. *Journal of Marriage and the Family, 57*: 153–164.

Lavner, J. A., & T. N. Bradbury. (2010). Patterns of change in marital satisfaction over the newlywed years. *Journal of Marriage and Family, 72*: 1171–1187.

———. (2012). Why do even satisfied newlyweds eventually go on to divorce? *Journal of Family Psychology, 26*: 1–10.

Looi, C., P. Seow, B. Zhang, H. So, W. Chen, & L. Wong. (2010). Leveraging mobile technology for sustainable seamless learning: A research agenda. *British Journal of Educational Technology, 41*: 154–169.

Lund, R., U. Christensen, C. J. Nilsson, M. Kriegbaum, & N. H. Rod. (2014, May). Stressful social relations and mortality: A prospective cohort study. *Journal of Epidemiology and Community Health.* Online.

Maatta, K., & S. Uusiautti. (2013). Silence is not golden: Review of studies of couple interaction. *Communication Studies, 64*: 33–48.

Marano, H. E. (1992, January/February). The reinvention of marriage. *Psychology Today,* 49.

Markman, H. J., G. K. Rhoades, S. M. Stanley, E. P. Ragan, & S. W. Whitton. (2010a). The premarital communication roots of marital distress and divorce: The first five years of marriage. *Journal of Family Psychology, 24*: 289–298.

Markman, H. J., S. M. Stanley, & S. L. Blumberg. (2010b). *Fighting for your marriage* (3rd ed.). San Francisco: Jossey-Bass.

Marshall, T. C., K. Chuong, & A. Aikawa. (2011). Day-to-day experiences of amae in Japanese romantic relationships. *Asian Journal of Social Psychology, 14*: 26–35.

Merolla, A. J., & S. Zhang. (2011). In the wake of transgressions: Examining forgiveness communication in personal relationships. *Personal Relationships, 18*: 79–95.

Moore, M. M. (2010). Human nonverbal courtship behavior—A brief historical review. *Journal of Sex Research, 47*: 171–180.

Norton, A. M, & J. Baptist. (2012, November). Couple boundaries for social networking: Impact of trust and satisfaction. Paper presented at the National Council on Family Relations, Phoenix, AZ.

Parker, M., D. Knox, & B. Easterling. (2011, February 24–26). SEXTING: Sexual content/images in romantic relationships. Poster presented at the Eastern Sociological Society, Philadelphia.

Perry, M. S., & R. J. Werner-Wilson. (2011). Couples and computer-mediated communication: A closer look at the affordances and use of the channel. *Family & Consumer Sciences Research Journal, 40*: 120–134.

Pettigrew, J. (2009). Text messaging and connectedness within close interpersonal relationships. *Marriage & Family Review, 45*: 697–716.

Purkett, T. (2014, Fall). Sexually transmitted infections. Presented to Courtship and Marriage class, Department of Sociology, East Carolina University.

Rappleyea, D. L., A. C. Taylor, & X Fang. (2014). Gender differences and communication technology use among emerging adults in the initiation of dating relationships. *Marriage & Family Review, 50*: 269–284.

Reich, S. M., K. Subrahmanyam, & G. Espionoza. (2012). Friending, IMing, and hanging out face-to-face: Overlap in adolescents' online and offline social networks. *Developmental Psychology, 48*: 356–368.

Sanford, K., & A. J. Grace. (2011). Emotion and underlying concerns during couples' conflict: An investigation of within-person change. *Personal Relationships, 18*: 96–109.

Schade, L. C., J. Sandberg, R. Bean, D. Busby, & S. Coyne. (2013). Using technology to connect in romantic relationships: Effects on attachment, relationship satisfaction, and stability in emerging adults. *Journal of Couple & Relationship Therapy: Innovations in Clinical and Educational Interventions, 12*: 314–338.

Sharpe, A. (2012). *Transgender marriage and the legal obligation to disclose gender history. Modern Law Review, 75*: 33–53.

South, S. C., B. D. Doss, & A. Christensen. (2010). Through the eyes of the beholder: The mediating role of relationship acceptance in the impact of partner behavior. *Family Relations, 59*: 611–622.

Strassberg, D. S., R. K. McKinnon, M. A. Sustaita, & J. Rullo. (2013). Sexting by high school students: An exploratory and descriptive study. *Archives of Sexual Behavior, 42*: 15–21.

Strickler, B. L., & J. D. Hans. (2010, November 3–5). Defining infidelity and identifying cheaters: An inductive approach with a factorial design. Poster presented at the annual meeting of the National Council on Family Relations, Minneapolis, MN.

Tan, R., N. C. Overall, & J. K. Taylor. (2012). Let's talk about us: Attachment, relationship-focused disclosure, and relationship quality. *Personal Relationships, 19*: 521–534.

Tannen, D. (1990). *You just don't understand: Women and men in conversation.* London: Virago.

———. (1998). *The argument culture.* New York: Random House.

———. (2006). *You're wearing that? Understanding mothers and daughters in conversation.* New York: Random House.

Taylor, A. C., D. L. Rappelea, X. Fang, & D. Cannon. (2013). Emerging adults' perceptions of acceptable behaviors prior to forming a committed, dating relationship. *Journal of Adult Development, 20*: 173–184.

Temple, J. R., J. A. Paul, P. van den Berg, V. D. Le, Am McElhany, & B. W. Temple. (2012). Teen sexting and its association with sexual behaviors. *Archives of Pediatrics and Adolescent Medicine, 166:* 828–833.

Vail-Smith, K., L. MacKenzie, & D. Knox. (2010). The illusion of safety in "monogamous" undergraduates. *American Journal of Health Behavior, 34:* 15–20.

Waller, W., & R. Hill. (1951). *The family: A dynamic interpretation.* New York: Holt, Rinehart and Winston.

Walsh, S. (2013). Match's 2012 singles in America survey. Retrieved from http://www.hookingupsmart.com/2013/02/07/hookinguprealities/matchs-2012-singles-in-america-survey/

Webley, K. (2012, September 17). Cheating Harvard. *Time,* 22.

Weisskirch, R. S., & R. Delevi. (2011). "Sexting" and adult romantic attachment. *Computers in Human Behavior, 27:* 1697–1701.

Wheeler, L. A., K. A. Updegraff, & S. M. Thayer. (2010). Conflict resolution in Mexican-origin couples: Culture, gender, and marital quality. *Journal of Marriage and the Family, 72:* 991–1005.

White, S. S., N. El-Bassel, L. Gilbert, E. Wu, & M. Chang. (2010). Lack of awareness of partner STD risk among heterosexual couples. *Perspectives on Sexual & Reproductive Health, 42:* 49–55.

Wickrama, K. A. S., C. M. Bryant, & T. K. Wickrama. (2010). Perceived community disorder, hostile marital interactions, and self-reported health of African American couples: An interdyadic process. *Personal Relationships, 17:* 515–531.

Woszidlo, A., & C. Segrin. (2013). Negative affectivity and educational attainment as predictors of newlyweds' problem solving communication and marital quality. *Journal of Psychology, 147:* 49–73.

Xu, K. (2013). Theorizing difference in intercultural communication: A critical dialogic perspective. *Communication Monographs, 80:* 379–397.

Chapter 6

Ali, L. (2011, November). Sexual self disclosure predicting sexual satisfaction. Poster presented at the National Council on Family Relations, Orlando, FL.

Ballard, S. M., C. Sugita, & K. H. Gross. (2011, November 17). A qualitative investigation of the sexual socialization of emerging adults. Paper presented at the 73rd Conference of the National Council on Family Relations, Orlando, FL.

Bersamin, M. M., B. L. Zamboanga, S. J. Schwartz, M. B. Donnellan, M. Hudson, R. S. Weisskirch, S. Y. Kim, V. B. Agocha, S. K. Whitbourne, & J. Caraway. (2014). Risky business: Is there an association between casual sex and mental health among emerging adults? *The Journal of Sex Research, 51:* 43–51.

Brown, C. C., J. S. Carroll, D. M. Busby, B. J. Willoughby, & RELATE Institute Brigham Young University. (2013, November). The pornography gap: Differences in men's and women's pornography patterns in couple relationships. Poster presented at the annual meeting of the National Council on Family Relations, San Antonio, TX.

Brucker, H., & P. Bearman. (2005). After the promise: The STD consequences of adolescent virginity pledges. *Journal of Adolescent Health, 36:* 271–278.

DeLamater, J., & M. Hasday. (2007). The sociology of sexuality. In Clifton D. Bryant & Dennis L. Peck (Eds.), *21st century sociology: A reference handbook* (pp. 254–264). Thousand Oaks, CA: Sage.

Dotson-Blake, K. P., D. Knox, & M. Zusman. (2012). Exploring social sexual scripts related to oral sex: A profile of college student perceptions. *The Professional Counselor, 2:* 1–11. Online journal at http://tpcjournal.nbcc.org/

Fennell, J. (2014, March 18). Dungeon rules: Normalizing kinky desires. Presentation to Sociology of Human Sexuality class, Department of Sociology, East Carolina University, Greenville, NC.

Fielder, R. L., J. L. Walsh, K. B. Carey, & M. P. Carey. (2014). Sexual hookups and adverse health outcomes: A longitudinal study of first year college women. *The Journal of Sex Research, 51:* 131–144.

Galinsky, A. M., & F. L. Sonenstein. (2013). Relationship commitment, perceived equity, and sexual enjoyment among young adults in the United States. *Archives of Sexual Behavior, 42:* 93–104.

Hall, S., & D. Knox. (2015). Relationship and sexual behaviors of a sample of 4,689 university students. Unpublished data collected for this text. Department of Family and Consumer Sciences, Ball State University, Muncie, IN and Department of Sociology, East Carolina University, Greenville, NC.

Hawes, Z. C., K. Wellings, & J. Stephenson. (2010). First heterosexual intercourse in the United Kingdom: A review of the literature. *Journal of Sex Research, 47:* 137–152.

Hendrickx, L. G., & P. Enzlin. (2014). Associated distress in heterosexual men and women: Results from an internet survey in Flanders. *The Journal of Sex Research, 51:* 1–12.

Herbenick, D., M. Reece, V. Schick, K. N. Jozkowski, S. E. Middelstadt, S. A. Sanders, B. S. Dodge, A. Ghassemi & J. D. Fortenberry (2011). Beliefs about women's vibrator use: Results from a nationally representative probability survey in the United States. *Journal of Sex & Marital Therapy, 37:* 329–345.

Herbenick, D., M. Reece, S. A. Sanders, B. Dodge, A. Ghassemi, & D. Fortenberry. (2010). Women's vibrator use in sexual partnerships: Results from a nationally representative survey in the United States. *Journal of Sex & Marital Therapy, 36:* 49–65.

Horowitz, A. D., & L. Spicer. (2013). Definitions among heterosexual and lesbian emerging adults in the U. K. *Journal of Sex Research, 50:* 139–150.

Humphreys, T. P. (2013). Cognitive frameworks of virginity and first intercourse. *Journal of Sex Research, 50:* 664–675.

Kimberly, C., & J. Hans. (2012, November). Sexual self-disclosure and communication among swinger couples. Poster presented at the National Council on Family Relations, Phoenix, AZ.

Landor, A., & L. G. Simons. (2010, November 3–6). The impact of virginity pledges on sexual attitudes and behaviors among college students. Paper presented at the annual meeting of the National Council on Family Relations, Minneapolis, MN.

Lee, J. T., C. L. Lin, G. H. Wan, & C. C. Liang. (2010). Sexual positions and sexual satisfaction of pregnant women. *Journal of Sex & Marital Therapy, 36:* 408–420.

Lehmiller, J. J., L. E. VanderDrift, & J. R. Kelly. (2014). Sexual communication, satisfaction, and condom use behavior in friends with benefits and romantic partners. *Journal of Sex Research, 51:* 74–85.

Levine, S. B. (2010). What is sexual addiction? *Journal of Sex & Marital Therapy, 36:* 261–275.

Malacad, B., & G. Hess. (2010). Oral sex: Behaviours and feelings of Canadian young women and implications for sex education. *European Journal of Contraception & Reproductive Health Care, 15:* 177–185.

Masters, W. H., & V. E. Johnson. (1970). *Human sexual inadequacy.* Boston, MA: Little, Brown.

McBride, K., & J. D. Fortenberry. (2010). Heterosexual anal sexuality and anal sex behaviors: A review. *Journal of Sex Research, 47:* 123–136.

McCabe, M. P., & D. L. Goldhammer. (2012). Demographic and psychological factors related to sexual desire among heterosexual women in a relationship. *Journal of Sex Research, 49:* 78–87.

Michael, R. T., J. H. Gagnon, E. O. Laumann, & G. Kolata. (1994). *Sex in America.* Boston, MA: Little, Brown.

Miller, S., A. Taylor, & D. Rappleyea. (2011, April 4–8). The influence of religion on young adult's attitudes of dating events. Poster presented at the Fifth Annual Research & Creative Achievement Week, East Carolina University.

Mongeau, P. A., K. Knight, J. Williams, J. Eden, & C. Shaw. (2013). Identifying and explicating variation among friends with benefits relationships. *Journal of Sex Research, 50:* 37–47.

Montesi, J., B. Conner, E. Gordon, R. Fauber, K. Kim, & R. Heimberg. (2013). On the relationship among social anxiety, intimacy, sexual communication, and sexual satisfaction in young couples. *Archives of Sexual Behavior*, 42: 81–91.

Paik, A. (2010). The contexts of sexual involvement and concurrent sexual partnerships. *Perspectives on Sexual and Reproductive Health*, 42: 33–43.

Penhollow, T. M., A. Marx, & M. Young. (2010, March 31). Impact of recreational sex on sexual satisfaction and leisure satisfaction. *Electronic Journal of Human Sexuality, 13*.

Porter, Nora. (2014, April). Slut-shaming: The double standard on campus. Paper presented at the annual meeting of the Southern Sociological Society, Charlotte, NC.

Reece, M., D. Herbenick, B. Dodge, S. A. Sanders, A. Ghassemi, & D. Fortenberry. (2010). Vibrator use among heterosexual men varies by partnership status: Results from a nationally representative study in the United States. *Journal of Sex & Marital Therapy*, 36: 389–407.

Reissing, E. D., H. L. Andruff, & J. J. Wentland. (2012). Looking back: The experience of first sexual intercourse and current sexual adjustment in young heterosexual adults. *Journal of Sex Research*, 49: 27–35.

Sanchez, D. T., J. C. Fetterolf, & L. A. Rudman. (2012). Eroticizing inequality in the United States: The consequences and determinants of traditional gender role adherence in intimate relationships. *Journal of Sex Research*, 49: 168–183.

Sandberg-Thoma, S. E., & C. M. Kamp Dush. (2014). Casual sexual relationships and mental health in adolescence and emerging adulthood. *Journal of Sex Research*, 51: 121–130.

Scimeca, G., A. Bruno, G. Pandolfo, U. Mico, V. M. Romeo, E. Abenavoli, A. Schjimmenti, R. Zoccali, & M. R. A. Muscatello. (2013). Alexithymia, negative emotions, and sexual behavior in heterosexual university students from Italy. *Archives of Sexual Behavior*, 42:117–127.

Simms, D. C., & E. S. Byers. (2013). Heterosexual dater's sexual initiation behaviors: Use of the theory of planned behavior. *Archives of Sexual Behavior*, 42: 105–116.

Sprecher, S. (2014, March10). Evidence of change in men's versus women's emotional reactions to first sexual intercourse: A 23-year study in a human sexuality course at a midwestern university. *The Journal of Sex Research*. Online.

Stephenson, K. R., A. H. Rellini, & C. M. Meston. (2013). Relationship satisfaction as a predictor of treatment response during cognitive behavioral sex therapy. *Archives of Sexual Behavior*, 42: 143–152.

Stulhofer, A., & D. Ajdukovic. (2011). Should we take anodyspareunia seriously? A descriptive analysis of pain during receptive anal intercourse in young heterosexual women. *Journal of Sex and Marital Therapy*, 37: 346–358.

Stulhofer, A., B. Traeen, & A. Carvalheira. (2013). Job-related strain and sexual health difficulties among heterosexual men from three European countries: The role of culture and emotional support. *Journal of Sexual Medicine*, 10: 747–756.

Symons, K., H. Vermeersch, and M. Van Houtte. (2014). The emotional experiences of early first intercourse: A multi-method study. *Journal of Adolescent Research*, 29: 533–560.

Thomsen, D., & I. J. Chang. (2000, November). Predictors of satisfaction with first intercourse: A new perspective for sexuality education. Poster presented at the 62nd Annual Conference of the National Council on Family Relations, Minneapolis, MN.

True Love Waits. (2014). Retrieved April 29 from http://www.lifeway.com/ArticleView?storeId=10054&catalogId=10001&langId=-1&article=true-love-waits

Van den Brink, F., M. A. M. Smeets, D. J. Hessen, J. G. Talens, & L. Woertman. (2013). Body satisfaction and sexual health in Dutch female university students. *Journal of Sex Research*, 50: 786–794.

Vannier, S. A., & L. F. O'Sullivan. (2010). Sex without desire: Characteristics of occasions of sexual compliance in young adults' committed relationships *Journal of Sex Research*, 47: 429–439.

Vasilenko, S. A., E. S. Lefkowitz, & J. L. Maggs. (2012). Short-term positive and negative consequences of sex based on daily reports among college students. *Journal of Sex Research*, 49: 558–569.

Vazonyi, A. I., & D. D. Jenkins. (2010). Religiosity, self-control, and virginity status in college students from the "Bible belt": A research note. *Journal for the Scientific Study of Sex*, 49: 561–568.

Walsh, S. (2013). Match's 2012 singles in America survey. Retrieved from http://hookingupsmart.com/2013/02/07/hookinguprealities/matchs-2012-singles-in-america-survey/

Walsh, J. L., R. L. Fielder, K. B. Carey, & M. P. Carey. (2013). Changes in women's condom use over the first year of college. *Journal of Sex Research*, 50: 128–138.

Walsh, J. L., & L. M. Ward. (2010). Magazine reading and involvement and young adults' sexual health knowledge, efficacy, and behaviors. *Journal of Sex Research*, 47: 285–300.

Wester, K. A., & A. E. Phoenix. (2013, April 8–12). Are there really rules and expectations in talking relationships? Gender differences in relationship formation among young adults. Research and Creative Week, East Carolina University, Greenville, NC.

Wetherill, R. R., D. J. Neal, & K. Fromme. (2010). Parents, peers, and sexual values influence sexual behavior during the transition to college. *Journal Archives of Sexual Behavior*, 39: 682–694.

Woertman, L., & F. Van den Brink. (2012). Body image and female sexual functioning and behavior: A review. *Journal of Sex Research*, 49: 184–211.

Wright, P. J., A. K. Randall, & J. G. Hayes. (2012). Predicting the condom assertiveness of collegiate females in the United States from the expanded health belief model. *International Journal of Sexual Health*, 24: 137–153.

Chapter 7

Adolfsen, A., J. Iedema, & S. Keuzenkamp. (2010). Multiple dimensions of attitudes about homosexuality: Development of a multifaceted scale measuring attitudes toward homosexuality. *Journal of Homosexuality*, 57: 1237–1257.

Amato, Paul R. (2004). Tension between institutional and individual views of marriage. *Journal of Marriage and Family*, 66: 959–965.

Becker, A. B., & M. E. Todd. (2013). A new American Family? Public opinion toward family studies and perceptions of the challenges faced by children of same-sex parents. *Journal of GLBT Family Studies*, 9: 425–448.

Bergman, K., R. J. Rubio, R. J. Green, & E. Padron. (2010). Gay men who become fathers via surrogacy: The transition to parenthood. *Journal of GLBT Family Studies*, 6: 111–141.

Biblarz, T. J., & E. Savci. (2010). Lesbian, gay, bisexual, and transgender families. *Journal of Marriage and Family*, 72: 480–497.

Biblarz, T. J., & J. Stacey. (2010). How does the gender of parents matter? *Journal of Marriage and Family*, 72: 3–22.

Blackwell, C. W., & S. F. Dziegielewski. (2012). Using the Internet to meet sexual partners: Research and practice implications. *Journal of Social Service Research*, 38: 46–55.

Bos, H., & T. G. M. Sandfort. (2010). Children's gender identity in lesbian and heterosexual two-parent families. *Journal of Sex Roles*, 62: 114–126.

Centers for Disease Control and Prevention. (2012). HIV incidence. Retrieved March 21, 2012, from http://www.cdc.gov/hiv/topics/surveillance/incidence.htm

Chase, L. M. (2011). Wives' tales: The experience of trans partners. *Journal of Gay & Lesbian Social Services*, 23: 429–451.

Chonody, J. M., K. S. Smith, & M. A. Litle. (2012). Legislating unequal treatment: An exploration of public policy on same-sex marriage. *Journal of GLBT Family Studies*, 8: 270–286.

Clunis, D. M., & G. Dorsey Green. (2003). *The lesbian parenting book*, 2nd ed. Emeryville, CA: Seal Press.

Crowl, A., S. Ahn, & J. Baker. (2008). A meta-analysis of developmental outcomes for children of same-sex and heterosexual parents. *Journal of GLBT Family Studies, 4:* 385–407.

D'Amico, E., & D. Julien. (2012). Disclosure of sexual orientation and gay, lesbian, and bisexual youth's adjustment: Associations with past and current parental acceptance and rejection. *Journal of GLBT Family Studies, 8:* 215–242.

Davis-Delano, L. R. (2014, June 2). Characteristics of activities that affect the development of women's same-sex relationships. *Journal of Homosexuality.* Online.

Denes, A., & T. D. Afifi. (2014). Coming out again: Exploring GLBQ individuals' communication with their parents after the first coming out. *Journal of GLBT Family Studies, 10:* 298–325.

Doaring, C. (2014, February 10). Justice department extends benefits. *USA Today,* p. A1.

Ducharme, J. K., & M. M. Kollar. (2012). Does the "marriage benefit" extend to same-sex union?: Evidence from a sample of married lesbian couples in Massachusetts. *Journal of Homosexuality, 59:* 580–591.

Esmaila, A. (2010). Negotiating fairness: A study on how lesbian family members evaluate, construct, and maintain "fairness" with the division of household labor. *Journal of Homosexuality, 57:* 591–609.

Finneran, C., & R. Stephenson. (2014). Intimate partner violence, minority stress, and sexual risk-taking among U.S. men who have sex with men. *Journal of Homosexuality, 61:* 288–306.

Gartrell, N., H. M. W. Bos, H. Peyser, A. Deck, & C. Rodas. (2012). Adolescents with lesbian mothers describe their own lives. *Journal of Homosexuality, 59:* 1211–1229.

Gates, G. J. (2011). How many people are lesbian, gay, bisexual and transgender? The Williams Institute, UCLA School of Law. Retrieved from http://williamsinstitute.law.ucla.edu/wp-content/uploads/Gates-How-Many-People-LGBT-Apr-2011.pdf

Gilla, D. L., R. G. Morrow, K. E. Collinsc, A. B. Lucey, & A. M. Schultze. (2010). Perceived climate in physical activity settings. *Journal of Homosexuality, 57:* 895–913.

Glass, V. Q. (2014). "We are with family": Black lesbian couples negotiate rituals with extended families. *Journal of GLBT Family Studies, 10:* 79–100.

Goldberg, A. E., & K. R. Allen. (2013). Same-sex relationship dissolution and LGB stepfamily formation: Perspectives of young adults with LGB parents. *Family Relations, 62:* 529–544.

Goldberg, A. E., J. B. Downing, & A. M. Moyer. (2012). Why parenthood, and why now? Gay men's motivations for pursuing parenthood. *Family Relations, 61:* 157–174.

Goldberg, A. E., J. Z. Smith, & D. A. Kashy. (2010). Preadoptive factors predicting lesbian, gay, and heterosexual couples' relationship quality across the transition to adoptive parenthood. *Journal of Family Psychology, 24:* 221–232.

Gonzales, G. (2014). Same-sex marriage: A prescription for better health. *New England Journal of Medicine, 370:* 1373–1376.

Gonzalez, K. A., S. S. Rostosky, R. D. Odom, & E. D. B. Riggle. (2013). The positive aspects of being the parent of an LGBTQ child. *Family Process, 52:* 325–337.

Gottman, J. M., R. W. Levenson, C. Swanson, K. Swanson, R. Tyson, & D. Yoshimoto. (2003). Observing gay, lesbian, and heterosexual couples' relationships: Mathematical modeling of conflict interaction. *Journal of Homosexuality, 45:* 65–91.

Grafsky, E. L. (2014). Becoming the parent of a GLB son or daughter. *Journal of GLBT Family Studies, 10:* 36–57.

Hall, S., & D. Knox. (2015). Relationship and sexual behaviors of a sample of 4,711 university students. Unpublished data collected for this text. Department of Family and Consumer Sciences, Ball State University, Muncie, IN and Department of Sociology, East Carolina University, Greenville, NC.

Human Rights Campaign. (2014). Retrieved from http://www.hrc.org/Iantiffi, A., & W. O. Bockting. (2011). View from both sides of the bridge? Gender, sexual legitimacy and transgender people's experiences of relationships. *Culture, Health & Sexuality, 13:* 355–370.

Israel, T., & J. J. Mohr. (2004). Attitudes toward bisexual women and men: Current research, future directions. In R. C. Fox (Ed.), *Current research on bisexuality* (pp. 117–134). New York: Harrington Park Press.

Johnson, C. W., A. A. Singh, & M. Gonzalez. (2014). "It's complicated": Collective memories of transgender, queer, and questioning youth in high school. *Journal of Homosexuality, 61:* 419–434.

Jones, S. L., & M. A. Yarhouse. (2011). A longitudinal study of attempted religiously mediated sexual orientation change. *Journal of Sex & Marital Therapy, 37:* 404–427.

Karten, E. Y., & J. C. Wade. (2010). Sexual orientation change efforts in men: A client perspective. *Journal of Men's Studies, 18:* 84–102.

Kinsey, A. C., W. B. Pomeroy, & C. E. Martin. (1948). *Sexual behavior in the human male.* Philadelphia, PA: Saunders.

Kinsey, A. C., W. B. Pomeroy, C. E. Martin, & P. H. Gebhard. (1953). *Sexual behavior in the human female.* Philadelphia, PA: Saunders.

Kuper, L. E., R. Nussbaum, & B. Mustanski. (2004). Gay men and lesbians: The family context. In M. Coleman & L. H. Ganong (Eds.), *Handbook of contemporary families: Considering the past, contemplating the future* (pp. 96–115). Thousand Oaks, CA: Sage Publications.

_____. (2005). What do we know about gay and lesbian couples? *Current Directions in Psychological Science, 14:* 251–254.

_____. (2008). Change in relationship quality for partners from lesbian, gay male, and heterosexual couples. *Journal of Family Psychology, 22:* 701–711.

_____. (2012). Exploring the diversity of gender and sexual orientation identities in an online sample of transgender individuals. *Journal of Sex Research, 49:* 244–254.

Kurdek, L. A. (1994). Conflict resolution styles in gay, lesbian, heterosexual nonparent, and heterosexual parent couples. *Journal of Marriage and the Family, 56:* 705–722.

Lalicha, J., & K. McLaren. (2010). Inside and outcast: Multifaceted stigma and redemption in the lives of gay and lesbian Jehovah's Witnesses. *Journal of Homosexuality, 57:* 1303–1333.

Leddy, A., N. Gartrell, & H. Bos. (2012). Growing up in a lesbian family: The life experience of adult daughters and sons of lesbian mothers. *Journal of GLBT Family Studies, 8:* 243–257.

Lee, T., & G. R. Hicks. (2011). An analysis of factors affecting attitudes toward same-sex marriage: Do the media matter? *Journal of Homosexuality, 58:* 1391–1408.

Lehman, A. D. (2010). Inappropriate injury: The case for barring consideration of a parent's homosexuality in custody actions. *Family Law Quarterly, 44:* 115–123.

Lever, J. (1994, August 23). The 1994 *Advocate* survey of sexuality and relationships: The men. *The Advocate,* pp. 16–24.

Lyons, M., A. Lynch, G. Brewer, & D. Bruno. (2014). Detection of sexual orientation ("gaydar") by homosexual and heterosexual women. *Archives of Sexual Behavior, 43:* 345–352.

McIntyre, S. L., E. A. Antonucci, & S. C. Haden. (2014). Being white helps: Intersections of self-concealment, stigmatization, identity formation, and psychological distress in racial and sexual minority women. *Journal of Lesbian Studies, 18:* 158–173.

McLaren, S., P. M. Gibbs, & E. Watts. (2013). The interrelations between age, sense of belonging, and depressive symptoms among Australian gay men and lesbians. *Journal of Homosexuality, 60:* 1–15.

McLean, K. (2004). Negotiating (non)monogamy: Bisexuality and intimate relationships. In R. C. Fox (Ed.), *Current research on bisexuality* (pp. 82–97). New York: Harrington Park Press.

Mena, J. A., & A. Vaccaro. (2013). Tell me you love me no matter what: Relationships and self-esteem among GLBQ young adults. *Journal of GLBT Family Studies, 9:* 3–23.

Mock, S. E., & R. P. Eibach. (2012). Stability and change in sexual orientation identity over a ten-year period in adulthood. *Archives of Sexual Behavior, 41:* 641–648.

Moss, A. R. (2012). Alternative families, alternative lives: Married women doing bisexuality. *Journal of GLBT Family Studies, 8:* 405–427.

Muraco, J. A., S. T. Russell, M. A. Curran, & E. A. Butler. (2012, November). Sexual orientation and romantic relationship quality as moderated by gender, age, and romantic relationship history. Paper presented at the National Council on Family Relations, Phoenix, AZ.

National Coalition of Anti-Violence Programs. (2012). *2011 National hate crimes report: Anti-lesbian, gay, bisexual and transgender violence in 2011.* New York: National Coalition of Anti-Violence Programs.

Newcomb, M. E., M. Birkett, H. L. Corliss, & B. Mustanski. (2014). Sexual orientation, gender, and racial differences in illicit drug use in a sample of US high school students. *American Journal of Public Health, 104:* 304–310.

Ocobock, A. (2013). The power and limits of marriage: Married gay men's family relationships. *Journal of Marriage and Family, 75:* 191–205.

Oswalt, S. B., & T. J. Wyatt. (2011). Sexual orientation and differences in mental health, stress, and academic performance in a national sample of U.S. college students. *Journal of Homosexuality, 58:* 1255–1280.

———. (2013). Sexual health behaviors and sexual orientation in a U.S. national sample of college students. *Archives of Sexual Behavior, 42:* 1561–1572.

Overby, L. Marvin. (2014). Etiology and attitudes: Beliefs about the origins of homosexuality and their implications for public policy. *Journal of Homosexuality, 61:* 568–587.

Padilla, Y. C., C. Crisp, & D. L. Rew. (2010). Parental acceptance and illegal drug use among gay, lesbian, and bisexual adolescents: Results from a national survey. *Social Work, 55:* 265–276.

Parsons, J., T. Starks, S. DuBois, C. Grov, & S. Golub. (2013). Alternatives to monogamy among gay male couples in community survey: Implications for mental health and sexual risk. *Archives of Sexual Behavior, 42:* 303–312.

Pearcey, M. (2004). Gay and bisexual married men's attitudes and experiences: Homophobia, reasons for marriage, and self-identity. *Journal of GLBT Family Studies, 1:* 21–42.

Peter, T., & C. Taylor. (2014). Buried above ground: A university-based study of risk/protective factors for suicidality among sexual minority youth in Canada. *Journal of LGBT Youth, 11:* 125–149.

Pew Research Center. (2012, May 14). Half say view of Obama not affected by gay marriage decision: Independents mostly unmoved.

Pinello, D. R. (2008). Gay marriage: For better or for worse? What we've learned from the evidence. *Law & Society Review, 42:* 227–230.

Pizer, J. C., B. Sears, C. Mallory, & N. D. Hunter. (2012). Evidence of persistent and pervasive workplace discrimination against GLBT people: The need for Federal legislation prohibiting discrimination and providing for equal employment benefits. *Loyola of Los Angeles Law Review, 45:* 715–779.

Potok, M. (2010). Anti-gay hate crimes: Doing the math. *Intelligence Report, 140* (Winter): 29.

Power, J. J., A. Perlesz, R. Brown, M. J. Schofield, M. K. Pitts, R. McNair, & A. Bickerdike. (2013). Bisexual parents and family diversity: Findings from the work, love, play study. *Journal of Bisexuality, 12:* 519–538.

Priebe, G., & C. G. Svedin. (2013). Operationalization of three dimensions of sexual orientation in a national survey of late adolescents. *Journal of Sex Research, 50:* 727–738.

Reinhardt, R. U. (2011). Bisexual women in heterosexual relationships: A study of psychological and sociological patterns: A reflective paper. *Journal of Bisexuality, 11:* 439–447.

Resource Guide to Coming Out for African Americans. (2011). Human Rights Campaign.

Ricks, J. L. (2012). Lesbians and alcohol abuse: Identifying factors for future research. *Journal of Social Service Research, 38,* 37–45.

Roberts, A. L., M. M. Glymour, & K. C. Koenen. (2013). Does maltreatment in childhood affect sexual orientation in adulthood? *Archives of Sexual Behavior, 42:* 161–171.

Rosenberger, J., D. Herbenick, D. Novak, & M. Reece. (2014). What's love got to do with it? Examinations perception and sexual behaviors among gay and bisexual men in the United States. *Archives of Sexual Behavior, 43:* 119–128.

Rothman, E. F., M. Sullivan, S. Keyes, & U. Boehmer. (2012). Parents' supportive reactions to sexual orientation disclosure associated with better health: Results from a population-based survey of LGB adults in Massachusetts. *Journal of Homosexuality, 59:* 186–200.

Rubinstein, G. (2010). Narcissism and self-esteem among homosexual and heterosexual male students. *Journal of Sex & Marital Therapy, 36:* 24–34.

Russell, S. T. (2013). LGBTQ youth well-being: The role of parents and policy. Poster presented at the annual meeting of the National Council on Family Relations, San Antonio, TX.

Russell, S. T., C. Ryan, R. B. Toomey, R. M. Diaz, & J. Sanchez. (2011). Lesbian, gay, bisexual and transgender adolescent school victimization: Implications for young adult health and adjustment. *Journal of School Health, 81:* 223–230.

Rust, Paula. (1993). Neutralizing the political threat of the marginal woman: Lesbians' beliefs about bisexual women. *Journal of Sex Research, 30*(3): 214–218.

Schrimshaw, E. W., M. J. Downing, Jr., & K. Siegel. (2013). Sexual venue selection and strategies for concealment of same-sex behavior among non-disclosing men who have sex with men and women. *Journal of Homosexuality, 60:* 120–145.

Serovich, J. M., S. M. Craft, P. Toviessi, R. Gangamma, et al. (2008). A systematic review of the research base on sexual reorientation therapies. *Journal of Marital and Family Therapy, 34:* 227–239.

Sharpe, A. (2012). Transgender marriage and the legal obligations to disclose gender history. *Modern Law Review, 75:* 33–53.

SIECUS (Sexuality Information and Education Council of the United States). (2014). Retrieved June 28, 2014, from http://www.siecus.org/index.cfm?fuseaction=Page.viewPage&pageId=591&parentID=477

Singh, A. A., D. G. Hays, & L. S. Watson. (2011). Strength in the face of adversity: Resilience strategies of transgender individuals. *Journal of Counseling & Development, 89:* 20–27.

Starks, T., & J. Parsons. (2014). Adult attachment among partnered gay men: Patterns and associations with sexual relationship quality. *Archives of Sexual Behavior, 43:* 107–117.

Sullivan, A. (1997). The conservative case. In A. Sullivan (Ed.), *Same-sex marriage: Pro and con* (pp. 146–154). New York: Vintage Books.

Sutphina, S. T. (2010). Social exchange theory and the division of household labor in same-sex couples. *Marriage and Family Review, 46:* 191–206.

Svab, A., & R. Kuhar. (2014). The transparent and family closets: Gay men and lesbians and their families of origin. *Journal of GLBT Family Studies, 10:* 15–35.

Todd, M., L. H. Rogers, & C. R. Boyer. (2013, November 5–9). Sexual vs. spiritual? Poster presented at the annual meeting of the National Council on Family Relations, San Antonio, TX.

U.S. Department of Health and Human Services Releases Report on LGBT Health. (2012).

Van Bergen, D. D., H. M. W. Bos, J. V. Lisdonk, S. Keuzenkamp, & T. G. M. Sandfort. (2013). Victimization and suicidality among Dutch lesbian, gay, and bisexual youths. *American Journal of Public Health, 103:* 70–72.

Van Eeden-Moorefield, B., C. R. Martell, M. Williams, & M. Preston. (2011). Same-sex relationships and dissolution: The connection between heteronormativity and homonormativity. *Family Relations, 60:* 562–571.

Wagner, C. G. (2006). Homosexual relationships. *Futurist, 40:* 6.

Wright, P. J., & S. Bae. (2013). Pornography consumption and attitudes toward homosexuality: A national longitudinal study. *Human Communication Research, 39*: 492–513.

Wright, R. G., A. J. LeBlanc, & L. Badgett. (2013). Same-sex legal marriage and psychological well-being: Findings from the California health interview survey. *American Journal of Public Health, 103*: 339–346.

Yost, M., & G. Thomas. (2012). Gender and binegativity: Men's and women's attitudes toward male and female bisexuals. *Archives of Sexual Behavior, 41*: 691–702.

Chapter 8

Aducci, A. J., J. A. Baptist, J. George, P. M. Barros, & B. S. Nelson Goff. (2012, November). Military wives' experience during OIF/OEF deployment. Paper presented at the National Council on Family Relations, Orlando, FL.

Amato, P. R., A. Booth, D. R. Johnson, & S. F. Rogers. (2007). *Alone together: How marriage in America is changing.* Cambridge, MA: Harvard University Press.

Anderson, J. R., M. J. Van Ryzin, & W. J. Doherty. (2010). Developmental trajectories of marital happiness in continuously married individuals: A group-based modeling approach. *Journal of Family Psychology, 24*: 587–596.

Arnold, A. L., C. W. O'Neal, C. Bryant, K. A. S. Wickrama, & C. Cutrona. (2011, November). Influences of intra and interpersonal factors in marital success. Presentation at the National Council on Family Relations, Orlando, FL.

Ashwin, S. & O. Isupova. (2014). "Behind every great man….": The male marriage wage premium examined qualitatively. *Journal of Marriage and Family, 76*: 37–55.

Averett, S. L., L. M. Argys, and J. Sorkin. (2013). In sickness and in health: An examination of relationship status and health using data from the Canadian National Public Health Survey. *Review of Economics of the Household, 2*: 599–633.

Ballard, S. M., & A. C. Taylor. (2011). *Family life education with diverse populations.* Thousand Oaks, CA: Sage.

Barzoki, M. H., N. Seyedroghani, & T. Azadarmaki. (2012, July). Sexual dissatisfaction in a sample of married Iranian women. *Sexuality and Culture.* Online.

Berge, J. M., K. W. Bauer, R. MacLehose, M. E. Eisenberg, & D. Neumark-Sztainer. (2014). Associations between relationship status and day-to-day health behaviors and weight among diverse young adults. *Families, Systems, and Health, 32*(1): 67–77.

Burrows, K. (2013). Age preferences in dating advertisements by homosexuals and heterosexuals: From sociobiological to sociological explanations. *Archives of Sexual Behavior, 42*: 203–211.

Campbell, K., J. C. Kaufman, T. D. Ogden, T. T. Pumaccahua, & H. Hammond. (2011a). Wedding rituals: How cost and elaborateness relate to marital outcomes. Poster presented at the National Council on Family Relations, Orlando, FL.

Campbell, K., L. C. Silva, & D. W. Wright. (2011b). Rituals in unmarried couple relationships: An exploratory study. *Family and Consumer Sciences Research Journal, 40*: 45–57.

Carroll, J. S., L. R. Dean, L. L. Call, & D. M. Busby. 2011. Materialism and marriage: Couple profiles of congruent and incongruent spouses. *Journal of Couple & Relationship Therapy, 10*: 287–308.

Choi, H., & N. F. Marks. (2013). Marital quality, socioeconomic status, and physical health. *Journal of Marriage and Family, 75*: 903–919.

Cottle, N. R., R. Hammond, K. Yorgason, K. Stookey, & B. Mallet. (2013, November 5–9). Marital quality among current and former college students. Poster presented at the annual meeting of the National Council on Family Relations, San Antonio, TX.

Covin, T. (2013, February 21–23). Personal communication at the annual meeting of the Southeastern Council on Family Relations, Birmingham, AL.

DeMaris, D. (2010). The 20-year trajectory of marital quality in enduring marriages: Does equity matter? *Journal of Social and Personal Relationships, 27*: 449–471.

Dew, J. (2008). Debt change and marital satisfaction change in recently married couples. *Family Relations, 57*: 60–71.

Dollahite, D. C., & L. D. Marks. (2012). The Mormon American family. In R. Wright, C. H. Mindel, T. Van Tran, & R. W. Habenstein (Eds.), *Ethnic families in America: Patterns and variations,* 5th ed. (pp. 461–486). Boston, MA: Pearson.

Dowd, D. A., M. J. Means, J. F. Pope, & J. H. Humphries. (2005). Attributions and marital satisfaction: The mediated effects of self-disclosure. *Journal of Family and Consumer Sciences, 97*: 22–27.

Duba, J. D., A. W. Hughey, T. Lara, & M. G. Burke. (2012). Areas of marital dissatisfaction among long-term couples. *Adultspan: Theory Research & Practice, 11*: 39–54.

Easterling, B. A., & D. Knox. (2010, July 20). Left behind: How military wives experience the deployment of their husbands. *Journal of Family Life.* Internet link no longer operative. Email Knoxd@ecu.edu for copy of paper.

Field, C. J., S. R. Kimuna, & M. A. Straus. (2013). Attitudes toward interracial relationships among college students: Race, class, gender and perceptions of parental views. *Journal of Black Studies, 44*: 741–776.

Foran, H. M., K. M. Wright, & M. D. Wood. (2013). Do combat exposure and post-deployment mental health influence intent to divorce? *Journal of Social and Clinical Psychology, 32*: 917–938.

Gibson, V. (2002). *Cougar: A guide for older women dating younger men.* Boston, MA: Firefly Books.

Gottman, J., & S. Carrere. (2000, September/October). Welcome to the love lab. *Psychology Today,* 42.

Hall, S. S., & R. Adams. (2011). Newlyweds' unexpected adjustments to marriage. *Family and Consumer Sciences Research Journal, 39*: 375–387.

Hall, S., & D. Knox. (2015). Relationship and sexual behaviors of a sample of 4,694 university students. Unpublished data collected for this text. Department of Family and Consumer Sciences, Ball State University, Muncie, IN and Department of Sociology, East Carolina University, Greenville, NC.

Huyck, M. H., & D. L. Gutmann. (1992). Thirty something years of marriage: Understanding experiences of women and men in enduring family relationships. *Family Perspective, 26*: 249–265.

Jackson, J. B., R. B. Miller, M. Oka, & R. G. Henry. (2014). Gender differences in marital satisfaction. *Journal of Marriage and Family, 76*: 105–129.

Jorgensen, B. L., J. Yorgason, R. Day, & J. A. Mancini. (2011, November 18). The influence of religious beliefs and practices on marital commitment. Poster presented at the annual meeting of the National Council on Family Relations, Orlando, FL.

Kilmann, P. R., H. Finch, M. M. Parnell, & J. T. Downer. (2013). Partner attachment and interpersonal characteristics. *Journal of Sex and Marital Therapy, 39*: 144–159.

Klos, L. A., & J. Sobal. (2013). Weight and weddings. Engaged men's body weight ideals and wedding weight management behaviors. *Appetite, 60*: 133–139.

Lacks, M. H., A. L. Lamson, A. Ivanescu, M. B. White, & C. Russoniello. (2013, April 8–12). An exploration of marital status and stress among military couples. Research and Creative Week, East Carolina University.

LeBaron, C. D. L., R. B. Miller, & J. B. Yorgason. (2014). A longitudinal examination of women's perceptions of marital power and marital happiness in midlife marriages. *Journal of Couple and Relationship Therapy, 13*: 93–113.

Levchenko, P., & C. Solheim. (2013). International marriages between Eastern European-born women and U.S. born men. *Family Relations, 62*: 30–41.

Lundquist, J. H. (2007). A comparison of civilian and enlisted divorce rates during the early all-volunteer force era. *Journal of Political and Military Sociology, 35*(2): 199–217.

Mitchell, B. A. (2010). Midlife marital happiness and ethnic culture: A life course perspective. *Journal of Comparative Family Studies, 41*: 167–183.

Murdock, G. P. (1949). *Social structure*. New York: Free Press.

Musick, K., & L. Bumpass. (2012). Reexamining the case for marriage: Union formation and changes in well-being. *Journal of Marriage and Family, 74*: 1–18.

Passel, J. S., W. Wang, & P. Taylor. (2010, June 4). Marrying out: One-in-seven new U.S. marriages is interracial or interethnic. Pew Research Center. Retrieved from http://pewresearch.org/pubs/1616/american-marriage-interracial-interethnic

Patrick, S., J. N. Sells, F. G. Giordano, & T. Tollerud. (2007). Intimacy, differentiation, and personality variables as predictors of marital satisfaction. *The Family Journal: Counseling and Therapy for Couples and Families, 15*(4): 359–367.

Perry, S. L. (2013). Racial composition of social settings, interracial friendship, and whites' attitudes toward interracial marriage. *Social Science Journal, 50*: 13–22.

Pew Research: Pew Forum on Religion and Public Life. (2008). The U.S. religious landscape survey. Retrieved from http://pewresearch.org/pubs/743/united-states-religion

Plagnol, A. C., & R. A. Easterlin. (2008, July). Aspirations, attainments, and satisfaction: Life cycle differences between American women and men. *Journal of Happiness Studies*. Online.

Proulx, C. M., & L. A. Snyder. (2009). Families and health: An empirical resource guide for researchers and practitioners. *Family Relations, 58*: 489–504.

Qian, Z., & D. T. Lichter. (2011). Changing patterns of interracial marriage in a multiracial society. *Journal of Marriage and Family, 73*: 1065–1084.

Riela, S., G. Rodriguez, A. Aron, X. Xu, & B. P. Acevedo. (2010). Experiences of falling in love: Investigating culture, ethnicity, gender, and speed. *Journal of Social and Personal Relationships, 27*: 473–493.

Shaley, O., N. Baum, & H. Itzhaki. (2013). "There's a man in my bed": The first sex experience among modern-orthodox newlyweds in Israel. *Journal of Sex & Marital Therapy, 39*: 40–55.

Solomon, Z., S. Debby-Aharon, G. Zerach, & D. Horesh. (2011). Marital adjustment, parental functioning, and emotional sharing in war veterans. *Journal of Family Issues, 32*: 127–147.

Spencer, J., & P. Amato. (2011, November). Marital quality across the life course: Evidence from latent growth curves. Presentation at the annual meeting of the National Council on Family Relations, Orlando, FL.

Stanley, S. M., E. P. Ragan, G. K. Rhoades, & H. J. Markman. (2012). Examining changes in relationship adjustment and life satisfaction in marriage. *Journal of Family Psychology, 26*: 165–170.

Statistical Abstract of the United States, 2012–2013, 131th ed. Washington, DC: U.S. Bureau of the Census.

Teten, A. L., J. A. Schumacher, C. T. Taft, M. A. Stanley, T. A. Kent, S. D. Bailey, N. Jo Dunn, & D. L. White. (2010). Intimate partner aggression perpetrated and sustained by male Afghanistan, Iraq, and Vietnam veterans with and without posttraumatic stress disorder. *Journal of Interpersonal Violence, 25*: 1612–1630.

Totenhagen, C., M. Katz, E. Butler, & M. Curran. (2011, November). Daily variability in relational experiences. Presentation at the National Council on Family Relations, Orlando, FL.

United States Census Bureau. (2014). Retrieved from http://quickfacts.census.gov/qfd/states/00000.html

Wallerstein, J., & S. Blakeslee. (1995). *The good marriage*. Boston, MA: Houghton-Mifflin.

Weisfeld, G. E., N. T. Nowak, T. Lucas, C. C. Weisfeld, E. O. Imamoglu, M. Butovskaya, et al. (2011). Do women seek humorousness in men because it signals intelligence? A cross-cultural test. *International Journal of Humor Research, 24*: 435–462.

Wheeler, B., S. Bertagnolli, & J. Yorgason. (2012, November). Marriage: Exploring predictors of marital quality in husband-older, wife-older, and same-age marriage. Poster presented at the National Council on Family Relations, Phoenix, AZ.

Wick, S., & B. S. Nelson Goff. (2014). A qualitative analysis of military couples with high and low trauma symptoms and relationship distress levels. *Relationship Therapy: Innovations in Clinical and Educational Interventions, 13*: 63–88.

Wilson, A. C., & T. L Huston. (2011, November). Shared reality in courtship: Does it matter for marital success? Paper presented at the annual meeting of the National Council on Family Relations, Orlando, FL.

Woldarsky, M., & L. S. Greenberg. (2014). Interpersonal forgiveness in emotion-focused couples' therapy: Relating process to outcome. *Journal of Marital and Family Therapy, 40*: 49–67.

Woszidlo, A., & C. Segrin. (2013). Negative affectivity and educational attainment as predictors of newlyweds' problem solving and marital quality. *Journal of Psychology, 147*: 49–73.

Wright, R. G., A. J. LeBlanc, & L. Badgett. (2013). Same-sex legal marriage and psychological well-being: Findings from the California health interview survey. *American Journal of Public Health, 103*: 339–346.

Wright, R., C. H. Mindel, T. Van Tran, & R. W. Habenstein. (2012). *Ethnic families in America: Patterns and variations*, 5th ed. Boston, MA: Pearson.

Yodanis, C., S. Lauer, & R. Ota. (2012). Interethnic romantic relationships: Enacting affiliative ethnic identities. *Journal of Marriage and Family, 74*: 1021–1037.

Chapter 9

Bauer, K. W., M. O. Hearst, K. Escoto, J. M. Berge, & D. Neumark-Sztainer. (2012). Parental employment and work-family stress: Associations with family food environments. *Social Science & Medicine, 75*: 496–504.

CareerBuilder.com. (2014). Office romance survey. Retrieved April 30 from http://www.careerbuilder.com/share/aboutus/pressreleasesdetail.aspx?sd=2%2F13%2F2014&id=pr803&ed=12%2F31%2F2014

Carey, A. R., & P. Trapp. (2014a, May 10). How much are "mom jobs" worth? *USA Today*, p. A-1.

_____. (2014b, July 22). Stay-at-home dads. *USA Today*, p. A-1.

Cohn, D., G. Livingston, & W. Wang. (2014, April 8). After decades of decline, a rise in stay-at-home mothers. Pew Research Center. Retrieved from http://www.pewresearch.org/fact-tank/2014/04/08/7-key-findings-about-stay-at-home-moms/

De Schipper, J. C., L. W. C. Tavecchio, & M. H. Van IJzendoorn. (2008). Children's attachment relationships with day care caregivers: Associations with positive caregiving and the child's temperament. *Social Development, 17*: 454–465.

Etzioni, A. (2011). The new normal. *Sociological Forum, 26*: 779–789.

Fiona, C. S., A. W. Chau, & K. Y. Chan. (2012). Financial knowledge and aptitudes: Impacts on college students' financial well-being. *College Student Journal, 46*: 114–132.

Fry, R., & A. Caumont. (2014, May 14). 5 key findings about student debt. Pew Research Center. Retrieved from http://www.pewresearch.org/fact-tank/2014/05/14/5-key-findings-about-student-debt/

Gunnar, M. R., E. Kryzer, M. J. Van Ryzin, & D. A. Phillips. (2010). The rise in cortisol in family day care: Associations with aspects of care quality, child behavior, and child sex. *Child Development, 81*: 851–886.

Hardie, J. H., & A. Lucas. (2010). Economic factors and relationship quality among young couples: Comparing cohabitation and marriage. *Journal of Marriage and the Family, 72*: 1141–1154.

Healy, M., & P. Trap. (2012, May 7). Desired changes in one's life. *USA Today*, p. 1D.

Helms, H. M., J. K. Walls, A. C. Crouter, & S. M. McHale. (2010). Provider role attitudes, marital satisfaction, role overload, and housework: A dyadic approach. *Journal of Family Psychology, 24*: 568–577.

Hendrix, J. A., & T. L. Parcel. (2014). Parental nonstandard work, family processes, and delinquency during adolescence. *Journal of Family Issues, 35*: 1363–1393.

Hochschild, A. R. (1989). *The second shift.* New York: Viking.

_____. (1997). *The time bind.* New York: Metropolitan Books.

Hoffnung, M., & M. Williams. (2013). Balancing act: Career and family during college-educated women's 30s. *Sex Roles, 68*: 321–334.

Hoser, N. (2012). Making it a dual-career family in Germany: Exploring what couples think and do in everyday life. *Marriage and Family Review, 48*: 643–666.

Insure.com. (2014). Annual Father's Day Index using Bureau of Labor Statistics.

Johnson, S., J. Li, G. Kendall, L. Strazdin, & P. Jacoby. (2013). Mothers' and fathers' work hours, child gender, and behavior in middle childhood. *Journal of Marriage & Family, 75*: 56–74.

Kahn, J. R., J. Garcia-Manglano, & S. M. Bianchi. (2014). The motherhood penalty at midlife: Long-term effects of children on women's careers. *Journal of Marriage and Family, 76*: 56–72.

Kahneman, D., & A. Deaton. (2010, September 6). High income improves evaluation of life but not emotional well-being. Proceedings of the National Academy of Sciences, early edition.

Kornrich, S., J. Brines, & K. Leupp. (2013). Egalitarianism, housework, and sexual frequency in marriage. *American Sociological Review, 78*: 26–50.

Lam, C. B., S. M. McHale, & A. C. Crouter. (2012). The division of household labor: Longitudinal changes and within-couple variation. *Journal of Marriage and Family, 74*: 944–952.

Ma a tta, K., & S. Uusiautti. (2012). Seven rules on having a happy marriage along with work. *The Family Journal, 20*: 267–271.

McBridge, M. C., & K. M. Bergen. (2014). Voices of women in commuter marriages: A site of discursive struggle. *Journal of Social and Personal Relationships, 31*: 554-557.

Meisenbach, M. J. (2010). The female breadwinner: Phenomenological experience and gendered identity in work/family spaces. *Journal of Sex Roles, 62*: 2–19.

Merrill, J., & D. Knox. (2010). *Finding love from 9 to 5.* New York: Praeger.

Meteyer, K., & M. Perry-Jenkins. (2012). Father involvement among working-class, dual-earner couples. *Fathering: A Journal of Theory, Research, & Practice about Men as Fathers, 8*: 379–403.

Minnotte, K. L., M. C. Minnotte, & D. E. Pedersen. (2013). Marital satisfaction among dual-earner couples: Gender ideologies and family-to-work conflict. *Family Relations, 62*: 686–698.

Minnottea, K. L., D. E. Pedersena, S. E. Mannonb, & G. Kiger. (2010). Tending to the emotions of children: Predicting parental performance of emotion work with children. *Marriage & Family Review, 46*: 224–241.

Neppl, T. (2012, November 1). The effects of economic pressure on family and child outcomes. Poster presented at the annual meeting of the National Council on Family Relations. Phoenix, AZ.

Olsen, K. M., & S. Dahl. (2010). Working time: implications for sickness absence and the work–family balance. *International Journal of Social Welfare, 19*: 45–53.

Opree, S. J., & M. Kalmijn. (2012). Exploring causal effects of combining work and intergenerational support on depressive symptoms among middle-aged women. *Ageing & Society, 32*: 130–146.

Parker, K., & W. Wang. (2013, March 14). Modern parenthood: Roles of moms and dads converge as they balance work and family. *Pew Research: Social & Demographic Trends.* Retrieved from http://www.pewsocialtrends.org/2013/03/14/modern-parenthood-roles-of-moms-and-dads-converge-as-they-balance-work-and-family/

Poortman, A., & T. Van der Lippe. (2009). Attitudes toward housework and child care and the gendered division of labor. *Journal of Marriage and the Family, 71*: 526–541.

Poverty line. (2014). Retrieved February 23, 2014 from http://aspe.hhs.gov/poverty/14poverty.cfm

Sandberg, J. G., J. M. Harper, E. J. Hill, R. B. Miller, J. B. Yorgason, & R. D. Day. (2013). "What happens at home does not necessarily stay at home": The relationship of observed negative couple interaction with physical health, mental health, and work satisfaction. *Journal of Marriage and the Family, 75*: 808–821.

Sasaki, T., N. L. Hazen, & W. B. Swann Jr. (2010). The supermom trap: Do involved dads erode moms' self-competence? *Personal Relationships, 17*: 71–79.

Schellekens, J., & D. Gliksberg. (2013). Inflation and marriage in Israel. *Journal of Family History, 38*: 78–93.

Schoen, R., N. M. Astone, K. Rothert, N. J. Standish, & Y. J. Kim. (2002). Women's employment, marital happiness, and divorce. *Social Forces, 81*: 643–662.

Schoen, R., S. J. Rogers, & P. R. Amato. (2006). Wives' employment and spouses' marital happiness: Assessing the direction of influence using longitudinal couple data. *Journal of Family Issues, 27*: 506–528.

Stanfield, J. B. (1998). Couples coping with dual careers: A description of flexible and rigid coping styles. *Social Science Journal, 35*: 53–62.

Statistical Abstract of the United States, 2012–2013, 131th ed. Washington, DC: U.S. Bureau of the Census.

Stulhofer, A., B. Traeen, & A. Carvalheira. (2013). Job-related strain and sexual health difficulties among heterosexual men from three European countries: The role of culture and emotional support. *Journal of Sexual Medicine, 10*: 747–756.

Treas, J., T. Van der Lippe, & T. C. Tai. (2011). The happy homemaker? Married women's well-being in cross-national perspective. *Social Forces, 90*: 111–132.

Vandell, D. L., J. Belsky, M. Burchinal, L. Steinberg, N. Vandergrift, & NICHD Early Child Care Research Network. (2010). Do effects of early child care extend to age 15 years? Results from the NICHD study of early child care and youth development. *Child Development, 81*: 737–756.

Vanderkam, L. (2010). *168 hours: You have more time than you think to achieve your dreams.* Portfolio Press (online).

Vault Office Romance Survey. (2014, February 12). Retrieved February 22 from http://www.vault.com/blog/workplace-issues/love-is-in-the-air-vaults-2014-office-romance-survey/

Wagner, S. L., B. Forer, I. L. Cepeda, H. Goelman, S. Maggi, A. D'Angiulli, J. Wessel, C. Hertzman & R. E. Grunau. (2013). Perceived stress and Canadian early childcare educators. *Child and Youth Care Forum, 42*: 53–70.

Walsh, S. (2013). Match's 2012 singles in America survey. Retrieved from http://www.hookingupsmart.com/2013/02/07/hookinguprealities/matchs-2012-singles-in-america-survey/

Wang, W., K. Parker, & P. Taylor. (2013, May 29). Breadwinner moms: Mothers as the sole or primary provider in four-in-ten households with children. *Pew Social & Demographic Trends.* Retrieved from http://www.pewsocialtrends.org/2013/05/29/breadwinner-moms/

Wheeler, B., & J. Kerpelman. (2013, November 5–9). Change in frequency of disagreements about money: A "gateway" to poorer marital outcomes among newlywed couples over the first five years of marriage. Poster presented at annual meeting of the National Council on Family Relations, San Antonio, TX.

Yang, J., & S. Ward. (2010, December 7). Working wives/Ernst & Young Survey. *USA Today,* Sec. B.

Yang, J., & A. Gonzalez. (2013, February 25). Virtual versus office work. *USA Today,* Sec. B.

Chapter 10

Alexander, P. C., A. Tracy, M. Radek, & C. Koverola. (2009). Predicting stages of change in battered women. *Journal of Interpersonal Violence, 24*: 1652–1672.

Anderson, K. L. (2010). Conflict, power, and violence in families. *Journal of Marriage and Family, 72*: 726–742.

Azziz-Baumgartner, E., L. McKeown, P. Melvin, D. Quynh, & J. Reed. (2011). Rates of femicide in women of different races, ethnicities, and places of birth: Massachusetts, 1993–2007. *Journal of Interpersonal Violence, 26*: 1077–1090.

Bartholomew, K., K. V. Regan, M. A. White, & D. Oram. (2008). Patterns of abuse in male same-sex relationships. *Violence and Victims, 23*: 617–637.

Becker, K. D., J. Stuewig, & L. A. McCloskey. (2010). Traumatic stress symptoms of women exposed to different forms of childhood victimization and intimate partner violence. *Journal of Interpersonal Violence, 25*: 1699–1715.

Bennett, S., V. L. Banyard, & L. Garnhart. (2014). To act or not to act, that is the question? Barriers and facilitators of bystander intervention. *Journal of Interpersonal Violence, 29*: 476–496.

Brownridge, D. A. (2010). Does the situational couple violence-intimate terrorism typology explain cohabitors' high risk of intimate partner violence? *Journal of Interpersonal Violence, 25*: 1264–1283.

Burke, S., M. Wallen, K. Vail-Smith, & D. Knox. (2011). Using technology to control intimate partners: An exploratory study of college undergraduates. *Computers in Human Behavior, 27*: 1162–1167.

Chavis, A., J. Hudnut-Beumler, M. W. Webb, J. A. Neely, L. Bickman, M. S. Dietrich, & S. J. Scholer. (2013). A brief intervention affects parents' attitudes toward using less physical punishment. *Child Abuse & Neglect, 37*: 1192–1201.

Clements, C. M., & R. L. Ogle. (2009). Does acknowledgment as an assault victim impact post assault psychological symptoms and coping? *Journal of Interpersonal Violence, 24*: 1595–1614.

Cohn, A., H. M. Zinzow, H. S. Resnick, & D. G. Kilpatrick. (2013). Correlates of reasons for not reporting rape to police: Results from a national telephone household probability sample of women with forcible or drug-or-alcohol facilitated/incapacitated rape. *Journal of Interpersonal Violence, 28*: 455–473.

Diem, C., & J. M. Pizarro. (2010). Social structure and family homicides. *Journal of Family Violence, 25*: 521–532.

Dominguez, M. M., J. High, E. Smith, B.Cafferky, P. Dharnidharka, & S. Smith. (2013, November 5–9). The intergenerational transmission of family violence: A meta-analytic review. Poster presented at the annual meeting of the National Council on Family Relations, San Antonio, TX.

Duntley, J. D., & D. M. Buss. (2012). The evolution of stalking. *Sex Roles, 66*: 311–327.

Eaton, L., S. Kalichman, D. Skinner, M. Watt, D. Pieterse, & E. Pitpitan. (2012). Pregnancy, alcohol intake, and intimate partner violence among men and women attending drinking establishments in a Cape Town, South Africa, township. *Journal of Community Health, 37*: 208–216.

Eke, A., N. Hilton, G. Harris, M. Rice, & R. Houghton. (2011). Intimate partner homicide: Risk assessment and prospects for prediction. *Journal of Family Violence, 26*: 211–216.

Farris, C., R. J. Viken, & T. Treat. (2010). Alcohol alters men's perceptual and decisional processing of women's sexual interest. *Journal of Abnormal Psychology, 119*: 427–432.

Fernandez-Montalvo, J., J. J. Lopez-Goni, & A. Arteaga. (2012). Violent behaviors in drug addiction: Differential profiles of drug-addicted patients with and without violence problems. *Journal of Interpersonal Violence, 27*: 142–157.

Few, A. L., & K. H. Rosen. (2005). Victims of chronic dating violence: How women's vulnerabilities link to their decisions to stay. *Family Relations, 54*: 265–279.

Follingstad, D. R., & M. Edmundson. (2010). Is psychological abuse reciprocal in intimate relationships? Data from a national sample of American adults. *Journal of Family Violence, 25*: 495–508.

Graham, A. M., H. K. Kim, & P. A. Fishcr. (2012). Patner aggression in high-risk families from birth to age 3 years: Associations with harsh parenting and child maladjustment. *Journal of Family Psychology, 26*: 105–114.

Hall, S., & D. Knox. (2012). Double victims: Sexual coercion by a dating partner and a stranger. *Journal of Aggression, Maltreatment & Trauma, 22*: 145–158.

————. (2015). Relationship and sexual behaviors of a sample of 4,722 university students. Unpublished data collected for this text. Department of Family and Consumer Sciences, Ball State University and Department of Sociology, East Carolina University.

Halligan, C., D. Knox, & J. Brinkley. (2013). TRAPPED: Technology as a barrier to leaving an abusive relationship. *College Student Journal, 47*: 644–648.

Halpern-Meekin, S., W. D. Manning, P. C. Giordano, & M. A. Longmore. (2013). Relationship churning, physical violence, and verbal abuse in young adult relationships. *Journal of Marriage & Family, 75*: 2–12.

Heath, N. M., S. M. Lynch, A. M. Fritch, & M. M Wong. (2013). Rape myth acceptance impacts the reporting of rape to the police: A study of incarcerated women. *Violence Against Women, 19*: 1065–1078.

Henning, K., & J. Connor-Smith. (2011). Why doesn't he leave? Relationship continuity and satisfaction among male domestic violence offenders. *Journal of Interpersonal Violence, 26*: 1366–1387.

Jacobsen, N., & J. Gottman. (2007). *When men batter women: New insights into ending abusive relationships.* New York: Simon & Schuster.

Johnson, D. W., C. Zlotnick, and S. Perez. (2008). The relative contribution of abuse severity and PTSD severity on the psychiatric and social morbidity of battered women in shelters. *Behavior Therapy, 39*: 232–247.

Katz, J., & J. Moore. (2013). Bystander education training for campus sexual assault prevention: An initial meta-analysis. *Violence and Victims, 28*: 1054–1067.

Klipfel, K. M., S. E. Claxton, & M. H. M. Van Dulmen. (2014). Interpersonal aggression victimization within casual sexual relationships and experiences. *Journal of Interpersonal Violence, 29*: 557–569.

Kothari, C. L., K. V. Rhodes, J. A. Wiley, J. Fink, S. Overholt, M. E. Dichter, S. C. Marcus, & C. Cerulli. (2012). Protection orders protect against assault and injury: A longitudinal study of police-involved women victims of intimate partner violence. *Journal of Interpersonal Violence, 27*: 2845– 2868.

Kress, V. E., J. J. Protivnak, & L. Sadlak. (2008). Counseling clients involved with violent intimate partners: The mental health counselor's role in promoting client safety. *Journal of Mental Health Counseling, 30*: 200–211.

Kulkarni, M., S. Graham-Bermann, S. Rauch, & J. Seng. (2011). Witnessing versus experiencing direct violence in childhood as correlates of adulthood PTSD. *Journal of Interpersonal Violence, 26*: 1264–1281.

Larsen, C. D., J. G. Sandberg, J. M. Harper, & R. Bean. (2011). The effects of childhood abuse on relationship quality: Gender differences and clinical implications. *Family Relations, 60*: 435–445.

Lawrence, Q., & M. Penaloza. (2013, March 21). Sexual violence victims say military justice system is broken. NPR.

Lawyer, S., H. Resnick, V. Bakanic, T. Burkett, & D. Kilpatrick. (2010). Forcible, drug-facilitated, and incapacitated rape and sexual assault among undergraduate women. *Journal of American College Health, 58*: 453–460.

Littleton, H., A. Grills-Taquechel, & D. Axsom. (2009). Impaired and incapacitated rape victims: Assault characteristics and post-assault experiences. *Violence & Victims, 24*: 439–457.

Liu, S., M. M. Dore, & I. Amrani-Cohen. (2013). Treating the effects of interpersonal violence: A comparison of two group models. *Social Work with Groups, 36*: 59–72.

Lyndon, A. E. H., C. Sinclair, J. MacArthur, B. Fay, E. Ratajack, & K. E. Collier. (2012). An introduction to issues of gender in stalking research. *Sex Roles, 66*: 299–310.

Ma, J., Y. Han, A. Grogan-Kaylor, J. Delva, & M. Castillo. (2012). Corporal punishment and youth externalizing behavior in Santiago, Chile. *Child Abuse & Neglect, 36*: 481–490.

MacMillan, H. L., C. N. Wathen, & C. M. Varcoe. (2013). Intimate partner violence in the family: Considerations for children's safety. *Child Abuse & Neglect, 37*: 1186–1191.

Mahr, K. (2013, January 14). India's shame. *Time*, 12.

Maneta, E., S. Cohen, M. Schulz, & R. J. Waldinger. (2012). Links between childhood physical abuse and intimate partner aggression: The mediating role of anger expression. *Violence and Victims, 27*: 315–328.

Marcus, R. E. (2012). Patterns of intimate partner violence in young adult couples: Nonviolent, unilaterally violent, and mutually violent couples. *Violence & Victims, 27*: 299–314.

Mathes, E. W. (2013). Why is there a strong positive correlation between perpetration and being a victim of sexual coercion? An exploratory study. *Journal of Family Violence, 28*: 783–796.

McMahon, S. (2010). Rape myth beliefs and bystander attitudes among incoming college students. *Journal of American College Health, 59*: 3–11.

Miller, J. D., A. Zeichner, & L. F. Wilson. (2012). Personality correlates of aggression: Evidence from measures of the five-factor model, UPPS model of impulsivity, and BIS/BAS. *Journal of Interpersonal Violence, 27*: 2903–2919.

Mouilso, E. R., & K. S. Calhoun. (2013). The role of rape myth acceptance and psychopathy in sexual assault perpetration. *Journal of Aggression, Maltreatment & Trauma, 22*: 159–174.

Mullen, P. E., M. Pathe, R. Purcell, & G. W. Stuart. (1999). A study of stalkers. *American Journal of Psychiatry, 56*: 1244–1249.

O'Leary, D. K., H. Foran, & S. Cohen. (2013). Validation of fear of partner scale. *Journal of Marital and Family Therapy, 39*: 502–514.

Oswald, D. L., & B. L. Russell. (2006). Perceptions of sexual coercion in heterosexual dating relationships: The role of aggressor gender and tactics. *The Journal of Sex Research, 43*: 87–98.

Pendry, P., F. Henderson, J. Antles, & E. Conlin. (2011, November 18). Parents' use of everyday conflict tactics in the presence of children: Predictors and implications for child behavior. Poster presented at the annual meeting of the National Council on Family Relations, Orlando, FL.

Porter, J., & L. M. Williams. (2011). Intimate violence among underrepresented groups on a college campus. *Journal of Interpersonal Violence, 26*, 3210–3224.

Powers, R. A., & S. S. Simpson. (2012). Self-protective behaviors and injury in domestic violence situations: Does it hurt to fight back? *Journal of Interpersonal Violence, 27*: 3345–3365.

Rhatigan, D. L., & A. M. Nathanson. (2010). The role of female behavior and attributions in predicting behavioral responses to hypothetical male aggression. *Violence Against Women, 16*: 621–637.

Rhodes, K. V., C. Cerulli, M. E. Dichter, C. L. Kothari, & F. K. Barg. (2010). "I didn't want to put them through that": The influence of children on victim decision making in intimate partner violence cases. *Journal of Family Violence, 25*: 485–493.

Rothman, E. F., D. Exner, & A. L. Baughman. (2011). The prevalence of sexual assault against people who identify as gay, lesbian, or bisexual in the United States: A systematic review. *Violence & Abuse, 12*: 55–66.

Russella, D., K. W. Springerb, & E. A. Greenfield. (2010). Witnessing domestic abuse in childhood as an independent risk factor for depressive symptoms in young adulthood. *Child Abuse & Neglect, 34*: 448–453.

Sarkar, N. N. (2008). The impact of intimate partner violence on women's reproductive health and pregnancy outcome. *Journal of Obstetrics and Gynecology, 28*: 266–278.

Shorey, R. C., J. Febres, H. Brasfield, & G. L. Stuart. (2012, August 12). The prevalence of mental health problems in men arrested for domestic violence. *Journal of Family Violence*. Online. Retrieved from http://www.springerlink.com.jproxy.lib.ecu.edu/content /0g67243666512182/fulltext.pdf

Sidebotham, P. (2013). Rethinking filicide. *Child Abuse Review, 22*: 305–310.

Simons, L. G., R. L. Simons, M. Lei, & T. E. Sutton. (2012). Exposure to harsh parenting and pornography as explanations for male's sexual coercion and female's sexual victimization. *Violence and Victims, 27*: 378–395.

Smith, P. H., C. E. Murray, & A. L. Coker. (2010). The Coping Window: A contextual understanding of the methods women use to cope with battering. *Violence and Victims, 25*: 18–28.

Spitzberg, B. H., & W. R. Cupach. (2007). Cyberstalking as (mis) matchmaking. In M. T. Whitty, A. J. Baker, & J. A. Inman (Eds.), *Online matchmaking* (pp. 127–146). New York: Palgrave Macmillan.

Swan, S. C., L. J. Gambone, J. E. Caldwell, T. P. Sullivan, & D. L Snow. (2008). A review of research on women's use of violence with male intimate partners. *Violence and Victims, 23*: 301–315.

Teten, A. L., B. Ball, L. A. Valle, R. Noonan, & B. Rosenbluth. (2009). Considerations for the definition, measurement, consequences, and prevention of dating violence victimization among adolescent girls. *Journal of Women's Health, 18*: 923–927.

Walby, S. (2013). Violence and society: Introduction to an emerging field of sociology. *Current Sociology, 61*: 95–111.

Ward, J. T., M. D. Krohn, & C. L. Gibson. (2014). The effects of police on trajectories of violence: A group-based, propensity score matching analysis. *Journal of Interpersonal Violence, 29*: 440–475.

Whitaker, M. P. (2014). Motivational attributions about intimate partner violence among male and female perpetrators. *Journal of Interpersonal Violence, 29*: 517–535.

Witte, T. H., & R. Kendra. (2010). Risk recognition and intimate partner violence. *Journal of Interpersonal Violence, 25*: 2199–2216.

Zinzow, H. M., H. S. Resnick, A. B. Amstadter, J. L. McCauley, K. J. Ruggiero, & D. G. Kilpatrick. (2010). Drug- or alcohol-facilitated, incapacitated, and forcible rape in relationship to mental health among a national sample of women. *Journal of Interpersonal Violence, 25*: 2217–2236.

Chapter 11

Allen, R. E. S., & J. L. Wiles. (2013). How older people position their late-life childlessness: A qualitative study. *Journal of Marriage and the Family, 75*: 206–220.

Ames, C. M., & W. V. Norman (2012). Preventing repeat abortion in Canada: Is the immediate insertion of intrauterine devices post abortion a cost-effective option associated with fewer repeat abortions? *Contraception, 85*: 51–55.

Baldur-Felskov, B., S. K. Kjaer, V. Albieri, M. Steding-Jessen, T. Kjaer, C. Johansen, S. O. Dalton, & A. Jensen. (2013). Psychiatric disorders in women with fertility problems: Results from a large Danish register-based cohort study. *Human Reproduction, 28*: 683–690.

Brandes, M., C. Hamilton, J. van der Steen, J. de Bruin, R. Bots, W. Nelen, & J. Kremer. (2011). Unexplained infertility: Overall ongoing pregnancy rate and mode of conception. *Human Reproduction, 26*: 360–368.

Canario, C., B. Figueiredo, & M. Ricou. (2013). Women and men's psychological adjustment after abortion: A six-months' perspective pilot study. *Journal of Reproductive and Infant Psychology, 29*: 262–275.

Curtin, S. C., J. C. Abma, & S. J. Ventura. (2013, December). Pregnancy rates for U.S. women continue to drop. U.S. Department of Health and Human Services. NCHS Data # 136.

Department of Commerce et al. (2011). *Women in America: Indicators of social and economic well-being.* Retrieved March 15 from http://www.whitehouse.gov/sites/default/files/rss_viewer/Women _in_America.pdf

Dougall, K. M., Y. Beyene, & R. D. Nachtigall. (2013). Age shock: Misperceptions of the impact of age on fertility before and after IVF in women who conceived after age 40. *Human Reproduction, 28*: 350–356.

Falcon, M., F. Valero, M. Pellegrini, M. Rotolo, G. Scaravelli, J. Joya, O. Vall, et al. (2010). Exposure to psychoactive substances in women who request voluntary termination of pregnancy assessed by serum and hair testing. *Forensic Science International, 196:* 22–26.

Feng, W., Y. Cai, & B. Gu. (2013). Population, policy, and politics: How will history judge China's one-child policy? *Population and Development Review, 38:* 115–129.

Fergusson, D. M., L. J. Horwood, & J. M. Boden. (2013). Does abortion reduce the mental health risks of unwanted or unintended pregnancy? A reappraisal of the evidence. *Australian and New Zealand Journal of Psychiatry, 47:* 819–827.

Finer, L. B., L. F. Frohwirth, L. A. Dauphinne, S. Singh, & A. M. Moore. (2005). Reasons U.S. women have abortions: Quantitative and qualitative reasons. *Perspectives on Sexual and Reproductive Health, 37:* 110–118.

Finer, L. B., & M. R. Zolna. (2014). Shifts in intended and unintended pregnancies in the United States, 2001–2008. *American Journal of Public Health, 104:* 43–48.

Foster, D. G., H. Gould, J. Taylor, & T. A. Weitz. (2012). Attitudes and decision making among women seeking abortions in one U.S. clinic. *Perspectives on Sexual & Reproductive Health, 44:* 117–124.

Foster, D. G., K. Kimport, H. Gould, S. C. M. Roberts, & T. A. Weitz. (2013). Effect of abortion protesters on women's emotional response to abortion. *Contraception, 87:* 81–87.

Frisco, M. L., & M. Weden. (2013). Early adult obesity and U.S. women's lifetime childbearing experiences. *Journal of Marriage and Family, 75:* 920–932.

Garrett, T. M., H. W. Baillie, & R. M. Garrett. (2001). *Health care ethics,* 4th ed. Upper Saddle River, NJ: Prentice Hall.

Geller, P., C. Psaros, & S. L. Kornfield. (2010). Satisfaction with pregnancy loss aftercare: Are women getting what they want? *Archives of Women's Mental Health, 13:* 111–124.

Goldberg, A. E., L. A. Kinkler, H. B. Richardson, & J. B. Downing. (2011). Lesbian, gay, and heterosexual couples in open adoption arrangements: A qualitative study. *Journal of Marriage and Family, 73:* 502–518.

Guttmacher Institute. (2014). Abortion facts. Retrieved May 31 from http://www.guttmacher.org/media/presskits/abortion-US/statsandfacts.html.

Hall, S., & D. Knox. (2015). Relationship and sexual behaviors of a sample of 4,700 university students. Unpublished data collected for this text. Department of Family and Consumer Sciences, Ball State University and Department of Sociology, East Carolina University.

Healy, M. (2013, December 17). The web has transformed adoption, for good and bad. *USA Today,* p. 3D.

Hoffnung, M., & M. Williams. (2013). Balancing act: Career and family during college-educated women's 30s. *Sex Roles, 68:* 321–334.

Jones, R. K., & K. Kooistra. (2011). Abortion incidence and access to services in the United States, 2008. *Perspectives on Sexual & Reproductive Health, 43:* 41–50.

Jones, R. K., A. M. Moore, & L. F. Frohwirth. (2011). Perceptions of male knowledge and support among U.S. women obtaining abortions. *Women's Health Issues, 21:* 117–123.

Juffer, F., M. van Ijzendoorn, & J. Palacios. (2011). Children's recovery after adoption. *Infancia y Aprendizaje, 34:* 3–18.

Katz, P. J., et al. (2011). Costs of infertility treatment: Results from an 18-month prospective cohort study. *Fertility and Sterility, 95:* 915–921.

Kondapalli, L. A., & A. Perales-Puchalt. (2013). Low birth weight: Is it related to assisted reproductive technology or underlying infertility? *Fertility and Sterility, 99:* 303–310.

Lee, D. (2006). Device brings hope for fertility clinics. Retrieved February 22, 2006, from http://www.indystar.com/apps/pbcs.dll/article?AID=/20060221/BUSINESS/602210365/1003

Lee, K., & A. M. Zvonkovic. (2014). Journeys to remain childless: A grounded theory examination of decision-making processes among voluntarily childless couples. *Journal of Social and Personal Relationships, 31:* 535–553.

MacLeod, C. (2012). Forced abortion ignites online outrage. *USA Today,* p. 5A.

Major, B., M. Appelbaum, & C. West. (2008, August 13). Report of the APA task force on mental health and abortion.

McQuillan, J., A. L. Greil, K. M. Shreffler, P. A. Wonch-Hill, K. C. Gentzler, & J. D. Hathcoat. (2012). Does the reason matter? Variations in childlessness concerns among U.S. women. *Journal of Marriage and Family, 74:* 1166–1181.

Meyer, A. S., L. M. McWey, W. McKendrick, & T. L. Henderson. (2010). Substance-using parents, foster care, and termination of parental rights: The importance of risk factors for legal outcomes. *Children & Youth Services Review, 32:* 639–649.

Pritchard, K. M., & L. Kort-Butler. (2012, November). Multiple motherhoods: The interactive effects of importance of life satisfaction. Poster presented at the National Council on Family Relations.

Puri, S., & R. D. Nachtigall. (2010). The ethics of sex selection: A comparison of the attitudes and experiences of primary care physicians and physician providers of clinical sex selection services. *Fertility and Sterility, 93:* 2107–2114.

Roby, J., & H. White. (2010). Adoption activities on the Internet: A call for regulation. *Social Work, 55:* 203–212.

Sandler, L. (2010, July 19). One and done. *Time,* 34–41.

Sandlow, J. I. (2013). Size does matter: Higher body mass index may mean lower pregnancy rates for microscopic testicular sperm extraction. *Fertility and Sterility, 99:* 347.

Sankar, S. (2012, November 1). Adoption of older children: Factors influencing decision to adopt. Poster presented at the annual meeting of the National Council on Family Relations, Phoenix, AZ.

Schmidt, L., T. Sobotka, J. G. Bentzen, & A. Nyboe Andersen. (2012). Demographic and medical consequences of the postponement of parenthood. *Human Reproduction Update, 18:* 29–43.

Scott, L. S. (2009). *Two is enough.* Berkeley, CA: Seal Press.

Shapiro, C. H. (2012, November). Decade of change: New interdisciplinary needs of people with infertility. Paper presented at the National Council on Family Relations, Phoenix, AZ.

Singer, E. (2010). The "W.I.S.E. Up!" tool: Empowering adopted children to cope with questions and comments about adoption. *Pediatric Nursing, 36:* 209–212.

Smock, P. J., & F. R. Greenland. (2010). Diversity in pathways to parenthood: Patterns, implications, and emerging research directions. *Journal of Marriage and Family, 72:* 576–593.

Statistical Abstract of the United States, 2012–2013, 131th ed. Washington, DC: U.S. Bureau of the Census.

Steinberg, J. R. (2013). Another response to: "Does abortion reduce the mental health risks of unwanted or unintended pregnancy?" Fergusson et al. (2013). *Australian and New Zealand Journal of Psychiatry, 47:* 1204–1205.

Sugiura-Ogasawara, M., S. Suzuki, Y. Ozaki, K. Katano, N. Suzumori, & T. Kitaori. (2013). Frequency of recurrent spontaneous abortion and its influence on further marital relationship and illness: The Okazaki Cohort Study in Japan. *Journal of Obstetrics and Gynaecology Research, 39:* 126–131.

Tal, G., J. Lafortune, & C. Low. (2014). What happens the morning after? The costs and benefits of expanding access to emergency contraception. *Journal of Policy Analysis and Management, 33:* 70–93.

Teskereci, G., & S. Oncel. (2013). Effect of lifestyle on quality of life of couples receiving infertility treatment. *Journal of Sex & Marital Therapy, 39:* 476–492.

U.S. Department of Agriculture. (2013, August 1). Parents projected to spend $241,080 to raise a child born in 2012. USDA report.

Van Geloven, N., F. Van der Veen, P. M. M. Bossuyt, P. G. Hompes, A. H. Zwinderman, & B. W. Mol. (2013). Can we distinguish between infertility and subfertility when predicting natural conception in couples with an unfulfilled child wish? *Human Reproduction, 28:* 658–665.

Waterman, J., E. Nadeem, E. Paczkowski, J. Foster, J. Lange, T. Belin, & J. Miranda. (2011, November). Behavior problems and parenting stress for children adopted from foster care over the first five years of placement. National Council on Family Relations, Orlando, FL.

Weitz, T. A., D. Taylor, S. Desai, U. D. Upadhyay, J. Waldman, M. F. Battistelli, & E. A. Drey. (2013). Safety of aspiration abortion performed by nurse practitioners, certified nurse midwives, and physician assistants under a California legal waiver. *American Journal of Public Health, 103:* 454–461.

Chapter 12

Ali, M. M., & D. S. Dwyer. (2010). Social network effects in alcohol consumption among adolescents. *Addictive Behaviors, 35:* 337–342.

Armstrong, A. B., & E. F. Clinton. (2012). Altering positive/negative interaction ratios of mothers and young children. *Child Behavior and Family Therapy, 34:* 231–242.

Bastaits, K., K. Ponnet, & D. Mortelmans. (2014). Do divorced fathers matter? The impact of parenting styles on the well-being of children. *Journal of Divorce & Remarriage, 55:* 363–390.

Baumrind, D. (1966). Effects of authoritative parental control on child behavior. *Child Development, 37:* 887–907.

Becker, K. (2010). Risk behavior in adolescents: What we need to know. *European Psychiatry, 25:* 85–95.

Beekman, D. (1977). *The mechanical baby: A popular history of the theory and practice of child rearing.* Westport, CT: Lawrence Hill & Company.

Biehle, S. N., & K. D. Michelson. (2012). First-time parent's expectations about the division of childcare and play. *Journal of Family Psychology, 26:* 36–45.

Bisakha, S. (2010). The relationship between frequency of family dinner and adolescent problem behaviors after adjusting for other family characteristics. *Journal of Adolescence, 33:* 187–196.

Byrd-Craven, J., B. J. Auer, D. A. Granger, & A. R. Massey. (2012). The father-daughter dance: The relationship between father-daughter relationship quality and daughters' stress response. *Journal of Family Psychology, 26:* 87–94.

Carr, K., & T. R. Wang. (2012). "Forgiveness isn't a simple process: It's a vast undertaking": Negotiating and communicating forgiveness in nonvoluntary family relationships. *Journal of Family Communication, 12:* 40–56.

Clarke, J. I. (2004, November). The overindulgence research literature: Implications for family life educators. Poster presented at the annual meeting of the National Council on Family Relations, Orlando, FL.

Copeland, L. (2012, October 23). Tech keeps tabs on teen drivers. *USA Today,* p. A3.

Coyne, S. M., L. M. Padilla-Walker, L. Stockdale, & A. Fraser. (2011, November 18). Media and the family: Associations between family media use and family connections. Poster presented at the annual meeting of National Council on Family Relations, Orlando, FL.

Crisp, B., & D. Knox. (2009). *Behavioral family therapy: An evidence based approach.* Durham, NC: Carolina Academic Press.

Crosnoe, R., & S. E. Cavanagh. (2010). Families with children and adolescents: A Review, critique, and future agenda. *Journal of Marriage and Family, 72:* 594–611.

Crutzen, R., E. Nijhuis, & S. Mujakovic. (2012). Negative associations between primary school children's perception of being allowed to drink at home and alcohol use. *Mental Health and Substance Use, 5:* 64–69.

Dearden, K., B. Crookston, H. Madanat, J. West, M. Penny, & S. Cueto. (2013). What difference can fathers make? Early paternal absence compromises Peruvian children's growth. *Maternal & Child Nutrition, 9:* 143–154.

Dumka, L. E., N. A. Gonzales, L. A. Wheeler, & R. E. Millsap. (2010). Parenting self-efficacy and parenting practices over time in Mexican American families. *Journal of Family Psychology, 24:* 522–531.

Durtschi, J. A., & K. L. Soloski. (2012, November). The dyadic effects of coparenting and parental stress on relationship quality. Presentation at the National Council on Family Relations, Phoenix, AZ.

El Nasser, H. (2012, August 1). No place like home for adult kids. *USA Today,* p. 1A.

Flood, M. (2009). The harms of pornography exposure among children and young people. *Child Abuse Review, 18:* 384–400.

Forbes, E. E., & Dahl, R. E. (2012). Research review: Altered reward function in adolescent depression: What, when and how? *Journal of Child Psychology and Psychiatry, 53:* 3–15.

Freeman, D., & J. Temple. (2010). Social factors associated with history of sexual assault among ethnically diverse adolescents. *Journal of Family Violence, 25:* 349–356.

Galinsky, E. (2010). *Mind in the making.* New York: Harper Collins.

Gelabert, E., S. Subirà, L. García-Esteve, P. Navarro, A. Plaza, E. Cuyàs, et al. (2012). Perfectionism dimensions in major postpartum depression. *Journal of Affective Disorders, 136:* 17–25.

Gibbs, N. (2012, August 27). Your life is fully mobile: Time Mobility Survey. *Time,* 32 and following.

Goldberg, A. E., J. Z. Smith, & D. A. Kashy. (2010). Preadoptive factors predicting lesbian, gay, and heterosexual couples' relationship quality across the transition to adoptive parenthood. *Journal of Family Psychology, 24:* 221–232.

Green, M. (2014, May 26). My life has completely changed. *People,* 53–58.

Haag, P. (2011). *Marriage Confidential.* New York: Harper Collins.

Hall, S., & D. Knox. (2015). Relationship and sexual behaviors of a sample of 4,700 university students. Unpublished data collected for this text. Department of Family and Consumer Sciences, Ball State University and Department of Sociology, East Carolina University.

Holmes, E. K., T. Sasaki, & N. L. Hazen. (2013). Smooth versus rocky transitions to parenthood: Family systems in developmental context. *Family Relations, 62:* 824–837.

Hong, R., & A. Welch. (2013). The lived experiences of single Taiwanese mothers being resilient after divorce. *Journal of Transcultural Nursing, 24:* 51–59.

Hyde, A., J. Drennan, M. Butler, E. Howlett, M. Carney, & M. Lohan. (2013). Parents' constructions of communication with their children about safer sex. *Journal of Clinical Nursing, 22:* 3438–3446.

Kaiser Family Foundation. (2010). Media use among teens. Retrieved February 7, 2010, from http://www.kff.org/entmedia/entmedia012010nr.cfm

Keim, B., & A. L. Jacobson. (2011). *Wisdom for parents: Key ideas from parent educators.* Toronto: de Sitter Publications of Canada.

Killewald, A. (2013). A reconsideration of the fatherhood premium: Marriage, coresidence, biology, and fathers' wages. *American Sociological Review, 78:* 96–116.

Knog, G., D. Camenga, & S. Krishnan-Sarin (2012). Parental influence on adolescent smoking cessation: Is there a gender difference? *Addictive Behaviors, 37:* 211–216.

Knox, D., C. Schacht, & M. Whatley. (2014). *Human sexuality: Making informed decisions,* 4th rev. ed. Reddington, CA: Best Value Publishers.

Kotila, L. E., S. J. Schoppe-Sullivan, & C. M. Kamp Dush. (2013). Time parenting activities in dual-earner families at the transition to parenthood. *Family Relations, 62:* 795–807.

Livingston, G. (2014). Growing number of dads home with the kids. *Pew Research Center.* Published June 5, 2014, from http://www.pewsocialtrends.org/2014/06/05/growing-number-of-dads-home-with-the-kids/

Maier, C., & C. R. McGeorge. (2013). Positive perceptions of single mothers and fathers: Implications for therapy. Paper presented at the annual meeting of the National Council on Family Relations, San Antonio, TX.

Mashoa, S. W., D. Chapmana, & M. Ashbya. (2010). The impact of paternity and marital status on low birth weight and preterm births. *Marriage & Family Review, 46:* 243–256.

Mayer, L., & W. W. Blome. (2013). The importance of early, targeted intervention: The effect of family, maternal, and child characteristics on the use of physical discipline. *Journal of Human Behavior in the Social Environment, 23:* 144–158.

McClain, L. R. (2011). Better parents, more stable partners: Union transitions among cohabiting parents. *Journal of Marriage and Family, 73:* 889–901.

McDaniel, B. T., L. E. Philbrook, & D. M. Teti. (2012, November). Becoming parents: Coparenting quality and salivary cortisol. Paper presented at the National Council on Family Relations, Phoenix, AZ.

McKinney, C., & K. Renk. (2008). Differential parenting between mothers and fathers: Implications for late adolescents. *Journal of Family Issues, 29:* 806–827.

McLanahan, S. S. (1991). The long-term effects of family dissolution. In Brice J. Christensen (Ed.), *When families fail: The social costs* (pp. 5–26). New York: University Press of America for the Rockford Institute.

McLanahan, S. S., & K. Booth. (1989). Mother-only families: Problems, prospects, and politics. *Journal of Marriage and the Family, 51:* 557–580.

Minnottea, K. L., D. E. Pedersena, S. E. Mannonb, & G. Kiger. (2010). Tending to the emotions of children: Predicting parental performance of emotion work with children. *Marriage & Family Review, 46:* 224–241.

Morawska, A., & M. Sanders. (2011). Parental use of time out revisited: A useful or harmful parenting strategy? *Journal of Child & Family Studies, 20:* 1–8.

Mortensen, O., T. Torsheim, O. Melkevik, & F. Thuen. (2012). Adding a baby to the equation: Married and cohabiting women's relationship satisfaction in the transition to parenthood. *Family Process, 51:* 122–139.

Nelson, M. K. (2010). *Parenting out of control.* New York: New York University Press.

Nomaguchi, K., S. L. Brown, & T. Leyman. (2012, November). Father involvement and maternal parenting stress: The role of relationship status. Paper presented at the National Council on Family Relations, Phoenix, AZ.

Offer, S. (2013). Family time activities and adolescents' emotional well-being. *Journal of Marriage & Family, 75:* 26–41.

Panetta, S. M., C. L. Somers, A. R. Ceresnie, S. B. Hillman, & R. T. Partridge. (2014). Maternal and paternal parenting style patterns and adolescent emotional and behavioral outcomes. *Marriage & Family Review, 50:* 342–359.

Parker, K. (2012). The boomerang generation. *Pew Research Center.* Retrieved on March 31 from http://www.pewsocialtrends.org/2012/03/15/the-boomerang-generation/2/#who-are-the-boomerang-kids

Payne, K., & P. Trapp. (2013, April 30). What makes parenting tougher today? Survey by Survey.com. *USA Today,* p. D1.

Pritchard, K. M., & L. Kort-Butler. (2012, November). Multiple motherhoods: The interactive effects of importance of life satisfaction. Poster presented at the National Council on Family Relations, Phoenix, AZ.

ProQuest Statistical Abstract of the United States, 2014, 2nd ed., Table 112. Bethesda, MD.

Quing, M., Z. Li-xia, & S. Xiao-yin. (2011). A comparison of postnatal depression and related factors between Chinese new mothers and fathers. *Journal of Clinical Nursing, 20:* 645–652.

Rayne, K. (2013). Based of workshop presentations. Email Dr. Rayne at karen@karenrayne.com for information about workshops/further use of this material.

Ready, B., L. Asare, & E. Long. (2011, November). Factors that impact a father's philosophy on parenting. Presentation at the National Council on Family Relations, Orlando, FL.

Sassler, S., D. Ciambrone, & G. Benway. (2008). Are they really mamma's boys/daddy's girls? The negotiation of adulthood upon returning home to the parental home. *Sociological Forum, 23:* 670–698.

Schoppe-Sullivan, S. J., G. L. Brown, E. A. Cannon, S. C. Mangelsdorf, & M. S. Sokolowski. (2008). Maternal gate keeping, coparenting quality, and fathering behavior in families with infants. *Journal of Family Psychology, 22:* 389–397.

Slinger, M. R., & D. J. Bredehoft. (2010, November 3–5). Relationships between childhood overindulgence and adult attitudes and behavior. Poster presented at the annual meeting of the National Council on Family Relations, Minneapolis, MN.

Solomon, A. (2012). *Far from the tree.* New York: Scribner.

Stanley, S. M., & H. J. Markman. (1992). Assessing commitment in personal relationships. *Journal of Marriage and Family, 54:* 595–608.

Su, J. H. (2012). Pregnancy intentions and parents' psychological well-being. *Journal of Marriage and the Family, 74:* 1182–1196.

Surjadi, F. F., F. O. Lorenz, R. D. Conger, & K. A. S. Wickrama. (2013). Harsh, inconsistent parental discipline and romantic relationships: Mediating processes of behavioral problems and ambivalence. *Journal of Family Psychology, 27:* 762–772.

Trifan, T. A., H. Stattin, & L. Tilton-Weaver. (2014). Have authoritarian parenting practices and roles changed in the last 50 years? *Journal of Marriage and Family, 76:* 744–761.

Uji, M., A. Sakamoto, K. Adachi, & T. Kitamura. (2014). The impact of authoritative, authoritarian and permissive parenting styles on children's later mental health in Japan: Focusing on parent and child gender. *Journal of Child & Family Studies, 23:* 293–302.

Walcheski, M. J., & D. J. Bredehoft. (2010, November 3–5). Exploring the relationship between overindulgence and parenting styles. Poster presented at the annual meeting of the, National Council on Family Relations, Minneapolis, MN.

Wickel, K. (2012, November). Partner support and postpartum depression: A review of the literature. Poster presented at the National Council on Family Relations, Phoenix, AZ.

Wilson, E. K., B. T. Dalberth, H. P. Koo, & J. C. Gard. (2010). Parents' perspectives on talking to preteenage children about sex. *Perspectives on Sexual and Reproductive Health, 42:* 56–64.

Wilson, K. R., S. S. Havighurst, & A. E. Harley. (2012). Tuning in to kids: An effectiveness trial of a parenting program targeting emotion socialization of preschoolers. *Journal of Family Psychology, 26:* 56–65.

Wiseman, R. (2013). *Masterminds and wingmen.* New York: Harmony Books.

Chapter 13

Abrahamson, I., H. Rafat, K. Adeel, & M. J. Schofield. (2012). What helps couples rebuild their relationship after infidelity. *Journal of Family Issues, 33:* 1494–1519.

Alam, R., M. Barrera, N. D'Agostino, D. B. Nicholas, & G. Schneiderman. (2012). Bereavement experiences of mothers and fathers over time after the death of a child due to cancer. *Death Studies, 36:* 1–22.

Backhans, M. C., & T. Hemmingsson. (2012). Unemployment and mental health Who is not affected? *European Journal of Public Health, 22:* 429–433.

Baer, R. A., J. Carmody, & M. Hunsinger. (2012). Weekly change in mindfulness and perceived stress reduction program. *Journal of Clinical Psychology, 68:* 755–765.

Bagarozzi, D. A. (2008). Understanding and treating marital infidelity: A multidimensional model. *The American Journal of Family Therapy, 36:* 1–17.

Barnes, H., D. Knox, & J. Brinkley. (2012, March). CHEATING: Gender differences in reactions to discovery of a partner's cheating. Paper presented at the annual meeting of the Southern Sociological Society, New Orleans, LA.

Barton, A. L., & M. S. Kirtley. (2012). Gender differences in the relationships among parenting styles and college student mental health. *Journal of American College Health, 60,* 21–26.

Bermant, G. (1976). Sexual behavior: Hard times with the Coolidge Effect. In M. H. Siegel & H. P. Zeigler (Eds.), *Psychological research: The inside story.* New York: Harper and Row.

Bodenmann, G., D. C. Atkins, M. Schär, & V. Poffet. (2010). The association between daily stress and sexual activity. *Journal of Family Psychology, 24:* 271–279.

Boss, P. (2013, November). Myth of closure: What is normal grief after loss, clear or ambiguous? Paper presented at the annual meeting of the National Council on Family Relations, San Antonio, TX.

Burke, M. L., G. G. Eakes, & M. A. Hainsworth. (1999). Milestones of chronic sorrow: Perspectives of chronically ill and bereaved persons and family caregivers. *Journal of Family Nursing, 5:* 387–384.

Burr, W. R., & S. R. Klein. (1994). *Reexamining family stress: New theory and research.* Thousand Oaks, CA: Sage.

Butterworth, P., & B. Rodgers. (2008). Mental health problems and marital disruption: Is it the combination of husbands and wives' mental health problems that predicts later divorce? *Social Psychiatry and Psychiatric Epidemiology, 43:* 758–764.

Cavaglion, G., & E. Rashty. (2010). Narratives of suffering among Italian female partners of cybersex and cyber-porn. *Sexual Addiction & Compulsivity, 17:* 270–287.

Cohen, S., & D. Janicki-Deverts. (2012). Who's stressed? Distributions of psychological stress in the United States in probability samples from 1983, 2006, and 2009. *Journal of Applied Social Psychology, 42:* 1320–1334.

Confer, J. C., & M. D. Cloud. (2011). Sex differences in response to imagining a partner's heterosexual or homosexual affair. *Personality & Individual Differences, 50:* 129–134.

Cox, R. B., J. S. Ketner, & A. J. Blow. (2013). Working with couples and substance abuse: Recommendations for clinical practice. *The American Journal of Family Therapy, 41:* 160–172.

Cramer, R. E., R. E. Lipinski, J. D. Meteer, & J. A. Houska. (2008). Sex differences in subjective distress to unfaithfulness: Testing competing evolutionary and violation of infidelity expectations hypotheses. *The Journal of Social Psychology, 148:* 389–406.

Cravens, J. D., & J. B. Whiting. (2014, May 28). Clinical implications of Internet infidelity: Where Facebook fits in. *The American Journal of Family Therapy, 42:* 325–339. Online.

Crisp, B., & D. Knox. (2009). *Behavioral family therapy.* Durham, NC: Carolina Academic Press.

Derby, K., D. Knox, & B. Easterling. (2012). Snooping in romantic relationships. *College Student Journal, 46:* 333–343.

Doria, M. V., H. Kennedy, C. Strathie, & S. Strathie. (2014). Explanations for the success of video interaction guidance (VIG): An emerging method in family psychotherapy. *The Family Journal, 22:* 78–87.

Dotson-Blake, K., D. Knox, & A. R. Holman. (2010, Fall). Reaching out: College student perceptions of counseling. *Professional Issues in Counseling.* Retrieved from http://www.shsu.edu/~piic /CollegeStudentPerceptions.htm

Eliason, M. (2012). Lost jobs, broken marriages. *Journal of Population Economics, 25:* 1365–1397.

Ellison, C. G., A. K. Henderson, N. D. Glen, & K. E. Harkrider. (2011). Sanctification, stress, and marital quality. *Family Relations, 60:* 404–420.

Enright, E. (2004, July/August). A house divided. *AARP The Magazine,* 60.

Euser, E. M., M. H. van Ijzendoorn, P. Prinzie, & M. J. Bakermans-Kranenburg. (2010). Prevalence of child maltreatment in the Netherlands. *Child Maltreatment, 15:* 5–17.

Field, T., M. Diego, M. Pelaez, O. Deeds, & J. Delgado. (2012). Depression and related problems of university students. *College Student Journal, 46:* 193–202.

Goode, E. (1999, July 17). New study finds middle age is prime of life. *New York Times,* p. D6.

Goodman, M. A., D. C. Dollahite, L. D. Marks, & E. Layton. (2013). Religious faith and transformational processes in marriage. *Family Relations, 62:* 808–823.

Grases, G., C. Sanchez-Curto, E. Rigo, & D. Androver-Roi. (2012). Relationship between positive humor and state and trait anxiety. *Ansiedad y Estrés, 18:* 79–93.

Green, M., & M. Elliott. (2010). Religion, health, and psychological well-being. *Journal of Religion & Health, 49:* 149–163.

Hall, J. H., W. Fals-Stewart, & F. D. Fincham. (2008). Risky sexual behavior among married alcoholic men. *Journal of Family Psychology, 22:* 287–299.

Hall, S., & D. Knox. (2015). Relationship and sexual behaviors of a sample of 4,590 university students. Unpublished data collected for this text. Department of Family and Consumer Sciences, Ball State University and Department of Sociology, East Carolina University.

Hertlein, K. M., & F. P. Piercy. (2012). Essential elements of Internet infidelity treatment. *Journal of Marital & Family Therapy, 38:* 257–270.

Higgins, C. A., L. E. Duxbury, & S. T. Lyons. (2010). Coping with overload and stress: Men and women in dual-earner families. *Journal of Marriage and Family, 72:* 847–859.

Hill, E. W. (2010). Discovering forgiveness through empathy: Implications for couple and family therapy. *Journal of Family Therapy, 32:* 169–185.

Hines, J. (2012). A gender comparison of perceptions of offline and online sexual cheating in middle-aged adults. Dissertation, Department of Psychology, Walden University.

Hudson, C. G. (2012). Declines in mental illness over the adult years: An enduring finding or methodological artifact? *Aging & Mental Health, 16:* 735–752.

Hughes, M. (2011). Hey, we love animals! *Industrial Engineer, 43:* 6.

Jeanfreau, M. M., A. P. Jurich, & M. D. Mong. (2014). An examination of potential attractions of women's marital infidelity. *American Journal of Family Therapy, 42:* 14–28.

Jones, K. E., & A. E. Tuttle. (2012). Clinical and ethical considerations for the treatment of cybersex addiction for marriage and family therapists. *Journal of Couple & Relationship Therapy, 11:* 274–290.

Lane, C. D., P. S. Meszaros, & T. Savla. (2012, November). Developing the family resilience measure. Paper presented at the annual meeting of the National Council on Family Relations, Phoenix, AZ.

Lindblad-Goldberg, M., & W. Northey. (2013). Ecosystemic structural family therapy. Theoretical and clinical foundations. *Contemporary Family Therapy: An International Journal, 35:* 147–160.

Lipscomb, R. (2009). Person-first practice: Treating patients with disabilities. *Journal of the American Dietetic Association, 109:* 21–25.

Lotspeich-Younkin, F., & S. Bartle-Haring. (2012, November). Differentiation and relationship satisfaction: Mediating effects of emotional intimacy and alcohol/substance use. Paper presented at annual meeting of the National Council on Family Relations, Phoenix, AZ.

Maddox Shaw, A. M., G. K. Rhoades, E. S. Allen, S. M. Stanley, & H. J. Markman. (2013). Predictors of extradyadic sexual involvement in unmarried opposite-sex relationships. *Journal of Sex Research, 50:* 598–610.

Magaziner, J. (2010, September/October). The new technologies of change. *Psychology Networker,* 42–47.

Maple, M., H., E. Edwards, V. Minichiello, & D. Plummer. (2013). Still part of the family: The importance of physical, emotional and spiritual memorial places and spaces for parents bereaved through the suicide death of their son or daughter. *Mortality, 18:* 54–71.

Marshall, A., J. M. Bell, & N. J. Moules. (2010). Beliefs, suffering, and healing: A clinical practice model for families experiencing mental illness. *Perspectives in Psychiatric Care, 46:* 197–208.

Mata, J., C. L. Hogan, J. Joormann, C. E. Waugh, & I. H. Gotlib. (2013). Acute exercise attenuates negative affect following repeated sad mood inductions in persons who have recovered from depression. *Journal of Abnormal Psychology, 122:* 45–50.

Mendenhall, R., P. Bowman, & L. Zhang. (2013). Single black mothers' role strain and adaptation across the life course. *Journal of African American Studies, 17:* 74–89.

Merrill, J., & D. Knox. (2010). *Finding love from 9 to 5: Trade secrets of an office love.* Santa Barbara, CA: Praeger.

Morell, V. (1998). A new look at monogamy. *Science, 281:* 1982.

Mutch, K. (2010). In sickness and in health: Experience of caring for a spouse with MS. *British Journal of Nursing, 19:* 214–219.

Negash, S., M. Cui, F. D. Fincham, & K. Pasley. (2014). Extradyadic involvement and relationship dissolution in heterosexual women university students. *Archives of Sexual Behavior, 43:* 531–539.

Nelson, Eve-Lynn, & Thao Bui. (2010). Rural telepsychology services for children and adolescents. *Journal of Clinical Psychology: In Session, 66:* 490–501.

Omarzu, J., A. N. Miller, C. Shultz, & A. Timmerman. (2012, June). Motivations and emotional consequences related to engaging in extramarital relationships. *International Journal of Sexual Health, 24.* Online.

Phillips, L. J., J. Edwards, N. McMurray, & S. Francey. (2012). Comparison of experiences of **stress** and coping between young people at risk of psychosis and a non-clinical cohort. *Behavioural & Cognitive Psychotherapy, 40:* 69–88.

Pietrzak, R. H., C. A. Morgan, & S. M. Southwick. (2010). Sleep quality in treatment-seeking veterans of Operations Enduring Freedom and Iraqi Freedom: The role of cognitive coping strategies and unit cohesion. *Journal of Psychosomatic Research, 69:* 441–448.

Russell, V. M, L. R. Baker, & J. K. Mcnulty. (2013). Attachment insecurity and infidelity in marriage: Do studies of dating relationships really inform us about marriage? *Journal of Family Psychology, 27:* 241–251.

Schmitt, J., & J. Jones. (2012). America's "New Class": A profile of the long-term unemployed. *New Labor Forum, 21:* 57–65.

Silveira, H., H. Moraesa, N. Oliveira, E. Freire Coutinho, & A. Jerson. (2013). Physical exercise and clinically depressed patients: A systematic review and meta-analysis. *Neuropsychobiology, 67:* 61–68.

Smith, A., & J. G. Grzywacz. (2013, November). Health and well-being in midlife parents of children with special needs. Paper presented at the annual meeting of the National Council on Family Relations, San Antonio, TX.

South, S. C., B. D. Doss, & A. Christensen. (2010). Through the eyes of the beholder: The mediating role of relationship acceptance in the impact of partner behavior. *Family Relations, 59:* 611–622.

Stephenson, K. R., & C. M. Meston. (2012). The young and the restless? Age as a moderator of the association between sexual desire and sexual distress in women. *Journal of Sex & Marital Therapy, 38:* 445–457.

Vina, J., F. Sanchis-Gomar, V. Martinez-Bello, & M. C. Gomezx-Cabrera. (2012). Exercise acts as a drug; the pharmacological benefits of exercise. *British Journal of Pharmacology, 167:* 1–12.

Walters, A. S., & B. D. Burger. (2013). "I love you, and I cheat": Investigating disclosures of infidelity to primary romantic partners. *Sexuality & Culture, 17:* 20–49.

Williamson, H. C., B. R. Karney, & T. N. Bradbury. (2013). Financial strain and stressful events predict newlyweds' negative communication independent of relationship satisfaction. *Journal of Family Psychology, 27:* 65–75.

Wright, P. J. (2013). U.S. males and pornography, 1973–2010: Consumption, predictors, correlates. *Journal of Sex Research, 50:* 60–71.

Chapter 14

Allen, E. S., & D. C. Atkins. (2012). The association of divorce and extramarital sex in a representative U.S. sample. *Journal of Family Issues, 33:* 1477–1493.

Amato, P. R. (2010). Research on divorce: Continuing trends and new developments. *Journal of Marriage and Family, 72:* 650–666.

Amato, P. R., J. B. Kane, & S. James. (2011). Reconsidering the 'Good Divorce.' *Family Relations, 60:* 511–524.

Arkes, J. (2013). The temporal effects of parental divorce on youth substance use. *Substance Use & Misuse, 48:* 290–297.

Baker, A. J. L., & D. Darnall. (2007). A construct study of the eight symptoms of severe parental alienation syndrome: A survey of parental experiences. *Journal of Divorce & Remarriage, 47:* 55–62.

Benjamin, Le, N. L. Dove, C. R. Agnew, M. S. Korn, & A. A. Musso. (2010). Predicting nonmarital romantic relationship dissolution: A meta-analytic synthesis. *Personal Relationships, 17:* 377–390.

Benton, S. D. (2008, November 10). Divorce mediation. Lecture. East Carolina University.

Brenner, R. E., & D. G. Vogal. (2014, August 7). Positive thought content valence after a breakup: Development of the positive and negative ex-relationship thoughts scale. Poster presented at the annual meeting of the American Psychological Association, Washington, DC.

Brown, S. L., & I. Fen Lin. (2012). The gray divorce revolution: Rising divorce among middle aged and older adults, 1990–2009. National Center for Marriage and Family Research, Bowling Green State University, Working Paper Series WP-12-04 March.

Brown, S. M., & J. Porter. (2013). The effects of religion on remarriage among American women: Evidence from the National Survey of Family Growth. *Journal of Divorce and Remarriage, 54:* 142–162.

Carr, D., & K. W. Springer. (2010). Advances in families and health research in the 21st century. *Journal of Marriage and Family, 72:* 743–761.

Cherlin, A. J. (2010). Demographic trends in the United States: A review of research in the 2000s. *Journal of Marriage and Family, 72:* 403–419.

Chiu, H., & D. Busby. (2010, November 3–5). Parental influence in adult children's marital relationship. Poster presented at the annual meeting of the National Council on Family Relations, Minneapolis, MN.

Clark, S., & C. Kenney. (2010). Is the United States experiencing a "matrilineal tilt"?: Gender, family structures and financial transfers to adult children. *Social Forces, 88:* 1753–1776.

Clarke, S. C., & B. F. Wilson. (1994). The relative stability of remarriages: A cohort approach using vital statistics. *Family Relations, 43:* 305–310.

Crawford, J. (2012, November). Changes among parents attending a divorce education program at 6 and 12 month follow up. Paper presented at the National Council on Family Relations, Phoenix, AZ.

Crosnoe, R., & S. E. Cavanagh. (2010). Families with children and adolescents: A review, critique, and future agenda. *Journal of Marriage and Family, 72:* 594–611.

Cui, M., F. D. Fincham, & J. A. Durtschi. (2011). The effect of parental divorce on young adults' romantic relationship dissolution: What makes a difference? *Personal Relationships, 18:* 410–426.

Dixon, L. J., K. C. Gordon, N. N. Frousakis, & J. A. Schumm. (2012). Expectations and the marital quality of participants of a marital enrichment seminar. *Family Relations, 61:* 75–89.

Djamba, Y. K., L. C. Mullins, K. P. Brackett, & N. J. McKenzie. (2012). Household size as a correlate of divorce rate: A county-level analysis. *Sociological Spectrum, 32:* 436–448.

Ebert, R. (2011). *Life itself.* New York: Grand Central Publishing.

Faircloth, M., D. Knox, & J. Brinkley. (2012, March 21–24). The good, the bad and technology mediated communication in romantic relationships. Paper presented at the Southern Sociological Society, New Orleans, LA.

Fox, W. E., & M. Shriner. (2014). Remarried couples in premarital education: Does the content match participant needs? *Journal of Divorce & Remarriage, 55:* 276–299.

Frimmel, W., M. Halla, & R. Winter-Ebmer. (2013). Assortative mating and divorce: Evidence from Austrian register data. *Journal of the Royal Statistical Society, Series A, Statistics in Society, 176*: 907–920.

Ganong, L. H., & M. Coleman. (1999). *Changing families, changing responsibilities: Family obligations following divorce and remarriage.* New York: Lawrence Erlbaum.

Gardner, R. A. (1998). *The parental alienation syndrome,* 2d ed. Cresskill, NJ: Creative Therapeutics.

Gershon, I. (2010). *The breakup 2.0.* New York: Cornell University Press.

Godbout, E., & C. Parent. (2012). The life paths and lived experiences of adults who have experienced parental alienation: A retrospective study. *Journal of Divorce and Remarriage, 53*: 34–54.

Goetting, A. (1982). The six stations of remarriage: The developmental tasks of remarriage after divorce. *The Family Coordinator, 31*: 213–222.

Guinart, M., & M. Grau. (2014). Qualitative analysis of the short-term and long-term impact of family breakdown on children: Case study. *Journal of Divorce & Remarriage, 55*: 408–422.

Hall, S., & D. Knox. (2015). Relationship and sexual behaviors of a sample of 4,730 university students. Unpublished data collected for this text. Department of Family and Consumer Sciences, Ball State University and Department of Sociology, East Carolina University.

Halligan, C., J. Chang, & D. Knox. (2014). Positive effects of parental divorce on undergraduates. Submitted for publication. Email knoxd@ecu.edu for publication reference.

Hawkins, A. J., B. J. Willoughby, & W. J. Doherty. (2012). Reasons for divorce and openness to marital reconciliation. *Journal of Divorce & Remarriage, 53*: 453–463.

Healy, M., & V. Salazar. (2010, April 27). Dealbreakers. *USA Today,* p. D1.

Hetherington, E. M. (2003). Intimate pathways: Changing patterns in close personal relationships across time. *Family Relations, 52*: 318–331.

Jayson, S. (2013, September 13). Remarriage in an age of cohabitation. *USA Today,* p. 2A.

———. (2014, June 14). Dads fighting for equal rights. *USA Today,* p. 1A.

Kesselring, R. G., & D. Bremmer. (2006). Female income and the divorce decision: Evidence from micro data. *Applied Economics, 38*: 1605–1617.

Kim, H. (2011). Exploratory study on the factors affecting marital satisfaction among remarried Korean couples. *Families in Society, 91*: 193–200.

Lavner, J. A., & T. N. Bradbury. (2010). Patterns of change in marital satisfaction over the newlywed years. *Journal of Marriage and Family, 72*: 1171–1187.

Lucier-Greer, M., F. Adler-Baeder, S. A. Ketring, K. T. Harcourt, & T. Smith. (2012). Comparing the experiences of couples in first marriages and remarriages in couple and relationship education. *Journal of Divorce & Remarriage, 53*: 55–75.

Manning, W. D., & P. J. Smock. (2000). "Swapping" families: Serial parenting and economic support for children. *Journal of Marriage and the Family, 62*: 111–122.

Martin-Uzzi, M., & D. Duval-Tsioles. (2013). The experience of remarried couples in blended families. *Journal of Divorce & Remarriage, 54*: 43–57.

Masheter, C. (1999). Examples of commitment in post divorce relationships between spouses. In J. M. Adams & W. H. Jones (Eds.), *Handbook of interpersonal commitment and relationship stability* (pp. 293–306). New York: Academic/Plenum Publishers.

McIntosh, J. E., Y. D. Wells, B. M. Smyth, & C. M. Long. (2008). Child-focused and child-inclusive divorce mediation: Comparative outcomes from a prospective study of post separation adjustment. *Family Court Review, 46*: 105–115.

McNamee, C. B., P. Amato, & V. King. (2014). Nonresident father involvement with children and divorced women's likelihood of remarriage. *Journal of Marriage and Family, 76*: 862–874.

Meier, J. S. (2009). A historical perspective on parental alienation syndrome and parental alienation. *Journal of Child Custody, 6*: 232–257.

Mercadante, C., M. F. Taylor, & J. A. Pooley. (2014). "I wouldn't wish it on my worst enemy": Western Australian fathers' perspectives on their marital separation experiences. *Marriage & Family Review, 50*: 318–341.

Mirecki, R. M., J. L. Chou, M. Elliott, & C. M. Schneider. (2013). What factors influence marital satisfaction? Differences between first and second marriages. *Journal of Divorce & Remarriage, 54*: 78–93.

Mitchell, B. A. (2010). Midlife marital happiness and ethnic culture: A life course perspective. *Journal of Comparative Family Studies, 41*: 167–183.

Morgan, E. S. (1944). *The Puritan family.* Boston: Public Library.

Murray, C. (2012). *Coming apart: The state of white America, 1960–2010.* New York: Crown Forum.

Nielsen, L. (2004). *Embracing your father: How to build the relationship you always wanted with your dad.* New York: McGraw-Hill.

Northrup, J. C., & S. Shumway. (2014). Gamer widow: A phenomenological study of spouses of online video game addicts. *The American Journal of Family Therapy, 42*: 269–281.

Nunley, J. M., & A. Seals. (2010). The effects of household income volatility on divorce. *The American Journal of Economics and Sociology, 69*: 983–1011.

Nuru, A. K., & T. R. Wang. (2014). "She was stomping on everything that we used to think of as family": Communication and turning points in cohabiting (step) families. *Journal of Divorce & Remarriage, 55*: 145–163.

Park, H., & J. M. Raymo. (2013). Divorce in Korea: Trends and educational differentials. *Journal of Marriage and the Family, 75*: 110–126.

Parker-Pope, T. (2010). *For better: The science of a good marriage.* New York: E. P. Dutton.

Reck, K., & B. Higginbotham. (2012, November 1). No longer newlyweds: Difficulties experienced by remarried couples over time. Poster presented at the annual meeting of the National Council on Family Relations, Phoenix, AZ.

Rhoades, G. K., S. M. Stanley, & H. J. Markman. (2010). Should I stay or should I go? Predicting dating relationship stability from four aspects of commitment. *Journal of Family Psychology, 24*: 543–550.

Scarf, M. (2013). *The remarriage blueprint: How remarried couples and their families succeed or fail.* New York: Scribner.

Schacht, T. E. (2000). Protection strategies to protect professionals and families involved in high-conflict divorce. *UALR Law Review, 22*(3): 565–592.

Soria, K. M., & S. Linder. (2014). Parental divorce and first-year college students' persistence and academic achievement. *Journal of Divorce & Remarriage, 55*: 103–116.

South, A. L. (2013). Perceptions of romantic relationships in adult children of divorce. *Journal of Divorce & Remarriage, 54*: 126–141.

Sprecher, S., C. Zimmerman, & B. Fehr. (2014). The influence of compassionate love on strategies used to end a relationship. *Journal of Social and Personal Relationships, 31*: 697–705.

Strochschein, L. (2012). Parental divorce and child mental health: Accounting for predisruption differences. *Journal of Divorce & Remarriage, 53*: 489–502.

Sweeney, M. M. (2010). Remarriage and stepfamilies: Strategic sites for family scholarship in the 21st century. *Journal of Marriage and the Family, 72*: 667–684.

Teich, M. (2007). A divided house. *Psychology Today, 40*: 96–102.

Toren, P., B. L. Bregman, E. Zohar-Reich, G. Ben-Amitay, L. Wolmer, & N. Laor. (2013). Sixteen-session group treatment for children and adolescents with parental alienation and their parents. *American Journal of Family Therapy, 41:* 187–197.

Walsh, S. (2013). Match's 2012 singles in America survey. Retrieved from http://www.hookingupsmart.com/2013/02/07/hookinguprealities/matchs-2012-singles-in-america-survey/

Warrener, C., J. M. Koivunen, & J. L. Postmus. (2013). Economic self-sufficiency among divorced women: Impact of depression, abuse, and efficacy. *Journal of Divorce & Remarriage, 54:* 163–175.

Whitton, S. W., S. M. Stanley, H. J. Markman, & C. A. Johnson. (2013). Attitudes toward divorce, commitment, and divorce proneness in first marriages and remarriages. *Journal of Marriage and the Family, 75:* 276–287.

Yazedjian, A., & M. L. Toews. (2010, November 3–5). Breakups, depression, and self-esteem as predictors of college adjustment. Poster presented at annual meeting of the National Council on Family Relations, Minneapolis, MN.

Chapter 15

Ahmend, A. A., G. A. H. Van den Elsen, M. A. Van der Marck, & M. G. M. Olde Rickkert. (2014). Medicinal use of cannabis and cannabinoids in older adults: Where is the evidence? *American Journal of the American Geriatrics Society, 62:* 410–411.

Alterovitz, S. R., & G. A. Mendelsohn. (2013). Relationship goals of middle-aged, young-old, and old-old Internet daters: An analysis of online person ads. *Journal of Aging Studies, 27:* 159–165.

Aoyagi, Y., & R. Shephard. (2011). Habitual physical activity and health in the elderly: The Nakanojo study. *Geriatrics & Gerontology International, 10:* 236–243.

Barovick, H. (2012, March 22). Niche aging. *Time,* 84–87.

Ben-Zur, H. (2012). Loneliness, optimism, and well-being among married, divorced, and widowed individuals. *Journal of Psychology, 146:* 23–36.

Boyes, M. (2013). Outdoor adventure and successful ageing. *Ageing and Society, 33:* 644–665.

Brown, S. L., & I. F. Lin. (2012). The Gray Divorce Revolution: Rising divorce among middle aged and older adults, 1990–2009. National Center for Marriage and Family Research. Bowling Green State University. Working Paper Series, WP-12-04 March.

Brown, S. L., & S. K. Shinohara. (2013). Dating relationships in older adulthood: A national portrait. *Journal of Marriage and Family, 75:* 1194–1202.

Bryant, C. D. (2007). The sociology of death and dying. In Clifton D. Bryant & Dennis L. Peck (Eds.), *21st century sociology: A reference handbook* (pp. 156–156). Thousand Oaks, CA: Sage.

Cagney, K. A., C. R. Browning, J. Iveniuk, & N. English. (2014). The onset of depression during the great recession: Foreclosure and older adult mental health. *American Journal of Public Health,* 104: 498–505.

Carr, D., & K. W. Springer. (2010). Advances in families and health research in the 21st century. *Journal of Marriage and Family, 72:* 743–761.

Clarke, L. H., & E. Bennett. (2013). "You learn to live with all the things that are wrong with you": Gender and the experience of multiple chronic conditions in later life. *Ageing and Society, 33:* 342–360.

Doyle, M., C. O'Dywer, & V. Timonen. (2010). "How can you just cut off a whole side of the family and say move on?" The reshaping of paternal grandparent-grandchild relationships following divorce or separation in the middle generation. *Family Relations, 59:* 587.

Drew, E. N., & J. E. King. (2012, March 26–30). Young adult's perspectives on the difference between grandmother and grandfather involvement. Poster presented during Research and Creative Achievement Week, East Carolina University.

Dunifon, R., & A. Bajracharya. (2012). The role of grandparents in the lives of youth. *Journal of Family Issues, 9:* 1168–1194.

Felding, R. A., W. J. Rejeski, S. Blair, T. Church, M. A. Espeland, et al. (2011). The lifestyle interventions and independence for

elders study: Design and methods. *Journals of Gerontology, Series A: Biological Sciences & Medical Sciences, 66A:* 1226–1237.

Ford, P. (2011). *Glenn Ford: A life.* Madison, WI: University of Wisconsin Press.

Haboush, A., C. S. Warren, & L. Benuto. (2012). Beauty, ethnicity, and age: Does internalization of mainstream media ideals influence attitudes towards older adults? *Sex Roles, 66:* 3–20.

Hall, S., & D. Knox. (2015). Relationship and sexual behaviors of a sample of 4,690 university students. Unpublished data collected for this text. Department of Family and Consumer Sciences, Ball State University and Department of Sociology, East Carolina University.

Hensley, B., P. Martin, J. A. Margrett, M. MacDonald, I. C. Siegler, & L.W. Poon. (2012). Life events and personality predicting loneliness among centenarians: Findings from the Georgia Centenarian Study. *Journal of Psychology, 146:* 173–188.

Hogsnes, L., C. Melin-Johansson, K. Gustaf Norbergh, & E. Danielson. (2014). The existential life situations of spouses of persons with dementia before and after relocating to a nursing home. *Aging & Mental Health, 18:* 152–160.

Iveniuk, J., L. J. Waite, E. Laumann, M. K. McClintock, & A. D. Tiedt. (2014). Marital conflict in older couples: Positivity, personality, and health. *Journal of Marriage and Family, 76:* 130–144.

Johnson, C. L., & B. M. Barer. (1997). *Life beyond 85 years: The aura of survivorship.* New York: Springer Publishing.

Johnson, J. D., C. J. Whitlatch, & H. L. Menne. (2014). Activity and well-being of older adults: Does cognitive impairment play a role? *Research on Aging, 36:* 147–160.

Karraker, A., & J. DeLamater. (2013). Past year inactivity among older married persons and their partners. *Journal of Marriage and Family, 75:* 142–163.

Kelly, S. (2012). Dead bodies that matter: Toward a new ecology of human death in American culture. *The Journal of American Culture, 35:* 37–51.

Lee, J. A., P. Foos, & C. Clow. (2010). Caring for one's elders and family-to-work conflict. *Psychologist-Manager Journal, 13:* 15–39.

Leopold, T., M. Raab, & H. Engelhardt. (2014). The transition to parent care: Costs, commitments, and caregiver selection among children. *Journal of Marriage and the Family, 76:* 300–318.

Madden, M. (2010). Older adults and social media. Pew Research Internet Project. Posted and retrieved August 2 from http://pewresearch.org/pubs/1711/older-adults-social-networking-facebook-twitter

Mansson, D. H. (2012). A qualitative analysis of grandparents' expressions of affection for their young adult grandchildren. *North American Journal of Psychology, 14:* 207–219.

McGowan, C. M. (2011). Legal aspects of end of life care. *Critical Care Nurse, 31:* 64–69.

Meilaender, G. (2013). *Should we live forever? The ethical ambiguities of aging.* Grand Rapids, MI: Eerdmans.

Nadorff, M. R., A. Fiske, J. A. Sperry, R. Petts, and J. J. Gregg. (2013). Insomnia symptoms, nightmares, and suicidal ideation in older adults. *The Journals of Gerontology: Series, 68:* 145–152.

Nelson, Todd D. (2011). Ageism: The strange case of prejudice against the older you. In R. L. Wiener & S. L. Willborn (Eds.), *Disability and aging discrimination* (p. 37). New York: Springer Science + Business Media.

North, M. S., & S. T. Fiske. (2012). An inconvenienced youth: Ageism and its potential intergenerational roots. *Psychological Bulletin,* 138: 982–997.

Park, N. S., Y. Jang, B. S. Lee, & D. A. Chiriboga. (2013). Racial/ethnic differences in predictors of self-rated health: Findings from the survey of older Floridians. *Research on Aging, 35:* 201–219.

Park-Lee, E., M. Sengupta, A. Bercovitz, & C. Caffrey. (2013). Oldest old long-term care recipients: Findings from the national center for health statistics' long-term care surveys. *Research on Aging,* 35: 296–321.

Pilkauskas, N. V. (2012). Three-generation family households: Differences by family structure at birth. *Journal of Marriage and Family,* 74: 931–943.

Pond, R. L., C. Stephens, & F. Alpass. (2010). How health affects retirement decisions : Three pathways taken by middle-older aged New Zealanders. *Ageing & Society*, 30: 527–545.

Popham, L. E., S. M. Kennison, & K. I. Bradley. (2011). Ageism and risk-taking in young adults: Evidence for a link between death and anxiety and ageism. *Death Studies*, 35: 751–763.

Price, C. A., & O. Nesteruk. (2013, November 5–8). Being married in retirement: Can it be a double-edged sword? Paper presented at the annual meeting of the National Council on Family Relations, San Antonio, TX.

ProQuest Statistical Abstract of the United States, 2014, 2nd ed., Table 112. Bethesda, MD, 2013.

Proulx, C. M., & A. E. Ermer. (2013, November 5–8). Marital quality and health: Longitudinal associations across two generations. Paper presented at the annual meeting of the National Council on Family Relations, San Antonio, TX.

Rauer, A. (2013, February 22). From golden bands to the golden years: The critical role of marriage in older adulthood. Paper presented at the Southeastern Council on Family Relations, Birmingham, AL.

Roy, M., & H. Payette. (2012). The body image construct among Western seniors: A systematic review of the literature. *Archives of Gerontology and Geriatrics*, 55: 505–521.

Scheetz, L. T., P. Martin, & L. W. Poon. (2012). Do centenarians have higher levels of depression? Findings from the Georgia Centenarian Study. *Journal of the American Geriatrics Society,* 60: 238–242.

Simeonova, E. (2013). Marriage, bereavement and mortality: he role of health care utilization. *Journal of Health Economics*, 32: 33–50.

Stahl, L. (2014, May 4). Living to 90 and beyond. *60 Minutes*. CBS.

Stone, E. (2008). The last will and testament in literature: Rupture, rivalry, and sometimes rapprochement from Middlemarch to Lemony Sniket. *Family Process*, 47: 425–439.

Syme, M. L., E. A. Klonoff, C. A. Macera, & S. K. Brodine. (2013). Predicting sexual decline and dissatisfaction among older adults: The role of partnered and individual physical and mental health factors. *The Journals of Gerontology: Series B*, 68: 323–332.

Torres, S., & G. Hammarström. (2009). Successful aging as an oxymoron: Older people with and without home-help care. *International Journal of Ageing & Later Life*, 4: 23–54.

Vickers, R. (2010, March 22). Sexuality and the elderly. Presentation for the Sociology of Human Sexuality class, Department of Sociology, East Carolina University.

Walker, R. B., & M. A. Luszcz. (2009). The health and relationship dynamics of late-life couples: A systematic review of the literature. *Ageing and Society,* 29: 455–481.

Wang, M., K. Henkens, & H. Solinge. (2011). Retirement adjustment: A review of theoretical and empirical advancements. *American Psychologist,* 66: 204–213.

Wendell, C. R., J. Gunstad, S. R. Waldstein, J. G. Wright, L. Ferrucci, & A. B. Zonderman. (2014). Cardiorespiratory fitness and accelerated cognitive decline with aging. *Journal of Gerontology: A Biological Medicine*, 69: 455–462.

NAME INDEX

Dir, A. L., 102
Dixon, L. J., 272
Djamba, Y. K., 273
Doaring, C., 148
Dodd, M. D., 53
Dodge, B., 133
Doherty, W. J., 174, 272, 273
Dollahite, D. C., 163, 250
Dominguez, M. M., 199
Dore, M. M., 191
Doria, M., 264
Dorsey Green, G., 152
Doss, B. D., 113, 264
Dotson-Blake, K. P., 119, 263
Dougall, K. M., 217
Dove, N. L., 272
Dowd, D. A., 172
Downer, J. T., 171
Downing, J. B., 151, 219
Downing, M. J., Jr., 147
Doyle, M., 32, 306
Drennan, J., 241
Drew, E. N., 305
Drey, E. A., 223
Driessnack, M., 102
Duba, J. D. A., 172
Dubbs, S. L., 93
DuBois, S., 146
Ducharme, J. K., 149
Duggan, M., 35, 37, 38
Dula, C., 21
Dumka, L. E., 237
Dunifon, R., 305
Duntley, J. D., 194
Durtschi, J. A., 232, 278
DuToit, N. C., 31
Duval-Tsioles, D., 286
Duxbury, L. E., 249
Dwyer, D. S., 234
Dziegielewski, S. F., 146

E

Eagan, K., 3, 22
Eakes, G. G., 262
East, L., 60
Easterlin, R. A., 174
Easterling, B., 5, 62, 79, 83, 102, 109, 165, 258
Eaton, L., 198
Eaves, L. J., 85
Ebert, R., 286
Eck, B. A., 32
Eden, J., 121
Edmundson, M., 193, 194
Edwards, A. B., 103
Edwards, C., 64
Edwards, E., 262
Edwards, J., 250
Edwards, K., 58, 61
Edwards, T. M., 88
Eggermont, S., 46
Eibach, R. P., 135, 138
Eisenberg, M. E., 161
Eke, A., 199
El-Bassel, N., 111
El Nasser, H., 240
Eliason, M., 260
Elliott, L., 5, 79
Elliott, M., 251, 287
Ellison, C. G., 85, 251
Ellison, N. B., 36, 37
Engelhardt, H., 295

Engels, R. C., 58, 87
England, P., 35, 58
English, N., 300
English, S., 87
Enzlin, P., 129
Erickson, L., 54
Erikson, E., 70
Ermer, A. E., 299
Escoto, K., 185
Esmaila, A., 147
Espeland, M. A., 299
Espionoza, G., 99
Etzioni, A., 178
Eubanks Fleming, C. J., 62
Euser, E. M., 260
Exner, D., 200

F

Faircloth, M., 101, 269
Falcon, M., 221
Fals-Stewart, W., 256
Fang, X., 99
Farris, C., 201
Fauber, R., 133
Febres, J., 199
Fehr, B., 73, 270
Feistman, R., 99
Felding, R. A., 299
Feng, W., 213
Fennel, J., 127
Fergusson, D. M., 223
Fernandez-Lansac, V., 297
Fernandez-Montalvo, J., 198
Fetterolf, J. C., 130
Few, A. L., 204
Field, C. J., 166
Field, T., 252
Fielder, R. L., 122
Fields, H., 88
Figueiredo, B., 224
Finch, H., 171
Fincham, F. D., 35, 82, 256, 258, 278
Finer, L. B., 208, 221
Fink, J., 206
Finkenauer, C., 82
Finneran, C., 146
Fiona, C. S., 189
Fischer, J. L., 89
Fish, J., 275
Fisher, H., 66, 76, 79
Fisher, M. L., 91
Fisher, P. A., 199
Fiske, A., 300
Fiske, S. T., 295
Fleischman, D. S., 93
Fleming, K. M., 63
Flood, M., 238
Follingstad, D. R., 193, 194
Foos, P., 297
Foran, H., 165, 206
Forbes, E. E., 242
Forer, B., 185
Fortenberry, D., 133
Fortenberry, J. D., 53, 127
Foster, D. G., 222
Foster, J., 69, 88, 219
Fox, W. E., 287
Fraken, I. H. A., 69
Francey, S., 250
Fraser, A., 238
Freeman, D., 245
Freire Coutinho, E., 250

Freud, S., 56, 77
Freysteinsdottir, F. J., 170
Frimmel, W., 271
Frisco, M. L., 217
Fritch, A. M., 192
Frohwirth, L. F., 221, 224
Fromme, K., 123
Frousakis, N. N., 272
Fry, R., 178
Frye, N. E., 113
Frye-Cox, N., 81
Furstenberg, F. F., 16
Furukawa, R., 102

G

Gagnon, J. H., 128
Gaidarov, T. M., 35
Galen Buckwalter, J., 85
Galinsky, A. M., 130
Galinsky, E., 228
Gallmeier, C. P., 107
Gambone, L. J., 194
Gangamma, R., 138
Ganong, L. H., 99, 286
García-Esteve, L., 231
Garcia-Manglano, J., 180
Gardner, R. A., 277
Garfield, R., 63, 107
Garnhart, L., 206
Garrett, R. M., 221
Garrett, T. M., 221
Gartrell, N., 151
Gates, G. J., 136
Gelabert, E., 231
Geller, P., 220
Gentzler, K. C., 212
George, J., 165
George, L. J., 91
George, R. P., 14
Gerding, A., 58
Gershon, I., 99, 269
Gervais, S. J., 53
Ghassemi, A., 133
Gibbs, J. L., 36, 37
Gibbs, N., 99, 100, 238
Gibbs, P. M., 139
Gibson, C. L., 206
Gibson, L., 107
Gilbert, L., 111
Gilla, D. L., 144
Giordano, F. G., 171
Girgis, S., 14
Giuliano, T. A., 78
Glass, V. Q., 143
Glen, N. D., 251
Gliksberg, D., 177
Glymour, M. M., 137
Go, M., 79
Godbout, E., 277, 278
Goelman, H., 185
Goetting, A., 286
Goldberg, A. E., 151, 153, 219, 232
Goldberg, J. S., 9
Goldhammer, D. L., 132
Golub, S., 146
Gonzaga, G. C., 85
Gonzales, N. A., 237
Gonzalez, A., 186
Gonzalez, K. A., 141
Gonzalez, M., 136
Goode, E., 254
Goodman, M. A., 250

Jones, J., 260
Jones, K. E., 255
Jones, R. K., 220, 224
Jones, S. L., 138
Joormann, J., 250
Jorgensen, B. L., 172
Jose, A., 43
Joya, J., 221
Juffer, F., 218
Julien, D., 145
Jurich, A. P., 256

K

Kahn, J. R., 180
Kahneman, D., 179
Kaiser Family Foundation (2010), 238
Kalichman, S., 198
Kalish, R., 35
Kamiya, Y., 32
Kamp Dush, C. M., 122, 230
Kane, J. B., 281
Karney, B. R., 260
Karraker, A., 302
Karremans, J., 87
Karten, E. Y., 138
Kashy, D. A., 151, 232
Katono, K., 220
Katz, J., 206
Katz, M., 161
Katz, P. J., 217
Kauffman, L., 124
Kaufman, T. D., 158, 159
Kaysen, D. L., 34
Kefalas, M., 16
Keim, B., 237
Kelley, K., 109, 110
Kelly, J. R., 121
Kelly, S., 308
Kelton Research, 177
Kem, J., 78
Kendall, G., 184
Kendra, R., 192
Kennedy, D. P., 79
Kennedy, H., 264
Kenney, C., 276
Kenney, S. R., 35
Kennison, S. M., 295
Kent, T. A., 165
Kerestes, A. M., 105
Kesselring, R. G., 270
Ketner, J. S., 261
Ketring, S. A., 290
Keuzenkamp, S., 138, 140, 143
Keyes, S., 143, 145
Kiger, G., 62, 187, 228
Killewald, A., 231
Kilmann, P. R., 171
Kilpatrick, D., 192, 201
Kim, H., 199, 286
Kim, K., 133
Kimberly, C., 122
Kimmel, M., 35
Kimmes, J. G., 103
Kimport, K., 222
Kimuna, S. R., 166
King, A. L. S., 99
King, J. E., 305
King, V., 285
Kinkler, L. A., 219
Kinsey, A. C., 136
Kirtley, M. S., 246
Kitamura, T., 236

Kitaori, T., 220
Kjaer, S. K., 217
Kjaer, T., 217
Klamijn, M., 187
Klein, S. R., 249, 251
Klinenberg, E., 32
Klipfel, K. M., 192
Klonoff, E. A., 302
Kloosterman, M., 58
Klos, L. A., 158
Knight, K., 121
Knivickas, K., 61
Knog, G., 229
Knox, D., 5, 33, 53, 60, 62, 63, 70, 72, 79,
     81, 83, 85, 86, 91, 101, 102, 106, 107,
     109, 111, 119, 121–123, 126, 127,
     138, 156, 161, 165, 167, 170, 180,
     181, 191, 194, 195, 199, 200, 205,
     208, 222, 238, 239, 241, 250, 256,
     257, 258, 260, 263, 264, 269, 271,
     272, 275, 278, 306
Koch, J., 71
Koenen, K. C., 137
Kohlberg, L., 56
Koivunen, J. M., 276
Kolata, G., 128
Kollar, M. M., 149
Kondapalli, L. A., 218
Kooistra, K., 220
Korn, M. S., 272
Kornfield, S. L., 220
Kornich, S., 184
Kort-Butler, L., 217, 230
Kothari, C. L., 204, 206
Kotila, L. E., 230
Kotlyar, I., 37
Koverola, C., 204
Kraft, J. M., 99
Kremer, J., 217
Kress, V. E., 206
Kretz, S., 85
Kriegbaum, M., 112
Krishnan-Sarin, S., 229
Krohn, D., 206
Kryzer, M. J., 185
Kuhar, R., 141, 143
Kulkarni, M., 203
Kuntsman, J., 70
Kuper, L. E., 135, 136
Kuperberg, A., 43
Kurdek, L. A., 106, 114, 145

L

LaBrie, J. W., 35
Lac, A., 35
Lacks, M. H., 164
Lalicha, J., 139
Lam, C. B., 184
Lambert, N. M., 82
Lamson, A. L., 164
Landor, A., 119
Lane, C. D., 249
Lange, J., 219
Langeslag, S. J. E., 69
Langlois, J-P, 296
Laor, N., 278
Lara, T., 172
Larsen, C. D., 197
Lauer, S., 83, 166
Laumann, E. O., 128, 299
Lavner, J. A., 96, 112, 272
Lawrence, Q., 197

Lawyer, S., 201
Lax, E., 18
Layton, E., 250
Le, V. D., 102
Leaper, C., 64
Lease, S. H., 63
LeBaron, C. D. L., 162
LeBlanc, A. J., 149
Leddy, A., 151
Lee, B. S., 299
Lee, D., 218
Lee, J., 71
Lee, J. A., 297
Lee, J. T., 128
Lee, K., 212
Lee, R. M., 83
Lee, T., 139
Lefkowitz, E. S., 126
Lehman, A. D., 152
Lehmiller, J. J., 121
Lei, M., 197
Leno, J., 86, 276
Leopold, T., 295
Leung, L., 100
Leupp, K., 184
Levchenko, P., 167
Levenson, R. W., 146
Lever, J., 137
Levine, S. B., 122
Lewis, M. A., 34
Leyman, T., 231
Li, J., 184
Li-xia, Z., 231
Liang, C. C., 128
Liat, K., 43
Liben, L. S., 58
Lichter, D. T., 166
Lin, C. L., 128
Lin, I. F., 271, 300
Lincoln, A. E., 58
Lindblad-Goldberg, M., 263
Linder, S., 278
Lipinski, R. E., 255
Lipscomb, R., 252
Lisdonk, J. V., 140, 143
Litle, M. A., 149
Littleton, H., 201
Liu, S., 191
Lively, K., 61
Livingston, G., 189, 231
Lo, S. K., 37
Lobel, M., 60
Locker, L., 83
Lohan, R., 241
Long, C. M., 283
Long, E., 232
Looi, C., 99
Lopez-Goni, J. J., 198
Lorber, J., 17
Lorenz, F. O., 235
Lotspeich-Younkin, F., 261
Lozano, J. B., 3, 22
Lucas, A., 178
Lucas, T., 172
Lucey, A. B., 144
Lucier-Greer, M., 66, 290
Lund, R., 112
Luszcz, M. A., 304
Lydon, J., 40, 87
Lynch, A., 137
Lynch, S. M., 192
Lyons, H., 31
Lyons, M., 137

Sternberg, R. J., 74
Stewart, W., 93
Stillman, T. F., 82
Stinehart, M. A., 78
Stocco, C., 54
Stockdale, D., 100
Stockdale, L., 238
Stone, E., 308
Stookey, K., 170
Strachman, A., 104
Strassberg, D. S., 102
Strathie, C., 264
Strathie, S., 264
Straus, M. A., 166
Strazdin, L., 184
Strickler, B. L., 111
Strochschein, K., 278
Strong, G., 79
Stroud, C. B., 79
Stuart, G. L., 199
Stuart, G. W., 194
Stuewig, J., 203
Stulhofer, A., 127, 132, 186
Su, J. H., 230, 231
Subirà, S., 231
Subrahmanyam, K., 99
Sugita, C., 123
Sugiura-Ogasawara, M., 220
Sullivan, A., 150
Sullivan, M., 143, 145
Sullivan, T. P., 194
Surjadi, F. F., 235
Sustaita, M. A., 102
Sutphina, S. T., 147
Sutton, T. E., 197
Suzuki, S., 220
Suzumori, N., 220
Svab, A., 141, 143
Svedin, C. G., 137
Swan, S. C., 194
Swann, W. B., Jr., 187
Swanson, C., 146
Swanson, K., 146
Sweeney, J., 80
Sweeney, M. M., 285, 287, 288
Syme, M. L., 302
Symons, K., 126

T

Taft, C. T., 165
Tai, T. C., 180
Talens, J. G., 129, 132
Tan, R., 108
Tannen, D., 107, 108
Tavecchio, W. C., 185
Taylor, A. C., 99, 162
Taylor, C., 140
Taylor, D., 223
Taylor, J., 222
Taylor, J. K., 108
Taylor, M. F., 277
Taylor, P., 166, 167, 178
Teich, M., 277
Teidt, A. D., 299
Temple, B. W., 102
Temple, J., 102, 245
Ter Gogt, T. F. M., 58
Teskereci, G., 217
Teten, A. L., 165, 192
Teti, D. M., 232
Thayer, S. M., 106
Thomas, G., 140

Thomas, R. J., 36
Thomsen, D., 126
Thuen, F., 232
Tillman, K. H., 31
Tilton-Weaver, L., 236
Timmerman, A., 256
Timonen, V., 306
Todd, M. E., 139, 150, 159
Toews, M. L., 269
Tollerud, T., 171
Toomey, R. B., 66, 144
Toren, P. B., 278
Torres, S., 303
Torsheim, T., 232
Totenhagen, C., 161
Toufexis, A., 76
Toviessi, P., 138
Tracy, A., 204
Traeen, B., 132, 186
Trail, T. E., 77
Trap, P., 177
Trapp, P., 180–182, 238, 240
Treas, J., 180
Treat, T., 201
Trella, D., 31
Trifan, T. A., 236
Trucco, E. M., 57
True Love Waits (2014), 118
Tsunokai, G. T., 83
Tucker, J. S., 79
Tuttle, A. E., 255
Tyson, R., 146
Tzeng, O. C. S., 71

U

Uecker, J., 35
Uji, M., 236
Umbach, P., 59
United Nations (2011), 31
Upadhyay, U. D., 223
Updegraff, K. A., 9, 106
U.S. Department of Agriculture (2013), 211
U.S. Department of Commerce (2011), 211
Utley, E. A., 16
Uusiautti, S., 103, 179

V

Vaccaro, A., 143, 144
Vail-Smith, K., 102, 111
Valenca, A. M., 99
Valero, F., 221
Vall, O., 221
Valle, G., 31
Valle, L. A., 192
Van Bergen, D. D., 140, 143
van den Berg, P., 102
Van den Brink, F., 129, 132
Van den Elsen, G. A. H., 300
Van der Lippe, T., 180, 183
Van der Marck, M. A., 300
van der Steen, J., 217
Van der Veen, F., 217
Van Dulmen, M. H. M., 192
Van Eeden-Moorefield, B., 146
Van Geloven, N., 217
Van Houtte, M., 126
van Ijzendoorn, M. H., 185, 218, 260
Van Ryzin, M. J., 174, 185
Van Tran, T., 162

Vandell, D. L., 185
Vandello, J. A., 62
VanderDrift, E., 121
Vandergrift, N., 185
Vanderkam, L., 188
Vannier, S., 132
Varcoe, C. M., 206
Varner, F., 56
Vasilenko, S. A., 126
Vault Office Romance Survey (2014), 180
Vazonyi, A. I., 119
Velez-Blasini, C. J., 35
Vennum, A., 92
Ventura, S. J., 224
Verkamp, J. M., 105
Vermeersch, H., 126
Vespa, J., 13, 46
Vickers, R., 302
Viera, J., 88
Viken, R. J., 201
Vina, J., 250
Vogal, D. G., 284

W

Wade, J. C., 138
Wagner, C. G., 149
Wagner, S. L., 185
Waite, L. J., 299
Walby, S., 196
Walcheski, M. J., 235, 236
Waldinger, R. J., 197
Waldman, J., 223
Walker, A., 9
Walker, R. B., 304
Waller, W., 106
Wallerstein, J., 174
Wallis, C., 58
Walls, J. K., 184
Walmer, L., 278
Walsh, J. L., 122, 123
Walsh, S., 32, 39, 68, 110, 120, 178, 268
Walters, A. S., 258
Wan, G. H., 128
Wang, M. K., 301
Wang, T. R., 241, 288
Wang, W., 26, 28, 31, 166, 167, 178, 179–181, 189
Ward, J. T., 206
Ward, L. M., 123
Ward, R., 34, 35
Ward, S., 184
Warren, C. S., 295
Warren, J. T., 79
Warrener, C., 276
Waterman, J., 219
Wathen, C. N., 206
Watson, L. S., 140
Watt, M., 198
Watts, E., 139
Waugh, C. E., 250
Way, N., 62
Webb, M. W., 206
Weber, M., 296
Webley, K., 111
Weden, M., 217
Weisfeld, C. C., 172
Weisfeld, G. E., 172
Weisgram, E. S., 58
Weisskirch, R. S., 102
Weitz, T. A., 222, 223
Welch, A., 244
Wellings, K., 127, 127 126

SUBJECT INDEX

demandingness, 234
dementia, 300
dental dam, 126
Deol, Amon, 199
dependent personality disorder, 89
Depo-Provera, 215
depression, 252–253
developmental task, 288
diaphragm, 215
Dickens, Charles, 51, 108
DiFranco, Ani, 17
Dion, Celine, 168
disability, 252
discriminatory male-only clubs, 48
disenchantment, 161
disengagement theory of aging, 296
dishonesty, 110–112
Disney Corporation, 8
displacement, 115
Divided and Conquered (Mokhonoana), 254
divorce, 12, 266–290
    changing values, 273
    characteristics of those most likely to divorce, 273–275
    children, effect of, on, 278–281
    coparenting, 284
    custody of children, 277, 284
    deciding to end a relationship, 268–269
    defined, 266
    elderly people, 300, 306
    extramarital affair, 272
    falling out of love, 272
    financial consequences, 276–277
    growing apart, 272
    macro factors, 270–271
    making most out of divorce, 281–285
    mediation, 282–283, 284
    micro factors, 271–276
    moving on, 284
    no-fault, 270
    parental alienation, 277–278
    postnuptial agreement, 277
    prenuptial agreement, 271
    race, 271
    satiation, 273
    social class, 271
    trends, 290
divorce mediation, 282–283, 284
divorced singles, 32
divorcism, 290
DOMA. *See* Defense of Marriage Act (DOMA)
domestic partnership, 8, 43
domestic violence. *See* abusive relationships
dominance, 106
double victims, 200
Douglas, Michael, 168
Down, 239
down low, 147, 258
Dreschler, Fran, 147
Drug-Induced Rape Prevention and Punishment Act, 201
Drugfree.org, 238
dual-career marriage, 180–182
Dubreuil, Doreen, 179

E

Ebert, Robert, 83, 88
EC pills, 216
economic pressures, 178–179

economic values/money management, 85–86
EDI. *See* extradyadic involvement (EDI)
Edison, Thomas, 19
educational homogamy, 84
Edwards, John, 255
egalitarian marriage, 157
eight-minute date, 38–39
8minutedating.com, 38
Einstein, Albert, 13, 26
Einstein, Elsa, 13
elderly people, 292–309
    age discrimination, 298
    ageism, 294–295
    caregiving for frail elderly, 295–298
    dating, 303–304
    death of one's spouse, 307
    divorce, 300, 306
    grandparenthood, 305–306
    income, 298–299
    life expectancy, 294
    mental health, 300
    physical health, 299–300
    physiological sexual changes, 302
    preparing for death, 308–309
    relationship with spouse, 304
    retirement, 300–301
    sandwich generation, 295–298
    sexuality, 302–303
    successful aging, 303
    technology, 304, 307
    theories of aging, 295
    trends, 309
emergency contraception, 215, 216
Emerson, Ralph Waldo, 304
emotion work, 187
emotional abuse, 193–194
emotional competence, 241
emotional remarriage, 286
empathy, 103
employed wives, 179–180
empty love, 74
"End of Men, The" (Rosin), 28
endogamy, 82
engaged persons, 41
entrapped, 205
erectile dysfunction, 129, 302
eros love style, 71
erotic melancholy, 78
erotomania, 78
escapism, 114
evangelicals, 233
Everything I Never Wanted To Be (Kucera), 261
evolutionary theory of love, 76, 78
exaggerated sense of oneself, 88
exchange theory
    defined, 86
    marital success, 174
    mate selection, 86–87
    relationship communication, 112
exercise, 250
exogamy, 82
expected family stressors, 248
Experience Works, 301
experimental group, 22
expression of emotions, 63
extended family, 10
external crisis, 248
external genitalia, 50
extradyadic involvement (EDI), 256, 258
extramarital affair, 254–260
    alienation of affection lawsuits, 258–260

    divorce, and, 272
    reactions to discovering the affair, 259
    reasons for, 256–258
    revealing one's affair, 258
    successful recovery from infidelity, 260, 261
    types, 254–255
extrarelational involvement, 256
eye contact, 98, 103

F

Facebook infidelity, 255
Facebookcheating.com, 255
Fafafini (Fafa), 57
falling out of love, 272
familism, 11, 12, 271
family
    agent of socialization, as, 56–57
    changes (1950-2015), 12
    cultural differences, 163–164
    defined, 8
    functions, 14–15
    marriage, contrasted, 10
    overview, 10
    pets, 8
    pronatalism, 209
    sexual values, and, 124
    trends, 25
    types, 8–10
family caregiver, 219
family caregiving, 295
family crisis, 248
family development theory, 174
family homicide, 192
family housing, 12
family life course development, 13
family life cycle, 14
family meals, 240
family of orientation, 8–9
family of origin, 9
family of procreation, 9
family relationship values, 12
family resiliency, 249
family secrets, 110
family stress model, 249
family systems framework, 17
Far from the Tree (Solomon), 234
Farrell, Warren, 63
fatuous love, 74
FBW. *See* female breadwinner (FBW)
fear of intimacy, 63
fellatio, 126
female attraction to "bad boys," 89–90
female breadwinner (FBW), 178, 184
female condom, 215
female-dominated occupations, 58
female rape myths, 200
female role socialization
    body image, 60–61
    children, bonding with, 62
    feminization of poverty, 60
    income, 59–60
    life expectancy, 61–62
    personal/marital satisfaction, 61
    relationship focus, 62
    "relationship talk," 62
    STIs, 60
female-to-male transsexual (FtM), 52, 136, 147
feminist framework, 17
feminization of poverty, 60
Fertell (at-home fertility kit), 217

hypersensitivity, 88
hypothesis, 21

I

"I" statements, 105
Iceland
    parental protective function, 230
    undergraduate attitudes towards
        relationships, 169–170
Ildan, Mehmet Murat, 268
IM addiction, 100
Implanon, 215
impulse control, 88
In a Lonely Place (film), 70
in vitro fertilization (IVF), 217–218
incest taboos, 82
incompetent suitor, 193
India
    arranged marriage, 75
    Google Baby, 224
    sexual abuse, 199
Indian Princess role, 232
individualism, 11, 12, 271
Indonesia, arranged marriage, 75
induced abortion, 220
Industrial Revolution, 11
infatuation, 69, 74
infertility, 216–217
infidelity, 111–112, 170. *See also* extra-
    marital affair
inflated ego, 88
insecurity, 88
instant messaging, 100
Institutional Review Board (IRB), 21–22
integrative behavioral couple therapy
    (IBCT), 264
internal sex organs, 50
internalized homophobia, 139
international dating, 39
international marriage, 167–168
Internet, 234
Internet adoption, 219–220
Internet dating, 35–38
Internet generation, 20
Internet use, 255
interpersonal theory of aging, 296
interracial marriage, 166–167, 169
interracial relationships, 83
interreligious marriage, 167, 169–170
intersex development, 50
intersexed individuals, 50
intimacy stalker, 193
intimate partner homicide, 192
intimate partner violence (IPV), 191
intimate terrorism (IT), 192
intrauterine device (IUD), 215
investment model of commitment, 81
Invitation, The (Mountain Dreamer), 188
Involved Couple's Inventory, 91
IPV. *See* intimate partner violence (IPV)
IRB approval, 21–22
isolation, 193
Italy, corporal punishment, 196
IVF. *See* in vitro fertilization (IVF)

J

Jacobson, Peter Marc, 147
Japan
    amae, 114
    authoritative parents, 236
Jdate.com, 36

jealousy, 198
Jekyll-and-Hyde personality, 198
job satisfaction, 132
Jobs, Paul and Clara, 90
Jobs, Steve, 19
Johansson, Scarlett, 32
Johnson, Dave, 231
Johnson, Samuel, 287
Johnson, Virginia, 86–87
Jolie, Angelina, 85, 218, 226, 257
Jordan, honor killing, 197
Joseph, Jenny, 254
Journal of Marriage and the Family,
    25
judgmental statements, 105

K

Kafir tribe, 221
Kardashian, Kim, 166
Kawas, Claudia, 303
Kazantzakis, Nikos, 250
Keaton, Diane, 31
Kennedy, John F., 18
Kennedy, Ted, 309
Kik Messenger, 239
kink, 127–128
Kinsch, Amanda, 183
Kinsey scale, 136–137
Kintz, Jarod, 179
kissing, 125
Krauss, Alison, 96

L

Lady Gaga, 64
Lahad, Kinneret, 29
Lambert, Miranda, 85
LAT. *See* living apart together (LAT)
latchkey parents, 281
later years. *See* elderly people
Latinos/as. *See* Hispanic Americans
laughter and play, 251
LDDR. *See* long-distance dating relation-
    ship (LDDR)
learning theory of love, 76, 78
Lebowitz, Frances Ann, 103
Lennox, Annie, 64
Leno, Jay, 86, 276
lesbian, 135. *See also* GLBTQ
    relationships
Lesser, Elizabeth, 299
Letterman, David, 194
Levitra, 302
Lewis, Pam, 276
Life, The Truth, and Being Free
    (Maraboli), 247
life expectancy, 61–62, 63, 294
*Life of Unlearning: One Man's Journey to
    Find the Truth, A* (Venn-Brown), 144
liking, 74
limiting commitments, 187–188
listening, 104
literature review, 21
litigation, 283
living apart together (LAT)
    advantages, 44–45
    conditions precedent, 44
    defined, 43
    disadvantages, 45–46
    stepfamily, 289
living together, 41. *See also* cohabitation

long-distance dating relationship (LDDR),
    39–41
longitudinal study, 21
looking-glass self, 16, 112
love, 68–82
    companionate, 71
    compassionate, 73
    marriage, and, 155
    problems associated with, 78–80
    psychological conditions, 80–81
    realistic, 70
    risky, dangerous choices, 78–79
    romantic, 70
    social conditions, 80
    social control of, 75–76
    stress, and, 250
    triangular view, 74
    unrequited, 78
    what makes love last, 81–82
love addiction, 79
love at first sight, 33, 53
love junkies, 77
love languages, 74–75
love styles, 71–74
love theories
    attachment theory, 77–78
    biochemical theory, 77, 78
    evolutionary theory, 76, 78
    learning theory, 76, 78
    overview (table), 78
    psychosexual theory, 77, 78
    sociological theory, 76–77, 78
lovesickness, 78
ludic love style, 71
LuLu, 38
Lundquist, Anne, 260
lust, 69
lying, 110–112

M

MacArthur, Douglas, 293
Macbeth (Shakespeare), 108, 131
Maclaine, Shirley, 131
Madoff, Bernie, 110
Madoff, Mark, 110
Maher, Bill, 213
male-biased language, 54
male-dominated occupations, 58
male rape myths, 200
male role socialization
    available pool of potential partners, 63
    custody disadvantages, 63
    expression of emotions, 63
    fear of intimacy, 63
    freedom of movement, 63
    identity synonymous with occupation,
        62
    initiation of relationships, 63–64
    life expectancy, 63
male-to-female transsexual (MtF), 52,
    136, 147
mandatory education, 11
Manet, Edouard, 203
mania love style, 71
marijuana, 130, 262
marital rape, 203
marital satisfaction, 61
marital sex, 128
marital success
    commitment, 172
    common interests, 172
    communication, 171–172

LEARNING OUTCOMES

1-1 Marriage?

Marriage is a system of binding a man and a woman together for the reproduction, care (physical and emotional), and socialization of offspring. Marriage in the United States is a legal contract between a couple and their state that regulates their economic and sexual relationship. The federal government supports marriage education in the public school system with the intention of reducing divorce (which is costly to both individuals and society). The various types of marriage are polygyny, polyandry, polyamory, pantagamy, and domestic partnerships.

1-2 Family?

In recognition of the diversity of families, the definition of family is increasing beyond the U.S. Census Bureau's definition to include two adult partners whose interdependent relationship is long-term and characterized by an emotional and financial commitment. Types of family include nuclear, extended, and blended. There are also traditional, modern, postmodern, and binuclear families. See the table to the right for differences between marriage and family.

Differences between Marriage and the Family in the United States

Marriage	Family
Usually initiated by a formal ceremony.	Formal ceremony not essential.
Involves two people.	Usually involves more than two people.
Ages of the individuals tend to be similar.	Individuals represent more than one generation.
Individuals usually choose each other.	Members are born or adopted into the family.
Ends when spouse dies or is divorced.	Continues beyond the life of the individual.
Sex between spouses is expected and approved.	Sex between near kin is neither expected nor approved.
Requires a license.	No license needed to become a parent.
Procreation expected.	Consequence of procreation.
Spouses are focused on each other.	Focus changes with addition of children.
Spouses can voluntarily withdraw from marriage.	Parents cannot divorce themselves from obligations to children via divorce.
Money in unit is spent on the couple.	Money is used for the needs of children.
Recreation revolves around adults.	Recreation revolves around children.

Reprinted by permission of Dr. Lee Axelson.

1-3 Changes in marriage and family

The advent of industrialization, urbanization, and mobility involved the demise of familism and the rise of individualism. When family members functioned together as an economic unit, they were dependent on one another for survival and were concerned about what was good for the family. The shift from familism to individualism is only one change; others include divorce replacing death as the endpoint for the majority of marriages, marriage and relationships emerging as legitimate objects of scientific study, the rise of feminism, changes in gender roles, increasing marriage age, the emergence of living apart together (LAT), technological advances in communication, a growing openness surrounding discussion of sexuality and relationships, and the acceptance of singlehood, cohabitation, heterogamous mating, homosexuality, and childfree marriages.

1-4 Theoretical frameworks for viewing marriage and the family

Theoretical frameworks provide a set of interrelated principles designed to explain a particular phenomenon and provide a point of view. The following table gives an overview of the frameworks used in this text.

1-5 Choices in relationships: View of the text

A central theme of this text is to encourage you to be proactive—to make conscious, deliberate relationship choices to enhance your own well-being and the well-being of those in your intimate groups. Though global, structural, cultural, and media influences are operative, a choices framework emphasizes that individuals have some control over their relationships. Important issues to keep in mind about a choices framework for viewing marriage and the family are that (1) not to decide is to decide, (2) action must follow a choice, (3) some choices require correcting, (4) all choices involve trade-offs, (5) choices include

Theoretical Frameworks for Marriage and the Family

Theory	Description	Concepts	Level of Analysis	Strengths	Weaknesses
Social Exchange	In their relationships, individuals seek to maximize their benefits and minimize their costs.	Benefits, Costs, Profit, Loss	Individual, Couple, Family	Provides explanations of human behavior based on outcome.	Assumes that people always act rationally and all behavior is calculated.
Family Life Course Development	All families have a life course that is composed of all the stages and events that have occurred within the family.	Stages, Transitions, Timing	Institution, Individual, Couple, Family	Families are seen as dynamic rather than static. Useful in working with families who are facing transitions in their life courses.	Difficult to adequately test the theory through research.
Structure-Function	The family has several important functions within society; within the family, individual members have certain functions.	Structure, Function	Institution	Emphasizes the relation of family to society, noting how families affect and are affected by the larger society.	Families with nontraditional structures (single-parent, same-sex couples) are seen as dysfunctional.
Conflict	Conflict in relationships is inevitable, due to competition over resources and power.	Conflict, Resources, Power	Institution	Views conflict as a normal part of relationships and as necessary for change and growth.	Sees all relationships as conflictual, and does not acknowledge cooperation.
Symbolic Interaction	People communicate through symbols and interpret the words and actions of others.	Definition of the situation, Looking-glass self, Self-fulfilling prophecy	Couple	Emphasizes the perceptions of individuals, not just objective reality or the viewpoint of outsiders.	Ignores the larger social interaction context and minimizes the influence of external forces.
Family Systems	The family is a system of interrelated parts that function together to maintain the unit.	Subsystem, Roles, Rules, Boundaries, Open system, Closed system	Couple, Family	Very useful in working with families who are having serious problems (violence, alcoholism). Describes the effect family members have on each other.	Based on work with systems, troubled families, and may not apply to nonproblem families.
Feminism	Women's experience is central and different from men's experience of social reality.	Inequality, Power, Oppression	Institution, Individual, Couple, Family	Exposes inequality and oppression as explanations for frustrations women experience.	Multiple branches of feminism may inhibit central accomplishment of increased equality.

selecting a positive or negative view, and (6) some choices are not revocable. Generation Yers (born in the early 1980s) are relaxed about relationship choices. Rather than pair bond, they "hang out," "hook up," and "live together." They are in no hurry to find "the one," to marry, and to begin a family.

1-6 Research: Process and evaluation

Caveats that are factors to be used in evaluating research include a random sample (the respondents providing the data reflect those who were not in the sample), a control group (the group not subjected to the experimental design for a basis of comparison), terminology (the phenomenon being studied should be objectively defined), researcher bias (present in all studies), time lag (takes two years from study to print), and distortion or deception (although rare, some researchers distort their data). Few studies avoid all research problems.

1-7 Trends in marriage and family

Although there is a decline in marriage among middle Americans, marriage remains the dominant choice for most Americans. Delaying marriage has increased in popularity, but marriage continues to be a life goal. Same-sex marriage is on the rise as well, as a result of increasing legalization of same-sex marriage and federal tax codes permitting the filing of joint returns by same-sex couples.

Potential Research Problems in Marriage and Family

Weaknesses	Consequences	Examples
Sample not random	Cannot generalize findings	Opinions of college students do not reflect opinions of other adults.
No control group	Inaccurate conclusions	Study on the effect of divorce on children needs control group of children whose parents are still together.
Age differences between groups of respondents	Inaccurate conclusions	Effect may be due to passage of time or to cohort differences.
Unclear terminology	Inability to measure what is not clearly defined	What is the definition of cohabitation, marital happiness, sexual fulfillment, good communication, and quality time?
Researcher bias	Slanted conclusions	A researcher studying the value of a product (e.g., Atkins Diet) should not be funded by the organization being studied.
Time lag	Outdated conclusions	Often-quoted Kinsey sex research is over 65 years old.
Distortion	Invalid conclusions	Research subjects exaggerate, omit information, and/or recall facts or events inaccurately. Respondents may remember what they wish had happened.
Deception	Public misled	Dr. Anil Potti (Duke University) changed data on research reports and provided fraudulent results (Darnton, 2012).

KEY TERMS

binuclear family family in which the members live in two households.

blended family a family created when two individuals marry and at least one of them brings a child or children from a previous relationship or marriage. Also referred to as a stepfamily.

civil union a pair-bonded relationship given legal significance in terms of rights and privileges.

common-law marriage a marriage by mutual agreement between cohabitants without a marriage license or ceremony (recognized in some, but not all, states).

conflict framework view that individuals in relationships compete for valuable resources.

control group group used to compare with the experimental group that is not exposed to the independent variable being studied.

cross-sectional study means studying the whole population at one time (e.g., finding out from persons now living together about their experience).

domestic partnership a relationship in which individuals who live together are emotionally and financially interdependent and are given some kind of official recognition by a city or corporation so as to receive partner benefits.

experimental group the group exposed to the independent variable.

extended family the nuclear family or parts of it plus other relatives.

familism philosophy in which decisions are made in reference to what is best for the family as a collective unit.

family a group of two or more people related by blood, marriage, or adoption.

family life course development the stages and process of how families change over time.

family life cycle stages which identify the various challenges faced by members of a family across time.

family of orientation the family of origin into which a person is born.

family of origin the family into which an individual is born or reared, usually including a mother, father, and children.

family of procreation the family a person begins by getting married and having children.

family systems framework views each member of the family as part of a system and the family as a unit that develops norms of interaction.

feminist framework views marriage and the family as contexts for inequality and oppression.

functionalists structural functionalist theorists who view the family as an institution with values, norms, and activities meant to provide stability for the larger society.

Generation Y children of the baby boomers, typically born between 1979 and 1984. Also known as the Millennial or Internet Generation.

hypothesis a suggested explanation for a phenomenon.

individualism philosophy in which decisions are made on the basis of what is best for the individual.

IRB approval Institutional Review Board approval is the OK by one's college, university, or institution that the proposed research is consistent with research ethics standards and poses no undo harm to participants.

longitudinal study means studying the same group across time (e.g., follow several couples who are living together at one-year intervals for 10 years).

marriage-resilience perspective the view that changes in the institution of marriage are not indicative of a decline and do not have negative effects.

marriage a legal contract signed by a couple with the state in which they reside that regulates their economic and sexual relationship.

modern family the dual-earner family, in which both spouses work outside the home.

nuclear family family consisting of an individual, his or her spouse, and his or her children, or of an individual and his or her parents and siblings.

open relationship a stable relationship in which the partners regard their own relationship as primary but agree that each may have emotional and physical relationships with others.

pantagamy a group marriage in which each member of the group is "married" to the others.

polyamory multiple loves (poly = many; amorous = love) and is a lifestyle in which lovers embrace the idea of having multiple emotional and sexual partners.

polyandry a form of polygamy in which one wife has two or more husbands.

polygamy a generic term referring to a marriage involving more than two spouses.

polygyny a form of polygamy in which one husband has two or more wives.

postmodern family nontraditional families emphasizing that a healthy family need not be heterosexual or have two parents.

random sample sample in which each person in the population being studied has an equal chance of being included in the sample.

social exchange framework spouses exchange resources, and decisions are made on the basis of perceived profit and loss.

structure-function framework emphasizes how marriage and family contribute to the larger society.

symbolic interaction framework views marriage and families as symbolic worlds in which the various members give meaning to each other's behavior.

theoretical framework a set of interrelated principles designed to explain a particular phenomenon and to provide a point of view.

traditional family the two-parent nuclear family with the husband as breadwinner and wife as homemaker.

utilitarianism the doctrine holding that individuals rationally weigh the rewards and costs associated with behavioral choices.

LEARNING OUTCOMES

2-1 Singlehood

An increasing percentage of people are delaying marriage. Young Americans are not alone in their delay of marriage—individuals in France, Germany, and Italy are engaging in a similar pattern. The primary attraction of singlehood is the freedom to do as one chooses. As a result of the sexual revolution, the women's movement, and the gay liberation movement, social approval of being unmarried has increased. Single people are those who have never married as well as those who are divorced or widowed.

The term *singlehood* is most often associated with young, unmarried individuals. However, there are three categories of single people: the never married, the divorced, and the widowed.

2-2 Ways of finding a partner

Besides the traditional way of meeting people at work or school or through friends and going on a date, couples today may also "hang out," which may lead to "hooking up." Internet dating, video dating, speed-dating, and international dating are newer forms for finding each other. Wealthy, busy clients looking for marriage partners pay Selective Search $20,000 to find them a mate.

2-3 Long-distance relationships

Long-distance relationships may arise as the result of online dating, or separation that occurs between two individuals already in a relationship. There are several advantages and disadvantages to a long-distance relationship. Advantages include: positive labeling, keeping the relationship "high," and having freedom over one's personal time and space. Partners in long-distance relationships report that the primary disadvantages of their relationships are frustration over not being able to be with the partner and loneliness. Other disadvantages include: missing out on other activities and relationships, less physical intimacy, spending a lot of money on phone calls/travel, and not discussing important relationship topics.

How can long-distance relationships be maintained?
Methods of Maintaining Long-Distance Relationships

1. Maintain daily contact via text messaging.
2. Enjoy or use the time when apart.
3. Avoid arguing during phone conversations.
4. Stay monogamous.
5. Use Skype.
6. Be creative.

2-4 Cohabitation

Cohabitation, also known as living together, is becoming a "normative life experience," with almost 60% of American women reporting that they had cohabited before marriage. Reasons for the increase in cohabitation include career or educational commitments; increased tolerance of society, parents, and peers; improved birth control technology; desire for a stable emotional and sexual relationship without legal ties; avoiding loneliness (Kasearu, 2010) and greater disregard for convention. Some types of relationships in which couples live together include the here-and-now, testers, engaged couples, money savers, pension partners, alimony maintenance, security blanket cohabiters, rebellious cohabiters, and cohabitants forever.

KEY TERMS

cohabitation (living together) two unrelated adults (by blood or by law) involved in an emotional and sexual relationship who sleep in the same residence at least four nights a week for three months.

cohabitation effect multiple cohabitation experiences before marriage has negative effect on a subsequent marriage such as lower levels of happiness/higher divorce.

domestic partnership a relationship in which individuals who live together are emotionally and financially interdependent and are given some kind of official recognition by a city or corporation so as to receive partner benefits.

hanging out going out in groups where the agenda is to meet others and have fun.

hooking up having a one-time sexual encounter in which there are generally no expectations of seeing one another again.

living apart together (LAT) a long-term committed couple who does not live in the same dwelling.

long-distance dating relationship (LDDR) lovers are separated by a distance, usually 500 miles, which prevents weekly face-to-face contact.

satiation a stimulus loses its value with repeated exposure or people get tired of each other if they spend relentless amounts of time with each other.

singlehood state of being unmarried.

Living together before marriage does not ensure a happy, stable marriage. People commonly have more than one cohabitation experience, often resulting in the cohabitation effect. This means that those who have multiple cohabitation experiences prior to marriage are more likely to end up in marriages characterized by violence, lower levels of happiness, lower levels of positive communication, higher levels of depression, you name it (Booth et al., 2008).

Not all researchers have found negative effects of cohabitation on relationships. Kuperberg (2014) found that the link between cohabitation and divorce was a result of the age at which cohabitation began. She suggested that individuals delay their marriage into their midtwenties "when they are older and more established in the lives, goals and careers, whether married or not at the time of coresidence rather than avoiding premarital cohabitation altogether" (p. 368).

2-5 Living apart together

The definition of **living apart together (LAT)** is a committed couple who does not live in the same home. A couple defined as LAT must define themselves as a committed couple, must be defined by others as a committed couple, and must live in separate homes. Below are some of the advantages and disadvantages of LAT.

Advantages and Disadvantages of Living Apart Together

Advantages	Disadvantages
Space and privacy	Stigma or disapproval
Variable sleep needs are met	More expensive to maintain two residences
Variable social needs are met	Travel between residences may be inconvenient
Reduced satiation	Lack of shared history
Freedom of self-expression and comfort	No legal protection

2-6 Trends in singlehood

Individuals will remain single longer. Cohabitation will continue to increase. Fewer cohabitation relationships will transition into marriage.

LEARNING OUTCOMES

3-1 Terminology of gender roles

Sex refers to the biological distinction between females and males. One's biological sex is identified on the basis of one's chromosomes, gonads, hormones, internal sex organs, and external genitals, and exists on a continuum rather than being a dichotomy. *Gender* refers to the social and psychological characteristics often associated with being female or male. Other terms related to gender include *gender identity, gender role, gender role ideology, transgender,* and *transgenderism.*

3-2 Theories of gender role development

Biosocial theory emphasizes that social behaviors (e.g., gender roles) are biologically based and have an evolutionary survival function. Traditionally, women stayed close to shelter or gathered food nearby, whereas men traveled far to find food. Such a conceptualization focuses on the division of labor between women and men as functional for the survival of the species. The bioecological model emphasizes the importance of understanding bidirectional influences between an individual's development and his or her surrounding environmental contexts. Social learning theory emphasizes the roles of reward and punishment in explaining how children learn gender role behavior. Identification theory says that children acquire the characteristics and behaviors of their same-sex parent through a process of identification. Boys identify with their fathers; girls identify with their mothers. Cognitive-developmental theory emphasizes biological readiness, in terms of cognitive development, of the child's responses to gender cues in the environment. Once children learn the concept of gender permanence, they seek to become competent, proper members of their gender group.

3-3 Agents of socialization

Various socialization influences include parents and siblings (representing different races and ethnicities), peers, religion, the economy, education, and mass media. These factors shape individuals toward various gender roles and influence what people think, feel, and do in their roles as women or men. For example, the family is a gendered institution with female and male roles highly structured by gender. The names parents assign to their children, the clothes they dress them in, and the toys they buy them all reflect gender. Parents may also be stricter with female children, determining the age at which they are allowed to leave the house at night, setting a time for curfew, and giving directives such as "call when you get to the party."

3-4 Consequences of traditional gender role socialization

Traditional female role socialization may result not only in negative outcomes, such as less income, a negative body image, and less marital satisfaction, but also in positive outcomes, such as a longer life, a stronger relationship focus, keeping relationships on track, and a closer emotional bond with children. Traditional male role socialization may result in the fusion of self and occupation, a more limited expression of emotion, disadvantages in child custody disputes, and a shorter lifespan but also in higher income, greater freedom of movement, a greater available pool of potential partners, and greater acceptance in initiating relationships. A recent research study of 335 undergraduate men revealed that

KEY TERMS

androgyny a blend of traits that are stereotypically associated with masculinity and femininity.

benevolent sexism the belief that women are innocent creatures who should be protected and supported.

biosocial theory (sociobiology) emphasizes the interaction of one's biological or genetic inheritance with one's social environment to explain and predict human behavior.

cross-dresser a generic term for individuals who may dress or present themselves in the gender of the opposite sex.

Fafafini in Samoan society, these are effeminate males socialized and reared as females due to the lack of women to perform domestic chores.

feminization of poverty the idea that women disproportionately experience poverty.

gender dysphoria the condition in which one's gender identity does not match one's biological sex.

gender identity the psychological state of viewing oneself as a girl or a boy and, later, as a woman or a man.

gender role ideology the proper role relationships between women and men in a society.

gender roles behaviors assigned to women and men in a society.

gender role transcendence abandoning gender frameworks and looking at phenomena independent of traditional gender categories

gender the social and psychological behaviors associated with being female or male.

intersex development refers to congenital variations in the reproductive system, sometimes resulting in ambiguous genitals.

intersexed individuals people with mixed or ambiguous genitals.

occupational sex segregation the concentration of women in certain occupations and men in other occupations.

parental investment any investment by a parent that increases the chance that the offspring will survive and thrive.

purging for cross-dressers, the term means to destroy one's clothes of the other sex as a means of distancing one's self from cross-dressing. The practice is rarely effective as the cross-dresser acquires new clothes.

sex the biological distinction between being female and being male.

sexism an attitude, action, or institutional structure that subordinates or discriminates against an individual or group because of their sex.

sex roles behaviors defined by biological constraints.

socialization the process through which we learn attitudes, values, beliefs, and behaviors appropriate to the social positions we occupy.

sociobiology emphasizes the interaction of one's biological or genetic inheritance with one's social environment to explain and predict human behavior.

transgender a generic term for a person of one biological sex who displays characteristics of the opposite sex.

transsexual an individual who has the anatomical and genetic characteristics of one sex but the self-concept of the other.

about 31% of college males reported their preference for marrying a traditional wife (one who would stay at home to take care of children). These men believe that a wife's working outside the home weakens the marriage.

Consequences of Traditional Female Role Socialization

Negative Consequences	Positive Consequences
Less income (more dependent)	Longer life
Feminization of poverty	Stronger relationship focus
Higher STD/HIV infection risk	Keep relationships on track
Negative body image	Bonding with children
Less personal/marital satisfaction	Identity not tied to job

Consequences of Traditional Male Role Socialization

Negative Consequences	Positive Consequences
Identity tied to work role	Higher income and occupational status
Limited emotionality	More positive self-concept
Fear of intimacy; more lonely	Less job discrimination
Disadvantaged in getting custody	Freedom of movement; more partners to select from; more normative to initiate relationships
Shorter life	Happier marriage

3-5 Changing gender roles

Androgyny refers to a blend of traits that are stereotypically associated with both masculinity and femininity. The term may also imply flexibility of traits; for example, an androgynous individual may be emotional in one situation, logical in another, assertive in another, and so forth. The concept of gender role transcendence involves abandoning gender schema (i.e., becoming "gender aschematic"), so that personality traits, social and occupational roles, and other aspects of our lives become divorced from gender categories. However, such transcendence is not equal for women and men. Although females are becoming more masculine, partly because our society values whatever is masculine, men are not becoming more feminine.

3-6 Trends in gender roles

Progress toward equality has occurred, and continues very slowly. Characteristics such as strength, independence, logical thinking, and aggressiveness are no longer associated with maleness, just as passivity, dependence, emotions, intuitiveness, and nurturance are no longer associated with femaleness. Women are becoming increasingly independent, and outstretching men in terms of education.

LEARNING OUTCOMES

4-1 Ways of viewing love

Love remains an elusive and variable phenomenon. Researchers have conceptualized love as a continuum from romanticism to realism, as a triangle consisting of three basic elements (intimacy, passion, and commitment), and as a style (from playful ludic love to obsessive and dangerous manic love).

4-2 Social control of love

The society in which we live exercises considerable control over our love object or choice and conceptualizes it in various ways. Parents inadvertently influence the mate choice of their children by moving to certain neighborhoods, joining certain churches, and enrolling their children in certain schools. Doing so increases the chance that their offspring will "hang out" with, fall in love with, and marry people who are similar in race, education, and social class.

In the 1100s in Europe, marriage was an economic and political arrangement that linked two families. As aristocratic families declined after the French Revolution, love bound a woman and man together. Previously, Buddhists, Greeks, and Hebrews had their own views of love. Love in colonial America was also tightly controlled.

4-3 Love theories: Origins of love

There are several theories about how love originates. The evolutionary theory of love suggests that individuals are motivated to emotionally bond with a partner to ensure a stable relationship for producing and rearing children. Learning theory emphasizes that love feelings develop in responses to certain behaviors engaged in by the partner; partners learn love behaviors from each other. The sociological theory of love proposes that there are four stages of love development: (1) rapport, (2) self-revelation, (3) mutual dependency, and (4) fulfillment of personality needs. Psychosexual theory suggests that love results from blocked biological sexual desires. Biochemical theory notes the chemical changes that occur during social bonding—specifically the biochemistry of oxytocin and vasopressin. Attachment theory emphasizes that a primary motivation in life is to be emotionally connected with other people; love results from this need.

4-4 Love as a context for problems

Several factors create problems in a love context. Unrequited or unreciprocated love can hurt both partners in a relationship. Moreover, those in love tend to make risky, dangerous, or questionable decisions, like beginning to smoke (if one's partner is a smoker). Problems may arise if parents disapprove of a relationship, which may result in one partner ending the relationship with their parents. Too much love may cause problems as well—a partner that is in love with two or more people at the same time may cause serious problems. Emotional and/or physical abuse within a relationship is both psychologically and physically detrimental to the abused, and often continues on the basis of love for the abusive partner. When a relationship of any sort ends, it can cause profound sadness or depression for one or both partners.

4-5 How love develops in a new relationship

Love occurs under certain conditions. Social conditions include a society that promotes the pursuit of love, peers who enjoy it, and a set of norms that link love and marriage. Psychological conditions involve high self-esteem, a willingness to disclose one's self to others, and gratitude. Researchers have also found that one of the most important psychological factors associated with falling in love is the perception of reciprocal liking. This factor is especially powerful if the person feels strong physical attraction. Individuals seeking emotional relationships are not attracted to those with little affect, a condition known as alexithymia. Physiological and cognitive conditions imply that the individual experiences a stirred-up state and labels it "love." Other factors associated with love include appearance, common interests, and similar friends. Love sometimes provides a context for problems in that a young person in love will lie to parents and become distant from them so as to be with the lover. Also, lovers experience problems such as being in love with two people at the same time, being in love with someone who is abusive, and making risky, dangerous, or questionable choices while in love (for example, not using a condom) or reacting to a former lover who has become a stalker.

4-6 Cultural factors in relationship development

Two types of cultural influences in mate selection are endogamy (to marry someone inside one's own social group—race, religion, social class) and exogamy (to marry someone outside one's own family).

4-7 Sociological factors in relationship development

Sociological aspects of mate selection involve homogamy. Variables include race, age, religion, education, social class, personal appearance, career, marital status, circadian preference, desired marital roles, economic values, geographic background, attachment, personality, and open-mindedness. Couples who have a lot in common are more likely to have a happy and durable relationship.

4-8 Psychological factors in relationship development

Psychological aspects of mate selection

Personality Disorders Problematic in a Potential Partner

Disorder	Characteristics	Impact on Partner
Paranoid	Suspicious, distrustful, thin-skinned, defensive	Partners may be accused of everything.
Schizoid	Cold, aloof, solitary, reclusive	Partners may feel that they can never connect and that the person is not capable of returning love.
Borderline	Moody, unstable, volatile, unreliable, suicidal, impulsive	Partners will never know what their Jekyll-and-Hyde partner will be like, which could be dangerous.
Antisocial	Deceptive, untrustworthy, conscienceless, remorseless	Such a partner could cheat on, lie to, or steal from a partner and not feel guilty.
Narcissistic	Egotistical, demanding, greedy, selfish	Such a person views partners only in terms of their value. Don't expect such a partner to see anything from your point of view; expect such a person to bail in tough times.
Dependent	Helpless, weak, clingy, insecure	Such a person will demand a partner's full time and attention, and other interests will incite jealousy.
Obsessive-compulsive	Rigid, inflexible	Such a person has rigid ideas about how a partner should think and behave and may try to impose them on the partner.
Neurotic	Worries, obsesses about negative outcomes	This individual will impose negative scenarios on the partners and couple.

include complementary needs, exchange theory, and parental characteristics. Personality characteristics of a potential mate that are desired by both men and women include being warm, kind, and open and having a sense of humor. Negative personality characteristics to avoid in a potential mate include controlling behavior, narcissism, poor impulse control, hypersensitivity to criticism, inflated ego, perfectionism or insecurity, control by someone else (e.g., parents), and substance abuse. Paranoid, schizoid, and borderline personalities are also to be avoided.

4-9 Engagement

The engagement period is the time to ask specific questions about one's partner's values, goals, and marital agenda, to visit each other's parents to assess parental models, and to consider involvement in premarital educational programs or counseling or both.

4-10 Delay or call off the wedding if...

Getting married on the rebound is not a good reason to marry. One should wait until the negative memories of past relationships have been replaced by positive aspects of one's current relationship. One should not marry to escape an unhappy home. Getting married because a partner becomes pregnant is a bad idea. Factors suggesting that a couple may not be ready for marriage include being in their teens, having known each other less than two years, abuse in the relationship, high frequency of negative comments/low frequency of positive comments, having the relationship characterized as on-and-off, low sexual satisfaction, limited relationship knowledge, and having a relationship characterized by significant differences or dramatic parental disapproval or both. Some research suggests that partners with the greatest number of similarities in values, goals, and interests are most likely to have happy and durable marriages.

4-11 Trends in love relationships

As our society becomes more diverse, the range of potential love partners will widen to include those with demographic characteristics different from oneself. Romantic love will continue and love will maintain its innocence as those getting remarried love just as deeply and invest in the power of love all over again. The development of a new love relationship will involve the same cultural constraints and sociological and psychological factors identified in the chapter. Individuals are not "free" to select their partner but do so from the menu presented by their culture.

REVIEWcard

KEY TERMS

agape love style love style characterized by a focus on the well-being of the love object, with little regard for reciprocation; the love of parents for their children is agape love.

alexithymia a personality trait which describes a person with little affect.

arranged marriage mate selection pattern whereby parents select the spouse of their offspring.

attachment theory of mate selection emphasizes the drive toward psychological intimacy and a social and emotional connection.

circadian preference refers to an individual's preference for morningness–eveningness in regard to intellectual and physical activities.

compassionate love emotional feelings toward another that generate behaviors to promote the partner's well-being.

complementary-needs theory tendency to select mates whose needs are opposite and complementary to one's own needs.

conjugal (married) love the love between married people characterized by companionship, calmness, comfort, and security.

cyclical relationships when couples break up and get back together several times.

dark triad personality term identifying traits of "bad boys" including narcissistic, deceptive, and no empathy.

endogamy the cultural expectation to select a marriage partner within one's social group.

engagement period of time during which committed, monogamous partners focus on wedding preparations and systematically examine their relationship.

eros love style love style characterized by passion and romance.

evolutionary theory of love theory that individuals are motivated to emotionally bond with a partner to ensure a stable relationship for producing and rearing children.

exchange theory theory that emphasizes that relations are formed and maintained between individuals offering the greatest rewards and least costs to each other.

exogamy the cultural expectation that one will marry outside the family group.

five love languages identified by Gary Chapman and now part of American love culture, these are gifts, quality time, words of affirmation, acts of service, and physical touch.

homogamy tendency to select someone with similar characteristics.

infatuation emotional feelings based on little actual exposure to the love object.

ludic love style love style that views love as a game in which the love interest is one of several partners, is never seen too often, and is kept at an emotional distance.

lust sexual desire.

mania love style an out-of-control love whereby the person "must have" the love object; obsessive jealousy and controlling behavior are symptoms of manic love.

marriage squeeze the imbalance of the ratio of marriageable-age men to marriageable-age women.

mating gradient the tendency for husbands to be more advanced than their wives with regard to age, education, and occupational success.

open-minded being open to understanding alternative points of view, values, and behaviors.

oxytocin a hormone released from the pituitary gland during the expulsive stage of labor that has been associated with the onset of maternal behavior in lower animals.

polyamory open emotional and sexual involvement with three or more people.

pool of eligibles the population from which a person selects an appropriate mate.

pragma love style love style that is logical and rational; the love partner is evaluated in terms of assets and liabilities.

role theory of mate selection emphasizes that a son or daughter models after the parent of the same sex by selecting a partner similar to the one the parent selected as a mate.

romantic love an intense love whereby the lover believes in love at first sight, only one true love, and love conquers all.

spatial homogamy the tendency for individuals to marry who grew up in close physical proximity.

storge love style a love consisting of friendship that is calm and nonsexual.

swinging persons who exchange partners for the purpose of sex.

unrequited love a one-sided love where one's love is not returned.

Courtesy of Rachel Calisto

REVIEWcard

LEARNING OUTCOMES

5-1 Communication: Verbal and nonverbal

Communication is the exchange of information and feelings by two individuals. It involves both verbal and nonverbal messages. The nonverbal part of a message often carries more weight than the verbal part.

5-2 Technology-mediated communication in romantic relationships

Use of texting in the initiation, escalation, and maintenance of romantic relationships is increasing with positive and negative effect. While it allows for quick and constant communication, it encourages isolation in those that find personal interaction difficult. Male texting frequency was negatively associated with relationship and stability for both partners, while female texting frequency was positively associated with their own relationship stability. Use of video-mediated communication in relationships where the individuals are separated from one another is increasing.

Conflict is both inevitable and desirable. Unless individuals confront and resolve issues over which they disagree, one or both may become resentful and withdraw from the relationship. Conflict may result from one partner's doing something the other does not like, having different perceptions, or having different values. Sometimes it is easier for one partner to view a situation differently or alter a value than for the other partner to change the behavior causing the distress.

5-3 Principles of effective communication

Some basic principles and techniques of effective communication include making communication a priority, maintaining eye contact, asking open-ended questions, using reflective listening, using "I" statements, complimenting each other, and sharing power. Partners must also be alert to keeping the dialogue (process) going even when they don't like what is being said (content).

Judgmental and Nonjudgmental Responses to a Partner's Saying, "I'd Like to Go Out with My Friends One Night a Week"

Nonjudgmental, Reflective Statements	Judgmental Statements
You value your friends and want to maintain good relationships with them.	You only think about what you want.
You think it is healthy for us to be with our friends some of the time.	Your friends are more important to you than I am.
You really enjoy your friends and want to spend some time with them.	You just want a night out so that you can meet someone new.
You think it is important that we not abandon our friends just because we are involved.	You just want to get away so you can drink.
You think that our being apart one night each week will make us even closer.	You are selfish.

5-4 Gender, culture, and communication

Women and men differ in their approach to patterns of communication. Women are more communicative about relationship issues, view a situation emotionally, and initiate discussions about relationship problems. A woman's goal is to

KEY TERMS

amae expecting a close other's indulgence when one behaves inappropriately.

authentic speaking and acting in a manner according to what one feels.

brainstorming suggesting as many alternatives as possible without evaluating them.

branching in communication, going out on different limbs of an issue rather than staying focused on the issue.

catfishing process whereby a person makes up an online identity and an entire social facade to trick a person into becoming involved in an emotional relationship.

closed-ended questions questions that allow for a one-word answer and do not elicit much information.

communication the process of exchanging information and feelings between two or more people.

congruent message one in which verbal and nonverbal behaviors match.

defense mechanisms unconscious techniques that function to protect individuals from anxiety and minimize emotional hurt.

displacement shifting one's feelings, thoughts, or behaviors from the person who evokes them onto someone else.

empathy the ability to emotionally experience and cognitively understand another person and his or her experiences.

escapism the simultaneous denial of and withdrawal from a problem.

flirting showing another person romantic interest without serious intent.

nomophobia the individual is dependent on virtual environments to the point of having a social phobia.

nonverbal communication the "message about the message," using gestures, eye contact, body posture, tone, volume, and rapidity of speech.

open-ended questions questions that encourage answers that contain a great deal of information.

power the ability to impose one's will on one's partner and to avoid being influenced by the partner.

principle of least interest principle stating that the person who has the least interest in a relationship controls the relationship.

projection attributing one's own feelings, attitudes, or desires to one's partner while avoiding recognition that these are one's own thoughts, feelings, and desires.

rationalization the cognitive justification for one's own behavior that unconsciously conceals one's true motives.

reflective listening paraphrasing or restating what a person has said to indicate that the listener understands.

sexting sending erotic text and photo images via a cell phone.

texting text messaging (short typewritten messages—maximum of 160 characters sent via cell phone).

video-mediated communication (VMC) communication via computer between separated lovers, spouses, and family members.

---

preserve intimacy and avoid isolation. To men, conversations are about winning and achieving the upper hand. Mothers and fathers speak differently to children. Fathers use more assertive speech than mothers.

5-5 Self-disclosure and secrets

5-6 Dishonesty, lying, and cheating

The levels of self-disclosure and honesty influence intimacy in relationships. High levels of self-disclosure are associated with increased intimacy. Despite the importance of honesty in relationships, deception occurs frequently in interpersonal relationships. Partners sometimes lie to each other about previous sexual relationships, how they feel about each other, and how they experience each other sexually. Telling lies is not the only form of dishonesty. People exaggerate, minimize, tell partial truths, pretend, and engage in self-deception. Partners may withhold information or keep secrets to protect themselves or to preserve the relationship, or both. The top reason reported in a 2012 study of 431 undergraduates was "To avoid hurting the partner." Even in "monogamous" relationships, there is considerable cheating. However, the more intimate the relationship, the greater our desire to share our most personal and private selves with our partner and the greater the emotional consequences of not sharing. In intimate relationships, keeping secrets can block opportunities for healing, resolution, self-acceptance, and a deeper intimacy with your partner.

5-7 Theories of relationship communication

Symbolic interactionists examine the process of communication between two actors in terms of the meanings each attaches to the actions of the other. Definition of the situation, the looking-glass self, and taking the role of the other are all relevant to understanding how partners communicate.

Exchange theorists suggest that the partners' communication can be described as a ratio of rewards to costs. Rewards are positive exchanges, such as compliments, compromises, and agreements. Costs refer to negative exchanges, such as critical remarks, complaints, and attacks. When the rewards are high and the costs are low, the outcome is likely to be positive for both partners (profit). When the costs are high and the rewards low, neither may be satisfied with the outcome (loss).

5-8 Fighting fair: Steps in conflict resolution

The sequence of resolving conflict includes deciding to address recurring issues rather than suppressing them, asking the partner for help in resolving issues, finding out the partner's point of view, summarizing in a nonjudgmental way the partner's perspective, and finding alternative win-win solutions. Defense mechanisms that interfere with conflict resolution include escapism, rationalization, projection, and displacement.

5-9 Trends in communication and technology

The future of communication will increasingly involve technology in the form of texting, smart phones, Facebook, etc. Such technology will be used to initiate, enhance, and maintain relationships. Parental communication with children will also be altered.

"I" statements statements that focus on the feelings and thoughts of the communicator without making a judgment on others.

"you" statements statements that blame or criticize the listener and often result in increasing negative feelings and behavior in the relationship.

LEARNING OUTCOMES

6-1 Alternative sexual values

Sexual values are moral guidelines for sexual behavior in relationships. One's sexual values may be identical to one's sexual behavior, but sexual behavior does not always correspond with sexual values.

6-2 Sources of sexual values

The sources of sexual values include one's school, family, and religion as well as technology, television, social movements, and the Internet.

6-3 Sexual behaviors

Sexual behaviors include kissing, masturbation, oral sex, coitus, anal sex, cybersex, and kinks. Some individuals are asexual, which means there is an absence of sexual behavior with a partner and oneself. Social scripts dictate sexual behavior.

6-4 Sexuality in relationships

Never married and noncohabiting individuals report more sexual partners than do those who are married or living with a partner. Marital sex is distinctive for its social legitimacy, declining frequency, and satisfaction (both physical and emotional). Divorced individuals have a lot of sexual partners but are the least sexually fulfilled.

6-5 Sexual fulfillment: Some prerequisites

Fulfilling sexual relationships involve self-knowledge, self-esteem, health, a good nonsexual relationship, open sexual communication, safer sex practices, and making love with, not to, one's partner. Other variables include realistic expectations ("my partner will not always want what I want") and not buying into sexual myths ("masturbation is sick").

6-6 Trends in sexuality in relationships

The future of sexual relationships will involve continued individualism as the driving force in sexual relationships. Numerous casual partners with predictable negative outcomes will continue to characterize individuals in adolescence and early twenties. The goal of sexuality transitions to settling down and new sexual behaviors.

KEY TERMS

absolutism belief system based on unconditional allegiance to the authority of religion, law, or tradition.

anodyspareunia frequent, severe pain during receptive anal sex.

asceticism the belief that giving in to carnal lusts is wrong and that one must rise above the pursuit of sensual pleasure to a life of self-discipline and self-denial.

asexual an absence of sexual behavior with a partner and oneself.

coitus sexual union of a man and woman by insertion of the penis into the vagina.

concurrent sexual partnerships relationships in which the partners have sex with several individuals concurrently.

condom assertiveness the unambiguous messaging that sex without a condom is unacceptable.

cybersex any consensual sexual experience mediated by a computer that involves at least two people.

friends with benefits relationship (FWBR) a relationship between nonromantic friends who also have a sexual relationship.

hedonism belief that the ultimate value and motivation for human actions lie in the pursuit of pleasure and the avoidance of pain.

kink typically refers to BDSM (bondage and discipline/dominance and submission/sadism and masochism).

masturbation stimulating one's own body with the goal of experiencing pleasurable sexual sensations.

relativism belief system in which decisions are made in reference to the emotional, security, and commitment aspects of the relationship.

satiation the state in which a stimulus loses its value with repeated exposure.

sexual compliance an individual willingly agrees to participate in sexual behavior without having the desire to do so.

sexual double standard the view that encourages and accepts sexual expression of men more than women.

sexual readiness determining when one is ready for first intercourse in reference to contraception, autonomy of decision, consensuality, and absence of regret.

sexual values moral guidelines for sexual behavior in relationships.

social script the identification of the roles in a social situation, the nature of the relationship between the roles, and the expected behaviors of those roles.

spectatoring mentally observing one's own and one's partner's sexual performance.

STI sexually transmitted infection.

swinging relationships (open relationships) involve married/pair-bonded individuals agreeing that they may have sexual encounters with others.

LEARNING OUTCOMES

7-1 Language and identification

In order to understand GLTBQ relationships, it is imperative to define sexual orientation, heterosexuality, homosexuality, lesbian, gay, transgender, transsexual, cross-dresser, and queer. Sexual orientation is more dynamic than defining individuals as "gay" or "straight." The Kinsey scale suggests that heterosexuality and homosexuality represent two ends of a sexual-orientation continuum and that most individuals are neither entirely homosexual nor entirely heterosexual, but fall somewhere along this continuum. Individual ability to identify sexual orientation by looking at someone, also known as "gaydar," is only slightly better than chance.

7-2 Sexual orientation

Some individuals believe that homosexual people choose their sexual orientation and think that, through various forms of reparative therapy, homosexual people should change their sexual orientation. However, many professional organizations (including the American Psychiatric Association, the American Psychological Association, and the American Medical Association) agree that homosexuality is not a mental disorder, that it needs no cure, and that efforts to change sexual orientation do not work and may, in fact, be harmful. These organizations also support the notion that changing societal reaction to those who are homosexual is a more appropriate focus.

7-3 Heterosexism, homonegativity, etc.

Heterosexism is based on the belief that heterosexuality is superior to homosexuality. There are five dimensions of attitudes toward homosexuality, homonegativity, and homophobia: general attitude, equal rights, close quarters, public display, and modern homonegativity. Just as the term *homophobia* is used to refer to negative attitudes toward homosexuality, gay men, and lesbians, biphobia refers to a parallel set of negative attitudes toward bisexuality/those identified as bisexual. Transsexuals are targets of transphobia, a set of negative attitudes toward transsexuality or those who self-identify as transsexual.

7-4 Coming out

Coming out is a major decision with which GLBT individuals struggle. Parents of gay children often discover very positive outcomes from their child's "coming out," including personal growth, positive emotions, activism, social connection, and closer relationships. Coming out includes several risks. LGBT individuals worry about parental and family members' reactions, whether they might face harassment and discrimination at school or workplace, and potentially becoming a victim of a hate crime. However, there are several benefits of coming out as well. These benefits include higher levels of acceptance from parents, lower levels of alcohol and drug consumption, and fewer identity adjustment problems.

7-5 Mixed-orientation inside relationships

Research suggests that gay and lesbian couples tend to be more similar to than different from heterosexual couples. Both gay and straight couples tend to value long-term monogamous relationships, experience high relationship satisfaction early in the relationship that decreases over time, disagree about the same topics, and have the same factors linked to relationship satisfaction.

KEY TERMS

biphobia (binegativity) refers to a parallel set of negative attitudes toward bisexuality and those identified as bisexual.

bisexuality emotional and sexual attraction to members of both sexes.

coming out being open and honest about one's sexual orientation and identity.

conversion therapy (reparative therapy) therapy designed to change a person's homosexual orientation to a heterosexual orientation.

cross-dresser broad term for individuals who may dress or present themselves in the gender of the other sex.

Defense of Marriage Act (DOMA) legislation passed by Congress denying federal recognition of homosexual marriage and allowing states to ignore same-sex marriages licensed by other states.

down low non-gay-identifying men who have sex with men and women meet their partners out of town, not in predictable contexts, or on the Internet.

gay homosexual woman or man.

genderqueer the person does not identify as either male or female since he or she does not feel sufficiently like one or the other.

hate crime instances of violence against homosexuals.

hetero-gay family a heterosexual mother and a gay father conceive and raise a child together but reside separately.

heterosexism the denigration and stigmatization of any behavior, person, or relationship that is not heterosexual.

heterosexuality the predominance of emotional and sexual attraction to individuals of the other sex.

homonegativity a construct that refers to antigay responses such as negative feelings (fear, disgust, anger), thoughts ("homosexuals are HIV carriers"), and behavior ("homosexuals deserve a beating").

homophobia negative (almost phobic) attitudes toward homosexuality.

homosexuality predominance of emotional and sexual attraction to individuals of the same sex.

internalized homophobia a sense of personal failure and self-hatred among lesbians and gay men resulting from the acceptance of negative social attitudes and feelings toward homosexuals.

lesbian homosexual woman.

queer broad self-identifier term to indicate that the person has a sexual orientation other than heterosexual.

second-parent adoption (coparent adoption) a legal procedure that allows individuals to adopt their partner's biological or adoptive child without terminating the first parent's legal status as parent.

sexual orientation (sexual identity) classification of individuals as heterosexual, bisexual, or homosexual, based on their emotional, cognitive, and sexual attractions and self-identity.

transgender individuals who express their masculinity and femininity in nontraditional ways consistent with their biological sex.

transphobia negative attitudes toward transsexuality or those who self-identify as transsexual.

transsexual individual with the biological and anatomical sex of one gender but the self-concept of the other sex.

However, same-sex couples' relationships are different from heterosexual relationships. Same-sex couples have more concern about when and how to disclose their relationships, are more likely to achieve a fair distribution of household labor, argue more effectively and resolve conflict in a positive way, and face prejudice and discrimination without much government support.

7-6 Same-sex marriage

In 2014, the Justice Department affirmed that it will extend equal benefits to spouses in same-sex marriages. While upwards of 20 states have legalized same-sex marriage, it remains a contested issue in many states and political candidates are careful about their position in regard to same-sex relationships for fear of losing votes.

The website Infoplease provides a timeline about the gay rights movement in the United States from 1924 to the present (http://www.infoplease.com/ipa/A0761909.html).

7-7 Parenting issues

A growing body of credible, scientific research on gay and lesbian parenting concludes that children raised by gay and lesbian parents adjust positively and their families function well. Lesbian and gay parents are as likely as heterosexual parents to provide supportive and healthy environments for their children, and the children of lesbian and gay parents are as likely as those of heterosexual parents to flourish. Some research suggests that children raised by lesbigay parents develop in less gender-stereotypical ways, are more open to homoerotic relationships, contend with the social stigma of having gay parents, and have more empathy for social diversity than children of opposite-sex parents. There is no credible social science evidence that gay parenting negatively affects the well-being of children.

7-8 Trends

Moral acceptance and social tolerance/acceptance of gays, lesbians, bisexuals, and transsexuals as individuals, couples, and parents will come slowly. Heterosexism, homonegativity, biphobia, and transphobia are entrenched in American society. However, as more states recognize same-sex marriage, more GLBT individuals will come out, their presence will become more evident, and tolerance, acceptance, and support will increase slowly.

LEARNING OUTCOMES

8-1 Motivations, functions, and transition to egalitarian marriage

Individuals' motives for marriage include:
- personal fulfillment
- companionship
- legitimacy of parenthood
- emotional and financial security

Social functions include:
- continuing to provide society with socialized members
- regulating sexual behavior
- stabilizing adult personalities

Traditional versus Egalitarian Marriages

Traditional Marriage	Egalitarian Marriage
Limited expectation of husband to meet emotional needs of wife and children.	Husband is expected to meet emotional needs of his wife and children.
Wife is not expected to earn income.	Wife is expected to earn income.
Emphasis is on ritual and roles.	Emphasis is on companionship.
Couples do not live together before marriage.	Couples often live together before marriage.
Wife takes husband's last name.	Wife may keep her maiden name. In some cases, he will take her last name.
Husband is dominant; wife is submissive.	Neither spouse is dominant.
Roles for husband and wife are rigid.	Roles for spouses are flexible.
Husband initiates sex; wife complies.	Either spouse initiates sex.
Wife takes care of children.	Fathers more involved in child rearing.
Education is important for husband, not for wife.	Education is important for both spouses.
Husband's career decides family residence.	Career of either spouse may determine family residence.

8-2 Weddings and honeymoons

The wedding is a rite of passage signifying the change from the role of fiancé to the role of spouse. In general women, more than men, are invested in preparation for the wedding; the wedding is perceived to be more for the bride's family, and many women prefer a traditional wedding. The honeymoon is a time in which the couple recovers from the wedding and solidifies their new status as spouses.

KEY TERMS

artifact concrete symbol that reflects the existence of a cultural belief or activity.

connection rituals habits which occur daily in which the couple share time and attention.

cougar a woman, usually in her thirties or forties, who is financially stable and mentally independent and looking for a younger man with whom to have fun.

disenchantment the transition from a state of newness and high expectation to a state of mundaneness tempered by reality.

honeymoon the time following the wedding whereby the couple becomes isolated to recover from the wedding and to solidify their new status change from lovers to spouses.

marital success the quality of the marriage relationship measured in terms of stability and happiness.

marriage rituals deliberate repeated social interactions that reflect emotional meaning to the couple.

May–December marriage age-dissimilar marriage (ADM) in which the woman is typically in the spring of her life (May) and her husband is in the later years (December).

military contract marriage a military person will marry a civilian to get more money and benefits from the government.

rite of passage an event that marks the transition from one social status to another.

satiation a stimulus loses its value with repeated exposure.

8-3 Changes after marriage

Changes after the wedding are:

- legal
- personal
- social
- economic
- sexual
- parental

8-4 Diversity in marriage

Mixed marriages include interracial, international, interreligious, and age-discrepant. When age-discrepant and age-similar marriages are compared, there are no differences in regard to marital happiness. There are three main types of military marriages: (1) those in which the soldier falls in love with a high school sweetheart, marries the person, and subsequently joins the military; (2) those in which the partners meet and marry after one of them has signed up for the military; and (3) the contract marriage in which a soldier will marry a civilian to get more money and benefits from the government. Military contract marriages are not common, but they do exist. Military families cope with deployment, the double standard, and limited income. Military marriages are particularly difficult for women.

8-5 Marriage success

Marital success is defined in terms of both marital stability and marital happiness. Characteristics associated with marital success include commitment, common interests, communication, religiosity, trust, and nonmaterialism, and having positive role models, absence of negative attributions, health, and equitable relationships.

8-6 Trends in marriage relationships

Diversity will continue to characterize marriage relationships in the future. Openness to interracial, interreligious, cross-national, and age-discrepant relationships will increase.

LEARNING OUTCOMES

9-1 Money and relationships

Generally, the more money a partner makes, the more power that person has in the relationship. Males make considerably more money than females and generally have more power in relationships. Seventy percent of all U.S. wives are in the labor force. The stereotypical family consisting of a husband who earns an income and a wife who stays at home with two or more children is no longer the norm. Only 13% are "traditional" in the sense of consisting of a breadwinning husband, a stay-at-home wife, and their children. In contrast, most marriages may be characterized as dual earner. Employed wives in unhappy marriages are more likely to leave the marriage than unemployed wives.

9-2 Work and marriage

A couple's marriage is organized around the work of each spouse. Where the couple live is determined by where the spouses can get jobs. Jobs influence what time spouses eat, which family members eat with whom, when they go to bed, and when, where, and for how long they vacation. The workplace is also a very common location for romance to arise. Within marital context, wives are most likely to be employed when their children are teenagers. Many partnerships, though, are dual-career marriages, in which both partners pursue careers and may or may not have dependents.

> **What do you think?**
> Take the self-assessment for Chapter 9 in the self-assessment card deck to determine your attitudes toward maternal employment.

9-3 Effects of wife's employment on the spouses and marriage

Recent studies of employment patterns and marital satisfaction demonstrate that coprovider couples reported the highest marital satisfaction, and most equitable division of household labor. This work is contrasted by other research that suggests egalitarian marriages report less marital satisfaction. What can be derived from these works is that a wife's employment will not affect a happy marriage, but it can affect an unhappy one by providing means for the woman to take care of herself if she leaves the marriage.

9-4 Work and family: Effects on children

Children do not appear to suffer cognitively or emotionally from their parents' working as long as positive, consistent child care alternatives are in place. However, less supervision of children by parents is an outcome of having two-earner parents. Leaving children to come home to an empty house is particularly problematic.

Parents view quality time as structured, planned activities, talking with their children, or just hanging out with them. Day care is typically mediocre but day care workers who engage in high-frequency, positive behavior engender secure attachments with the children they work with. High-quality child care predicts

KEY TERMS

cognitive restructuring viewing a situation in positive terms.

commuter marriage a type of long-distance marriage where spouses live in different locations during the workweek (and sometimes for longer periods of time) to accommodate the careers of the respective spouses.

consumerism to buy everything and to have everything now.

dual-career marriage one in which both spouses pursue careers.

HER/HIS career marriage a wife's career is given precedence over a husband's career.

HIS/HER career marriage a husband's and wife's careers are given equal precedence.

mommy track stopping paid employment to spend time with young children.

poverty the lack of resources necessary for material well-being.

role compartmentalization separating the roles of work and home so that an individual does not dwell on the problems of one role while physically being at the place of the other role.

role conflict being confronted with incompatible role obligations.

role overload not having the time or energy to meet the demands of their responsibilities in the roles of wife, parent, and worker.

role strain the anxiety that results from being able to fulfill only a limited number of role obligations.

second shift housework and child care that are done when the parents return home after work.

shift work having one parent work during the day and the other parent work at night so that one parent can always be with the children.

spillover thesis work spreads into family life in the form of the worker/parent doing overtime, taking work home, attending seminars organized by the company, being "on call" on the weekend/ during vacation, and always being on the computer in reference to work.

superperson strategy involves working as hard and as efficiently as possible to meet the demands of work and family.

superwoman (supermom) a cultural label that allows a mother who is experiencing role overload to regard herself as particularly efficient, energetic, and confident.

THEIR career marriage a career shared by a couple who travel and work together (e.g., journalists).

third shift the emotional energy expended by a spouse or parent in dealing with various family issues.

higher cognitive-academic achievement at age 15, with increasing positive effects at higher levels of quality.

9-5 Balancing work and family life

Strategies used for balancing the demands of work and family include the superperson strategy, cognitive restructuring, delegation of responsibility, planning and time management, and role compartmentalization. Government and corporations have begun to respond to the family concerns of employees by implementing work-family policies and programs. These policies are typically inadequate and cosmetic.

9-6 Trends in money, work, and family life

Families will continue to be stressed by work. The number of female spouses working outside the home will increase.

LEARNING OUTCOMES

10-1 Types of relationship abuse

Violence or physical abuse may be defined as the deliberate infliction of physical harm by either partner on the other. Violence may occur over an issue on which the partners disagree, or it may be a mechanism of control. Female violence is as prevalent as male violence. The difference is that female violence is often in response to male violence, whereas male violence is more often to control the partner. Therefore, women tend to be striking back rather than throwing initial blows.

Emotional abuse is designed to denigrate the partner, reduce the partner's status, and make the partner vulnerable, thereby giving the abuser more control. Stalking is unwanted following or harassment that induces fear in the target person. The stalker is most often a heterosexual male who has been rejected by someone who fails to return his advances. Stalking is typically designed either to seek revenge or to win a partner back. Cyber victimization includes being sent threatening email, unsolicited obscene email, computer viruses, or junk mail (spamming). It may also include flaming (online verbal abuse) and leaving improper messages on message boards designed to get back at the person. Obsessive relationship intrusion (ORI) is the relentless pursuit of intimacy with someone who does not want it. Unlike stalking, the goal of which is to harm, ORI involves hyperintimacy (telling people that they are beautiful or desirable to the point of making them uncomfortable), relentless mediated contacts (flooding people with email messages, cell phone messages, or faxes), or interactional contacts (showing up at work or the gym, or joining the same volunteer groups as the pursued).

10-2 Reasons for violence and abuse in relationships

Cultural explanations for violence include violence in the media, corporal punishment in childhood, gender inequality, and stress. Community explanations involve social isolation of individuals and spouses from extended family, poverty, inaccessible community services, and lack of violence prevention programs. Individual factors include psychopathology of the person (antisocial), personality (dependency or jealousy), and alcohol or substance abuse. Family factors include experiencing child abuse by one's parents and observing parents who abuse each other.

10-3 Sexual abuse in undergraduate relationships

Some women experience sexual abuse in addition to a larger pattern of physical abuse, and the combination is associated with less general satisfaction, less sexual satisfaction, more conflict, and more psychological abuse from the partner. About 33% of undergraduates report being forced to have sex against their will. Acquaintance rape is defined as nonconsensual sex between adults who know each other.

One type of acquaintance rape is date rape, which refers to nonconsensual sex between people who are dating or on a date. Rohypnol, known as the date rape drug, causes profound sedation so that the person may not remember being raped. Most women do not report being raped by an acquaintance or date.

10-4 Abuse in marriage relationships

Abuse in marriage is born out of the need to control the partner and may include repeated rape. About half of the women raped by an intimate partner and two-thirds of the women physically assaulted by an intimate partner have been victimized multiple times.

KEY TERMS

acquaintance rape nonconsensual sex between adults (of same or other sex) who know each other.

battered woman syndrome (battered woman defense) legal term used in court that the person accused of murder was suffering from to justify their behavior. Therapists define battering as physical aggression that results in injury and accompanied by fear and terror.

corporal punishment the use of physical force with the intention of causing a child to experience pain, but not injury, for the purpose of correction or control of the child's behavior.

cyber control use of communication technology, such as cell phones, email, and social networking sites, to monitor or control partners in intimate relationships.

cyber victimization harassing behavior which includes being sent threatening email, unsolicited obscene email, computer viruses, or junk mail (spamming); can also include flaming (online verbal abuse) and leaving improper messages on message boards.

date rape one type of acquaintance rape which refers to nonconsensual sex between people who are dating or are on a date.

double victims individuals who report being a victim of forced sex by both a stranger and by a date or acquaintance.

emotional abuse nonphysical behavior designed to denigrate the partner, reduce the partner's status, and make the partner feel vulnerable to being controlled by the partner.

entrapped stuck in an abusive relationship and unable to extricate oneself from the abusive partner.

female rape myths beliefs that deny victim injury or cast blame on the woman for her own rape.

filicide murder of an offspring by a parent.

honor crime (honor killing) refers to unmarried women who are killed because they bring shame on their parents and siblings; occurs in Middle Eastern countries such as Jordan.

intimate partner homicide murder of a spouse.

intimate partner violence an all-inclusive term that refers to crimes committed against current or former spouses, boyfriends, or girlfriends.

intimate terrorism behavior designed to control the partner.

male rape myths beliefs that deny victim injury or make assumptions about his sexual orientation.

marital rape forcible rape by one's spouse—a crime in all states.

obsessive relational intrusion the relentless pursuit of intimacy with someone who does not want it.

parricide murder of a parent by an offspring.

periodic reinforcement reinforcement that occurs every now and then (unpredictable). The abused victim never knows when the abuser will be polite and kind again (e.g., flowers and candy).

revenge porn posting nude photos of ex; a form of emotional and sexual abuse; some states considering legislation against this behavior.

Rophypnol causes profound, prolonged sedation and short-term memory loss; also known as the date rape drug, roofies, Mexican Valium, or the "forget (me) pill."

siblicide murder of a sibling.

situational couple violence conflict escalates over an issue and one or both partners lose control.

stalking unwanted following or harassment of a person that induces fear in the victim.

uxoricide the murder of a woman by a romantic partner.

violence physical aggression with the purpose to control, intimidate, and subjugate another human being.

10-5 Effects of abuse

The effects of IPV (intimate partner violence) include symptoms of PTSD—loss of interest in activities and life in general, a feeling of detachment from others, inability to sleep, and irritability. Women with low self-esteem are more likely to take some responsibility for their boyfriends' abusive behavior. In effect, they excuse the rape as their fault. Violence on pregnant women significantly increases the risk for miscarriage and birth defects. Negative effects may also accrue to children who witness abuse. These children are more likely to be depressed as adults.

10-6 The cycle of abuse

The cycle of abuse begins when a person is abused and the perpetrator feels regret, asks for forgiveness, and starts acting nice (e.g., gives flowers). The victim, who perceives few options and feels guilty terminating the relationship with the partner who asks for forgiveness, feels hope for the relationship at the contriteness of the abuser and does not call the police or file charges. The couple usually experiences a period of making up or honeymooning, during which the victim feels good again about the abusing partner. However, tensions mount again and are released in the form of violence. Such violence is followed by the familiar sense of regret and pleadings for forgiveness, accompanied by being nice (a new bouquet of flowers, and so on).

The reasons people stay in abusive relationships include love, emotional dependency, commitment to the relationship, hope, view of violence as legitimate, guilt, fear, economic dependency, and isolation. The catalyst for breaking free combines the sustained aversiveness of staying, the perception that they and their children will be harmed by doing so, and the awareness of an alternative path or of help in seeking one. One must be cautious in getting out of an abusive relationship because an abuser is most likely to kill his partner when she actually leaves the relationship.

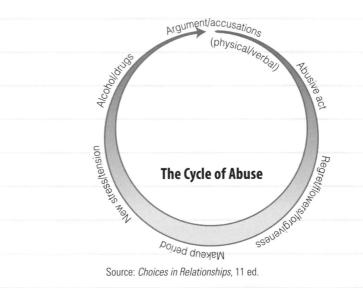

Source: *Choices in Relationships*, 11 ed.

10-7 Trends in abuse in relationships

Abuse in relationships will continue to occur behind closed doors, in private contexts where the abuse is undetected. Reducing such abuse will depend on prevention strategies focused at three levels: the general population, specific groups at high risk for abuse, and individuals/couples who have already experienced abuse. Public education and media campaigns aimed at the general population will continue to convey the criminal nature of domestic assault, suggest ways the abused might learn escape from abuse, and identify where abuse victims and perpetrators can get help.

LEARNING OUTCOMES

11-1 Do you want to have children?

Having children continues to be a major goal of most young adults (women more than men). Among youth between the ages of 18 and 29, almost three-fourths in a Pew Research Center report noted that they wanted to have children and most said that "being a good parent" was more important than "having a successful marriage." Social influences to have a child include family, friends, religion, government, favorable economic conditions, and cultural observances. The reasons people give for having children include love and companionship with one's own offspring, the desire to be personally fulfilled as an adult by having a child, and the desire to recapture one's youth. Having a child (particularly for women) reduces one's educational and career advancement. The cost for housing, food, transportation, clothing, health care, and child care for a child up to age 2 is more than $11,000 annually.

11-2 How many children do you want?

About 18% of women ages 40 to 44 do not have children. About 44% of these childfree women have chosen not to have children. Those who choose to be childfree are sometimes viewed with suspicion, avoidance, discomfort, rejection, and pity.

The most preferred type of family in the United States is the two-child family. Some of the factors in a couple's decision to have more than one child are the desire to repeat a good experience, the feeling that two children provide companionship for each other, and the desire to have a child of each sex.

11-3 Infertility

Infertility is defined as the inability to achieve a pregnancy after at least one year of regular sexual relations without birth control, or the inability to carry a pregnancy to a live birth. Forty percent of infertility problems are attributed to the woman, 40% to the man, and 20% to both of them. The causes of infertility in women include blocked fallopian tubes, endocrine imbalance that prevents ovulation, dysfunctional ovaries, chemically hostile cervical mucus that may kill sperm, and effects of sexually transmitted infections (STIs). The psychological reaction to infertility is often depression over having to give up a lifetime goal. Some of the more common causes of infertility in men include low sperm production, poor semen motility, effects of STIs, and interference with passage of sperm through the genital ducts due to an enlarged prostate.

A number of technological innovations are available to assist women and couples in becoming pregnant. These include hormonal therapy, artificial insemination, ovum transfer, in vitro fertilization, gamete intrafallopian transfer, and zygote intrafallopian transfer. Being infertile (for the woman) may have a negative lifetime effect, both personal and interpersonal (half the women in one study were separated from their husbands or partners or reported a negative effect on their sex lives).

11-4 Adoption

Motives for adoption include a couple's inability to have a biological child (infertility), their desire to give an otherwise unwanted child a permanent loving home, or their desire to avoid contributing to overpopulation by having more

KEY TERMS

abortion rate the number of abortions per thousand women aged 15 to 44.

abortion ratio refers to the number of abortions per 1,000 live births. Abortion is affected by the need for parental consent and parental notification.

antinatalism opposition to children.

childlessness concerns the idea that holidays and family gatherings may be difficult because of not having children or feeling left out or sad that others have children.

competitive birthing having the same number (or more) of children in reference to one's peers.

conception refers to the fusion of the egg and sperm. Also known as fertilization.

emergency contraception (postcoital contraception) refers to various types of morning-after pills.

foster parent neither a biological nor an adoptive parent but a person who takes care of and fosters a child taken into custody.

induced abortion the deliberate termination of a pregnancy through chemical or surgical means.

infertility the inability to achieve a pregnancy after at least one year of regular sexual relations without birth control, or the inability to carry a pregnancy to a live birth.

parental consent a woman needs permission from a parent to get an abortion if under a certain age, usually 18.

parental notification a woman has to tell a parent she is getting an abortion if she is under a certain age, usually 18, but she does not need parental permission.

pregnancy when the fertilized egg is implanted (typically in the uterine wall).

procreative liberty the freedom to decide to have children or not.

pronatalism cultural attitude which encourages having children.

spontaneous abortion (miscarriage) the unintended termination of a pregnancy.

therapeutic abortions abortions performed to protect the life or health of the woman.

biological children. Adoption is actually quite rare. Just over 1% of 18- to 44-year-old women reported having adopted a child.

Although those who typically adopt are currently White, educated, and of high income, adoptees are increasingly being placed in nontraditional families including with older, gay, and single individuals; it is recognized that these individuals may also be White, educated, and of high income. Most college students are open to transracial adoption.

11-5 Foster parenting
Some individuals seek the role of parent via foster parenting. A foster parent, also known as a family caregiver, is a person who, either alone or with a spouse, takes care of and fosters a child taken into custody in his or her home. A foster parent has made a contract with the state for the service, has judicial status, and is reimbursed by the state.

11-6 Abortion
An abortion may be either induced, which is the deliberate termination of a pregnancy through chemical or surgical means, or spontaneous (miscarriage), which is the unintended termination of a pregnancy. The most frequently cited reasons for induced abortion were that having a child would interfere with a woman's education, work, or ability to care for dependents (74%); that she could not afford a baby now (73%); and that she did not want to be a single mother or was having relationship problems (48%). Nearly 4 in 10 women said they had completed their childbearing, and almost one-third of the women were not ready to have a child. Less than 1% said their parents' or partner's desire for them to have an abortion was the most important reason. In regard to the psychological effects of abortion, the American Psychological Association reviewed the literature and concluded that "among women who have a single, legal, first-trimester abortion of an unplanned pregnancy for nontherapeutic reasons, the relative risks of mental health problems are no greater than the risks among women who deliver an unplanned pregnancy."

11-7 Trends in deciding about children
Most couples will continue to have children, but the number of couples choosing to remain childless will grow. Options for infertile couples who want children will continue to increase.

Remember

There is a self-assessment card for this chapter in the self-assessment card deck.

LEARNING OUTCOMES

12-1 Parenting: A matter of choices

Although both genetic and environmental factors are at work, the choices parents make have a dramatic impact on their children. Parents who don't make a choice about parenting have already made one. The five basic choices parents make include deciding (1) whether to have a child, (2) the number of children, (3) the interval between children, (4) methods of discipline and guidance, and (5) the degree to which they will be invested in the role of parent.

12-2 Roles of parents

Parenting includes providing physical care for children, loving them, being an economic resource, providing guidance as a teacher or model, and protecting them from harm. One of the biggest problems confronting parents today is the societal influence on their children. These influences include drugs and alcohol; peer pressure; TV, Internet, and movies; and crime or gangs.

12-3 Transition to parenthood

Transition to parenthood refers to that period of time from the beginning of pregnancy through the first few months after the birth of a baby. The mother, father, and couple all undergo changes and adaptations during this period. Most mothers relish their new role; some may experience the transitory feelings of baby blues; a few report postpartum depression. Fathers may also experience depression following the birth of a baby. The preference for a male baby is associated with the fathers' depression.

The father's involvement with his children is sometimes predicted by the quality of the parents' romantic relationship. If the father is emotionally and physically involved with the mother, he is more likely to take an active role in the child's life. In recent years, there has been a renewed cultural awareness of fatherhood.

A summary of almost 150 studies involving almost 50,000 respondents on the question of how children affect marital satisfaction revealed that parents (both women and men) reported lower marital satisfaction than nonparents. In addition, the higher the number of children, the lower the marital satisfaction; the factors that depressed marital satisfaction were conflict and loss of freedom.

Percentage of Couples Getting Divorced by Number of Children

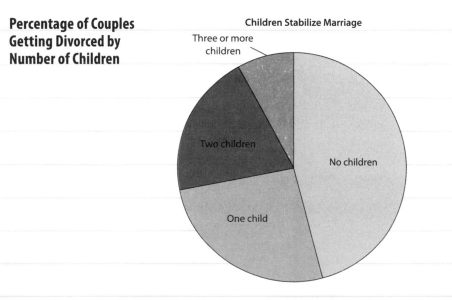

Children Stabilize Marriage

Three or more children

Two children

No children

One child

KEY TERMS

baby blues transitory symptoms of depression in a mother 24 to 48 hours after her baby is born.

boomerang generation adult children who return to live with their parents.

demandingness the manner in which parents place demands on children in regard to expectations and discipline.

emotional competence teaching the child to experience emotion, express emotion, and regulate emotion.

gatekeeper role term used to refer to the influence of the mother on the father's involvement with his children.

overindulgence defined as more than just giving children too much; includes overnurturing and providing too little structure.

oxytocin hormone from the pituitary gland during the expulsive stage of labor that has been associated with the onset of maternal behavior in lower animals.

parenting defined in terms of roles including caregiver, emotional resource, teacher, and economic resource.

parenting self-efficiency feeling competent as a parent.

postpartum depression a more severe reaction following the birth of a baby which occurs in reference to a complicated delivery as well as numerous physiological and psychological changes occurring during pregnancy, labor, and delivery; usually in the first month after birth but can be experienced after a couple of years have passed.

postpartum psychosis a reaction in which a woman wants to harm her baby.

reactive attachment disorder common among children who were taught as infants that no one cared about them; these children have no capacity to bond emotionally with others since they have no learning history of the experience and do not trust adults, caretakers, or parents.

responsiveness refers to the extent to which parents respond to and meet the needs of their children.

sextortion online sexual extortion.

single-parent family family in which there is only one parent and the other parent is completely out of the child's life through death, sperm donation, or abandonment, and no contact is made with the other parent.

single-parent household one parent has primary custody of the child/children with the other parent living outside of the house but still being a part of the child's family; also called binuclear family.

time-out a noncorporal form of punishment that involves removing the child from a context of reinforcement to a place of isolation.

transition to parenthood period from the beginning of pregnancy through the first few months after the birth of a baby during which the mother and father undergo changes.

12-4 Parenthood: Some facts

Parenthood will involve about 40% of the time a couple lives together, parents are only one influence on their children, each child is unique, and parenting styles differ. Research suggests that an authoritative parenting style, characterized by being both demanding and warm, is associated with positive outcomes. In addition, being emotionally connected to a child, respecting the child's individuality, and monitoring the child's behavior to encourage positive contexts have positive outcomes. Birth order effects include that firstborns are the first on the scene with parents and always have the "inside track." They want to stay that way so they are traditional and conforming to their parents' expectations. Children born later learn quickly that they entered an existing family constellation where everyone else is bigger and stronger. They cannot depend on having established territory, so they must excel in ways different from the firstborn. They are open to experience, adventurousness, and trying new things because their status is not already assured.

12-5 Principles of effective parenting

Giving time, love, praise, and encouragement; being realistic; avoiding overindulgence; monitoring activities and drug use; setting limits and disciplining children for inappropriate behavior; providing security; encouraging responsibility; teaching emotional competence; providing sex education; teaching nonviolence; establishing norms of forgiveness; keeping children connected with nature; engaging in leisure activity with children; and using technology to encourage safer driving are aspects of effective parenting.

12-6 Single-parenting issues

About 40% of births in the United States are to unmarried mothers. By adolescence, 20% of chldren have no contact with their father. The challenges of single parenthood for the parent include taking care of the emotional and physical needs of a child alone, meeting one's own adult emotional and sexual needs, earning money, and rearing a child without a father (the influence of whom can be positive and beneficial).

12-7 Trends in parenting

Parents will continue to be primarily responsible for child rearing, but with the financial need of both parents to earn income increasing, new parental norms involving a wider range of behaviors on the part of children will be accepted.

Remember

There is a self-assessment card for this chapter in the self-assessment card deck. For Chapter 12, there are two tools: the Traditional Motherhood Scale and the Spanking versus Time-Out Scale.

REVIEWcard

LEARNING OUTCOMES

13-1 Definitions and sources of stress and crisis

Stress is a reaction of the body to substantial or unusual demands (physical, environmental, or interpersonal). Stress is often accompanied by irritability, high blood pressure, and depression. Stress also has an effect on a person's relationships and sex life. Stress is a process rather than a state. A crisis is a situation that requires changes in normal patterns of behavior. A family crisis is a situation that upsets the normal functioning of the family and requires a new set of responses to the stressor. Sources of stress and crises can be external (e.g., a hurricane, a tornado, downsizing, military deployment) or internal (e.g., alcoholism, an extramarital affair, Alzheimer's disease).

Family resilience is displayed when family members successfully cope under adversity, which enables them to flourish with warmth, support, and cohesion. Key factors include a positive outlook, spirituality, flexibility, communication, financial management, shared recreation, routines or rituals, and support networks.

13-2 Positive stress-management strategies

Changing one's basic values and perspective is the most helpful strategy in reacting to a crisis. Viewing ill health as a challenge, bankruptcy as an opportunity to spend time with one's family, and infidelity as an opportunity to improve communication are examples. Other positive coping strategies are exercise, adequate sleep, love, religion, friends or relatives, humor, education, and counseling. Still other strategies include intervening early in a crisis, not blaming each other, keeping destructive impulses in check, and seeking opportunities for fun. Pets are also useful in helping individuals cope with stress. They are associated with reducing blood pressure, preventing heart disease, and fighting depression.

13-3 Harmful stress-management strategies

Some harmful strategies include keeping feelings inside, taking out frustrations on others, and denying or avoiding the problem.

Nicki Pardo/Getty Images

KEY TERMS

alienation of affection law which gives a spouse the right to sue a third party for taking the affections of a spouse away.

behavioral couple therapy therapeutic focus on behaviors the respective spouses want increased or decreased, initiated or terminated.

chronic sorrow grief-related feelings that occur periodically throughout the lives of those left behind.

coolidge effect term used to describe waning of sexual excitement and the effect of novelty and variety on increasing sexual arousal.

crisis a crucial situation that requires change in one's normal pattern of behavior.

down low term refers to African American married men who have sex with men and hide this from their spouse.

extradyadic involvement refers to sexual involvement of a pair-bonded individual with someone other than the partner; also called extrarelational involvement.

extramarital affair refers to a spouse's sexual involvement with someone outside the marriage.

family resiliency the successful coping of family members under adversity that enables them to flourish with warmth, support, and cohesion.

integrative behavioral couple therapy therapy which focuses on the cognitions or assumptions of the spouses, which impact the way spouses feel and interpret each other's behavior.

palliative care health care for the individual who has a life-threatening illness which focuses on relief of pain/suffering and support for the individual.

resiliency a family's strength and ability to respond to a crisis in a positive way.

sanctification viewing the marriage as having divine character or significance.

snooping investigating (without the partner's knowledge or permission) a romantic partner's private communication (e.g., text messages, email, and cell phone use) motivated by concern that the partner may be hiding something.

stress reaction of the body to substantial or unusual demands (physical, environmental, or interpersonal).

telerelationship therapy therapy sessions conducted online, often through Skype, where both therapist and couple can see and hear each other.

13-4 Five individual, couple, and family crisis events

Some of the more common crisis events that spouses and families face include physical illness and disability, mental illness, middle-age crazy (midlife crisis), an extramarital affair, unemployment, substance abuse, the death of a family member, and the suicide of a family member. When an affair is discovered, a sense of betrayal is pervasive for the partner. Surviving an affair involves forgiveness on the part of the offended spouse and the ability to grant a pardon to the offending spouse, to give up feeling angry, and to relinquish the right to retaliate against the offending spouse. In exchange, an offending spouse must take responsibility for the affair, agree not to repeat the behavior, and grant the partner the right to check up on the offending partner to regain trust. The best affair prevention is a happy and fulfilling marriage as well as avoidance of intimate conversations with members of the other sex and a context (e.g., where alcohol is consumed and being alone), which are conducive to physical involvement. The occurrence of a "midlife crisis" is reported by less than a quarter of adults in the middle years. Those who did experience a crisis were going through a divorce.

13-5 Marriage (relationship) therapy

There are around 50,000 marriage and family therapists in the United States. Whatever marriage therapy costs, it can be worth it in terms of improved relationships. If divorce can be averted, both spouses and children can avoid the trauma and thousands of dollars will be saved. Effective marriage therapy usually involves seeing the spouses together (conjoint therapy). Spouses may seek behavioral couple therapy, cognitive-behavioral therapy, or telerelationship (Skype) therapy.

13-6 Trends regarding stress and crisis in relationships

Stress and crisis will continue to be a part of relationships. No spouse, partner, marriage, family, or relationship is immune. Most relationship partners will also show resilience to rise above whatever crisis happens. The motivation to do so is strong, and having a partner to share one's difficulties reduces the sting. As noted, it is always one's perception of an event, not the event itself, which will determine the severity of a crisis and the capacity to cope with and overcome it.

> **Remember**
> Chapter 13 has self-assessment in the self-assessment card deck.

LEARNING OUTCOMES

14-1 Deciding whether to continue or end a relationship/get a divorce

Several factors are predictive of maintaining a relationship or letting it go. In addition, there is a process in terms of what should be considered before deciding to end a relationship/divorce. Specifically, four factors have been identified in whether a person continues or ends a relationship: dedication, perceived constraints, material constraints, and feeling trapped.

14-2 Macro factors contributing to divorce

Macro factors contributing to divorce include increased economic independence of women (women can afford to leave), changing family functions (companionship is the only remaining function), liberal divorce laws (it's easier to leave), prenuptial agreements, the Internet, fewer moral and religious sanctions (churches embrace single individuals), more divorce models (Hollywood models abound), mobility and anonymity, and social class, ethnicity, and culture. The United States has one of the highest divorce rates in the world, higher even than vanguard countries such as Sweden.

14-3 Micro factors contributing to divorce

Micro factors include having numerous differences, falling out of love, spending limited time together, decreasing positive behaviors, having an affair, having poor conflict resolution skills, changing values, experiencing the onset of satiation, and having the perception that one would be happier if divorced.

14-4 Consequences of divorce for spouses/parents

The psychological consequences for divorcing spouses depend on how unhappy the marriage was. Spouses who were miserable while in a loveless conflictual marriage often regard the divorce as a relief. Spouses who were left (e.g., a spouse leaves for another partner) may be devastated and suicidal. Women tend to fare better emotionally after separation and divorce than do men. Women are more likely than men not only to have a stronger network of supportive relationships but also to profit from divorce by developing a new sense of self-esteem and confidence because they are thrust into a more independent role. Divorced men are more likely than divorced women to date more partners sooner and to remarry more quickly.

A sample of 410 freshmen and sohomores at a large southeastern university revealed that though recovery was not traumatic for either men or women, men reported more difficulty than women did in adjusting to a breakup. Men more than women reported "a new partner" was more helpful in relationship recovery. Women more than men reported that "time" was more helpful in relationship recovery.

Getting divorced affects one financially, affects fathers' separation from children, and may result in shared parenting dysfunction and parental alienation.

KEY TERMS

arbitration third party listens to both spouses and makes a decision about custody, division of property, child support, and alimony.

binuclear family family that lives in two households as when parents live in separate households following a divorce.

blended family family wherein spouses in a remarriage bring their children to live with the new partner and at least one other child.

content valence positive or negative emotions associated with the content of a thought.

developmental task a skill that, if mastered, allows a family to grow as a cohesive unit.

divorce mediation meeting with a neutral professional who negotiates child custody, division of property, child support, and alimony directly with the divorcing spouses.

divorce the legal ending of a valid marriage contract.

divorcism the belief that divorce is a disaster.

latchkey parents divorcing parents who spend every other week with the children in the family home so the children do not have to alternate between residences.

litigation a judge hears arguments from lawyers representing the respective spouses and decides issues of custody, child support, division of property, etc.

negative commitment spouses who continue to be emotionally attached to and have difficulty breaking away from ex-spouses.

negotiation identifying both sides of an issue and finding a resolution that is acceptable to both parties.

no fault divorce neither party is identified as the guilty party or the cause of the divorce.

parental alienation estrangement of a child from a parent due to one parent turning the child against the other.

parental alienation syndrome an alleged disturbance in which children are obsessively preoccupied with deprecation and/or criticism of a parent; denigration that is unjustified and/or exaggerated.

physical custody the distribution of parenting time between divorced spouses.

postnuptial agreement an agreement about how money is to be divided should a couple later divorce, which is made after the couple marry.

satiation a stimulus loses its value with repeated exposure; also called habituation.

stepfamily family in which spouses in a new marriage bring children from previous relationships into the new home.

stepism the assumption that stepfamilies are inferior to biological families.

14-5 Negative and positive consequences of divorce for children

Although researchers agree that a civil, cooperative, coparenting relationship between ex-spouses is the greatest predictor of a positive outcome for children, researchers disagree on the long-term negative effects of divorce on children. However, there is no disagreement that most children do not experience long-term negative effects. Divorce mediation encourages civility between divorcing spouses who negotiate the issues of division of property, custody, visitation, child support, and spousal support.

14-6 Prerequisites for having a "successful" divorce

Divorce is an emotionally traumatic event for everyone involved, but there are some steps that spouses can take to minimize the pain and help each other and their children with the transition. Some of these steps include mediating the divorce, coparenting, sharing responsibility, creating positive thoughts, avoiding drugs and alcohol, being active, releasing anger, allowing time to heal, and progressing through the psychological stages of divorce.

14-7 Remarriage

Two-thirds of divorced females and three-fourths of divorced males remarry. Ninety percent of remarriages are of people who are divorced rather than widowed. Most divorced individuals select someone who is divorced to remarry just as widowed individuals select someone who is also widowed to remarry.

Two years is the recommended time from the end of one marriage to the beginning of the next. Older divorced women (over age 40) are less likely than younger divorced women to remarry. National data reflect that remarriages are more likely than first marriages to end in divorce in the early years of remarriage. After 15 years, however, second marriages tend to be more stable and happier than first marriages. The reason for this is that remarried individuals tend not to be afraid of divorce and would divorce if unhappy in a second marriage. First-time married individuals may be fearful of divorce and stay married even though they are unhappy.

Colin Oates

14-8 Stepfamilies

Stepfamilies represent the fastest-growing type of family in the United States. There is a movement away from the use of the term *blended* when referring to stepfamilies, because stepfamilies really do not blend. Although a stepfamily can be created when a never married or a widowed parent with children marries a person with or without children, most stepfamilies today are composed of spouses who were once divorced.

Stepfamilies differ from nuclear families: the children in nuclear families are biologically related to both parents, whereas the children in stepfamilies are biologically related to only one parent. Also, in nuclear families, both biological parents live with their children, whereas only one biological parent in stepfamilies lives with the children. In some cases, the children alternate living with each parent. Stepism is the assumption that stepfamilies are inferior to biological families. Stepism, like racism, heterosexism, sexism, and ageism, involves prejudice and discrimination.

Stepfamilies go through a set of stages. Newly remarried couples often expect instant bonding between the new members of the stepfamily. It does not often happen. The stages are fantasy (everyone will love everyone), reality (possible bitter conflict), assertiveness (parents speak their mind), strengthening pair ties (spouses nurture their relationship), and recurring change (stepfamily members know there will continue to be change). Involvement in stepfamily discussion groups such as the Stepfamily Enrichment Program provides enormous benefits.

Differences between Nuclear Families and Stepfamilies

Nuclear Families	Stepfamilies
1. Children are (usually) biologically related to both parents.	1. Children are biologically related to only one parent.
2. Both biological parents live together with children.	2. As a result of divorce or death, one biological parent does not live with the children. In the case of joint physical custody, children may live with both parents, alternating between them.
3. Beliefs and values of members tend to be similar.	3. Beliefs and values of members are more likely to be different because of different backgrounds.
4. The relationship between adults has existed longer than relationship between children and parents.	4. The relationship between children and parents has existed longer than the relationship between adults.
5. Children have one home they regard as theirs.	5. Children may have two homes they regard as theirs.
6. The family's economic resources come from within the family unit.	6. Some economic resources may come from an ex-spouse.
7. All money generated stays in the family.	7. Some money generated may leave the family in the form of alimony or child support.
8. Relationships are relatively stable.	8. Relationships are in flux: new adults adjusting to each other; children adjusting to a stepparent; a stepparent adjusting to stepchildren; stepchildren adjusting to each other.
9. No stigma is attached to nuclear family.	9. Stepfamilies are stigmatized.
10. Spouses had a childfree period.	10. Spouses had no childfree period.
11. Inheritance rights are automatic.	11. Stepchildren do not automatically inherit from stepparents.
12. Rights to custody of children are assumed if divorce occurs.	12. Rights to custody of stepchildren are usually not considered.
13. Extended family networks are smooth and comfortable.	13. Extended family networks become complex and strained.
14. Nuclear family may not have experienced loss.	14. Stepfamily has experienced loss.
15. Families experience a range of problems.	15. Stepchildren tend to be a major problem.

14-9 Trends in divorce and remarriage

Divorce remains stigmatized in our society. In view of this cultural attitude, a number of attempts will continue to be made to reduce divorce rates. Divorced fathers will also demand to be treated equally in custody decisions. The remarriage rate is dropping.

> **Remember**
> Take the Chapter 14 self-assessment, Children's Beliefs about Parental Divorce Scale, located in your self-assessment card deck.

REVIEWcard

LEARNING OUTCOMES

15-1 Age and ageism

Age is defined chronologically (by time), physiologically (by capacity to see, hear, and so on), psychologically (by self-concept), sociologically (by social roles), and culturally (by the value placed on the elderly). Ageism is the denigration of the elderly, and gerontophobia is the fear or dread of the elderly. Theories of aging range from disengagement (individuals and societies mutually disengage from each other) to continuity (the habit patterns of youth are continued in old age), as you can see in the following table. Life course is the aging theory currently in vogue. This approach examines differences in aging across cohorts by emphasizing that "individual biography is situated within the context of social structure and historical circumstance."

Theories of Aging

Name of Theory	Level of Theory	Theorists	Basic Assumptions	Criticisms
Disengagement	Macro	Elaine Cumming, William Henry	The gradual and mutual withdrawal of the elderly and society from each other is a natural process. It is also necessary and functional for society that the elderly disengage so that new people can be phased in to replace them in an orderly transition.	Not all people want to disengage; some want to stay active and involved. Disengagement does not specify what happens when the elderly stay involved.
Activity	Macro	Robert Havighurst	People continue the level of activity they had in middle age into their later years. Though high levels of activity are unrelated to living longer, they are related to reporting high levels of life satisfaction.	Ill health may force people to curtail their level of activity. The older a person, the more likely the person is to curtail activity.
Conflict	Macro	Karl Marx, Max Weber	The elderly compete with youth for jobs and social resources such as government programs (Medicare).	The elderly are presented as disadvantaged. Their power to organize and mobilize political resources such as the American Association of Retired Persons is underestimated.
Age stratification	Macro	M. W. Riley	The elderly represent a powerful cohort of individuals passing through the social system that both affect and are affected by social change.	Too much emphasis is put on age, and little recognition is given to other variables within a cohort such as gender, race, and socioeconomic differences.
Modernization	Macro	Donald Cowgill	The status of the elderly is in reference to the evolution of the society toward modernization. The elderly in premodern societies have more status because what they have to offer in the form of cultural wisdom is more valued. The elderly in modern technologically advanced societies have low status because they have little to offer.	Cultural values for the elderly, not level of modernization, dictate the status of the elderly. Japan has high respect for the elderly and yet is highly technological and modernized.
Symbolic	Micro	Arlie Hochschild	The elderly socially construct meaning in their interactions with others and society. Developing social bonds with other elderly can ward off being isolated and abandoned. Meaning is in the interpretation, not in the event.	The power of the larger social system and larger social structures to affect the lives of the elderly is minimized.
Continuity	Micro	Bernice Neugarten	The earlier habit patterns, values, and attitudes of the individual are carried forward as a person ages. The only personality change that occurs with aging is the tendency to turn one's attention and interest on the self.	Other factors than one's personality affect aging outcomes. The social structure influences the life of the elderly rather than vice versa.
Interpersonal	Micro	Julian Palmore III Jean-Pierre Langlois	Negative assumptions based on physical apperance (droopy eyes means sad person).	Some elderly are in good physical condition.

KEY TERMS

age term which may be defined chronologically (number of years), physiologically (physical decline), psychologically (self-concept), sociologically (roles for the elderly/retired), and culturally (meaning age in one's society).

age discrimination a situation where older people are often not hired and younger workers are hired to take their place.

ageism the systematic persecution and degradation of people because they are old.

ageism by invisibility when older adults are not included in advertising and educational materials.

blurred retirement an individual working part-time before completely retiring or taking a "bridge job" that provides a transition between a lifelong career and full retirement.

dementia loss of brain function that occurs with certain diseases. It affects memory, thinking, language, judgment, and behavior.

family caregiving adult children providing care for their elderly parents.

frail term used to define elderly people if they have difficulty with at least one personal care activity (feeding, bathing, toileting).

gerontology the study of aging.

gerontophobia fear or dread of the elderly, which may create a self-fulfilling prophecy.

phased retirement an employee agreeing to a reduced work load in exchange for reduced income.

sandwich generation generation of adults who are "sandwiched" between caring for their elderly parents and their own children.

thanatology the examination of the social dimensions of death, dying, and bereavement.

15-2 Caregiving for the frail elderly: The "sandwich generation"

Eldercare combined with child care is becoming common among the sandwich generation—adult children responsible for the needs of both their parents and their children. Two levels of eldercare include help with personal needs such as feeding, bathing, and toileting as well as instrumental care such as going to the grocery store, managing bank records, and so on. Members of the sandwich generation report feelings of exhaustion over the relentless demands, guilt over not doing enough, and resentment over feeling burdened.

Most adult children want to take care of their aging parents either in the parents' own home or in the adult child's home. Parents are usually resistant to full-time nursing home care, but become resigned to such care when there is no other option. But professional caregiving is expensive—$70,000 is the average annual cost. Home health care can reduce the strain of caring for an elderly parent.

15-3 Issues confronting the elderly

Issues of concern to the elderly include housing, health, retirement, and sexuality. Most elderly live in their own homes, which they have paid for. Most elderly housing is adequate, although repair becomes a problem when people age. Health concerns are paramount for the elderly. Good health is the single most important factor associated with an elderly person's perceived life satisfaction.

When elders' vision, hearing, physical mobility, and strength are markedly diminished, their sense of well-being is often significantly negatively impacted. Driving accidents also increase with aging. Mental problems may also occur with mood disorders; depression is the most common. Dementia, which includes Alzheimer's disease, is the mental disorder most associated with aging; however, only 3% of the aged population experience severe cognitive impairment.

Individuals who have a positive attitude toward retirement are those who have a pension waiting for them, are married, have planned for retirement, are in good health, and have high self-esteem.

For most elderly women and men, sexuality involves lower reported interest, activity, and capacity. Fear of the inability to have an erection is the sexual problem elderly men most frequently report (Viagra, Levitra, and Cialis have helped allay this fear). The absence of a sexual partner is the sexual problem elderly women most frequently report.

15-4 Successful aging

Factors associated with successful aging include not smoking (or quitting early), developing a positive view of life and life's crises, avoiding alcohol and substance abuse, maintaining healthy weight, exercising daily, continuing to educate oneself, and having a happy marriage. Indeed, those who were identified as "happy and well" were six times more likely to be in good marriages than those who were identified as "sad and sick." Success in one's career is also associated with successful aging.

15-5 Relationships and the elderly

Marriages that survive into old age (beyond age 85) tend to have limited conflict, considerable companionship, and mutual supportiveness. Marital satisfaction is related to equality of roles and marital communication. Some widowed or divorced elderly try to find and renew an earlier love relationship. Older adults and senior citizens are increasing their use of technology to stay connected by using networking sites such as Facebook and LinkedIn. Relationships with siblings are primarily emotional rather than functional. Relationships with children are emotional and expressive. Actual caregiving by one's children is rare.

15-6 The end of one's life

Thanatology is the examination of the social dimensions of death, dying, and bereavement. The end of life can involve adjusting to the death of one's spouse and to the gradual decline of one's health. Most elderly are satisfied with their lives, relationships, and health. Declines begin when people reach their 80s. Most fear the process of dying more than death itself.

15-7 Trends and the elderly in the United States

The percentage of the U.S. population over the age of 55 will increase from 21% to 30% by 2030.

NOTES

NOTES

NOTES

NOTES

Attitudes toward Marriage Scale

The purpose of this survey is to assess the degree to which you view marriage positively. Read each item carefully and consider what you believe. There are no right or wrong answers. After reading each statement, select the number that best reflects your level of agreement, using the following scale:

1	2	3	4	5	6	7
Strongly Disagree						Strongly Agree

____ **1.** I am married or plan to get married.

____ **2.** Being single and free is not as good as people think it is.

____ **3.** Marriage is NOT another word for being trapped.

____ **4.** Single people are more lonely than married people.

____ **5.** Married people are happier than single people.

____ **6.** Most of the married people I know are happy.

____ **7.** Most of the single people I know think marriage is better than singlehood.

____ **8.** The statement that singles are more lonely and less happy than marrieds is mostly true.

____ **9.** It is better to be married than to be single.

____ **10.** Married people enjoy their lifestyle more than single people.

____ **11.** Marrieds have more close intimate relationships than singles.

____ **12.** Marrieds have a greater sense of joy than singles.

____ **13.** Being married is a more satisfying lifestyle than being single.

____ **14.** People who think that married people are happier than single people are correct.

____ **15.** Single people struggle with avoiding loneliness.

____ **16.** Married people are not as lonely as single people.

____ **17.** The companionship of marriage is a major advantage of the lifestyle.

____ **18.** Married people have better sex than singles.

____ **19.** The idea that singlehood is a happier lifestyle than being married is nonsense.

____ **20.** Singlehood as a lifestyle is overrated.

Scoring

After assigning a number from 1 (strongly disagree) to 7 (strongly agree), add the numbers. The higher your score (140 is the highest possible score), the more positive your view of marriage. The lower your score (20 is the lowest possible score), the more negatively you view marriage. The midpoint is 60 (scores lower than 60 suggest a more negative view of marriage; scores higher than 60 suggest a more positive view of marriage.

Norms

The norming sample of this self-assessment was based on 32 males and 174 females at East Carolina University. The average score of the males was 92 and the average score of the females was 95, suggesting a predominantly positive view of marriage (with females more positive than males).

Source: "Attitudes toward Marriage Scale" was developed for this text by David Knox. It is to be used for general assessment and is not designed to be a clinical diagnostic tool or as a research instrument.

Living Apart Together Scale

This scale will help you assess the degree to which you might benefit from living in a separate residence from your spouse or partner with whom you have a lifetime commitment. There are no right or wrong answers. After reading each sentence carefully, circle the number that best represents the degree to which you agree or disagree with the sentence.

1	2	3	4	5
Strongly Agree	Mildly Agree	Undecided	Mildly Disagree	Strongly Disagree

		SA	MA	U	MD	SD
1.	I prefer to have my own place (apart from my partner) to live.	1	2	3	4	5
2.	Living apart from my partner feels "right" to me.	1	2	3	4	5
3.	Too much togetherness can kill a relationship.	1	2	3	4	5
4.	Living apart can enhance your relationship.	1	2	3	4	5
5.	By living apart you can love your partner more.	1	2	3	4	5
6.	Living apart protects your relationship from staleness.	1	2	3	4	5
7.	Couples who live apart are just as happy as those who don't.	1	2	3	4	5
8.	Couples who LAT are just as much in love as those who live together in the same place.	1	2	3	4	5
9.	People who LAT probably have less relationship stress than couples who live together in the same place.	1	2	3	4	5
10.	LAT couples are just as committed as couples who live together in the same residence.	1	2	3	4	5

Scoring

Add the numbers you circled. The lower your total score (10 is the lowest possible score), the more suited you are to the Living Apart Together lifestyle. The higher your total score (50 is the highest possible score), the least suited you are to the Living Apart Together lifestyle. A score of 25 places you at the midpoint between being the extremes. One-hundred and thirty undergraduates completed the LAT scale with an average score of 28.92, which suggests that both sexes view themselves as less rather than more suited (30 is the midpoint between the lowest score of 10 and the highest score of 50) for a LAT arrangement with females reporting less suitability than males.

Source: Email Dr. Knox at knoxd@scu.edu for scale use.

The Beliefs about Women Scale (BAWS)

The following statements describe different attitudes toward men and women. There are no right or wrong answers, only opinions. Indicate how much you agree or disagree with each statement, using the following scale: (A) strongly disagree, (B) slightly disagree, (C) neither agree nor disagree, (D) slightly agree, or (E) strongly agree.

_____	**1.**	Women are more passive than men.
_____	**2.**	Women are less career-motivated than men.
_____	**3.**	Women don't generally like to be active in their sexual relationships.
_____	**4.**	Women are more concerned about their physical appearance than are men.
_____	**5.**	Women comply more often than men.
_____	**6.**	Women care as much as men do about developing a job or career.
_____	**7.**	Most women don't like to express their sexuality.
_____	**8.**	Men are as conceited about their appearance as are women.
_____	**9.**	Men are as submissive as women.
_____	**10.**	Women are as skillful in business-related activities as are men.
_____	**11.**	Most women want their partner to take the initiative in their sexual relationships.
_____	**12.**	Women spend more time attending to their physical appearance than men do.
_____	**13.**	Women tend to give up more easily than men.
_____	**14.**	Women dislike being in leadership positions more than men.
_____	**15.**	Women are as interested in sex as are men.
_____	**16.**	Women pay more attention to their looks than most men do.
_____	**17.**	Women are more easily influenced than men.
_____	**18.**	Women don't like responsibility as much as men.
_____	**19.**	Women's sexual desires are less intense than men's.
_____	**20.**	Women gain more status from their physical appearance than do men.

The BAWS consists of fifteen separate subscales; only four are used here. The items for these four subscales and coding instructions are as follows:

1. Women are more passive than men (items 1, 5, 9, 13, 17).
2. Women are interested in careers less than men (items 2, 6, 10, 14, 18).
3. Women are less sexual than men (items 3, 7, 11, 15, 19).
4. Women are more appearance conscious than men (items 4, 8, 12, 16, 20).

Score the items as follows: strongly agree = +2; slightly agree = +1; neither agree nor disagree = 0; slightly disagree = −1; strongly disagree = −2.

Scores range from 0 to 40; subscale scores range from 0 to 10. The higher your score, the more traditional your gender beliefs about men and women.

Source: William E. Snell, Jr., Ph.D. (1997). College of Liberal Arts, Department of Psychology, Southeast Missouri State University. Reprinted with permission. Contact Dr. Snell for further use: wesnell@semo.edu

The Love Attitudes Scale (LAS)

This scale is designed to assess the degree to which you are romantic or realistic in your attitudes toward love. There are no right or wrong answers. After reading each sentence carefully, circle the number that best represents the degree to which you agree or disagree with the sentence.

1	2	3	4	5
Strongly Agree	*Mildly Agree*	*Undecided*	*Mildly Disagree*	*Strongly Disagree*

	SA	MA	U	MD	SD
1. Love doesn't make sense. It just is.	1	2	3	4	5
2. When you fall "head over heels" in love, it's sure to be the real thing.	1	2	3	4	5
3. To be in love with someone you would like to marry but can't is a tragedy.	1	2	3	4	5
4. When love hits, you know it.	1	2	3	4	5
5. Common interests are really unimportant; as long as each of you is truly in love, you will adjust.	1	2	3	4	5
6. It doesn't matter if you marry after you have known your partner for only a short time as long as you know you are in love.	1	2	3	4	5
7. If you are going to love a person, you will "know" after a short time.	1	2	3	4	5
8. As long as two people love each other, the educational differences they have really do not matter.	1	2	3	4	5
9. You can love someone even though you do not like any of that person's friends.	1	2	3	4	5
10. When you are in love, you are usually in a daze.	1	2	3	4	5
11. Love "at first sight" is often the deepest and most enduring type of love.	1	2	3	4	5
12. When you are in love, it really does not matter what your partner does because you will love him or her anyway.	1	2	3	4	5
13. As long as you really love a person, you will be able to solve the problems you have with the person.	1	2	3	4	5
14. Usually you can really love and be happy with only one or two people in the world.	1	2	3	4	5
15. Regardless of other factors, if you truly love another person, that is a good enough reason to marry that person.	1	2	3	4	5
16. It is necessary to be in love with the one you marry to be happy.	1	2	3	4	5
17. Love is more of a feeling than a relationship.	1	2	3	4	5
18. People should not get married unless they are in love.	1	2	3	4	5
19. Most people truly love only once during their lives.	1	2	3	4	5
20. Somewhere there is an ideal mate for most people.	1	2	3	4	5
21. In most cases, you will "know it" when you meet the right partner.	1	2	3	4	5
22. Jealousy usually varies directly with love; that is, the more you are in love, the greater your tendency to become jealous will be.	1	2	3	4	5
23. When you are in love, you are motivated by what you feel rather than by what you think.	1	2	3	4	5
24. Love is best described as an exciting rather than a calm thing.	1	2	3	4	5
25. Most divorces probably result from falling out of love rather than failing to adjust.	1	2	3	4	5
26. When you are in love, your judgment is usually not too clear.	1	2	3	4	5

	SA	MA	U	MD	SD
27. Love comes only once in a lifetime.	1	2	3	4	5
28. Love is often a violent and uncontrollable emotion.	1	2	3	4	5
29. When selecting a marriage partner, differences in social class and religion are of small importance compared with love.	1	2	3	4	5
30. No matter what anyone says, love cannot be understood.	1	2	3	4	5

Scoring

Add the numbers you circled. 1 (strongly agree) is the most romantic response and 5 (strongly disagree) is the most realistic response. The lower your total score (30 is the lowest possible score), the more romantic your attitudes toward love. The higher your total score (150 is the highest possible score), the more realistic your attitudes toward love. Of 45 undergraduate males, 85.3 was the average score; of 193 undergraduate females, 85.5 was the average score. These scores reflect a slight lean toward romanticism with no gender difference.

A team of researchers (Medora et al., 2002) gave the scale to 641 young adults at three international universities in the United States, Turkey, and India. Female respondents in all three cultures had higher romanticism scores than male respondents (reflecting their higher value for, desire for, and thoughts about marriage). When the scores were compared by culture, American young adults were the most romantic, followed by Turkish students, with Indians having the lowest romanticism scores.

Reference: Medora, N. P., J. H. Larson, N. Hortacsu, & P. Dave. (2002). Perceived attitudes towards romanticism: A cross-cultural study of American, Asian-Indian, and Turkish young adults. *Journal of Comparative Family Studies, 33:* 155–178.

Source: David Knox. Department of Sociology, East Carolina University. Email Dr. Knox for permission to use. Knoxd@ecu.edu

Supportive Communication Scale (SCS)

This scale is designed to assess the degree to which partners experience supportive communication in their relationships. After reading each item, circle the number that best approximates your answer.

0	1	2	3	4
Strongly Disagree	Disagree	Undecided	Agree	Strongly Agree

	SD	D	UN	A	SA
1. My partner listens to me when I need someone to talk to.	0	1	2	3	4
2. My partner helps me clarify my thoughts.	0	1	2	3	4
3. I can state my feelings without my partner getting defensive.	0	1	2	3	4
4. When it comes to having a serious discussion, it seems we have little in common (reverse scored).	0	1	2	3	4
5. I feel put down in a serious conversation with my partner (reverse scored).	0	1	2	3	4
6. I feel discussing some things with my partner is useless (reverse scored).	0	1	2	3	4
7. My partner and I understand each other completely.	0	1	2	3	4
8. We have an endless number of things to talk about.	0	1	2	3	4

Scoring

Look at the numbers you circled. Reverse-score the numbers for questions 4, 5, and 6. For example, if you circled a 0, give yourself a 4; if you circled a 3, give yourself a 1, and so on. Add the numbers and divide by 8, the total number of items. The lowest possible score would be 0, reflecting the complete absence of supportive communication; the highest score would be 4, reflecting complete supportive communication. One-hundred-and-eighty-eight individuals completed the scale. Thirty-nine percent of the respondents were married, 38% were single, and 23% were living together. The average age was just over 24. The average score of 94 male partners who took the scale was 3.01; the average score of 94 female partners was 3.07.

Source: Sprecher, S., S. Metts, B. Burelson, E. Hatfield, & A. Thompson. (1995). Domains of expressive interaction in intimate relationships: Associations with satisfaction and commitment. *Family Relations, 44:* 203–210. Published in 1995 by the National Council on Family Relations.

Conservative–Liberal Sexuality Scale (CLSS)

This scale is designed to assess the degree to which you are conservative or liberal in your attitudes toward sex. There are no right or wrong answers. After reading each sentence carefully, select the number that best represents the degree to which you agree or disagree with the sentence.

1	2	3	4	5
Strongly Agree	Mildly Agree	Undecided	Mildly Disagree	Strongly Disagree

_____ **1.** Abortion is wrong.

_____ **2.** Homosexuality is immoral.

_____ **3.** Couples should wait to have sexual intercourse until after they are married.

_____ **4.** Couples who are virgins at marriage have more successful marriages.

_____ **5.** Watching pornography is harmful.

_____ **6.** Kinky sex is something to be avoided.

_____ **7.** Having an extramarital affair is never justified.

_____ **8.** Masturbation is something an individual should try to avoid doing.

_____ **9.** One should always be in love when having sex with a person.

_____ **10.** Transgender people are screwed up and "not right."

_____ **11.** Sex is for youth, not for the elderly.

_____ **12.** There is entirely too much sex on TV today.

_____ **13.** The best use of sex is for procreation.

_____ **14.** Any form of sex that is not sexual intercourse is wrong.

_____ **15.** Our society is entirely too liberal when it comes to sex.

_____ **16.** Sex education gives youth ideas about sex they shouldn't have.

_____ **17.** Promiscuity is the cause of the downfall of an individual.

_____ **18.** Too much sexual freedom is promoted in our country today.

_____ **19.** The handicapped probably should not try to get involved in sex.

_____ **20.** The movies in America are too sexually explicit.

Scoring

Add the numbers you circled. 1 (strongly agree) is the most conservative response and 5 (strongly disagree) is the most liberal response. The lower your total score (20 is the lowest possible score), the more sexually conservative your attitudes toward sex. The higher your total score (100 is the highest possible score), the more liberal your attitudes toward sex. A score of 60 places you at the midpoint between being the ultimate conservative and the ultimate liberal about sex. Of 191 undergraduate females, the average score was 69.85. Of 39 undergraduate males the average score was 72.89. Hence, both women and men tended to be more sexually liberal than conservative with men more sexually liberal than women.

Source: Knox, D. (2014). "The Conservative–Liberal Sexuality Scale" was developed for this text. The scale is intended to be thought provoking and fun. It is not intended to be used as a clinical or diagnostic instrument.

Self-Report of Behavior Scale (SRBS)

This questionnaire is designed to examine which of the following statements most closely describes your behavior during past encounters with people you thought were homosexuals. Rate each of the following self-statements as honestly as possible using the following scale. Write each value in the provided blank.

1	2	3	4	5
Never	Rarely	Occasionally	Frequently	Always

_____ **1.** I have spread negative talk about someone because I suspected that the person was gay.

_____ **2.** I have participated in playing jokes on someone because I suspected that the person was gay.

_____ **3.** I have changed roommates and/or rooms because I suspected my roommate was gay.

_____ **4.** I have warned people who I thought were gay and who were a little too friendly with me to keep away from me.

_____ **5.** I have attended antigay protests.

_____ **6.** I have been rude to someone because I thought that the person was gay.

_____ **7.** I have changed seat locations because I suspected the person sitting next to me was gay.

_____ **8.** I have had to force myself to keep from hitting someone because the person was gay and very near me.

_____ **9.** When someone I thought to be gay has walked toward me as if to start a conversation, I have deliberately changed directions and walked away to avoid the person.

_____ **10.** I have stared at a gay person in such a manner as to convey my disapproval of the person being too close to me.

_____ **11.** I have been with a group in which one (or more) person(s) yelled insulting comments to a gay person or group of gay people.

_____ **12.** I have changed my normal behavior in a restroom because a person I believed to be gay was in there at the same time.

_____ **13.** When a gay person has checked me out, I have verbally threatened the person.

_____ **14.** I have participated in damaging someone's property because the person was gay.

_____ **15.** I have physically hit or pushed someone I thought was gay because the person brushed against me when passing by.

_____ **16.** Within the past few months, I have told a joke that made fun of gay people.

_____ **17.** I have gotten into a physical fight with a gay person because I thought the person had been making moves on me.

_____ **18.** I have refused to work on school and/or work projects with a partner I thought was gay.

_____ **19.** I have written graffiti about gay people or homosexuality.

_____ **20.** When a gay person has been near me, I have moved away to put more distance between us.

Scoring

Determine your score by adding your points together. The lowest score is 20 points, the highest 100 points. The higher the score, the more negative the attitudes toward homosexuals.

Comparison Data

Sunita Patel (1989) originally developed the Self-Report of Behavior Scale in her thesis research in her clinical psychology master's program at East Carolina University. College men (from a university campus and from a military base) were the original participants (Patel et al. 1995). The scale was revised by Shartra Sylivant (1992), who used it with a coed high school student population, and by Tristan Roderick (1994), who involved college students to assess its psychometric properties. The scale was found to have high internal consistency. Two factors were identifed: a passive avoidance of homosexuals and active or aggressive reactions.

In a study by Roderick et al. (1998), the mean score for 182 college women was 24.76. The mean score for 84 men was significantly higher, at 31.60. A similar-sex difference, although with higher (more negative) scores, was found in Sylivant's high school sample (with a mean of 33.74 for the young women, and 44.40 for the young men).

The following table provides the scores of the college students in Roderick's sample (from a mid-sized state university in the southeast):

	N	Mean	Standard Deviation
Women	182	24.76	7.68
Men	84	31.60	10.36
Total	266	26.91	9.16

Source: By Ms. Shartra Sylivant M.A., L.P.A. Clinical, H.S.P. The cognitive, affective, and behavioral components of adolescent homonegativity. Master's thesis, East Carolina University, 1992. The SBS-R is reprinted by the permission of S. Sylivant.

Attitudes toward Interracial Dating Scale

Interracial dating or marrying is the dating or marrying of two people from different races. The purpose of this survey is to gain a better understanding of what people think and feel about interracial relationships. Please read each item carefully, and in each space, score your response using the following scale. There are no right or wrong answers to any of these statements.

1	2	3	4	5	6	7
Strongly Disagree						*Strongly Agree*

____ **1.** I believe that interracial couples date outside their race to get attention.

____ **2.** I feel that interracial couples have little in common.

____ **3.** When I see an interracial couple, I find myself evaluating them negatively.

____ **4.** People date outside their own race because they feel inferior.

____ **5.** Dating interracially shows a lack of respect for one's own race.

____ **6.** I would be upset with a family member who dated outside our race.

____ **7.** I would be upset with a close friend who dated outside our race.

____ **8.** I feel uneasy around an interracial couple.

____ **9.** People of different races should associate only in nondating settings.

____ **10.** I am offended when I see an interracial couple.

____ **11.** Interracial couples are more likely to have low self-esteem.

____ **12.** Interracial dating interferes with my fundamental beliefs.

____ **13.** People should date only within their race.

____ **14.** I dislike seeing interracial couples together.

____ **15.** I would not pursue a relationship with someone of a different race, regardless of my feelings for that person.

____ **16.** Interracial dating interferes with my concept of cultural identity.

____ **17.** I support dating between people with the same skin color, but not with a different skin color.

____ **18.** I can imagine myself in a long-term relationship with someone of another race.

____ **19.** As long as the people involved love each other, I do not have a problem with interracial dating.

____ **20.** I think interracial dating is a good thing.

Scoring

First, reverse the scores for items 18, 19, and 20 by switching them to the opposite side of the spectrum. For example, if you selected 7 for item 18, replace it with a 1; if you selected 3, replace it with a 5, and so on. Next, add your scores and divide by 20. Possible final scores range from 1 to 7, with 1 representing the most positive attitudes toward interracial dating and 7 representing the most negative attitudes toward interracial dating.

Norms

The norming sample was based upon 113 male and 200 female students attending Valdosta State University. The participants completing the Attitudes toward Interracial Dating Scale (IRDS) received no compensation for their participation. All participants were U.S. citizens. The average age was 23.02 years (standard deviation [SD] = 5.09), and participants ranged in age from 18 to 50 years. The ethnic composition of the sample was 62.9% White, 32.6% Black, 1% Asian, 0.6% Hispanic, and 2.2% others. The classification of the sample was 9.3% freshmen, 16.3% sophomores, 29.1% juniors, 37.1% seniors, and 2.9% graduate students. The average score on the IRDS was 2.88 (SD = 1.48), and scores ranged from 1.00 to 6.60, suggesting very positive views of interracial dating. Men scored an average of 2.97 (SD = 1.58), and women, 2.84 (SD = 1.42). There were no significant differences between the responses of women and men.

Source: The Attitudes toward Interracial Dating Scale. (2004). Mark Whatley, Ph.D., Department of Psychology, Valdosta State University, Valdosta, GA 31698.

Satisfaction with Marriage Scale

Below are five statements with which you may agree or disagree. Using the 1–7 scale below, indicate your agreement with each item by circling the appropriate number on the line following that item. Please be open and honest in responding to each item.

1	2	3	4	5	6	7
Strongly Disagree	Disagree	Slightly Disagree	Neither Agree nor Disagree	Slightly Agree	Agree	Strongly Agree

	SD	D	SD	N	SA	A	SA
1. In most ways my married life is close to ideal	1	2	3	4	5	6	7
2. The conditions of my married life are excellent.	1	2	3	4	5	6	7
3. I am satisfied with my married life.	1	2	3	4	5	6	7
4. So far I have gotten the important things I want in my married life.	1	2	3	4	5	6	7
5. If I could live my married life over, I would change almost nothing.	1	2	3	4	5	6	7

Scoring

Add the numbers you circled. The marital satisfaction score will range from 5 to 35. For purposes of this study a couple's combined marital satisfaction score was calculated by summing both partners' scores, resulting in a possible score range of 10 to 70, with higher scores indicating greater marital satisfaction for the couple. The internal consistency of the SWML has been reported with a Cronbach's alpha of .92 along with some evidence of construct validity (Johnson et al., 2006).

Source: Johnson, H. A., Zabriskie, R. B. & Hill, B. (2006). The contribution of couple leisure involvement, leisure time, and leisure satisfaction to marital satisfaction. *Marriage and Family Review*, 40: 69–91. Publisher by Taylor & Francis Ltd.

Maternal Employment Scale (MES)

Directions

Using the following scale, please mark a number on the blank next to each statement to indicate how strongly you agree or disagree.

1	2	3	4	5	6
Disagree Very Strongly	Disagree Strongly	Disagree Slightly	Agree Slightly	Agree Strongly	Agree Very Strongly

_____ **1.** Children are less likely to form a warm and secure relationship with a mother who works full-time outside the home.

_____ **2.** Children whose mothers work are more independent and able to do things for themselves.

_____ **3.** Working mothers are more likely to have children with psychological problems than mothers who do not work outside the home.

_____ **4.** Teenagers get into less trouble with the law if their mothers do not work full-time outside the home.

_____ **5.** For young children, working mothers are good role models for leading busy and productive lives.

_____ **6.** Boys whose mothers work are more likely to develop respect for women.

_____ **7.** Young children learn more if their mothers stay at home with them.

_____ **8.** Children whose mothers work learn valuable lessons about other people they can rely on.

_____ **9.** Girls whose mothers work full-time outside the home develop stronger motivation to do well in school.

_____ **10.** Daughters of working mothers are better prepared to combine work and motherhood if they choose to do both.

_____ **11.** Children whose mothers work are more likely to be left alone and exposed to dangerous situations.

_____ **12.** Children whose mothers work are more likely to pitch in and do tasks around the house.

_____ **13.** Children do better in school if their mothers are not working full-time outside the home.

_____ **14.** Children whose mothers work full-time outside the home develop more regard for women's intelligence and competence.

_____ **15.** Children of working mothers are less well-nourished and don't eat the way they should.

_____ **16.** Children whose mothers work are more likely to understand and appreciate the value of a dollar.

_____ **17.** Children whose mothers work suffer because their mothers are not there when they need them.

_____ **18.** Children of working mothers grow up to be less competent parents than other children because they have not had adequate parental role models.

_____ **19.** Sons of working mothers are better prepared to cooperate with a wife who wants both to work and have children.

_____ **20.** Children of mothers who work develop lower self-esteem because they think they are not worth devoting attention to.

_____ **21.** Children whose mothers work are more likely to learn the importance of teamwork and cooperation among family members.

_____ **22.** Children of working mothers are more likely than other children to experiment with alcohol, other drugs, and sex at an early age.

_____ **23.** Children whose mothers work develop less stereotyped views about men's and women's roles.

_____ **24.** Children whose mothers work full-time outside the home are more adaptable; they cope better with the unexpected and with changes in plans.

Scoring

Items 1, 3, 4, 7, 11, 13, 15, 17, 18, 20, and 22 refer to "costs" of maternal employment for children and yield a Costs Subscale score. High scores on the Costs Subscale reflect strong beliefs that maternal employment is costly to children. Items 2, 5, 6, 8, 9, 10, 12, 14, 16, 19, 21, 23, and 24 refer to "benefits" of maternal employment for children and yield a Benefits Subscale score. To obtain a total score, reverse the score of all items in the Benefits Subscale so that 1 = 6, 2 = 5, 3 = 4, 4 = 3, 5 = 2, and 6 = 1. The higher one's total score, the more one believes that maternal employment has negative consequences for children.

Source: E. Greenberger, W. A. Goldberg, T. J. Crawford, and J. Granger, Beliefs about the consequences of maternal employment for children in *Psychology of Women Quarterly, Maternal Employment Scale, 12:* 35–59, 1988.

Abusive Behavior Inventory

Circle the number that best represents your closest estimate of how often each of the behaviors have happened in the relationship with your current or former partner during the previous six months.

1.	Called you a name and/or criticized you	1	2	3	4	5
2.	Tried to keep you from doing something you wanted to do (e.g., going out with friends or going to meetings)	1	2	3	4	5
3.	Gave you angry stares or looks	1	2	3	4	5
4.	Prevented you from having money for your own use	1	2	3	4	5
5.	Ended a discussion and made a decision without you	1	2	3	4	5
6.	Threatened to hit or throw something at you	1	2	3	4	5
7.	Pushed, grabbed, or shoved you	1	2	3	4	5
8.	Put down your family and friends	1	2	3	4	5
9.	Accused you of paying too much attention to someone or something else	1	2	3	4	5
10.	Put you on an allowance	1	2	3	4	5
11.	Used your children to threaten you (e.g., told you that you would lose custody or threatened to leave town with the children)	1	2	3	4	5
12.	Became very upset with you because dinner, housework, or laundry was not done when or how it was wanted	1	2	3	4	5
13.	Said things to scare you (e.g., told you something "bad" would happen or threatened to commit suicide)					
14.	Slapped, hit, or punched you	1	2	3	4	5
15.	Made you do something humiliating or degrading (e.g., begging for forgiveness or having to ask permission to use the car or do something)	1	2	3	4	5
16.	Checked up on you (e.g., listened to your phone calls, checked the mileage on your car, or called you repeatedly at work)	1	2	3	4	5
17.	Drove recklessly when you were in the car	1	2	3	4	5
18.	Pressured you to have sex in a way you didn't like or want	1	2	3	4	5
19.	Refused to do housework or child care	1	2	3	4	5
20.	Threatened you with a knife, gun, or other weapon	1	2	3	4	5
21.	Spanked you	1	2	3	4	5
22.	Told you that you were a bad parent	1	2	3	4	5
23.	Stopped you or tried to stop you from going to work or school	1	2	3	4	5
24.	Threw, hit, kicked, or smashed something	1	2	3	4	5
25.	Kicked you	1	2	3	4	5
26.	Physically forced you to have sex	1	2	3	4	5
27.	Threw you around	1	2	3	4	5
28.	Physically attacked the sexual parts of your body	1	2	3	4	5
29.	Choked or strangled you	1	2	3	4	5
30.	Used a knife, gun, or other weapon against you	1	2	3	4	5

Add the numbers you circled and divide the total by 30 to determine your score. The higher your score (5 is the highest score), the more abusive your relationship.

The inventory was given to 100 men and 78 women equally divided into groups of abusers or abused and nonabusers or nonabused. The men were members of a chemical dependency treatment program in a veterans' hospital and the women were partners of these men. Abusing or abused men earned an average score of 1.8; abusing or abused women earned an average score of 2.3. Nonabusing, abused men and women earned scores of 1.3 and 1.6, respectively.

Source: Shepard, M. F., & J. A. Campbell. The Abusive Behavior Inventory: A measure of psychological and physical abuse. *Journal of Interpersonal Violence, 7*(3):291–305, 1992.

Childfree Lifestyle Scale (CLS)

The purpose of this scale is to assess your attitudes toward having a childfree lifestyle. After reading each statement, select the number that best reflects your answer, using the following scale:

1	2	3	4	5	6	7
Strongly Disagree						Strongly Agree

_____ **1.** I do not like children.

_____ **2.** I would resent having to spend all my money on kids.

_____ **3.** I would rather enjoy my personal freedom than have it taken away by having children.

_____ **4.** I would rather focus on my career than have children.

_____ **5.** Children are a burden.

_____ **6.** I have no desire to be a parent.

_____ **7.** I am too "into me" to become a parent.

_____ **8.** I lack the nurturing skills to be a parent.

_____ **9.** I have no patience for children.

_____ **10.** Raising a child is too much work.

_____ **11.** A marriage without children is empty.

_____ **12.** Children are vital to a good marriage.

_____ **13.** You can't really be fulfilled as a couple unless you have children.

_____ **14.** Having children gives meaning to a couple's marriage.

_____ **15.** The happiest couples that I know have children.

_____ **16.** The biggest mistake couples make is deciding not to have children.

_____ **17.** Childfree couples are sad couples.

_____ **18.** Becoming a parent enhances the intimacy between spouses.

_____ **19.** A house without the "pitter patter" of little feet is not a home.

_____ **20.** Having a child means your marriage is successful.

Scoring

Reverse score items 11 through 20. For example, if you wrote a 7 for item 20, change this to a 1. If you wrote a 7, change to a 1. If you wrote a 2, change to a 6, etc. Add the numbers. The higher the score (140 is the highest possible score), the greater the value for a childfree lifestyle. The lower the score (20 is the lowest possible score), the less the desire to have a childfree lifestyle. The midpoint between the extremes is 80: Scores below 80 suggest less preference for a childfree lifestyle and scores above 80 suggest a desire for a childfree lifestyle. The average score of 52 male and 138 female undergraduates at Valdosta State University was below the midpoint (M = 68.78, SD = 17.06), suggesting a tendency toward a lifestyle that included children. A significant difference was found between males, who scored 72.94 (SD = 16.82), and females, who scored 67.21 (SD = 16.95), suggesting that males are more approving of a childfree lifestyle. There were no significant differences between Whites and Blacks or between students in different ranks (freshman, sophomore, junior, senior, etc.).

Source: "The Childfree Lifestyle Scale," 2010, by Mark A. Whatley, Ph.D., Department of Psychology, Valdosta State University, Valdosta, Georgia 31698-0100. Used by permission. Other uses of this scale by written permission of Dr. Whatley only (mwhatley@valdosta.edu). Information on the reliability and validity of this scale is available from Dr. Whatley.

Abortion Attitude Scale (AAS)

This is not a test. There are no wrong or right answers to any of the statements, so just answer as honestly as you can. The statements ask your feelings about legal abortion (the voluntary removal of a human fetus from the mother during the first three months of pregnancy by a qualified medical person). Tell how you feel about each statement by selecting only one response. Use the following scale for your answers:

5	4	3	2	1
Strongly Agree	Slightly Agree	Agree	Slightly Disagree	Strongly Disagree

____ 1. The Supreme Court should strike down legal abortions in the United States.

____ 2. Abortion is a good way of solving an unwanted pregnancy.

____ 3. A mother should feel obligated to bear a child she has conceived.

____ 4. Abortion is wrong no matter what the circumstances are.

____ 5. A fetus is not a person until it can live outside its mother's body.

____ 6. The decision to have an abortion should be the pregnant mother's.

____ 7. Every conceived child has the right to be born.

____ 8. A pregnant female not wanting to have a child should be encouraged to have an abortion.

____ 9. Abortion should be considered killing a person.

____ 10. People should not look down on those who choose to have abortions.

____ 11. Abortion should be an available alternative for unmarried pregnant teenagers.

____ 12. People should not have the power over the life or death of a fetus.

____ 13. Unwanted children should not be brought into the world.

____ 14. A fetus should be considered a person at the moment of conception.

Scoring and Interpretation

As its name indicates, this scale was developed to measure attitudes toward abortion. Sloan (1983) developed the scale for use with high school and college students. To compute your score, first reverse the point scale for items 1, 3, 4, 7, 9, 12, and 14. For example, if you selected a 5 for item one, this becomes a 0; if you selected a 1, this becomes a 4. After reversing the scores on the seven items specified, add the numbers you circled for all the items. Sloan provided the following categories for interpreting the results:

70–56   Strong proabortion

54–44   Moderate proabortion

43–27   Unsure

26–16   Moderate pro-life

15–0   Strong pro-life

Reliability and Validity

The AAS was administered to high school and college students, Right to Life group members, and abortion service personnel. Sloan (1983) reported a high total test estimate of reliability (0.92). Construct validity was supported in that the mean score for Right to Life members was 16.2; the mean score for abortion service personnel was 55.6; and other groups' scores fell between these values.

Source: Sloan, L. A. (1983, May/June). Abortion Attitude Scale. *Journal of Health Education, 14*(3). The *Journal of Health Education* is a publication of the American Alliance for Health, Physical Education, Recreation and Dance, 1900 Association Drive, Reston, VA 20191.

Spanking versus Time-Out Scale

Parents discipline their children to help them develop self-control and correct misbehavior. Some parents spank their children; others use time-out. Spanking is a disciplinary technique whereby a mild slap (i.e., a "spank") is applied to the buttocks of a disobedient child. Time-out is a disciplinary technique whereby, when a child misbehaves, the child is removed from the situation. The purpose of this survey is to assess the degree to which you prefer spanking versus time-out as a method of discipline. Please read each item carefully and select a number from 1 to 7, which represents your belief. There are no right or wrong answers; please give your honest opinion.

1	2	3	4	5	6	7
Strongly Disagree						Strongly Agree

_____ **1.** Spanking is a better form of discipline than time-out.

_____ **2.** Time-out does not have any effect on children.

_____ **3.** When I have children, I will more likely spank them than use a time-out.

_____ **4.** A threat of a time-out does not stop a child from misbehaving.

_____ **5.** Lessons are learned better with spanking.

_____ **6.** Time-out does not give a child an understanding of what the child has done wrong.

_____ **7.** Spanking teaches a child to respect authority.

_____ **8.** Giving children time-outs is a waste of time.

_____ **9.** Spanking has more of an impact on changing the behavior of children than time-out.

_____ **10.** I do not believe "time-out" is a form of punishment.

_____ **11.** Getting spanked as a child helps you become a responsible citizen.

_____ **12.** Time-out is only used because parents are afraid to spank their kids.

_____ **13.** Spanking can be an effective tool in disciplining a child.

_____ **14.** Time-out is watered-down discipline.

Scoring

If you want to know the degree to which you approve of spanking, reverse the number you selected for all odd-numbered items (1, 3, 5, 7, 9, 11, and 13) after you have selected a number from 1 to 7 for each of the 14 items. For example, if you selected a 1 for item 1, change this number to a 7 (1 = 7; 2 = 6; 3 = 5; 4 = 4; 5 = 3; 6 = 2; 7 = 1). Now add these 7 numbers. The lower your score (7 is the lowest possible score), the lower your approval of spanking; the higher your score (49 is the highest possible score), the greater your approval of spanking. A score of 21 places you at the midpoint between being very disapproving of or very accepting of spanking as a discipline strategy.

If you want to know the degree to which you approve of using time-out as a method of discipline, reverse the number you selected for all even-numbered items (2, 4, 6, 8, 10, 12, and 14). For example, if you selected a 1 for item 2, change this number to a 7 (i.e., 1 = 7; 2 = 6; 3 = 5; 4 = 4; 5 = 3; 6 = 2; 7 = 1). Now add these 7 numbers. The lower your score (7 is the lowest possible score), the lower your approval of time-out; the higher your score (49 is the highest possible score), the greater your approval of time-out. A score of 21 places you at the midpoint between being very disapproving of or very accepting of time-out as a discipline strategy.

Scores of Other Students Who Completed the Scale

The scale was completed by 48 male and 168 female student volunteers at East Carolina University. Their ages ranged from 18 to 34, with a mean age of 19.65 (SD = 2.06). The ethnic background of the sample included 73.1% White, 17.1% African American, 2.8% Hispanic, 0.9% Asian, 3.7% from other ethnic backgrounds; 2.3% did not indicate ethnicity. The college classification level of the sample included 52.8% freshman, 24.5% sophomore, 13.9% junior, and 8.8% senior. The average score on the spanking dimension was 29.73 (SD = 10.97), and the time-out dimension was 22.93 (SD = 8.86), suggesting greater acceptance of spanking than time-out.

Time-out differences. In regard to sex of the participants, female participants were more positive about using time-out as a discipline strategy (M = 33.72, SD = 8.76) than were male participants (M = 30.81, SD = 8.97; p = .05). In regard to ethnicity of the participants, White participants were more positive about using time-out as a discipline strategy (M = 34.63, SD = 8.54) than were non-White participants (M = 28.45, SD = 8.55; p = .05). In regard to year in school, freshmen were more positive about using spanking as a discipline strategy (M = 34.34, SD = 9.23) than were sophomores, juniors, and seniors (M = 31.66, SD = 8.25; p = .05).

Spanking differences. In regard to ethnicity of the participants, non-White participants were more positive about using spanking as a discipline strategy (M = 35.09, SD = 10.02) than were White participants (M = 27.87, SD = 10.72; p = .05). In regard to year in school, freshmen were less positive about using spanking as a discipline strategy (M = 28.28, SD = 11.42) than were sophomores, juniors, and seniors (M = 31.34, SD = 10.26; p = .05). There were no significant differences in regard to sex of the participants (p = .05) in the opinion of spanking.

Overall differences. There were no significant differences in overall attitudes to discipline in regards to sex of the participants, ethnicity, or year in school.

Source: "The Spanking vs. Time-Out Scale," 2004, by Mark Whatley, Ph.D., Department of Psychology, Valdosta State University, Valdosta, Georgia 31698-0100. Information on the reliability and validity of this scale is available from Dr. Whatley (mwhatley@valdosta.edu).

Traditional Motherhood Scale

The purpose of this survey is to assess the degree to which students possess a traditional view of motherhood. Read each item carefully and consider what you believe. There are no right or wrong answers, so please give your honest reaction and opinion. After reading each statement, select the number that best reflects your level of agreement, using the following scale:

1	2	3	4	5	6	7
Strongly Disagree						Strongly Agree

____ 1. A mother has a better relationship with her children than a father does.

____ 2. A mother knows more about her child than a father, thereby being the better parent.

____ 3. Motherhood is what brings women to their fullest potential.

____ 4. A good mother should stay at home with her children for the first year.

____ 5. Mothers should stay at home with the children.

____ 6. Motherhood brings much joy and contentment to a woman.

____ 7. A mother is needed in a child's life for nurturance and growth.

____ 8. Motherhood is an essential part of a female's life.

____ 9. I feel that all women should experience motherhood in some way.

____ 10. Mothers are more nurturing than fathers.

____ 11. Mothers have a stronger emotional bond with their children than do fathers.

____ 12. Mothers are more sympathetic to children who have hurt themselves than are fathers.

____ 13. Mothers spend more time with their children than do fathers.

____ 14. Mothers are more lenient toward their children than are fathers.

____ 15. Mothers are more affectionate toward their children than are fathers.

____ 16. The presence of the mother is vital to the child during the formative years.

____ 17. Mothers play a larger role than fathers in raising children.

____ 18. Women instinctively know what a baby needs.

Scoring

After assigning a number from 1 (strongly disagree) to 7 (strongly agree), add the numbers and divide by 18. The higher your score (7 is the highest possible score), the stronger the traditional view of motherhood. The lower your score (1 is the lowest possible score), the less traditional the view of motherhood.

Norms

The norming sample of this self-assessment was based upon 20 male and 86 female students attending Valdosta State University. The average age of participants completing the scale was 21.72 years (SD = 2.98), and ages ranged from 18 to 34. The ethnic composition of the sample was 80.2% White, 15.1% Black, 1.9% Asian, 0.9% American Indian, and 1.9% other. The classification of the sample was 16.0% freshmen, 15.1% sophomores, 27.4% juniors, 39.6% seniors, and 1.9% graduate students.

Participants responded to each of the 18 items according to the 7-point scale. The most traditional score was 6.33; the score reflecting the least support for traditional motherhood was 1.78. The midpoint (average score) between the top and bottom score was 4.28 (SD = 1.04); thus, people scoring above this number tended to have a more traditional view of motherhood and people scoring below this number tended to have a less traditional view of motherhood.

There was a significant difference (p = .05) between female participants' scores (mean = 4.19; SD = 1.08) and male participants' scores (mean = 4.68; SD = 0.73), suggesting that males had more traditional views of motherhood than females.

Source: Mark Whatley, Ph.D. (2004). The Traditional Motherhood Scale. Department of Psychology, Valdosta State University. Use of this scale is permitted only by prior written permission of Dr. Whatley (mwhatley@valdosta.edu).

Family Hardiness Scale (FHS)

This scale is designed to identify the degree to which your family has the characteristics of hardiness, which is defined as resistance to stress, having internal strength, and having a sense of control over life events and hardships. Read each statement and decide to what degree each describes your family. Choices include false, mostly false, mostly true, or totally true about your family. Write a 0 to 3 (or NA) next to each statement.

0	1	2	3	NA
False	*Mostly False*	*Mostly True*	*Totally True*	*Not Applicable*

In our family...

_____ **1.** Trouble results from mistakes we make.

_____ **2.** It is not wise to plan ahead and hope because things do not turn out anyway.

_____ **3.** Our work and efforts are not appreciated no matter how hard we try and work.

_____ **4.** In the long run, the bad things that happen to us are balanced by the good things that happen.

_____ **5.** We have a sense of being strong even when we face big problems.

_____ **6.** Many times I feel I can trust that even in difficult times things will work out.

_____ **7.** While we don't always agree, we can count on each other to stand by us in times of need.

_____ **8.** We do not feel we can survive if another problem hits us.

_____ **9.** We believe that things will work out for the better if we work together as a family.

_____ **10.** Life seems dull and meaningless.

_____ **11.** We strive together and help each other no matter what.

_____ **12.** When our family plans activities we try new and exciting things.

_____ **13.** We listen to each other's problems, hurts, and fears.

_____ **14.** We tend to do the same things over and over...it's boring.

_____ **15.** We seem to encourage each other to try new things and experiences.

_____ **16.** It is better to stay at home than go out and do things with others.

_____ **17.** Being active and learning new things are encouraged.

_____ **18.** We work together to solve problems.

_____ **19.** Most of the unhappy things that happen are due to bad luck.

_____ **20.** We realize our lives are controlled by accidents and luck.

Scoring

Reverse score items 1, 2, 3, 8, 10, 14, 16, 19, and 20. For example, if for number 1 you wrote down a 0, replace the 0 with a 3 and vice versa. Now add all the numbers from 1 to 20.

Norms

A "low" score indicating very low hardiness is 0 and a "high" score indicating very high hardiness is 60. The overall average of 304 families who took the scale scored 47.4 (SD = 6.7).

Source: The authors of the scale are Marilyn A. McCubbin, Hamilton I. McCubbin, and Anne I. Thompson. See McCubbin, H. I., & A. I. Thompson (Eds.). (1991). *Family assessment inventories for research and practice.* Madison, WI: University of Wisconsin.

Children's Beliefs about Parental Divorce Scale

The following are some statements about children and their separated parents. Some of the statements are **true** about how you think and feel, so you will want to check **yes**. Some are **not true** about how you think or feel, so you will want to check **no**. There are no right or wrong answers. Your answers will just indicate some of the things you are thinking now about your parents' separation.

1. It would upset me if other kids asked a lot of questions about my parents. ___ Yes ___ No
2. It was usually my father's fault when my parents had a fight. ___ Yes ___ No
3. I sometimes worry that both my parents will want to live without me. ___ Yes ___ No
4. When my family was unhappy, it was usually because of my mother. ___ Yes ___ No
5. My parents will always live apart. ___ Yes ___ No
6. My parents often argue with each other after I misbehave. ___ Yes ___ No
7. I like talking to my friends as much now as I used to. ___ Yes ___ No
8. My father is usually a nice person. ___ Yes ___ No
9. It's possible that both my parents will never want to see me again. ___ Yes ___ No
10. My mother is usually a nice person. ___ Yes ___ No
11. If I behave better, I might be able to bring my family back together. ___ Yes ___ No
12. My parents would probably be happier if I were never born. ___ Yes ___ No
13. I like playing with my friends as much now as I used to. ___ Yes ___ No
14. When my family was unhappy, it was usually because of something my father said or did. ___ Yes ___ No
15. I sometimes worry that I'll be left all alone. ___ Yes ___ No
16. Often I have a bad time when I'm with my mother. ___ Yes ___ No
17. My family will probably do things together just like before. ___ Yes ___ No
18. My parents probably argue more when I'm with them than when I'm gone. ___ Yes ___ No
19. I'd rather be alone than play with other kids. ___ Yes ___ No
20. My father caused most of the trouble in my family. ___ Yes ___ No
21. I feel that my parents still love me. ___ Yes ___ No
22. My mother caused most of the trouble in my family. ___ Yes ___ No
23. My parents will probably see that they have made a mistake and get back together again. ___ Yes ___ No
24. My parents are happier when I'm with them than when I'm not. ___ Yes ___ No
25. My friends and I do many things together. ___ Yes ___ No
26. There are a lot of things I like about my father. ___ Yes ___ No
27. I sometimes think that one day I may have to go live with a friend or relative. ___ Yes ___ No
28. My mother is more good than bad. ___ Yes ___ No
29. I sometimes think that my parents will one day live together again. ___ Yes ___ No
30. I can make my parents unhappy with each other by what I say or do. ___ Yes ___ No
31. My friends understand how I feel about my parents. ___ Yes ___ No
32. My father is more good than bad. ___ Yes ___ No
33. I feel my parents still like me. ___ Yes ___ No
34. There are a lot of things about my mother I like. ___ Yes ___ No
35. I sometimes think that my parents will live together again once they realize how much I want them to. ___ Yes ___ No
36. My parents would probably still be living together if it weren't for me. ___ Yes ___ No

Scoring

The Children's Beliefs about Parental Divorce Scale (CBAPS) identifies problematic responding. A **yes** response on items 1, 2, 3, 4, 6, 9, 11, 12, 14–20, 22, 23, 27, 29, 30, 35, and 36, and a **no** response on items 5, 7, 8, 10, 13, 21, 24–26, 28, and 31–34 indicate a problematic reaction to one's parents divorcing. A total score is derived by adding the number of problematic beliefs across all the items, with a total score of 36. The higher the score, the more problematic the beliefs about parental divorce.

Norms

A total of 170 schoolchildren whose parents were divorced completed the scale; of the children, 84 were boys and 86 were girls, with a mean age of 11. The mean for the total score was 8.20, with a standard deviation of 4.98.

Source: From Table 1 (adapted), p. 715, from Kurdek, L. A., & Berg, B. (1987). Children's Beliefs About Parental Divorce Scale: Psychometric characteristics and concurrent validity. *Journal of Consulting and Clinical Psychology, 55*(5), 712–718.

Family Member Well-Being (FMWB)

The purpose of this survey is to measure the adjustment of family members in terms of concern about health, tension, energy, cheerfulness, fear, anger, sadness, and general concerns. Read each of the eight statements and note that the words at each end of the 1 to 9 scale describe opposite feelings. Select the number along the bar which seems closest to how you have generally felt during the past month.

1. How concerned or worried about your health have you been? (during the past month)

   Not CONCERNED   *Very*
   at all   *CONCERNED*
   1   2   3   4   5   6   7   8   9

2. How relaxed or tense have you been? (during the past month)

   Very   *Very*
   RELAXED   *TENSE*
   1   2   3   4   5   6   7   8   9

3. How much energy, pep, and vitality have you felt? (during the past month)

   No energy at all   *Very energetic*
   LISTLESS   *DYNAMIC*
   1   2   3   4   5   6   7   8   9

4. How depressed or cheerful have you been? (during the past month)

   Very   *Very*
   DEPRESSED   *CHEERFUL*
   1   2   3   4   5   6   7   8   9

5. How afraid have you been? (during the past month)

   Not   *Very*
   AFRAID   *AFRAID*
   1   2   3   4   5   6   7   8   9

6. How angry have you been? (during the past month)

   Not ANGRY   *Very*
   at all   *ANGRY*
   1   2   3   4   5   6   7   8   9

7. How sad have you been? (during the past month)

   Not SAD   *Very*
   at all   *SAD*
   1   2   3   4   5   6   7   8   9

8. How concerned or worried about the health of another family member have you been? (during the past month)

   Not CONCERNED   *Very*
   at all   *CONCERNED*
   1   2   3   4   5   6   7   8   9

Scoring

Reverse score items 1, 2, 5, 6, 7, and 8 so that 1 = 9, 9 = 1, etc. Then, add each of the eight numbers. Higher scores reflect more positive family well-being. The scale was completed by large samples including 297 investment executives and 234 spouses of investment executives, midwest farm families involving 411 males and 389 females, 813 rural bank employees plus 448 of their spouses, and 524 male members of military families and 465 female members of military families. The means of the FMWB range from 37.5 (SD = 9.1) for Caucasian female members of military families to 45.5 (SD = 10.5) for spouses of investment executives.

Source: McCubbin, H. I., & A. I. Thompson (Eds). (1991). Family Member Well-Being (FMWB). In H. I. McCubbin, A. I. Thompson, & M. A. McCubbin (1996), *Family assessment resiliency, coping and adaptation.* Inventories for research and practice. Madison, WI: University of Wisconsin. Call the University of Wisconsin, Family Stress Coping, Coping and Health Project at 608-262-5070 for information.